NEUROMETHODS

Series Editor
Wolfgang Walz
University of Saskatchewan,
Saskatoon, Canada

For further volumes:
http://www.springer.com/series/7657

Pre-Clinical and Clinical Methods in Brain Trauma Research

Edited by

Amit K. Srivastava and Charles S. Cox, Jr.

Department of Pediatric Surgery, McGovern Medical School, The University of Texas Health Sciences Center at Houston, Houston, TX, USA

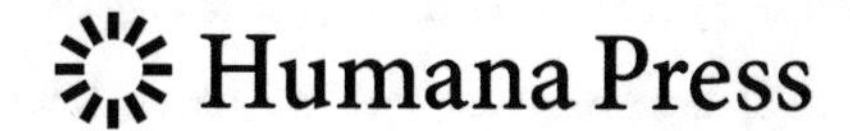

Editors
Amit K. Srivastava
Department of Pediatric Surgery
McGovern Medical School
The University of Texas
Health Sciences Center at Houston
Houston, TX, USA

Charles S. Cox, Jr.
Department of Pediatric Surgery
McGovern Medical School
The University of Texas
Health Sciences Center at Houston
Houston, TX, USA

ISSN 0893-2336 ISSN 1940-6045 (electronic)
Neuromethods
ISBN 978-1-4939-9327-7 ISBN 978-1-4939-8564-7 (eBook)
https://doi.org/10.1007/978-1-4939-8564-7

Printed on acid-free paper

This Humana Press imprint is published by the registered company Springer Science+Business Media, LLC part of Springer Nature.
The registered company address is: 233 Spring Street, New York, NY 10013, U.S.A.

Preface to the Series

Experimental life sciences have two basic foundations: concepts and tools. The *Neuromethods* series focuses on the tools and techniques unique to the investigation of the nervous system and excitable cells. It will not, however, shortchange the concept side of things as care has been taken to integrate these tools within the context of the concepts and questions under investigation. In this way, the series is unique in that it not only collects protocols but also includes theoretical background information and critiques which led to the methods and their development. Thus it gives the reader a better understanding of the origin of the techniques and their potential future development. The *Neuromethods* publishing program strikes a balance between recent and exciting developments like those concerning new animal models of disease, imaging, in vivo methods, and more established techniques, including, for example, immunocytochemistry and electrophysiological technologies. New trainees in neurosciences still need a sound footing in these older methods in order to apply a critical approach to their results.

Under the guidance of its founders, Alan Boulton and Glen Baker, the *Neuromethods* series has been a success since its first volume published through Humana Press in 1985. The series continues to flourish through many changes over the years. It is now published under the umbrella of Springer Protocols. While methods involving brain research have changed a lot since the series started, the publishing environment and technology have changed even more radically. Neuromethods has the distinct layout and style of the Springer Protocols program, designed specifically for readability and ease of reference in a laboratory setting.

The careful application of methods is potentially the most important step in the process of scientific inquiry. In the past, new methodologies led the way in developing new disciplines in the biological and medical sciences. For example, Physiology emerged out of Anatomy in the nineteenth century by harnessing new methods based on the newly discovered phenomenon of electricity. Nowadays, the relationships between disciplines and methods are more complex. Methods are now widely shared between disciplines and research areas. New developments in electronic publishing make it possible for scientists that encounter new methods to quickly find sources of information electronically. The design of individual volumes and chapters in this series takes this new access technology into account. Springer Protocols makes it possible to download single protocols separately. In addition, Springer makes its print-on-demand technology available globally. A print copy can therefore be acquired quickly and for a competitive price anywhere in the world.

Saskatoon, Canada *Wolfgang Walz*

Preface

Brain trauma is a silently expanding epidemic and a major socioeconomic and public health problem. Each year, a large number of people sustain some form of brain injury, which often becomes a cause of death. The survivors of brain trauma experience long-term debilitating changes in their physical, cognitive, and psychosocial status. According to the American Academy of Neurology, 40% of former National Football League (NFL) players suffer from brain trauma, and the US Department of Defense terms traumatic brain injury (TBI) as the "signature injury" among Iraq and Afghanistan war veterans. These new findings have fueled an increase in public interest in TBI. However, despite being a major cause of mortality and morbidity, there is limited success in the development of effective treatments for TBI. The pathophysiology is complex and involves both primary and secondary injury mechanisms. In clinical practice, aside from surgical intervention in a few selected cases, management of the disease is primarily supportive to prevent progressive secondary injury to the brain. Management of mild TBI and repeated subconcussive brain trauma is particularly difficult because of the lack of diagnostic guidelines and delayed-onset neurodegenerative nature of this condition. The burden of mortality and residual disability demands a more analytical approach to understand the complexity of TBI and factors that affect outcomes. The past few years have seen some extraordinary inventions and technological advancement in the TBI field. In recognition of growing advancement in the area of brain trauma research, there is a need to provide comprehensive information on the tools and techniques used in preclinical and clinical settings. This book has made an attempt to integrate chapters from experts across the field to address current perspectives and knowledge gaps in the field of TBI. The topics range from development of in vitro and animal TBI models to diagnostic imaging and disease monitoring in patients, as well as designing of preclinical and clinical trials. The book is edited to achieve a level of writing that is academically rigorous to address the needs of TBI researchers. We hope that this translational book will provide both basic scientists and clinical researchers with a comprehensive reference on the fundamental techniques and their potential application in the field of TBI.

Houston, TX, USA

Amit K. Srivastava
Charles S. Cox, Jr.

Contents

Contributors

PEETHAMBARAM ARUN • *Blast-Induced Neurotrauma Branch, Center for Military Psychiatry and Neuroscience, Walter Reed Army Institute of Research, Silver Spring, MD, USA*

ROJAHNE AZWOIR • *Department of Neuroscience, Johns Hopkins University School of Medicine, Baltimore, MD, USA*

SUPINDER BEDI • *Department of Pediatric Surgery, McGovern Medical School, The University of Texas Health Sciences Center at Houston, Houston, TX, USA*

DANIEL E. BONDER • *Department of Biomedical Engineering, Center for Injury Biomechanics, Materials and Medicine, New Jersey Institute of Technology, Newark, NJ, USA*

DEBORAH R. BOONE • *Department of Anesthesiology, University of Texas Medical Branch, Galveston, TX, USA*

PHILIP BOUGHTON • *School of Aerospace, Mechanical and Mechatronic Engineering, Faculty of Engineering and Information Technologies, University of Sydney, Sydney, NSW, Australia*

NAMAS CHANDRA • *Department of Biomedical Engineering, Center for Injury Biomechanics, Materials and Medicine, New Jersey Institute of Technology, Newark, NJ, USA*

LIAM CHEN • *Department of Pathology, Johns Hopkins University School of Medicine, Baltimore, MD, USA*

HUIMAHN ALEX CHOI • *Department of Neurosurgery, McGovern School of Medicine, The University of Texas Health Science Center at Houston, Houston, TX, USA*

CHARLES S. COX, JR. • *Department of Pediatric Surgery, McGovern Medical School, The University of Texas Health Science Center at Houston, Houston, TX, USA*

D. KACY CULLEN • *Department of Neurosurgery, Center for Brain Injury and Repair, University of Pennsylvania, Philadelphia, PA, USA; Center for Neurotrauma, Neurodegeneration and Restoration, Corporal Michael J. Crescenz VA Medical Center, Philadelphia, PA, USA*

YURI DANILOV • *Department of Kinesiology, University of Wisconsin, Madison, WI, USA*

PRAMOD DASH • *Department of Neurobiology and Anatomy, The University of Texas Health Science Center at Houston, Houston, TX, USA; University of Texas Graduate School of Biomedical Science, Houston, TX, USA*

C. EDWARD DIXON • *Department of Neurological Surgery, University of Pittsburgh, Pittsburgh, PA, USA; Safar Center for Resuscitation Research, University of Pittsburgh, Pittsburgh, PA, USA; V.A. Pittsburgh Healthcare System, Pittsburgh, PA, USA*

JOSH L. DUCKWORTH • *Uniformed Services University of the Health Sciences, Bethesda, MD, USA*

TIMOTHY Q. DUONG • *Department of Radiology, Stony Brook School of Medicine, Stony Brook, NY, USA*

MICHAEL F. W. FESTING • *Medical Research Toxicology Unit, University of Leicester, Leicester, UK*

JONATHAN A. N. FISHER • *Department of Physiology, New York Medical College, Valhalla, NY, USA*

VLADIMIR GERZANICH • *Department of Neurosurgery, University of Maryland School of Medicine, Baltimore, MD, USA*
THOMAS H. GOMEZ • *Center for Laboratory Animal Medicine and Care, The University of Texas Health Science Center at Houston, Houston, TX, USA*
MIRINDA GORMLEY • *Division of Epidemiology, Department of Family Medicine and Population Health, Virginia Commonwealth University, Richmond, VA, USA*
DANIEL GRIFFITHS • *Barrow Neurological Institute at Phoenix Children's Hospital, Phoenix, AZ, USA; Department of Child Health, University of Arizona College of Medicine—Phoenix, Phoenix, AZ, USA*
F. JAY HARAN • *Uniformed Services University of the Health Sciences, Bethesda, MD, USA; University of North Carolina at Greensboro, Greensboro, NC, USA*
ERIK HAYMAN • *Department of Neurosurgery, University of Maryland School of Medicine, Baltimore, MD, USA*
HELEN L. HELLMICH • *Department of Anesthesiology, University of Texas Medical Branch, Galveston, TX, USA*
SAMER M. JABER • *Department of Animal Medicine, University of Massachusetts Medical School, Worcester, MA, USA; Department of Pathology, University of Massachusetts Medical School, Worcester, MA, USA*
JENIFER JURANEK • *Department of Pediatrics, McGovern Medical School, The University of Texas Health Science Center at Houston, Houston, TX, USA*
KASPAR KALEDJIAN • *Department of Neurosurgery, University of Maryland School of Medicine, Baltimore, MD, USA*
CAROLYN E. KEATING • *Department of Neurosurgery, Center for Brain Injury and Repair, University of Pennsylvania, Philadelphia, PA, USA; Center for Neurotrauma, Neurodegeneration and Restoration, Corporal Michael J. Crescenz VA Medical Center, Philadelphia, PA, USA*
RYAN KITAGAWA • *Department of Neurosurgery, McGovern School of Medicine, The University of Texas Health Science Center at Houston, Houston, TX, USA*
NIKITA A. KUZNETSOV • *Louisiana State University, Baton Rouge, LA, USA*
JEREMY KWARCINSKI • *School of Aerospace, Mechanical and Mechatronic Engineering, Faculty of Engineering and Information Technologies, University of Sydney, Sydney, NSW, Australia*
JONATHAN LIFSHITZ • *Barrow Neurological Institute at Phoenix Children's Hospital, Phoenix, AZ, USA; Department of Child Health, University of Arizona College of Medicine—Phoenix, Phoenix, AZ, USA; Phoenix Veteran Affairs Healthcare System, Phoenix, AZ, USA*
YIN LIU • *Department of Neurobiology and Anatomy, The University of Texas Health Science Center at Houston, Houston, TX, USA; University of Texas Graduate School of Biomedical Science, Houston, TX, USA*
JOSEPH B. LONG • *Blast-Induced Neurotrauma Branch, Center for Military Psychiatry and Neuroscience, Walter Reed Army Institute of Research, Silver Spring, MD, USA*
JUAN LU • *Division of Epidemiology, Department of Family Medicine and Population Health, Virginia Commonwealth University, Richmond, VA, USA*
MARY F. MCGUIRE • *Department of Neurosurgery, McGovern School of Medicine, The University of Texas Health Science Center at Houston, Houston, TX, USA*
SCOTT D. OLSON • *Department of Pediatric Surgery, McGovern Medical School, The University of Texas Health Sciences Center at Houston, Houston, TX, USA*

NICOLE OSIER • *Holistic Adult Health Division, School of Nursing, University of Texas at Austin, Austin, TX, USA; Department of Neurology, Dell Medical School, University of Texas at Austin, Austin, TX, USA*
DAFNA PALTIN • *Department of Kinesiology, University of Wisconsin, Madison, WI, USA*
BRYAN J. PFISTER • *Department of Biomedical Engineering, Center for Injury Biomechanics, Materials and Medicine, New Jersey Institute of Technology, Newark, NJ, USA*
STACEY L. PIOTROWSKI • *Center for Laboratory Animal Medicine and Care, The University of Texas Health Science Center at Houston, Houston, TX, USA*
KARTHIK S. PRABHAKARA • *Department of Pediatric Surgery, McGovern Medical School, The University of Texas Health Sciences Center at Houston, Houston, TX, USA*
JOHN REDELL • *Department of Neurobiology and Anatomy, The University of Texas Health Science Center at Houston, Houston, TX, USA*
CHRISTOPHER K. RHEA • *Department of Kinesiology, University of North Carolina at Greensboro, Greensboro, NC, USA*
MARY A. ROBINSON • *Center for Laboratory Animal Medicine and Care, The University of Texas Health Science Center at Houston, Houston, TX, USA*
SCOTT E. ROSS • *University of North Carolina at Greensboro, Greensboro, NC, USA*
RACHEL K. ROWE • *Barrow Neurological Institute at Phoenix Children's Hospital, Phoenix, AZ, USA; Department of Child Health, University of Arizona College of Medicine—Phoenix, Phoenix, AZ, USA; Phoenix Veteran Affairs Healthcare System, Phoenix, AZ, USA*
ANDREW RUYS • *School of Aerospace, Mechanical and Mechatronic Engineering, Faculty of Engineering and Information Technologies, University of Sydney, Sydney, NSW, Australia*
KATHRYN E. SAATMAN • *Spinal Cord and Brain Injury Research Center, Department of Physiology, University of Kentucky, Lexington, KY, USA*
VENKATASIVASAI SUJITH SAJJA • *Blast-Induced Neurotrauma Branch, Center for Military Psychiatry and Neuroscience, Walter Reed Army Institute of Research, Silver Spring, MD, USA*
ANTHONY SAN LUCAS • *Department of Epidemiology, University of Texas M.D. Anderson Cancer Center, Houston, TX, USA*
JUDE P. J. SAVARRAJ • *Department of Neurosurgery, McGovern School of Medicine, The University of Texas Health Science Center at Houston, Houston, TX, USA*
DANIELLE SCOTT • *Spinal Cord and Brain Injury Research Center, Department of Physiology, University of Kentucky, Lexington, KY, USA*
STACY L. SELL • *Department of Anesthesiology, University of Texas Medical Branch, Galveston, TX, USA*
QIANG SHEN • *Research Imaging Institute, UT Health San Antonio, San Antonio, TX, USA; Department of Ophthalmology, UT Health San Antonio, San Antonio, TX, USA; Department of Radiology, UT Health San Antonio, San Antonio, TX, USA*
J. MARC SIMARD • *Department of Neurosurgery, University of Maryland School of Medicine, Baltimore, MD, USA*
AMIT K. SRIVASTAVA • *Department of Pediatric Surgery, McGovern Medical School, The University of Texas Health Science Center at Houston, Houston, TX, USA*
MIN D. TANG-SCHOMER • *Department of Pediatrics, UConn Health and Connecticut Children's Medical Center, Farmington, CT, USA; The Jackson Laboratory for Genomic Medicine, Farmington, CT, USA*

NAAMA E. TOLEDANO FURMAN • *Department of Pediatric Surgery, McGovern Medical School, The University of Texas Health Sciences Center at Houston, Houston, TX, USA*
STEPHEN A. VAN ALBERT • *The Geneva Foundation Supporting Walter Reed Army Institute of Research, Walter Reed Army Institute of Research, Silver Spring, MD, USA*
JAMES VAN GELDER • *Sydney Spine Institute, Burwood, NSW, Australia*
HARRIS A. WEISZ • *Department of Anesthesiology, University of Texas Medical Branch, Galveston, TX, USA*
CRISTIN G. WELLE • *Departments of Neurosurgery and Bioengineering, University of Colorado Denver, Aurora, CO, USA*
W. GEOFFREY WRIGHT • *Temple University, Philadelphia, PA, USA*

Chapter 1

Traumatic Brain Injury

Amit K. Srivastava and Charles S. Cox, Jr.

Abstract

Traumatic brain injury (TBI) is a common, complex, and costly condition. It is a multidimensional and highly complex condition and an important cause of disability and mortality all around the world. To date, there are no effective treatments available that are able to mitigate subacute injuries and improve long-term functional recovery in TBI. A major reason that several experimental treatments for TBI have failed in the past is the appreciation that TBI is not a single acute event but chronic and progressive tissue damage. Better understanding of the anatomy and pathophysiology of brain injuries, new biomarkers, advanced neuroimaging, and the reorganization of trauma systems have led to a significant reduction in deaths and disability resulting from TBI. In this chapter, we discuss the pathophysiology, classification, clinical presentation, and diagnosis of this condition.

Key words Brain injury, Trauma, TBI-classification, Diagnosis, Prediction

1 Introduction

The mention of brain trauma can be found as early as in ancient Greek literatures from the Age of Pericles. As translated by F. Adams in 1939 in his book "The genuine work of Hippocrates," Section 7 of the Aphorisms of Hippocrates (circa 415 B.C.) describes minor head injury as "*shaking or concussion of the brain produced by any cause inevitably leaves the patient with an instantaneous loss of voice*" (implying unconscious). The modern definition of traumatic brain injury (TBI) is "an alteration in brain function, or other evidence of brain pathology, caused by an external force" [1]. Often labeled as a silent epidemic, TBI is one of the most serious public health and socioeconomic problems. According to the 2010 report of the global burden of disease (GBD) study, 89% of trauma-related deaths occur in low- and middle-income countries. Within the spectrum of trauma-related injuries, TBI is one of the largest causes of death and disability [2]. Worldwide, more than 50 million TBIs occur each year [3]. In the United States, every year approximately 1.7 million people are assessed in

Amit K. Srivastava and Charles S. Cox, Jr. (eds.), *Pre-Clinical and Clinical Methods in Brain Trauma Research*, Neuromethods, vol. 139, https://doi.org/10.1007/978-1-4939-8564-7_1,

an emergency room after sustaining TBI of any severity, of whom 52,000 die, contributing to 30.5% of all injury-related deaths [4, 5]. Although high-quality data of TBI incidence and prevalence are scant, (mainly due to the absence of neurotrauma registries), TBI incidences vary substantially between countries with a significant rise in injuries noted in developing countries due to increased use of motor vehicles and road traffic accidents. On the other hand, the number of elderly people with TBI is increasing in developed countries mainly due to falls [6]. Approximately 5.3 million people in the USA, and 7.7 million people in European Union are currently living with some form of TBI-related disabilities such as impaired motor function, attention deficit, depression, aggressive behavior, and difficulty in decision-making [5, 7]. As a result of these chronic consequences, the financial burden in the USA has been estimated at over $60 billion per year, with 80% of the total lifetime cost per case of severe TBI (approximately $400,000) was attributable to loss of human potential and productivity [8, 9].

The pathophysiology of TBI is extremely complex and involves both primary and secondary injury mechanisms. The primary injury induces biochemical and cellular changes that leads to secondary injury that evolves over hours, days, months, or even years with perpetual neuronal damage and death overtime [5]. In addition to these complex and continuous pathophysiological events, TBI occurs frequently in association with multiple extracranial injuries. Multiple traumatic injuries have a conglomerate of clinical events such as hypotension, hypoxia, pyrexia, and coagulopathy that may adversely affect the brain and have long-term consequences. In addition, an increasing body of evidence indicates that TBI patients become susceptible to other neurological and psychiatric disorders such as chronic traumatic encephalopathy (CTE), dementia, stroke, parkinsonism, epilepsy, Alzheimer's disease (AD), and posttraumatic stress disorder (PTSD) [10–17]. In many cases, the manifestation of these associated conditions occurs several years after the injury. This level of disease complexity limits the development of effective treatment strategies and models to predict the outcome in TBI patients.

Few decades ago, it was commonly believed that aside from evacuating occasional hematomas or managing skull fractures, little could be done after TBI. Complications, such as the development of intracranial hematomas or increased intracranial pressure (ICP) and resulting clinical severity of injury were difficult to recognize, hence the treatment was delayed. In light of these concerns, in 1971, Graham Teasdale and Bryan Jennett developed the Glasgow Coma Scale (GCS) for the clinical assessment of posttraumatic unconsciousness [18]. Effectiveness of this tool made it the bedrock for all TBI studies and clinical trials. Since the introduction of GCS, TBI research has significantly progressed, advancing our knowledge and offering opportunities to limit processes involved in brain damage. Stratified TBI management approaches that incorporate emerging technologies have led to several breakthroughs in last decade. Both preclinical and clinical brain trauma research are now increasingly focused on understanding the disease pathophysiology, developing advanced diagnostic tools, and improving treatment and rehabilitation guidelines.

2 Pathophysiology of TBI

The knowledge of TBI-associated pathophysiology is crucial for adequate and patient-oriented treatment. Pathophysiology of TBI is a very complex process that involves both primary and secondary injuries. Primary injury is defined by direct mechanical damage to brain tissues caused by kinetic energy transfer. This results in shearing of white-matter tracts, focal contusions, hematomas, diffuse swelling, and cell death. Cell death due to primary injury is irreversible and not amenable to intervention. Secondary injury starts within seconds to minutes after the primary insult and may continue for days, weeks, and months, further contributing to brain damage [4]. It is believed that most of the brain dysfunction following TBI is attributed to secondary injury mechanisms [19], hence, most TBI treatment modalities are focused on the limitation of secondary injuries. Secondary injury can be influenced by impairment of cerebrovascular autoregulation, changes in cerebral blood flow (hypo- and hyperperfusion), cerebral metabolic dysfunction, and insufficient cerebral oxygenation [20]. Furthermore, at the molecular level, secondary injury is characterized by terminal membrane depolarization of the neurons with excessive release of excitatory neurotransmitters such as glutamate, aspartate, and activation of voltage-dependent Ca^{2+}- and Na^{+}-channels. Glutamate and other neurotransmitters' excitotoxicity and ion-channel leakage cause astrocytic swelling, and contribute to elevated ICP. This increased pressure can lead to shifts of brain structures or impair blood flow resulting in brain ischemia. Intracellular influx of Ca^{2+} and Na^{+} also trigger catabolic intracellular processes with the activation of lipid peroxidases, proteases, caspases, and release of free radicals. These events lead to membrane degeneration of vascular and cellular structures and ultimately necrosis and/or apoptosis. This cellular degeneration is also associated with a complex array of inflammatory responses and increased permeability of the blood brain barrier (BBB) [21–23]. While inflammatory responses play a crucial role in the acute phase of injury by clearing cellular debris at the site of injury, dysregulated and prolonged inflammation can be deleterious and further damages neurons that may have been salvageable. Several studies have indicated that the state of cerebral microenvironment and resultant secondary mechanism after brain injury determine the eventual clinical outcome. However, the heterogeneous pathology of TBI makes uniform treatment recommendations difficult. A better understanding of predominating pathophysiological mechanisms could lead to improved management of TBI patients and the development of therapies to limit secondary brain injury.

3 Clinical Classification System for TBI

It is well known that TBI is not just one disease, rather it is a heterogeneous collection of the outcomes determined by multiple factors. Detailed characterization of type and severity is essential to stratify patients for optimum treatment and to evaluate outcome. Traditionally, TBI is classified by physical mechanism (closed/blunt force, blast, and penetrating injuries); by clinical severity (mild, moderate, and sever); and, by structural damage (neuroimaging descriptions).

3.1 Classification by Physical Mechanism

Mechanistic classification of TBI has great utility in modeling injuries in research laboratories. Based on physical mechanism, TBI is classified as: (1) focal injury produced by direct contact loading (when the head is struck by a solid object at a tangible speed) and/or (2) diffused injury produced by noncontact loading (when the brain accelerates/decelerates within the skull). In direct contact loading TBI, an intense mechanical force (even the one that last less than 50 ms) can cause local areas of stretching or compression within brain tissue [24, 25]. One example of this type of injury is a motor vehicle striking a pedestrian's head. In contact loading injuries, intracranial hematomas occur in 25–35% of patients with severe impact and in 5–10% of moderate impact [26]. This type of injury can also lead to the induction of ischemia causing necrosis within the core of the injured brain tissue. Tissue necrosis further initiates multiple inflammatory cascades that cause additional local tissue damage [27]. On the other hand, noncontact loading TBI, such as exposure to explosive overpressure shock waves, often results in diffuse axonal injury (DAI) which is characterized by multiple small white-matter tracts lesions [28]. DAI is a progressive intra-axonal event that requires several hours after the injury for its complete evolution. It is often presented with profound coma and a poor outcome with enduring morbidity [26]. Overall, TBI classification based on physical mechanism is helpful in understanding the distinct pathophysiology produced by a specific force at specific magnitude and predict the pattern of injury.

3.2 Classification by Clinical Severity

Various trauma scores have been developed in the past to triage patients to assess the clinical severity of brain injury. Out of all of them, GCS has been the most commonly used scoring system. This scoring system allows bedside assessment of impaired level of

Table 1
The Glasgow Coma Scale

Eye response (E)	
4	Open
3	Open in response to voice
2	Open in response to pain
1	No response
Verbal response (V)	
5	Smiles, oriented to sounds
4	Cries but consolable
3	Inconsistently inconsolable
2	Inconsolable and agitated
1	No response
Motor response (M)	
6	Moves spontaneously
5	Withdrawal to touch
4	Withdrawal to painful stimuli
3	Decorticate flexion
2	Decerebrate extension
1	No response

Total score = $E + V + M$ [severe (3–8); moderate (9–12); mild (13–15)]

consciousness, the clinical hallmark of acute brain injury. It also helps in outcome prediction and treatment decisions [29, 30]. The GCS consists of the total score (range 3–15) of three components: (1) eye response (E); (2) verbal response (V); and (3) motor response (M). Each level of response is assigned a number—the worse the response, the lower the number. Based on total score ($E + V + M$), injuries are classified as severe (GCS 3–8), moderate (GCS 9–12), or mild (GCS 13–15) (Table 1). Often a point of confusion, the acronym GCS can be referred either the "Glasgow Coma Scale" (individual components) or the "Glasgow Coma Score" (total sum of components). It is important to note that the "scale" is applicable to the management of the individual patient, whereas the "score" summarizes information about groups of patients [29]. For an accurate classification of the clinical severity, the GCS should be assessed within a few hours of the injury and preferably before sedation and intubation. The GCS can also be affected by confounders such as intubation for airway protection, paralysis for medical control of raised ICP, sedation, alcohol

Table 2
The Pediatrics Glasgow Coma Scale

Eye response (*E*)	
4	Open
3	Open in response to voice
2	Open in response to pain
1	No response
Verbal response (*V*)	
5	Smiles, oriented to sounds
4	Cries but consolable
3	Inconsistently inconsolable
2	Inconsolable and agitated
1	No response
Motor response (*M*)	
6	Moves spontaneously
5	Withdrawal to touch
4	Withdrawal to painful stimuli
3	Decorticate flexion
2	Decerebrate extension
1	No response

Total score = $E + V + M$ [severe (3–8); moderate (9–12); mild (13–15)]

intoxication, shock, or low blood oxygen. For instance, blood alcohol concentrations greater than 240 mg/100 ml in TBI patients (highly intoxicated) has been found to be associated with 2–3 point reduction in GCS [31]. This scale has also been considered difficult to apply on preverbal children with head trauma. For these young patients, a slight modification of GCS, the Pediatric Glasgow Coma Scale (PGCS) was developed, where verbal responses were reported as appropriate words, social smiles, cries, irritability, and agitation [32] (Table 2). Another limitation of the GCS is that it does not account for a patient's transition from a vegetative-unresponsive state to a minimally conscious state. For this assessment a new coma score, the Full Outline of UnResponsiveness (FOUR) was developed. It consists of four components (eye, motor, brainstem, and respiration), and each component has a maximal score of 4. The FOUR tests for eye tracking or blinking to command also permits the early detection of locked-in syndrome [33]. Other neurological severity scales are the Brussels Coma Grades, Grady Coma Grades, and Innsbruck Coma Scale [34].

3.3 Classification by Structural Damage

TBI classification by structural damage determine the presence and extent of the injury by neuroimaging and plays an important role in the acute therapy of TBI. One of the advantages of this type of classification is that neuroimaging assessments of brain damage are not influenced by the confounders that affect the GCS [35–37]. For high-resolution neuroimaging, magnetic resonance imaging (MRI) is a very useful tool. However, the relatively long time period for image acquisition and its limited utility in ventilated patients limit its role in acute settings [38]. Therefore, amongst all available imaging tools, computerized tomography (CT) becomes the ideal option for assessment of acute structural damage following TBI. Marshall and colleagues, using the National Traumatic Coma Database, developed a CT-based classification of brain trauma [39]. This classification system utilizes the status of the mesencephalic cisterns, the degree of midline shift, and the presence or absence of one or more surgical masses and differentiates diffused injury in six categories as follows: (1) Diffuse injury I (no visible intracranial pathology); (2) Diffuse injury II (midline shift of 0–5 mm, basal cisterns remain visible, no high or mixed density lesions >25 cm^3); (3) Diffuse injury III (midline shift of 0–5 mm, basal cisterns compressed or completely effaced, no high or mixed density lesions >25 cm^3; (4) Diffuse injury IV (midline shift >5 mm, no high or mixed density lesions >25 cm^3; (5) Evacuated mass lesion (any lesion evacuated surgically); (6) Non-evacuated mass lesion (high or mixed density lesions >25 cm^3, not surgically evacuated). Patients with higher categories have a worse prognosis, which can be used in predicting mortality in adult patients. In a study with 634 traumatic neurosurgical patients, mortality in patients with Marshall score I and II was 0%, for score III was 40%, for score IV was 0%, for score V was 18.79% and for score VI was 95.66% [38]. Despite being broadly used and having strong predictive power, this classification system has certain limitations including the lack of specification of the type of mass lesion and difficulties in classifying patients with multiple injuries [34, 40]. The Rotterdam score is another CT-based classification system that utilizes some elements of the Marshall score to predict outcome [40]. This system differentiates between types of mass lesions and recognizes the more favorable prognosis [41, 42].

4 Diagnostic Procedures in TBI

The purpose of the TBI diagnostic procedures is to establish the type and severity of injury and to establish treatment goals (Fig. 1). A diagnosis of TBI is established based on clinical signs and symptoms such as risk factors, length of loss of consciousness (LOC), alteration of consciousness/mental state, and/or posttraumatic amnesia (PTA). Diagnosis of penetrating brain injury is self-evident

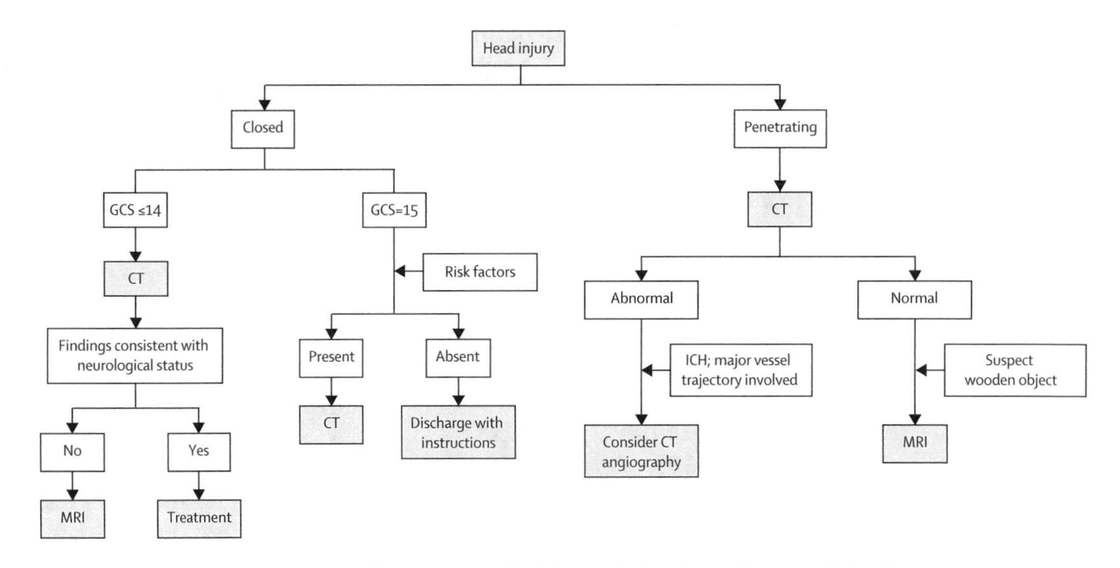

Fig. 1 Diagnostic approaches for TBI (reproduced with permission from Mass et al. [26])

and mostly straightforward. However, with closed-head injuries especially in cases of mild TBI, the symptoms are often not as straightforward or clear. In mild TBI PTA persists beyond loss of consciousness. This makes patients' self-reporting unreliable. False-negative diagnoses may also occur with patients who are fully oriented by the time the emergency personnel arrive but have no memory of the accident. It is recommended that in cases of mild TBI, a retrospective assessment of the patient's status should be performed [43]. Based on initial diagnosis, if certain signs and symptoms are present, the physician may further evaluate the injury by radioanatomical analysis of the brain [44]. CT examination is always recommended in patients with a GCS $\leq$14 and in patients with a GCS = 15 in the presence of risk factors [45, 46]. Traumatic intracranial lesions are frequently visible in CT-scan of severe and moderate TBI patients. However, the CT in Head Injury Patients (CHIP) study (2007) reported presence of intracranial lesions in 14% of patients with minor head injuries [47]. As discussed earlier in this chapter, CT is the preferred method of neuroimaging in TBI, especially for patients with acute injury and tracheal intubation. However, diffusion tensor imaging (DTI) has proved to be more informative in the subacute and chronic phases of injury. DTI quantifies isotropic and anisotropic water diffusion and provides better detection of white-matter brain lesions [48]. Because of the dynamic and progressive nature of TBI, a follow-up CT is advisable if lesions were present on the initial CT or there is indication of clinical deterioration or increasing ICP [26, 49]. There is also a minor chance of development of new lesions with diffuse injuries [50]. Computed tomography angiography (CTA) has also been increasingly used in TBI patients to uncover vascular lesions, especially in penetrating brain injuries. CTA has limited overall sensitivity in detecting arterial injuries but is more accurate in identifying traumatic intracranial aneurysms [51].

5 Long-Term Outcome Prediction After TBI

Approximately 30% sever TBI patients develop long-term neurological deficits and it is extremely difficult to identify those patients during the early phases of resuscitation [52, 53]. Models for the prediction of long-term outcomes are important in early decision making and interventions offered to patients. Several predictive models for patient outcomes after moderate to severe TBI have been proposed with varying methodological quality [54–57]. Specifically, International Mission for Prognosis and Analysis of Clinical trials in Traumatic brain injury database (IMPACT) models and Corticosteroid Randomization After Significant Head Injury (CRASH) models, were developed using large datasets and have good discriminatory power to predict long-term outcome [58, 59]. The predictions in IMPACT models (core, extended, and lab models) are based on age, GCS motor score, and pupillary reactivity for the core model; core model variables + CT classification, epidural hemorrhage (EDH) hypoxia, hypotension, and traumatic subarachnoid hemorrhage (tSAH) for extended model; and extended model + glucose and hemoglobin for lab model. These models predict mortality and unfavorable outcome (death, vegetative state or severe disability) at 6 months after injury (http://www.tbi-impact.org/?p=impact/calc). The CRASH models (basic model and CT model) were developed based on the CRASH trial dataset with patients who sustained mild, moderate and severe TBI [60, 61] for the prediction of mortality at 2 weeks after injury and unfavorable outcome at 6 months. The predictions in CRASH models are based on age, GCS motor score, pupillary reactivity, major extra-cranial injury, tSAH and CT classification (http://www.trialscoordinatingcentre.lshtm.ac.uk/Risk%20calculator/index.html). Both the IMPACT and CRASH models have been externally validated, indicating satisfactory generalizability and no relevant difference in performance [52]. However, a large observational study of 3626 patients concluded that although the IMPACT models demonstrated sufficient statistical performance, they fell below the level required to guide individual patient decision-making. The study suggests that management in a dedicated neurocritical care unit may be cost-effective compared with a combined neuro/general critical care unit and supports current recommendations that all patients with severe TBI would benefit from transfer to a tertiary care hospital, regardless of the need for surgery [62]. Large purpose-built neurotrauma registries may further improve these predictive models.

6 Future Directions

TBI management focuses on stabilizing the patient with prehospital and intensive clinical care. Once patients have been stabilized, they are often admitted to inpatient rehabilitation. These standardized TBI management protocols are based on weak evidence and there are weak attempts made for personalized treatment plans. Recent technological advances in neuroimaging and the development of tools for rapid detection and pathophysiological monitoring of injury, combined with informatics to integrate data from multiple sources, offer new avenues to improve TBI characterization and management approaches. New clinical trials such as the Collaborative European NeuroTrauma Effectiveness Research (CENTER-TBI) trial in adults and the Approaches and Decisions for Acute Pediatric TBI (ADAPT) trial in children, target the need for stronger evidence-based care. It is imperative that the knowledge gained from these clinical trials is swiftly translated to clinics for improved patient care. The development of Common Data Elements (CDE) for TBI research is another important step forward. TBI research projects and neurotrauma registries develop a huge amount of data and sharing of these data would greatly benefit future research. It is believed that TBI will likely remain the largest global contributor to neurological disability and although financial support for neurotrauma research is substantial, it is still very low when compared to other neurological conditions. Prospects for research funding can be improved by disease-awareness campaigns and a better explanation of TBI-associated socioeconomic burden education for the general public.

References

1. Menon DK, Schwab K, Wright DW, Maas AI, Demographics, Clinical Assessment Working Group of the I, Interagency Initiative toward Common Data Elements for Research on Traumatic Brain I, Psychological H (2010) Position statement: definition of traumatic brain injury. Arch Phys Med Rehabil 91 (11):1637–1640. https://doi.org/10.1016/j.apmr.2010.05.017
2. Rubiano AM, Carney N, Chesnut R, Puyana JC (2015) Global neurotrauma research challenges and opportunities. Nature 527(7578): S193–S197. https://doi.org/10.1038/nature16035
3. Feigin VL, Theadom A, Barker-Collo S, Starkey NJ, McPherson K, Kahan M, Dowell A, Brown P, Parag V, Kydd R, Jones K, Jones A, Ameratunga S, Group BS (2013) Incidence of traumatic brain injury in New Zealand: a population-based study. Lancet Neurol 12 (1):53–64. https://doi.org/10.1016/S1474-4422(12)70262-4
4. Jackson ML, Srivastava AK, Cox CS Jr (2017) Preclinical progenitor cell therapy in traumatic brain injury: a meta-analysis. J Surg Res 214:38–48. https://doi.org/10.1016/j.jss.2017.02.078
5. Cox CS (2017) Cellular therapy for traumatic neurological injury. Pediatr Res. https://doi.org/10.1038/pr.2017.253
6. Roozenbeek B, Maas AI, Menon DK (2013) Changing patterns in the epidemiology of traumatic brain injury. Nat Rev Neurol 9 (4):231–236. https://doi.org/10.1038/nrneurol.2013.22
7. Tagliaferri F, Compagnone C, Korsic M, Servadei F, Kraus J (2006) A systematic review of brain injury epidemiology in Europe. Acta

Neurochir 148(3):255–268.; discussion 268. https://doi.org/10.1007/s00701-005-0651-y

8. Rosenfeld JV, Maas AI, Bragge P, Morganti-Kossmann MC, Manley GT, Gruen RL (2012) Early management of severe traumatic brain injury. Lancet 380(9847):1088–1098. https://doi.org/10.1016/S0140-6736(12)60864-2
9. Stocchetti N, Zanier ER (2016) Chronic impact of traumatic brain injury on outcome and quality of life: a narrative review. Crit Care 20(1):148. https://doi.org/10.1186/s13054-016-1318-1
10. Stern RA, Riley DO, Daneshvar DH, Nowinski CJ, Cantu RC, McKee AC (2011) Long-term consequences of repetitive brain trauma: chronic traumatic encephalopathy. PM R 3 (10 Suppl 2):S460–S467. https://doi.org/10.1016/j.pmrj.2011.08.008
11. Van Den Heuvel C, Thornton E, Vink R (2007) Traumatic brain injury and Alzheimer's disease: a review. Prog Brain Res 161:303–316. https://doi.org/10.1016/S0079-6123(06)61021-2
12. Levin HS, Diaz-Arrastia RR (2015) Diagnosis, prognosis, and clinical management of mild traumatic brain injury. Lancet Neurol 14 (5):506–517. https://doi.org/10.1016/S1474-4422(15)00002-2
13. Fleminger S, Oliver DL, Lovestone S, Rabe-Hesketh S, Giora A (2003) Head injury as a risk factor for Alzheimer's disease: the evidence 10 years on; a partial replication. J Neurol Neurosurg Psychiatry 74 (7):857–862. http://dx.doi.org/10.1136/jnnp.74.7.857
14. Burke JF, Stulc JL, Skolarus LE, Sears ED, Zahuranec DB, Morgenstern LB (2013) Traumatic brain injury may be an independent risk factor for stroke. Neurology 81(1):33–39. https://doi.org/10.1212/WNL.0b013e318297eecf
15. Jafari S, Etminan M, Aminzadeh F, Samii A (2013) Head injury and risk of Parkinson disease: a systematic review and meta-analysis. Mov Disord 28(9):1222–1229. https://doi.org/10.1002/mds.25458
16. Crane PK, Gibbons LE, Dams-O'Connor K, Trittschuh E, Leverenz JB, Keene CD, Sonnen J, Montine TJ, Bennett DA, Leurgans S, Schneider JA, Larson EB (2016) Association of traumatic brain injury with late-life neurodegenerative conditions and neuropathologic findings. JAMA Neurol 73 (9):1062–1069. https://doi.org/10.1001/jamaneurol.2016.1948
17. Walsh S, Donnan J, Fortin Y, Sikora L, Morrissey A, Collins K, MacDonald D (2017) A systematic review of the risks factors associated with the onset and natural progression of epilepsy. Neurotoxicology 61:64–77. https://doi.org/10.1016/j.neuro.2016.03.011
18. Teasdale G, Jennett B (1974) Assessment of coma and impaired consciousness. A practical scale. Lancet 2(7872):81–84
19. Park E, Bell JD, Baker AJ (2008) Traumatic brain injury: can the consequences be stopped? CMAJ 178(9):1163–1170. https://doi.org/10.1503/cmaj.080282
20. Udomphorn Y, Armstead WM, Vavilala MS (2008) Cerebral blood flow and autoregulation after pediatric traumatic brain injury. Pediatr Neurol 38(4):225–234. https://doi.org/10.1016/j.pediatrneurol.2007.09.012
21. Werner C, Engelhard K (2007) Pathophysiology of traumatic brain injury. Br J Anaesth 99 (1):4–9. https://doi.org/10.1093/bja/aem131
22. McKee AC, Daneshvar DH (2015) The neuropathology of traumatic brain injury. Handb Clin Neurol 127:45–66. https://doi.org/10.1016/B978-0-444-52892-6.00004-0
23. Blennow K, Hardy J, Zetterberg H (2012) The neuropathology and neurobiology of traumatic brain injury. Neuron 76(5):886–899. https://doi.org/10.1016/j.neuron.2012.11.021
24. Pellman EJ, Viano DC, Tucker AM, Casson IR, Waeckerle JF (2003) Concussion in professional football: reconstruction of game impacts and injuries. Neurosurgery 53(4):799–812. discussion 812–794
25. Gurdjian ES, Roberts VL, Thomas LM (1966) Tolerance curves of acceleration and intracranial pressure and protective index in experimental head injury. J Trauma 6(5):600–604
26. Maas AI, Stocchetti N, Bullock R (2008) Moderate and severe traumatic brain injury in adults. Lancet Neurol 7(8):728–741. https://doi.org/10.1016/S1474-4422(08)70164-9
27. Gao TL, Yuan XT, Yang D, Dai HL, Wang WJ, Peng X, Shao HJ, Jin ZF, Fu ZJ (2012) Expression of HMGB1 and RAGE in rat and human brains after traumatic brain injury. J Trauma Acute Care Surg 72(3):643–649. https://doi.org/10.1097/TA.0b013e31823c54a6
28. Hicks RR, Fertig SJ, Desrocher RE, Koroshetz WJ, Pancrazio JJ (2010) Neurological effects of blast injury. J Trauma 68(5):1257–1263. https://doi.org/10.1097/TA.0b013e3181d8956d
29. Teasdale G, Maas A, Lecky F, Manley G, Stocchetti N, Murray G (2014) The Glasgow

Coma Scale at 40 years: standing the test of time. Lancet Neurol 13(8):844–854. https://doi.org/10.1016/S1474-4422(14)70120-6
30. Laureys S, Bodart O, Gosseries O (2014) The Glasgow coma scale: time for critical reappraisal? Lancet Neurol 13(8):755–757. https://doi.org/10.1016/S1474-4422(14)70152-8
31. Brickley MR, Shepherd JP (1995) The relationship between alcohol intoxication, injury severity and Glasgow Coma Score in assault patients. Injury 26(5):311–314
32. Reilly PL, Simpson DA, Sprod R, Thomas L (1988) Assessing the conscious level in infants and young children: a paediatric version of the Glasgow Coma Scale. Childs Nerv Syst 4 (1):30–33
33. Wijdicks EF, Bamlet WR, Maramattom BV, Manno EM, McClelland RL (2005) Validation of a new coma scale: the FOUR score. Ann Neurol 58(4):585–593. https://doi.org/10.1002/ana.20611
34. Saatman KE, Duhaime AC, Bullock R, Maas AI, Valadka A, Manley GT, Workshop Scientific T, Advisory Panel M (2008) Classification of traumatic brain injury for targeted therapies. J Neurotrauma 25(7):719–738. https://doi.org/10.1089/neu.2008.0586
35. Malec JF, Brown AW, Leibson CL, Flaada JT, Mandrekar JN, Diehl NN, Perkins PK (2007) The mayo classification system for traumatic brain injury severity. J Neurotrauma 24 (9):1417–1424. https://doi.org/10.1089/neu.2006.0245
36. Buechler CM, Blostein PA, Koestner A, Hurt K, Schaars M, McKernan J (1998) Variation among trauma centers' calculation of Glasgow Coma Scale score: results of a national survey. J Trauma 45(3):429–432
37. Moskopp D, Stahle C, Wassmann H (1995) Problems of the Glasgow Coma Scale with early intubated patients. Neurosurg Rev 18 (4):253–257
38. Munakomi S (2016) A comparative study between Marshall and Rotterdam CT scores in predicting early deaths in patients with traumatic brain injury in a major tertiary care hospital in Nepal. Chin J Traumatol 19(1):25–27
39. Marshall LF, Marshall SB, Klauber MR, Clark M v B, Eisenberg HM, Jane JA, Luerssen TG, Marmarou A, Foulkes MA (1991) A new classification of head injury based on computerized tomography. J Neurosurg Special Suppl 75 (1s):S14–S20. https://doi.org/10.3171/sup.1991.75.1s.0s14
40. Maas AI, Hukkelhoven CW, Marshall LF, Steyerberg EW (2005) Prediction of outcome in traumatic brain injury with computed tomographic characteristics: a comparison between the computed tomographic classification and combinations of computed tomographic predictors. Neurosurgery 57(6):1173–1182. discussion 1173–1182
41. Lobato RD, Cordobes F, Rivas JJ, de la Fuente M, Montero A, Barcena A, Perez C, Cabrera A, Lamas E (1983) Outcome from severe head injury related to the type of intracranial lesion. A computerized tomography study. J Neurosurg 59(5):762–774. https://doi.org/10.3171/jns.1983.59.5.0762
42. (2000) The Brain Trauma Foundation. The American Association of Neurological Surgeons. The Joint Section on Neurotrauma and Critical Care. Computed tomography scan features. J Neurotrauma 17 (6–7):597–627. https://doi.org/10.1089/neu.2000.17.597
43. Ruff RM, Iverson GL, Barth JT, Bush SS, Broshek DK, Policy NAN, Planning C (2009) Recommendations for diagnosing a mild traumatic brain injury: a National Academy of Neuropsychology education paper. Arch Clin Neuropsychol 24(1):3–10. https://doi.org/10.1093/arclin/acp006
44. Schmid KE, Tortella FC (2012) The diagnosis of traumatic brain injury on the battlefield. Front Neurol 3:90. https://doi.org/10.3389/fneur.2012.00090
45. Guidelines for minor head injured patients' management in adult age (1996) The study group on head injury of the Italian Society for Neurosurgery. J Neurosurg Sci 40(1):11–15
46. Vos PE, Battistin L, Birbamer G, Gerstenbrand F, Potapov A, Prevec T, Stepan Ch A, Traubner P, Twijnstra A, Vecsei L, von Wild K, European Federation of Neurological S (2002) EFNS guideline on mild traumatic brain injury: report of an EFNS task force. Eur J Neurol 9(3):207–219
47. Smits M, Dippel DW, Steyerberg EW, de Haan GG, Dekker HM, Vos PE, Kool DR, Nederkoorn PJ, Hofman PA, Twijnstra A, Tanghe HL, Hunink MG (2007) Predicting intracranial traumatic findings on computed tomography in patients with minor head injury: the CHIP prediction rule. Ann Intern Med 146 (6):397–405
48. Huisman TA, Schwamm LH, Schaefer PW, Koroshetz WJ, Shetty-Alva N, Ozsunar Y, Wu O, Sorensen AG (2004) Diffusion tensor imaging as potential biomarker of white matter injury in diffuse axonal injury. AJNR Am J Neuroradiol 25(3):370–376
49. Maas AI, Dearden M, Teasdale GM, Braakman R, Cohadon F, Iannotti F,

Karimi A, Lapierre F, Murray G, Ohman J, Persson L, Servadei F, Stocchetti N, Unterberg A (1997) EBIC-guidelines for management of severe head injury in adults. European Brain Injury Consortium. Acta Neurochir 139 (4):286–294

50. Servadei F, Murray GD, Penny K, Teasdale GM, Dearden M, Iannotti F, Lapierre F, Maas AJ, Karimi A, Ohman J, Persson L, Stocchetti N, Trojanowski T, Unterberg A (2000) The value of the "worst" computed tomographic scan in clinical studies of moderate and severe head injury. European Brain Injury Consortium. Neurosurgery 46 (1):70–75. discussion 75–77
51. Bodanapally UK, Shanmuganathan K, Boscak AR, Jaffray PM, Van der Byl G, Roy AK, Dreizin D, Fleiter TR, Mirvis SE, Krejza J, Aarabi B (2014) Vascular complications of penetrating brain injury: comparison of helical CT angiography and conventional angiography. J Neurosurg 121(5):1275–1283. https://doi.org/10.3171/2014.7.JNS132688
52. Roozenbeek B, Lingsma HF, Lecky FE, Lu J, Weir J, Butcher I, GS MH, Murray GD, Perel P, Maas AI, Steyerberg EW, International Mission on Prognosis Analysis of Clinical Trials in Traumatic Brain Injury Study G, Corticosteroid Randomisation After Significant Head Injury Trial C, Trauma A, Research N (2012) Prediction of outcome after moderate and severe traumatic brain injury: external validation of the International Mission on Prognosis and Analysis of Clinical Trials (IMPACT) and Corticoid Randomisation After Significant Head injury (CRASH) prognostic models. Crit Care Med 40(5):1609–1617. https://doi.org/10.1097/CCM.0b013e31824519ce
53. McHugh GS, Engel DC, Butcher I, Steyerberg EW, Lu J, Mushkudiani N, Hernandez AV, Marmarou A, Maas AI, Murray GD (2007) Prognostic value of secondary insults in traumatic brain injury: results from the IMPACT study. J Neurotrauma 24(2):287–293. https://doi.org/10.1089/neu.2006.0031
54. Perel P, Edwards P, Wentz R, Roberts I (2006) Systematic review of prognostic models in traumatic brain injury. BMC Med Inform Decis Mak 6:38. https://doi.org/10.1186/1472-6947-6-38
55. Hukkelhoven CW, Steyerberg EW, Habbema JD, Farace E, Marmarou A, Murray GD, Marshall LF, Maas AI (2005) Predicting outcome after traumatic brain injury: development and validation of a prognostic score based on admission characteristics. J Neurotrauma 22 (10):1025–1039. https://doi.org/10.1089/neu.2005.22.1025
56. Tasaki O, Shiozaki T, Hamasaki T, Kajino K, Nakae H, Tanaka H, Shimazu T, Sugimoto H (2009) Prognostic indicators and outcome prediction model for severe traumatic brain injury. J Trauma 66(2):304–308. https://doi.org/10.1097/TA.0b013e31815d9d3f
57. Mushkudiani NA, Hukkelhoven CW, Hernandez AV, Murray GD, Choi SC, Maas AI, Steyerberg EW (2008) A systematic review finds methodological improvements necessary for prognostic models in determining traumatic brain injury outcomes. J Clin Epidemiol 61(4):331–343. https://doi.org/10.1016/j.jclinepi.2007.06.011
58. Steyerberg EW, Mushkudiani N, Perel P, Butcher I, Lu J, McHugh GS, Murray GD, Marmarou A, Roberts I, Habbema JD, Maas AI (2008) Predicting outcome after traumatic brain injury: development and international validation of prognostic scores based on admission characteristics. PLoS Med 5(8):e165.; discussion e165. https://doi.org/10.1371/journal.pmed.0050165
59. Collaborators MCT, Perel P, Arango M, Clayton T, Edwards P, Komolafe E, Poccock S, Roberts I, Shakur H, Steyerberg E, Yutthakasemsunt S (2008) Predicting outcome after traumatic brain injury: practical prognostic models based on large cohort of international patients. BMJ 336 (7641):425–429. https://doi.org/10.1136/bmj.39461.643438.25
60. Roberts I, Yates D, Sandercock P, Farrell B, Wasserberg J, Lomas G, Cottingham R, Svoboda P, Brayley N, Mazairac G, Laloe V, Munoz-Sanchez A, Arango M, Hartzenberg B, Khamis H, Yutthakasemsunt S, Komolafe E, Olldashi F, Yadav Y, Murillo-Cabezas F, Shakur H, Edwards P, Ct c (2004) Effect of intravenous corticosteroids on death within 14 days in 10008 adults with clinically significant head injury (MRC CRASH trial): randomised placebo-controlled trial. Lancet 364 (9442):1321–1328. https://doi.org/10.1016/S0140-6736(04)17188-2
61. Edwards P, Arango M, Balica L, Cottingham R, El-Sayed H, Farrell B, Fernandes J, Gogichaisvili T, Golden N, Hartzenberg B, Husain M, Ulloa MI, Jerbi Z, Khamis H, Komolafe E, Laloe V, Lomas G, Ludwig S, Mazairac G, Munoz Sanchez Mde L, Nasi L, Olldashi F, Plunkett P, Roberts I, Sandercock P, Shakur H, Soler C, Stocker R, Svoboda P, Trenkler S, Venkataramana NK, Wasserberg J, Yates D,

Yutthakasemsunt S, Ct c (2005) Final results of MRC CRASH, a randomised placebo-controlled trial of intravenous corticosteroid in adults with head injury-outcomes at 6 months. Lancet 365(9475):1957–1959. https://doi.org/10.1016/S0140-6736(05)66552-X

62. Harrison DA, Prabhu G, Grieve R, Harvey SE, Sadique MZ, Gomes M, Griggs KA, Walmsley E, Smith M, Yeoman P, Lecky FE, Hutchinson PJ, Menon DK, Rowan KM (2013) Risk adjustment In Neurocritical care (RAIN)—prospective validation of risk prediction models for adult patients with acute traumatic brain injury to use to evaluate the optimum location and comparative costs of neurocritical care: a cohort study. Health Technol Assess 17(23):vii–viii., 1–350. https://doi.org/10.3310/hta17230

Chapter 2

Three-Dimensional In Vitro Brain Tissue Models

Min D. Tang-Schomer

Abstract

In vitro cell and tissue cultures are indispensable tools for brain research, including traumatic brain injury (TBI). Bioengineered three-dimensional (3D) tissue models are increasingly used as disease systems to understand complex cell–cell interactions and tissue functions. Here, we describe a bioengineered 3D in vitro brain tissue model that presents cortical tissue-like cell compartmentalization and mechanical property, long-term tissue growth and neurophysiological functions. The 3D model's brain-mimetic properties enabled recapitulation of dynamic tissue responses to TBI, as demonstrated with an experimental weight-drop injury setup. Here, we provide an overview of the design principles of the 3D brain tissue model and detailed instructions on constructing the model from raw materials (silkworm cocoons, hydrogel, cells), and on conducting mechanical testing and injury experiments. We describe downstream analytic assays for evaluation of the 3D tissue model and expected outcomes. Materials and methods described in this protocol can be adapted to other 3D culture systems.

Key words Brain cortex, Silk, Axon, Neuronal network, 3D tissue model, Tissue engineering

1 Introduction: 3D Brain Tissue Models

Brain research is performed at various levels, from studies of development to behavior. The brain's complexity requires sophisticated interdisciplinary methods and equipment for study. Basic brain research has traditionally relied on human studies, animal models or ex vivo brain slices. These in vivo models are highly variable, difficult to control and analyze, and cannot be scaled for high throughput drug testing. In vitro models reduce the complexity into elementary components and recapitulate fundamental features of the brain's behavior. By offering greater control of experimental conditions and more options for analytical evaluation, in vitro models have become indispensable tools to accompany in vivo studies for brain research. It is now widely recognized that 3D culture conditions are more physiologically relevant than conventional 2D cultures. There has been substantial progress in reconstructing native environments or niches for the cells by generating new materials to support the cells ex vivo, fabricating micro- and

Amit K. Srivastava and Charles S. Cox, Jr. (eds.), *Pre-Clinical and Clinical Methods in Brain Trauma Research*, Neuromethods, vol. 139, https://doi.org/10.1007/978-1-4939-8564-7_2,

nanoscale structural cues for guiding cell and tissue organization, and incorporating physiological flow conditions with microfluidic devices. Specialized systems have been developed to address specific research questions for the brain, such as hydrogel-based tumor niche, co-culture systems, and lab-on-chip designs for drug screening [1–3]. However, recreating functional features of the brain remains a significant challenge. Alternatively, advanced cell engineering technologies harness the endogenous programs of induced pluripotent cells and neural stem cells, and have generated cerebral organoids with features reminiscent of a developing brain [4]. These 3D organoids are used to study normal brain development, congenital malformation, and genetic disorders [5, 6]. However, these developmental brain tissue models have similar challenges as ex vivo systems regarding control of experimental conditions. The complementary advantages and limitations of different 3D in vitro models highlight the need for a versatile system that can be adapted for a wide range of research studies.

Here, we describe our design of a bioengineered 3D brain tissue model that presents a high degree of versatility regarding structural components and configuration, mechanical properties and assay options [7–11]. Much of the design focuses on its scalability for further improvement in complexity and functionality, and its adaptability to standard downstream biological assays, including dynamic and functional analysis. In this protocol, we provide detailed methods for model fabrication and assembly from raw materials, and assay options. We also describe the application of the 3D brain tissue model to traumatic brain injury (TBI) research. The methods developed for the 3D brain model are applicable to other 3D in vitro systems. Our goal is to provide the reader with sufficient information for reproducing the model, and guidelines for making adaptations to suit specific research questions.

1.1 3D Bioengineered Brain Models: Design Principles

3D in vitro model requires appropriate materials, tissue-mimetic structures and properties, and control of cells. The structural base of our model is a silk-collagen protein composite scaffold. Silk material is chosen due to its structural versatility and biocompatibility. The past research on silk materials has greatly advanced its application in tissue engineering and tissue-material interface controls [12, 13]. These studies have built a foundation for silk as a suitable base material for in vitro brain models. Specifically, silk material can be engineered with tunable mechanical property, and surface-coated to support neuronal growth. Moreover, the highly porous structure of the silk scaffold allows for free diffusion of nutrients and oxygen, and thus they prevent metabolic deficits in 3D tissue constructs and provide more relevant physiological conditions. Practically, silk scaffolds can be generated with low cost and in large quantities [14], which is a significant advantage compared to other biomaterials. Collagen is chosen as the base extracellular matrix

(ECM) component due to its stability and permissibility for neuronal growth. Other ECM components commonly used in brain research, such as Matrigel, fibrin or hyaluronic acid, lack long-term stability as stand-alone material for in vitro systems, but can be mixed into collagen base for additional growth support [11].

We chose a donut-shape design with a shell and a core region for a unit model. This design serves two purposes, structural mimicry of the brain's compartments, and controlled segregation of different cell components. The composite design allows neurons to be anchored to the scaffold shell region, and permits axon 3D growth in the gel matrix-filled core region, therefore, recapturing the compartmentalization of neuronal cell bodies and the axons into the grey and white matter of the brain. Additionally, it can be easily assembled into multiple concentric rings to represent separate structures incorporated with different cell populations. This feature could be used to mimic cortical layers.

The composite donut-shaped design has multiple practical advantages regarding experimental manipulation. The scaffold shell stabilizes hydrogel 3D volume, and prevents the gel from collapsing overtime that is a significant issue for hydrogel-based in vitro systems. The stiffer silk scaffold provides greater ease of handling 3D tissue models containing soft hydrogels, which is not easy with systems based solely on soft hydrogels. The relatively transparent gel-matrix core provides better optical clarity compared to the scaffold region, and permit 3D imaging of neurite processes and intracellular organelles.

We chose the donut shape to be 5–6 mm diameter by 2 mm height, with a center 2 mm diameter hole. The millimeter dimension combined with high porosity of the silk scaffold allows seeding of high cell numbers (in the range of millions) within a confined volume; a feature approximating the brain's high cellularity. The diameter can fit into a 96-well plate to enable microplate-based assays. Since most confocal microscopes are equipped with 10× lenses with a working distance up to 1 mm, the scaffold can be imaged from both sides to cover the entire *z*-range. These features maximize evaluable ranges of the 3D tissue model. In addition, the 3D tissue model's size is well suited to accommodate surgical manipulations developed for in vivo brains.

We have optimized these design parameters to generate 3D brain tissue models with rat embryonic cortical cells [7–11]. The neuronal networks can be observed after 3 days of culture, and they are maintained for weeks or months [7]. The 3D brain tissue model is highly reproducible, rapid to form, and consistent in outcome.

1.2 3D Bioengineered Brain Models: Application to TBI

TBI results from a direct impact to the brain followed by secondary injuries with lasting tissue damage, and encompasses a wide range of mechanical responses of the cortical tissue [15]. The complex sequelae after the initial injury event involves mechanical deformation, tissue structural damage and biochemical deregulation, inflammatory cascade, neuronal excitotoxicity and activity depression, and triggers for neurodegeneration [16]. Many cell culture-based systems have been devised to capture different aspects of TBI [17, 18]. Though useful for addressing specific questions especially at the cellular level, relating findings from these systems to functional outcomes at the tissue level remains a significant challenge. Animal models can provide behavioral outcomes, but direct observation of cellular responses cannot be achieved with sufficient temporal and spatial control. Therefore, 3D brain tissue models are necessary to bridge the gap between cell-level and tissue-level studies for TBI. Finally, the ability to capture real-time responses with in vitro systems can provide invaluable insights towards the understanding of TBI progression.

With our 3D brain tissue model, the silk scaffold can be tailored to match its mechanical properties to that of a brain, and the model design as described above (Sect. 1.1) permits comprehensive assessments including cellular damage, electrophysiological activity, and neurochemical changes in real time. We describe mechanical assessment of the 3D model in Sect. 3.2 and the adaptation of a weight-drop impact in vivo injury model to the in vitro system in Sect. 3.3. Responses of the model system to experimental TBI are investigated through the use of biochemical, immunological, morphological, and electrophysiological examinations, as described in Sects. 3.4–3.7.

2 Materials

2.1 Scaffold Biomaterials

2.1.1 Preparation of Silk Fibroin Solution from *Bombyx mori* (Silkworm) Cocoons

The schematics of silk fibroin solution preparation from silkworm cocoons is outlined in Fig. 1. We processed 5 g cocoons (~11 cocoons) per batch, and all the reagents were quantified for the batch size. Cocoons were cut into eight pieces each and soaked into boiling 2 L, 0.02 M Na_2CO_3 solution on a hot plate for 30 min. The de-gumming step dissolves the other undesirable silk fiber component, hydrophilic sericins, from insoluble silk fibroins. The fibroin were wrung out by hand and rinsed in distilled water at least three times to wash out any remaining sericin and chemicals [**Note 1**]. The fibroin extract was dried in a fume hood overnight, weighed and placed in a glass beaker. A 9.3 M LiBr solution was freshly prepared; and the required volume (mL) was calculated by multiplying the mass of dry fibroin by 4 [**Note 2**]. The LiBr solution was slowly poured over the silk fibroin, and a spatula was used to immerse all the fibroin fibers. The beaker was placed in a 60 °C oven for at least 4 h to allow the fibers to dissolve [**Note 3**]. The fibroin solution was collected with a syringe and 18G needle and loaded into MWCO 3500 dialysis cassettes (Thermo-Fisher). Dialysis against distilled water was performed for 48 h with water change every couple of hours. Afterwards, the fibroin

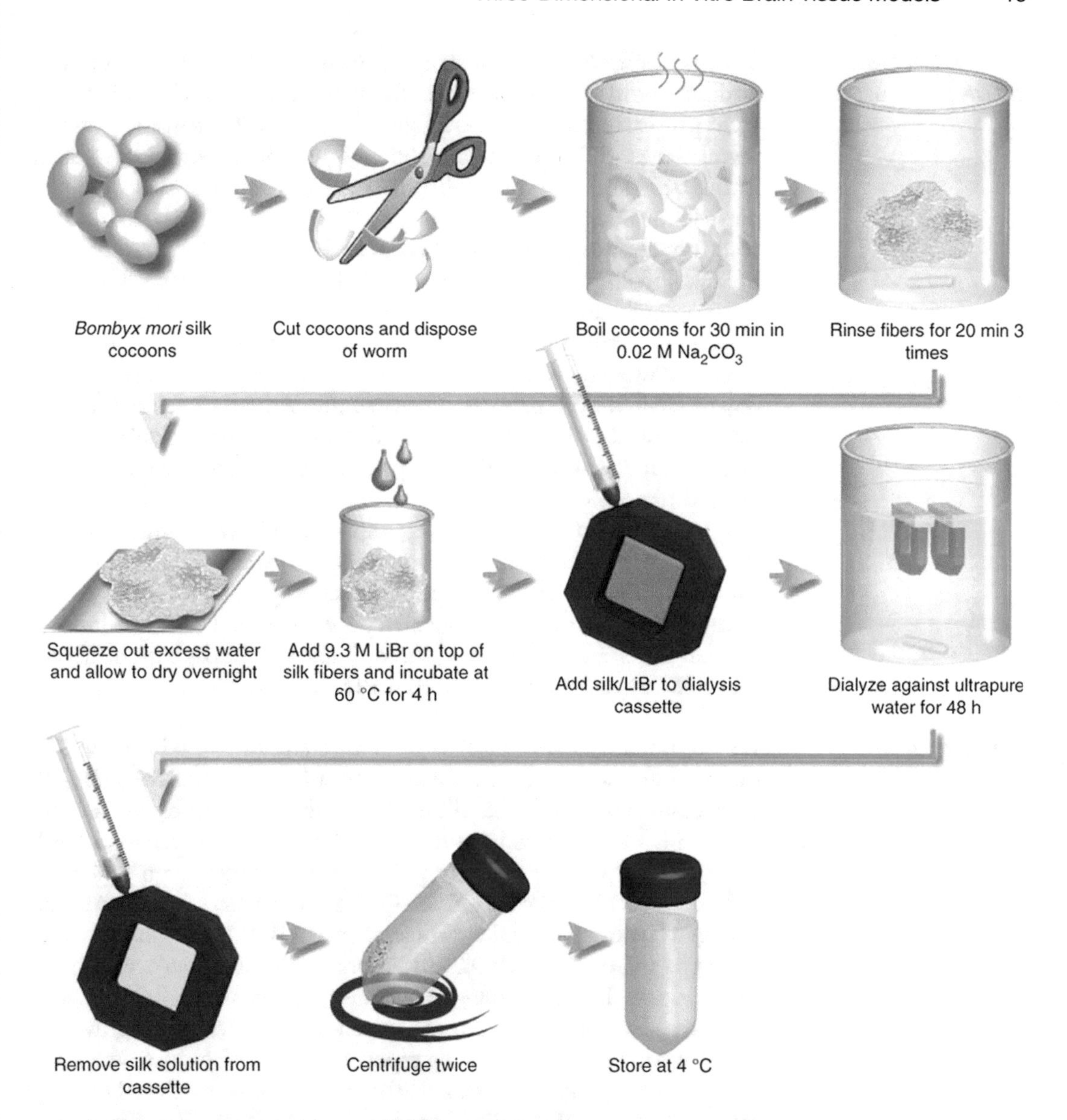

Fig. 1 Schematic of the silk fibroin extraction procedure. Reprinted by permission from Macmillan Publishers Ltd: [Nature Protocols] (Danielle N Rockwood, Rucsanda C Preda, Tuna Yücel, Xiaoqin Wang, Michael L Lovett, David L Kaplan. Materials fabrication from Bombyx mori silk fibroin. 2011, 6, 10, 1612–1631. copyright (2011)

solution was collected with a syringe from the cassettes into 50 mL conical tubes and centrifuged twice at maximum speed (~97,310 × *g*) at 4 °C for 20 min. After each centrifugation step, the supernatant was poured into a fresh tube and the pellet was discarded. The fibroin concentration was measured by estimating the dry weight of 1 mL solution after drying on a weight boat in a 60 °C oven for 2 h. Multiplying the dry weight by 100 gave the concentration of silk fibroin solution, typically 6–9% (w/v) per batch. The fibroin solution was adjusted to 6% (w/v) by diluting in distilled water [**Note 4**].

2.1.2 Preparation of Porous Silk Scaffolds

Porous scaffolds were constructed using a salt leaching method. Sodium chloride (salt) particles (Sigma-Aldrich, Natick, MA, USA) were pre-sorted into 500–600 μm-sized particles using test-grade metal sieves. Silk fibroin solution (6% w/v in water) was poured into a 100-mm petri dish, followed by dispensing of pre-sorted salt particles at a volume-to-weight ratio of 1:2 (i.e., 45 mL silk fibroin solution and 90 grams of salt) [**Note 5**]. At least 48 h was allowed for silk fibroin protein precipitation [**Note 6**]. After the silk fibroin is completely solidified, the mixture together with the dish was placed in distilled water under constant agitation for 2 days to dissolve away the salt particles; water was replaced at least once a day to facilitate the process. The final product of a porous silk mat can be stored at 4 °C or −20 °C for months. To prepare for the donut-shaped silk scaffold, biopsy punches were used to cut out the desired dimensions [**Note 7**].

To prepare for cell seeding, silk scaffolds were immersed in distilled water and autoclaved under a wet cycle (121 °C, 20 min). Before the planned cell seeding, the scaffolds were immersed in sterile 0.1 mg/mL poly-D-lysine (PDL) solution, incubated for 1 h and washed three times with phosphate buffered saline (PBS) to remove nonbound PDL [**Note 8**].

2.2 Hydrogels

We used collagen type I gel as the base matrix material for the 3D brain tissue model, and found that steps for handling collagen gel were easily adapted for handling many other types of hydrogels. In our hand, 1 mL liquid gel solution was sufficient to prepare 30–50 scaffolds. To prepare collagen gel, 10× M199 (containing phenol red) or 10× DMEM, 1 N NaOH solution and liquid rat tail collagen type I (in 0.02 N acetic acid, BD Biosciences) were mixed in the order at a 10:2:88 volume ratio in a microcentrifuge tube on ice. Because collagen gelation occurs immediately upon neutral pH at room temperature, care was taken to thoroughly mix the solution by slow and steady pipetting, and the solution was kept cold [**Note 9**]. Other hydrogels that were tested for the 3D tissue model include Matrigel (Corning) [**Note 10**] [7], fibrin gel (prepared by mixing 20 mg/mL fibrinogen and 10 U/mL thrombin at a volume ratio of 2:5) [7], Hydromatrix (Sigma-Aldrich) [11], Puramatrix (3D Matrix) [11], and HystemC (Sigma-Aldrich) [11] [**Note 11**].

2.3 Cells

We used primary cortical neurons from embryonic rats, with a standard dissociation protocol previously developed for neuroscience studies. Dissect cortices from embryonic day 18 (E18) Sprague-Dawley rats (Charles River) were obtained after the approval of Institutional Animal Care and Use Committee (IACUC) protocol. Ten cortices were incubated in 5 mL of 0.3% trypsin (Sigma) with 0.2% DNase I (Roche applied Science) for 20 min at 37 °C. Trypsin was inactivated by adding equal volume of

1 mg/mL soybean protein solution (Sigma). The cortices was titrated using 10 mL Pasteur pipette by pipetting up and down < 20 times till single cell suspension was generated [**Note 12**]. The cell suspension was centrifuged at $\sim 300 \times g$ for 5 min; and the cell pellet was resuspended in 10 mL of NeuroBasal media (Invitrogen) supplemented with B-27, penicillin/streptomycin (100 U/ml), and GlutaMax™ (2 mM) (Invitrogen). The expected cell concentration was about 2×10^7/mL.

3 Methods

In this section, we describe methods of the 3D brain model assembly with the base materials (scaffold, hydrogel, cells) (Sect. 3.1), mechanical testing (Sect. 3.2), TBI-like mechanical injury (Sect. 3.3), and downstream multimodal assessments (Sects. 3.4–3.7). The evaluation matrix includes nonterminal assays (Sect. 3.4), electrophysiological analysis (Sect. 3.5), immunostaining and morphological analysis (Sect. 3.6) and genetic analysis (Sect. 3.7). Figure 2 provides the outline of the methods (Fig. 2a) and examples of the 3D model assembly process (Fig. 2b). Each assay provides quantitative evaluation of a specific feature of the 3D brain tissue model. Combined, they provide comprehensive assessments of the 3D tissue model including genetic, biochemical, structural

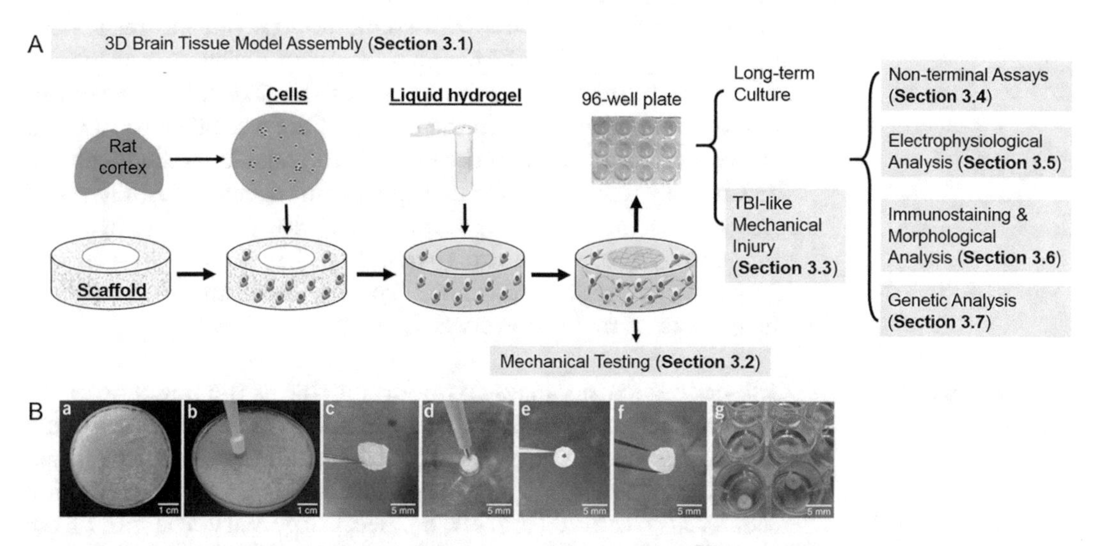

Fig. 2 Methods layout and examples. (**A**) Methods including 3D brain tissue model assembly and downstream assays. (**B**) Examples from silk fibroin scaffold to 3D neuronal cultures. (*a*) Aqueous-based silk sponge hydrated in water. (*b*) Cutting out 6-mm-diameter discs. (*c*) Porous silk scaffold. (*d*) Cutting out 2-mm-diameter central space. (*e*) Donut-shaped silk material scaffold. (*f*) Silk scaffold filled with collagen gel. (*g*) 3D bioengineered brain tissue constructs in tissue culture plate. **b** is reprinted by permission from Macmillan Publishers Ltd: [Nature Protocols] (Chwalek, K., Tang-Schomer, M.D., Omenetto, F.G., Kaplan, D.L. In vitro bioengineered model of cortical brain tissue, 2015; 10(9):1362–73. copyright (2015)

and functional characteristics. As mentioned in the Design Principles (Sect. 1.1), the 3D brain tissue model is constructed with maximum adaptability for standard biological assays; so the model requires minimal adaptation for easy assay. With one batch of cells, many replicates of 3D brain tissue models can be constructed and used for different assays at various time points. As many nonterminal assays can be used, users should consider performing more than one assay on one sample set, such as collecting medium supernatant for solution-based assays, followed by electrophysiological analysis, immunostaining, or nuclei acids or protein analysis.

3.1 3D Brain Tissue Model Assembly

3.1.1 Cell Seeding

Scaffold can be impregnated with single cell suspensions in two ways. For controlled seeding, scaffolds were placed into individual wells of a 96-well plate. Cell suspension of desired concentration was applied into the wells at 100 μL/well. For quick seeding with many scaffolds, cell suspension of 1×10^7 to 1×10^8/mL was applied to ~32 scaffolds in one well of a 6-well plate at 2 mL/well [**Note 13**]. The cells were incubated at 37 °C overnight to allow for cell attachment to the scaffolds. The following morning the cultures were aspirated off nonattached cells and replaced with fresh culture medium [**Note 14**].

3.1.2 Hydrogel Embedding

The cell-seeded silk scaffolds were embedded in hydrogel matrix between day 2 and 4 after seeding. The steps described here are for collagen gel, and are applicable to other types of hydrogels. The scaffold was first removed with sterile forceps onto a fresh, clean and dry surface [**Note 15**]. Apply 50 μL of collagen solution per scaffold by pipetting it in the center of the scaffold. This volume should be sufficient to fill in the sample. Care was taken to avoid air bubble formation. Return the embedded scaffold to the incubator for 1 h at 37 °C to allow for collagen gelation. A successfully gelled scaffold presents an opaque middle core [**Note 16**]. After hydrogel embedding, the 3D cultures were placed in a new 96-well and immersed in fresh, prewarmed culture medium. Half-medium change was carried out every 3–5 days.

3.1.3 Effects of Cell Densities

Cell seeding density plays an important role in the outcome of the final product, as demonstrated in Fig. 3. We had tried different seeding densities with the controlled seeding method, and found that cell attainment in the scaffold reached a plateau of ~2 million seeded cells/scaffold (Fig. 3A, B). Because primary cells suffer from initial dissociation trauma, and have ~50% cell loss within the first day, the number of cells retained in the 3D culture would be less than the initial seeded number. We used DNA measurement with Picogreen (Invitrogen) assay to quantify cell numbers. We first determined the linear DNA-cell number correlation with known cell numbers from 2D cultures of embryonic rat cortical neurons (Fig. 3c). Based on the standard curve, we estimated cell numbers

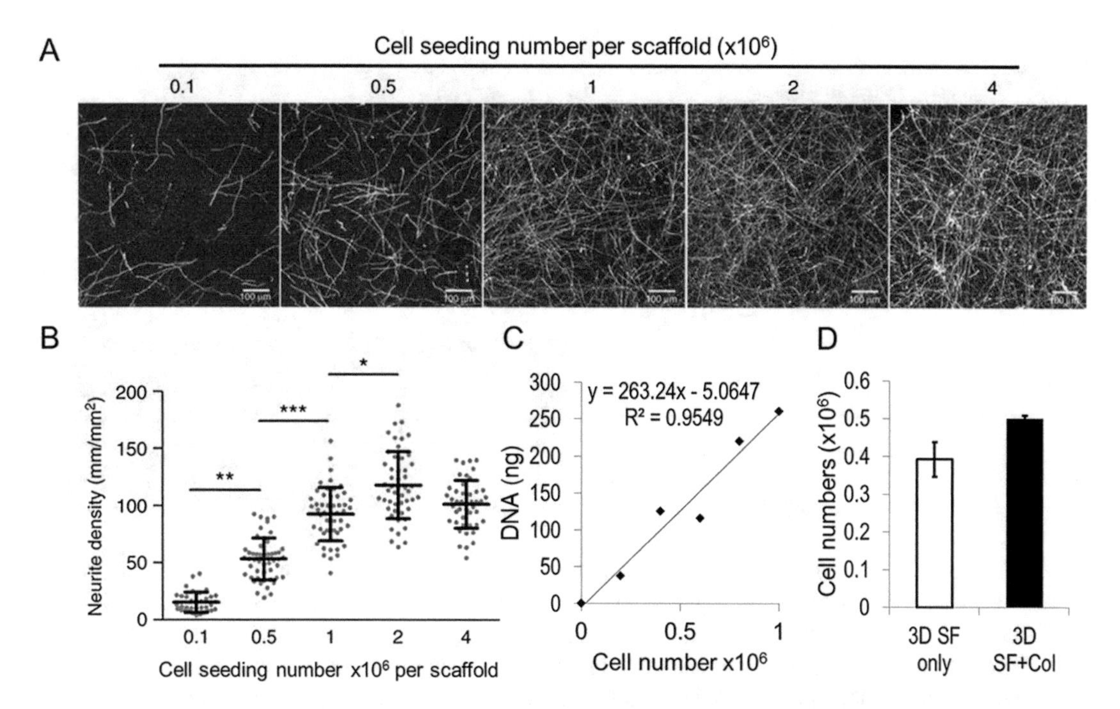

Fig. 3 Cell seeding densities and final cell numbers. (**A**) Axon networks in the center collagen gel region of 1 week-old 3D cultures of different cell seeding densities. Scale bars, 100 μm. (**B**) Neurite density quantification. (**C**) Linear correlation of DNA quantity with in vitro cultured neuronal numbers. (**D**) Cell numbers calculated according to the standard curve in **C** from extracted DNA quantities from 3D cultures. (**A** and **B**) are reprinted by permission from Macmillan Publishers Ltd: [Nature Protocols] (Chwalek, K., Tang-Schomer, M.D., Omenetto, F.G., Kaplan, D.L. In vitro bioengineered model of cortical brain tissue, 2015; 10(9):1362–73. copyright (2015). (**C** and **D**) are reprinted by permission from PNAS (Tang-Schomer MD, White JD, Tien LW, Schmitt LI, Valentin TM, Graziano DJ, Hopkins AM, Omenetto FG, Haydon PG, Kaplan DL. Bioengineered functional brain-like cortical tissue. 2014; 111(38):13811–6

of 3D tissue models using their extracted DNA counts. Cell attainment at the maximum seeding density (2–4 million cells/scaffold) was found to be ~500,000 cells/3D model (Fig. 3d, "3D SF+Col") [7].

3.2 Mechanical Testing

We characterized the 3D tissue model's Young's modulus and compared with fresh rat and mouse brain tissues. The 3D model's dimension and stiffness are comparable with the animal brain tissues, so the same apparatus and mechanical testing can be used for all these samples. We used confined compression test on an Instron mechanical tester (Instron 3366), and machined a customized chamber using aluminum stock and stainless steel porous discs (40 μm pore size) (McMaster-Carr). To perform a stress-relaxation test, the sample was compressed step-wise at 5% of its height, relaxed for 500 s to allow for equilibrium to be established as secreted water flowed out through the porous disc. The process was repeated for 2 h. The machine recorded the load-strain

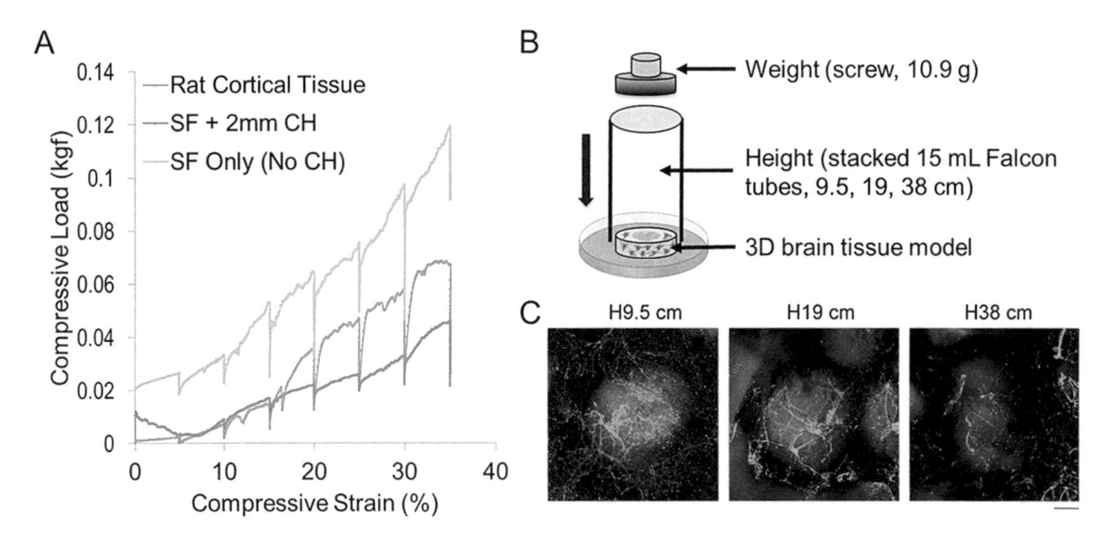

Fig. 4 Mechanical testing and TBI-like injury experiment. (**A**) Load-strain diagrams generated from Instron compression test. "*SF*" scaffold. "*CH*" center hole. "SF + 2 mm CH" is the 3D brain tissue model format. (**B**) Schematics of the weight-drop injury setup. (**C**) Fluorescence images of 3D brain tissue models fixed immediately after impact. Images are 2D sums of confocal image stacks of neurons immunostained with TUJ1 in *green* superposed with bright field images of silk structure in dark grey. Note the decrease of neuronal network density and staining intensity at increased impact forces. Scale bar, 100 μm. Figures are modified with permission from PNAS (Tang-Schomer MD, White JD, Tien LW, Schmitt LI, Valentin TM, Graziano DJ, Hopkins AM, Omenetto FG, Haydon PG, Kaplan DL. Bioengineered functional brain-like cortical tissue. 2014; 111(38):13811–6

diagram. Figure 4A provides example load-strain traces. We calculated the Young's modulus as the linear slope of the minima at equilibrium [**Note 17**].

3.3 TBI-Like Mechanical Injury

Many mechanical injury protocols have been developed for TBI models, including weight drop, fluid percussion, and blast injury [17, 18]. Because the weight drop method can easily control the injury severity by adjusting the height and induce a wide range of deficit [19], we adapted this method for the 3D in vitro brain model. Figure 4B shows the schematics of the injury setup; and Fig. 4C shows immunostained examples of tissue damage under different forces. We used a stainless steel screw with a head of 5 mm diameter as the weight of 10.9-g [**Note 18**], and fashioned a see-through tube of adjustable height by fitting longitudinally three 15 mL Falcon tubes with the bottoms cut-out. The tube gave three heights, 9.5, 19 and 38 cm. Calculations based on the compressive modulus of 3D brain-like tissues estimated that these conditions would generate compression distances of 0.10, 0.15, and 0.22 mm, respectively, comparable with the ~0.3-mm compression of the brain in vivo [19] [**Note 19**].

The test sample was placed into a 35-mm petri dish containing 1 mL artificial cerebral spinal fluid (CSF) solution [**Note 20**] before injury, and fitted inside the Falcon tube of the injury apparatus aligned with the weight. We removed the tube and the screw immediately after the screw hit the sample. The 3D tissue model was either returned to live electrophysiology recording (Sect. 3.5) or fixed for staining (Sect. 3.6). Solution samples were collected (100 μL) from the original petri dish before and after injury and preserved at −80 °C for downstream liquid analysis (Sect. 3.4).

3.4 Nonterminal Assays for 3D Brain Tissue Model

Nonterminal assays include solution-based assays that analyze the soluble components of the culture media, and microplate-based assays using nontoxic and cell permeable fluorescent indicators for cell activities. These assays provide quantitative measures of a cell culture's activity over time. Solution-based assays that had been successfully performed with our 3D brain tissue models include mass spectrometric measurements of neurotransmitters [7] and enzymatic assays [11]. Figure 5a shows mass spectrometric analysis of glutamate release from mechanically injured 3D brain tissue models. Microplate-based assays are especially valuable for 3D cultures due to the difficulty of direct visual assessment. We have adapted commercial assay kits for the 3D brain tissue models, as summarized in Table 1. Figure 5B shows AlamarBlue assay of 3D brain tissue model's viability in comparison with collagen gel-based cultures.

3.5 Electrophysiological Analysis of 3D Model

To assess the 3D brain tissue model's neuronal activities, we measured local field potential (LFP) (Fig. 6). Due to the comparable dimension and mechanical properties of the 3D brain tissue model with a rodent brain, a similar setup for brain slice extracellular recording was used for in vitro measurement [20]. The recording probe consisted of two parallel tungsten electrodes (50 μm dia. tip, 250 μm separation), as drawn in Fig. 6-*a*. The 3D model was transferred to HEPES-buffered artificial CSF in a 35-mm dish [**Notes 20** and **21**]. The reference electrode was placed in the liquid at a safe distance away from the tissue model. The tip of the recording probe was inserted into the tissue model with a micromanipulator. The electrode can be retrieved and reentered into a different spot [**Note 22**]. Signals were amplified with an AM-amplifier (AM-Systems) and digitized at 50 kHz. Acquisition was made using Clampex version 9.2 (Molecular Devices). Spontaneous activities were monitored using the Clampex software, and recorded once the trace became stable [**Note 23**].

For drug experiments, baseline signals were first obtained for 10 min, and 10 μL droplet of drug solution, such as 1 mM tetradotoxin (TTX) with a ~20 μM final concentration, was applied near the recording probe. After the signal stabilized in ~1 min, recording was resumed for another 10 min.

For mechanical injury experiments, baseline signals were first obtained for 10 min, and the probe was removed. After the weight

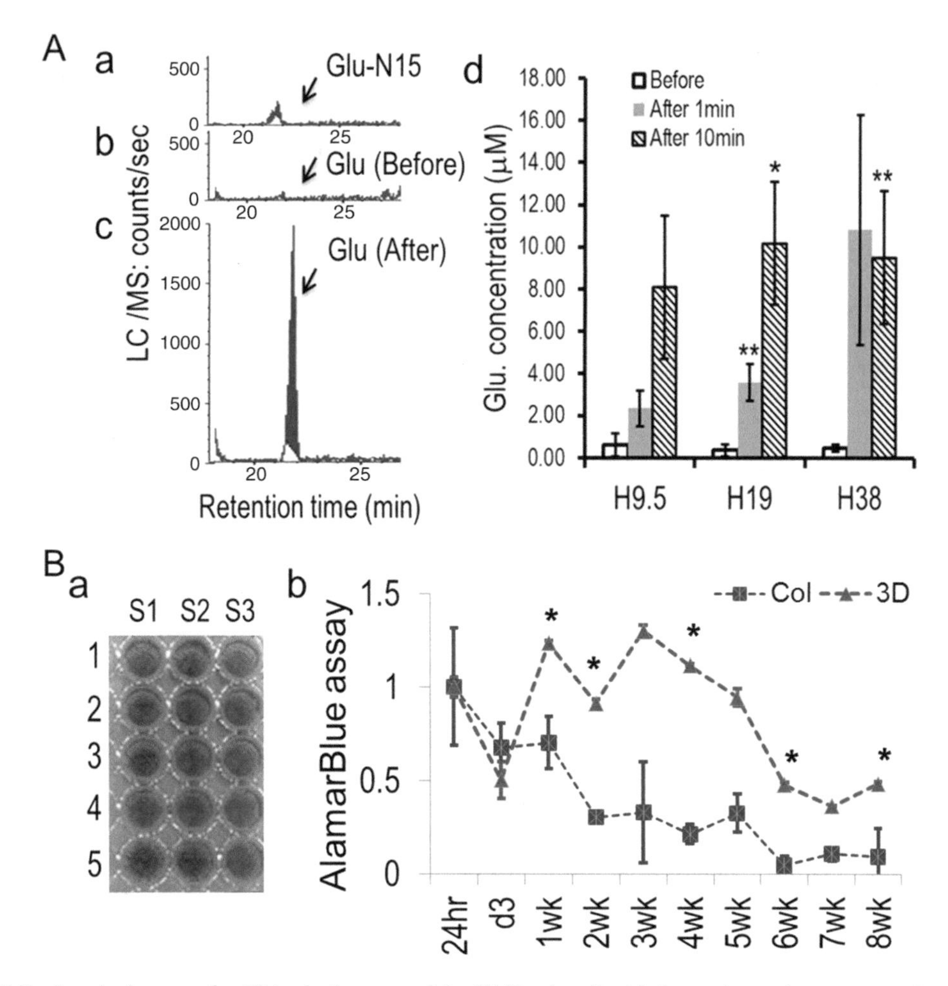

Fig. 5 Nonterminal assays for 3D brain tissue models. (**A**) Tandem liquid chromatography mass spectrometric (LC/MS) measurements of glutamate (Glu) release after mechanical injury of the 3D brain tissue model. (*a–c*) Representative LC/MS detection traces of the internal control Glu-N15 (*a*) and the Glu level at the baseline (*b*) and after impact (*c*). Glu peaks (arrow) at a retention time of ~21 min. (*d*) Quantification of Glu levels before and at 1 and 10 min after injury. Student *t* test. $^*P < 0.05$; $^{**}P < 0.01$ vs. before. (**B**) Microplate-based AlamarBlue assay to measure viability of 3D tissue cultures. (*a*) An example plate culture after alamarblue assay showing three samples (S1–S3) of five replicates per sample. (*b*) Brain-like tissues (*red*) and collagen (Col) gel-based cultures (*blue*) of similar cell seeding numbers were compared. Data was expressed as microplate reading relative to 24-h levels. Student *t* test. $^*P < 0.05$, 3D vs. Col. Figures are modified with permission from PNAS (Tang-Schomer MD, White JD, Tien LW, Schmitt LI, Valentin TM, Graziano DJ, Hopkins AM, Omenetto FG, Haydon PG, Kaplan DL. Bioengineered functional brain-like cortical tissue. 2014; 111 (38):13811–6

drop, the sample was immediately placed back to the recording stage and the probe was repositioned to approximately the same loci in the 3D model. The post-injury recording contained a ~2 min delay to exclude artifacts associated with probe reinsertion. The probe was rinsed with PBS between samples.

Table 1
Commercial assays adapted for the 3D brain tissue model

Assay	Source	Outcome	Action	Basic steps	Note	Reference
AlamarBlue	ThermoFisher	Viability/ cytotoxicity	The active ingredient of alamarBlue® (resazurin), a nontoxic, nonfluorescent, cell permeable compound, is converted by the natural reducing power of living cells to the fluorescent molecule, resorufin. This color change and increased fluorescence can be detected using absorbance (detected at 570 and 600 nm) or fluorescence (using an excitation between 530 and 560 and an emission at 590 nm).	The samples were incubated for 3 h with alamarBlue reagent diluted 1:10 (v:v) in culture medium, and followed by fluorescence reading of the medium at excitation/emission of 545/590 nm (SpectraMax M2™, Molecular Devices). 3D constructs without cells were used as control cultures. Minimum 4–8 replicates were used per condition.	AlamarBlue assayed 3D tissue models can be washed with fresh media and continued for culture. Samples can also be freezed and stored at −80 °C for nucleic acids and protein extraction.	Tang-Schomer et al. PNAS, 2014 [7]
WST-1	Roche	Viability/ mitochondria health	Tetrazolium salts (e.g., MTT, XTT, WST-1) are cleaved to soluble formazan by endogeneous reduction activities that occurs primarily at the cell surface. The amount of formazan dye formed directly correlates to the number of metabolically active cells in culture.	The samples were incubated for 1–2 h with WST-1 reagent diluted 1:10 (v:v) in culture medium, followed by absorbance reading of the medium at 450 nm with 600 nm as the reference wavelength on a microplate reader. The fresh medium was used as a baseline control and its average absorbance value was subtracted from the value of the samples.	Cultures cannot be continued after terazolium salt treatment.	Ren M, et al. Sci. Rep., 2016 [10]; Sood D et al. ACS Biomat. Sci. and Eng. 2016 [11]

(continued)

Table 1 (continued)

Assay	Source	Outcome	Action	Basic steps	Note	Reference
JC-1	Invitrogen (T3168)	Mitochondria (MT) membrane potential (Vmem)/ apoptosis	Changes of MT Vmem indicates mitochondrial disruption, and is an early sign of cell apoptosis. Probes that detect mitochondrial membrane potential are positively charged, causing them to accumulate in the electronegative interior of the mitochondrion. The membrane-permeant JC-1 dye exhibits potential-dependent accumulation in mitochondria, indicated by a fluorescence emission shift from green (~529 nm) to red (~590 nm), due to concentration-dependent formation of red fluorescent J-aggregates.	Cultures were washed once with HBSS prior to staining. JC1 dye was diluted in Hank's Balanced Salt Solution (HBSS) to a final concentration of 5 μM and added to the culture (100 μL/well). The cells were incubated at 37 °C for 30 min in the dark. The cells were washed once with HBSS to remove excess dye. Fluorescence intensities were measured using a top-read fluorescence microplate reader. Change of mitochondria membrane potential was expressed as the ratio of readings at two emission sites of the dye (red/green: 590 nm/ 529 nm, excitation: 488 nm) representing the ratio of JC-1 aggregates (red)/monomers (green).	Uneven JC-1 staining as a result of heterogeneous cellular responses in a 3D culture may affect global readings.	Ren M et al. Sci. Rep., 2016 [10]
Di-8-ANEPPS dye	Invitrogen (D-3167)	Plasma membrane potential/cell physiology including nerve impulses, cell	ANEP dyes are molecules that fluoresce in response to electrical potential changes in their environment. Their	Di-8-ANEPPS dye was diluted in HBSS to a final concentration of 2 μM and added to the cells (100 μL/well). The cells	Uneven dye staining as a result of heterogeneous cellular responses in a	Ren M et al. Sci. Rep., 2016 [10]

	signaling and ion-channel gating	optical response is sufficiently fast to detect transient (millisecond) potential changes in excitable cells. However, the magnitude of their potential-dependent fluorescence change is often small; typically a 2–10% fluorescence change per 100 mV. ANEP dyes are not fluorescent until bound to cell membrane. These dyes permit the quantitation of membrane potential using excitation ratio measurements.	were incubated at 37 °C for 30 min in the dark. The cells were washed once with HBSS to remove excess dye prior to measurement. Fluorescent intensities were measured using a top-read fluorescence microplate reader. Cell membrane potential was expressed as the ratio of readings at two excitation sites of the dye (450 nm/510 nm, emission: 640 nm).	3D culture may affect global readings.

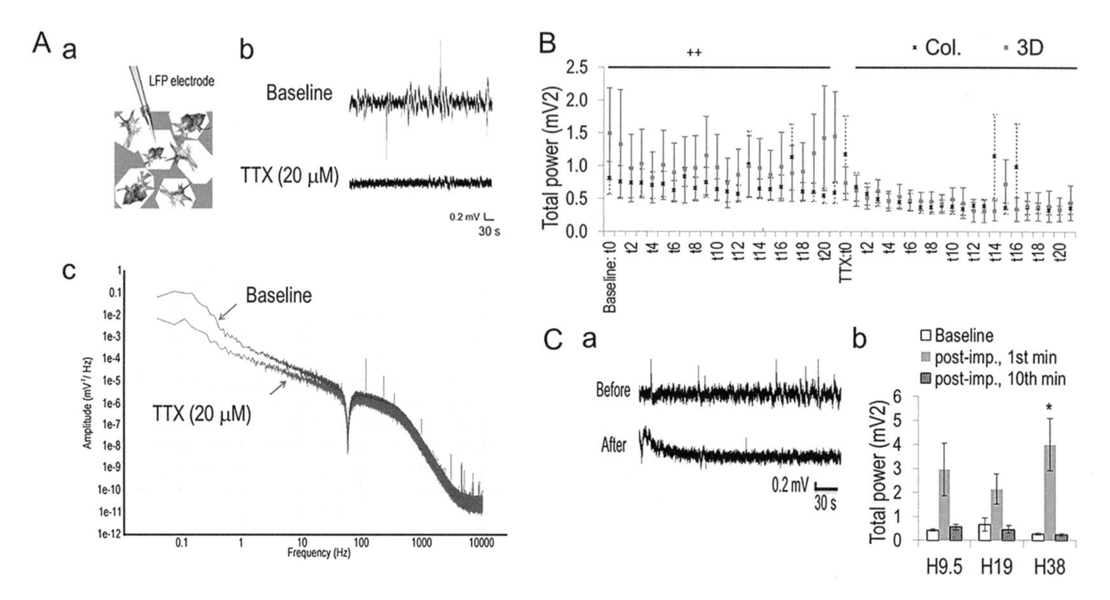

Fig. 6 Electrophysiological analysis. (**A**) LFP measurement. (*a*) Schematics. (*b*) Representative signal traces of baseline and after TTX (20 μM) treatment. (*c*) A representative power spectrum after fast Fourier transformation of raw signal traces. (**B**) Time-evolved changes of total power (millivolts^2; 0–50 Hz) over a 20-min duration (10 min of baseline and 10 min of TTX treatment). Each segment (t0–t20) represents a 27-s window. Col, collagen. (**C**) Injury-induced electrophysiological activity changes. (*a*) Representative signal traces of baseline (before) and after impact (after). The electrode was removed during injury and repositioned after a ~2-min delay to avoid artifacts from the impact. (*b*) Impact force-dependence and time-dependence of activity changes. Total power calculated from recordings of a 10-min baseline, the first 1-min post-impact, and tenth minute post-impact. Paired Student *t* test. *$P < 0.05$ vs. baseline. Figures are reprinted by permission from PNAS (Tang-Schomer MD, White JD, Tien LW, Schmitt LI, Valentin TM, Graziano DJ, Hopkins AM, Omenetto FG, Haydon PG, Kaplan DL. Bioengineered functional brain-like cortical tissue. 2014; 111 (38):13811–6

For analysis, the recording was imported into the Clampfit software (Molecular Devices). Power spectrum of a 10-min recording trace was obtained by fast Fourier transformation (FFT). Power at each time point was calculated by the sum (mV^2) of 0–50 Hz power (mV^2/Hz). For total power analysis assuming the 10-min trace is representative of the sample's homeostasis, averaging all time points' power gives the average total power. For activities that display time-dependent changes, such as an initial burst followed by slow depression, the recording time period needs to be divided into a series of time windows. We divided a 10-min trace into 22 segments (t0–t21) with a window size of 27 s. The power sums were plotted against the time points to demonstrate time-dependent changes.

Example recording traces and power spectrum after FFT are shown in Fig. 6A-*b* and -*c*, respectively. Time-evolved changes of the total power with TTX treatment are shown in Fig. 6B. Example recording traces and power analysis after mechanical injury are provided in Fig. 6C-*a* and -*b* respectively.

3.6 Immunostaining and Morphological Analysis of 3D Brain Tissue Models

Successful reconstitution of 3D brain tissue models can be assessed visually using bright-field microscopy by identifying the outgrowth of neurites within the central hydrogel. However, as the hydrogel is not fully transparent, the identification of thin neurites may be difficult for untrained eyed. Thus, fluorescent immunostaining need to be performed using anti-TUJ1 and anti-GFAP for neurons and astroglial cells, respectively. Figure 7A shows axon outgrowth into the collagen gel-filled center region of the 3D model overtime. The 3D tissue cultures were fixed with 4% paraformaldehyde (Fisher Scientific) for 20 min, washed, permeabilized with 0.1% Triton X-100 (Fisher Scientific) including 4% goat serum (Sigma) for 20 min. The fixed samples were incubated with primary antibodies overnight at 4 °C, followed by three 10 min washes and incubation with secondary antibodies for 1 h at room temperature and subsequent PBS washes.

3D tissue models were imaged with a confocal microscope (Leica SP2). In our system, a 10× lens captured a field of view of 1.5 mm by 1.5 mm which covered most of the center collagen gel

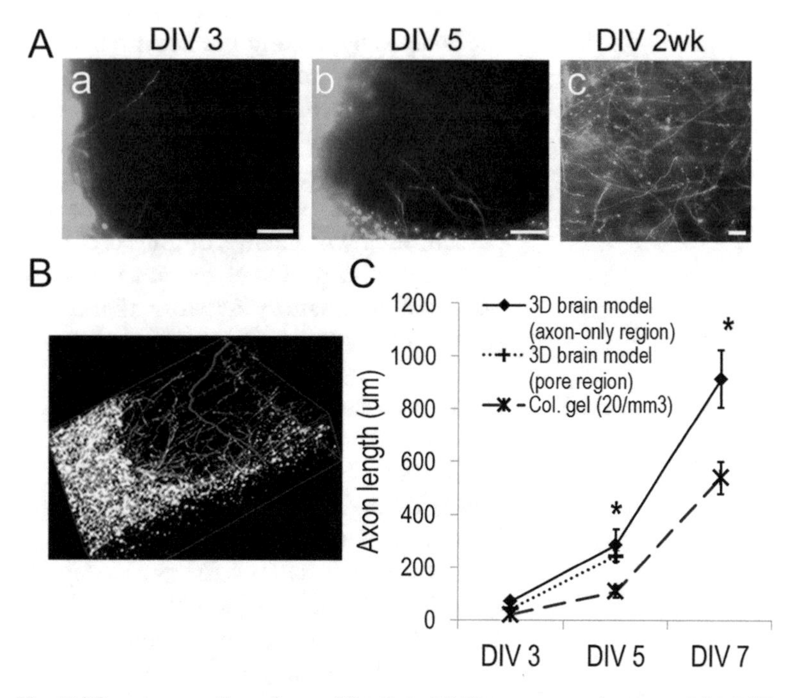

Fig. 7 3D axon growth and quantification. (**A**) Fluorescence images of (*a*) DIV3, (*b*) DIV5, and (*c*) 2-week axons immunostained with β3-tubulin in green. *DIV* day in vitro. Scale bar: 100 μm. (**B**) 3D neurite tracing with Fiji ImageJ. The traced process shown in pink. (**C**) Axon 3D length measurements. Figures are modified from reprints by permission from PNAS (Tang-Schomer MD, White JD, Tien LW, Schmitt LI, Valentin TM, Graziano DJ, Hopkins AM, Omenetto FG, Haydon PG, Kaplan DL. Bioengineered functional brain-like cortical tissue. 2014; 111 (38):13811–6

region. With a working distance reaching 1 mm, the model can be flipped and imaged again to cover the entire height of the 3D model.

Two quantitative measures were used for image analysis, axon 3D length growth and network density. Fiji ImageJ, "Simple Neurite Tracer" plugin was used for 3D axon tracing and measurements [21]. A start point and an end point of an axon in the 3D stack were manually picked. The tracer program automatically searched for a path between the two points approximating the axon trajectory. The process was repeated until a complete axon path was constructed. This method is useful to quantify axon length and suitable for 3D tissue models less than 1 week old. Figure 7b shows an example of 3D axon tracing, and Fig. 7c quantifications.

For 3D cultures with dense axon networks, network density analysis was performed on the z-stacks with a custom automated image analysis code. Each 2D plane within a z-stack was filtered to remove noise and cell bodies. The filtered images were then binarized to produce a neurite mask. Regions of low signal where neurite extensions were not observed were identified and excluded from the analysis by dilating the neurite mask with increasing radii until a minimal (≤ 5) number of connected components was achieved, giving a total area mask. The percent area covered by the neurites per 2D plane of the corresponding z-stack was determined by dividing the number of positive pixels in the neurite mask with the number of positive pixels in the total area mask. The mean percent area was finally computed over the entire z-stack for each sample resulting in average neurite network density. This method was used to quantify network density in Fig. 3B, and elsewhere [8, 9, 11].

3.7 Genetic Analysis

Analysis of the 3D brain model's genetic materials is necessary for evaluating the model's gene expression under controlled in vitro conditions. For extraction, we used Qiagene Mini AllPrep™ DNA/RNA/Protein kits and QIAshredder kit. The total DNA levels can be used to estimate the final cell numbers of 3D tissue cultures. Picogreen (Invitrogen) assay was used to measure DNA quantities. NanoDrop was used to measure RNA quantities, and BCA Protein Assay kit (Millipore) for protein quantification. Figure 8 shows expected nucleic acids and protein yields per 3D brain tissue model. RNAs were further purified with DNAse treatment (Sigma), and cDNA synthesized with an iScript Reverse Transcription Supermix kit (Bio-Rad). cDNAs were used with specific primers of targeted genes of interest for quantitative real-time PCR [7].

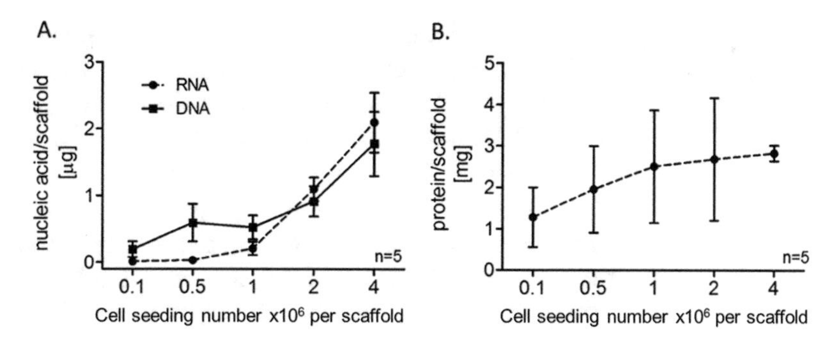

Fig. 8 Expected yield of nucleic acids and protein per construct in relation to the cell density. (**A**) DNA and RNA quantification. (**B**) Protein quantification. Figures are reprinted with permission from Journal of Visual Experiments: Chwalek, K., Sood, D., Cantley, W. L., White, J. D., Tang-Schomer, M., Kaplan, D. L. Engineered 3D Silk-collagen-based Model of Polarized Neural Tissue. J. Vis. Exp. (104), e52970, doi:https:/doi.org/10.3791/52970 (2015)

4 Conclusions

Here, we described a bioengineered in vitro 3D brain tissue model that is suitable for investigating the brain during homeostasis as well as brain injury. The model has exhibited in vivo-like behavior to TBI-like conditions including electrophysiological responses. The model's base materials (silk scaffold, hydrogel, cells) can be obtained from raw natural sources and commercially available; and the model assembly is a straightforward process. Following the design principles, the model can be expanded to other materials and structural forms. A wide range of biological assays have been adapted for the model to provide comprehensive assessment including structural, biochemical, genetic, and functional outcomes.

5 Notes

1. Fibroin purity is important to ensure the consistency of the final product. *Tips*: Silk boiling can be facilitated by stirring every couple of minutes with a spatula. Use force to wring and wash the fibroin.
2. LiBr purity is critical to ensure the concentration needed for complete dissolution of silk fibroin. *Tip*: LiBr absorbs vapor in ambient air. Use air-tight storage and keep fresh stocks.
3. Cover the beaker to prevent evaporation.
4. The liquid silk fibroin can be stored at 4 °C for up to 1 month in a closed container.
5. The ratio and mixing of the silk fibroin solution and salt are critical for the quality of the final product. Since salt-induced

fibroin precipitation occurs instantaneously, uneven salt distribution would result in pores of uneven sizes and density. *Tips*: (1) Handle the silk fibroin solution slowly and avoid bubbles. (2) Pour the salt particles steadily across the dish surface while rotating the dish to ensure even distribution.

6. Cover the dish with Para film to prevent water evaporation. *Tip*: Use a pipette tip to gently tap the surface of the mixture. If the tip pulls out viscous liquid, it means that fibroin precipitation is incomplete, and needs more time.
7. The silk mat's bottom face in contact with the petri dish is smoother than its top face, because the open surface has more defects. *Tip*: Trim off the top rough surface of the silk mat.
8. Silk scaffolds at the different stages of pre- or post-autoclave and pre-or post-PDL coating can be stored immersed in sterile water at 4 °C, or without water at −20 °C in a closed container for months. *Tips*: (1) Use care to handle dried silk scaffolds as they are brittle and light-weight when completely dehydrated. (2) Add water or PBS to rehydrate a dried scaffold. The scaffold can recover its shape without damage.
9. A neutral pH is critical for the collagen gel quality. Conditions too acidic or basic result in flimsy collagen strands and less homogeneous gel composition. *Tips*: Titration with NaOH at 1 μL a time is often needed to reach a neutral pH. Use the color change (from yellow to pinkish orange) when using M199 or a pH paper when using DMEM to gauge the gel's pH.
10. Matrigel solidifies at room temperature. *Tip:* Make frozen aliquots. Thaw on ice before use.
11. Hydrogels need to be prepared fresh right before use.
12. Be gentle and avoid air bubble formation.
13. The scaffolds were placed as one layer covering the well surface, maximizing cell access to the scaffolds.
14. The scaffold needs sufficient wash before hydrogel embedding to avoid accidental trapping of un-attached cells.
15. The scaffolds need to remain hydrated yet free of excess liquid to avoid diluting the gel solution during the next step. A hydrophobic surface helps contain the liquid gel within the scaffold; otherwise, a wet (or wetted) surface would lead the liquid gel to spread and unable to solidify in place. Process one sample at a time for this and the following step, to prevent the cells in the scaffold from drying. *Tip*: Dab the scaffold on a dry sterile surface a few times first to ensure that most of the excess liquid is removed, and set onto a new dish before adding the liquid gel.

16. We typically embed the scaffolds in batches of ten. At 30 min, we would check the samples for gel appearance. We would add a little more liquid gel to those with insufficient gel volume and allow for more gelation time.
17. We did not observe differences of un-fixed samples or samples fixed with 4% paraformaldehyde. Accordingly, 3D cultures after all the necessary biological assays can be fixed, stored in PBS at 4 °C and reserved for mechanical testing at later time points.
18. The screw head needs to be flat and smooth, with a size to match that of the 3D tissue model.
19. The weight and height of the injury apparatus can be adjusted based on Newton's gravitational formula, $mgh = 1/2kx^2{}_2{}^2$, where m is the weight, g is gravitational constant (9.91 m/s^2), h is the height. k is the spring constant, given as Young's modulus (E) multiplied by the original area (A_0) divided by the original length (L_0) of the test sample. x is the compressive distance.
20. The artificial CSF consists of (in mM): 125 NaCl, 5 KCl, 5 glutamate, 10 HEPES, 3.1 $CaCl_2$, and 1.3 $MgCl_2$, titrated to pH 7.4 using 1 M NaOH, supplemented with 10 mM fresh glucose before use.
21. It is important to keep the construct stationary in the liquid. The liquid level should submerge the construct; however, do not let it float.
22. Visual inspection is sufficient to determine the entry of the electrode into the construct. A slight movement of the electrode would move the construct if it is securely positioned.
23. Environmental noises need to be prevented by shielding the recording area with a Faraday cage, by turning off all electronic devices including lights within the cage, and by avoiding cell phone use near the cage.

Acknowledgments

We thank David Kaplan and the Kaplan laboratory at Tufts University for support.

References

1. Heffernan JM, Overstreet DJ, Le LD, Vernon BL, Sirianni RW (2015) Bioengineered scaffolds for 3D analysis of glioblastoma proliferation and invasion. Ann Biomed Eng 43:1965–1977
2. Ivanov DP, Parker TL, Walker DA, Alexander C, Ashford MB, Gellert PR, Garnett MC (2015) In vitro co-culture model of medulloblastoma and human neural stem cells for drug delivery assessment. J Biotechnol 205:3–13

3. Fan Y, Nguyen DT, Akay Y, Xu F, Akay M (2016) Engineering a brain cancer chip for high-throughput drug screening. Sci Rep 6:25062
4. Lancaster MA, Renner M, Martin CA, Wenzel D, Bicknell LS, Hurles ME, Homfray T, Penninger JM, Jackson AP, Knoblich JA (2013) Cerebral organoids model human brain development and microcephaly. Nature 501:373–379
5. Chamberlain SJ, Chen PF, Ng KY, Bourgois-Rocha F, Lemtiri-Chlieh F, Levine ES, Lalande M (2010) Induced pluripotent stem cell models of the genomic imprinting disorders Angelman and Prader-Willi syndromes. Proc Natl Acad Sci U S A 107:17668–17673
6. Mariani J, Simonini MV, Palejev D, Tomasini L, Coppola G, Szekely AM, Horvath TL, Vaccarino FM (2012) Modeling human cortical development in vitro using induced pluripotent stem cells. Proc Natl Acad Sci U S A 109:12770–12775
7. Tang-Schomer MD, White JD, Tien LW, Schmitt LI, Valentin TM, Graziano DJ, Graziano DJ, Hopkins AM, Omenetto FG, Haydon PG, Kaplan DL (2014) Bioengineered functional brain-like cortical tissue. Proc Natl Acad Sci U S A 111:13811–13816
8. Chwalek K, Sood D, Cantley WL, White JD, Tang-Schomer M, Kaplan DL (2015) Engineered 3D silk-collagen-based model of polarized neural tissue. J Vis Exp 105:e52970. https://doi.org/10.3791/52970
9. Chwalek K, Tang-Schomer MD, Omenetto FG, Kaplan DL (2015) In vitro bioengineered model of cortical brain tissue. Nat Protoc 10:1362–1373
10. Ren M, Du C, Herrero Acero E, Tang-Schomer MD, Ozkucur N (2016) A biofidelic 3D culture model to study the development of brain cellular systems. Sci Rep 6:24953
11. Sood D, Chwalek K, Stuntz E, Pouli D, Du C, Tang-Schomer MD, Georgakoudi I, Black LD, Kaplan DL (2016) Fetal brain extracellular matrix boosts neuronal network formation in 3D bioengineered model of cortical brain tissue. ACS Biomater Sci Eng 2(1):131–140
12. Altman GH, Diaz F, Jakuba C, Calabro T, Horan RL, Chen J, Lu H, Richmond J, Kaplan DL (2003) Silk-based biomaterials. Biomaterials 24:401–416
13. Omenetto FG, Kaplan DL (2010) New opportunities for an ancient material. Science 329:528–531
14. Rockwood DN, Preda RC, Yucel T, Wang X, Lovett ML, Kaplan DL (2011) Materials fabrication from Bombyx mori silk fibroin. Nat Protoc 6:1612–1631
15. Saatman KE, Duhaime AC, Bullock R, Maas AI, Valadka A, Manley GT, Workshop Scientific Team and Advisory Panel Members (2008) Classification of traumatic brain injury for targeted therapies. J Neurotrauma 25:719–738
16. Farkas O, Povlishock JT (2007) Cellular and subcellular change evoked by diffuse traumatic brain injury: a complex web of change extending far beyond focal damage. Prog Brain Res 161:43–59
17. Morrison B 3rd, Saatman KE, Meaney DF, McIntosh TK (1998) In vitro central nervous system models of mechanically induced trauma: a review. J Neurotrauma 15:911–928
18. Potts MB, Adwanikar H, Noble-Haeusslein LJ (2009) Models of traumatic cerebellar injury. Cerebellum 8:211–221
19. Marmarou A, Foda MA, van den Brink W, Campbell J, Kita H, Demetriadou K (1994) A new model of diffuse brain injury in rats. Part I: Pathophysiology and biomechanics. J Neurosurg 80:291–300
20. Schmitt LI, Sims RE, Dale N, Haydon PG (2012) Wakefulness affects synaptic and network activity by increasing extracellular astrocyte-derived adenosine. J Neurosci 32:4417–4425
21. Longair MH, Baker DA, Armstrong JD (2011) Simple neurite tracer: open source software for reconstruction, visualization and analysis of neuronal processes. Bioinformatics 27:2453–2454

Chapter 3

Modeling Traumatic Brain Injury In Vitro

Daniel E. Bonder, Carolyn E. Keating, Namas Chandra, D. Kacy Cullen, and Bryan J. Pfister

Abstract

Traumatic brain injury (TBI) is unique among neurological afflictions in that it is induced by a discrete physical event. To understand the relationship between mechanical loading and the evolution of structural and functional alterations of neural cells, TBI researchers have utilized in vitro models. These models were engineered to mimic loading conditions relevant for clinical TBI and to allow for the microscopic study of the cellular responses in real time. Collectively, this high degree of experimental control has resulted in robust platforms that enable the exploration of biological mechanisms involved in the progression of neural cellular injury. This chapter presents detailed background and methodology pertaining to two established in vitro models used in the field: (1) "stretch" injury to two-dimensional (2-D) cultures, and (2) simple shear deformation applied to three-dimensional (3-D) cell-containing matrices. The stretch injury paradigm uses a rapid pressure-pulse to stretch an elastic silicone membrane on which neural cells are cultured. The resulting deformation can be either biaxial or uniaxial, and is commonly applied to 2-D neuronal cultures with isolated axonal projections to model tensile loading in aligned axonal tracts, believed to be a proximal cause of diffuse axonal injury, the "hallmark" pathology of closed-head TBI. Rapid shear deformation to 3-D neural cellular constructs is applied using a linear actuator and is designed to replicate the complex loading conditions experienced by brain cells during inertial loading, with shear being the dominant mode of deformation in the nearly incompressible brain. This model has been utilized to study the acute and longer-term responses of 3-D neuronal cultures or 3-D neuronal-astrocytic cocultures to heterogeneous strain fields representative of loading patterns in vivo.

Key words Traumatic brain injury, Stretch injury, Axonal injury, Neuronal deformation, Astrocyte reactivity, Shearing injury, Plasma membrane permeability, In vitro model

1 Introduction

Traumatic brain injury (TBI) is unique among neurological afflictions in that it is induced by a discrete physical event, causing immediate effects on neural cells and vasculature while initiating a series of pathophysiological responses that may last over a protracted time frame.

The injury to neurons has been linked to the rapid deformation of brain tissue as a result of the damaging head motions [1, 2].

Amit K. Srivastava and Charles S. Cox, Jr. (eds.), *Pre-Clinical and Clinical Methods in Brain Trauma Research*, Neuromethods, vol. 139, https://doi.org/10.1007/978-1-4939-8564-7_3,

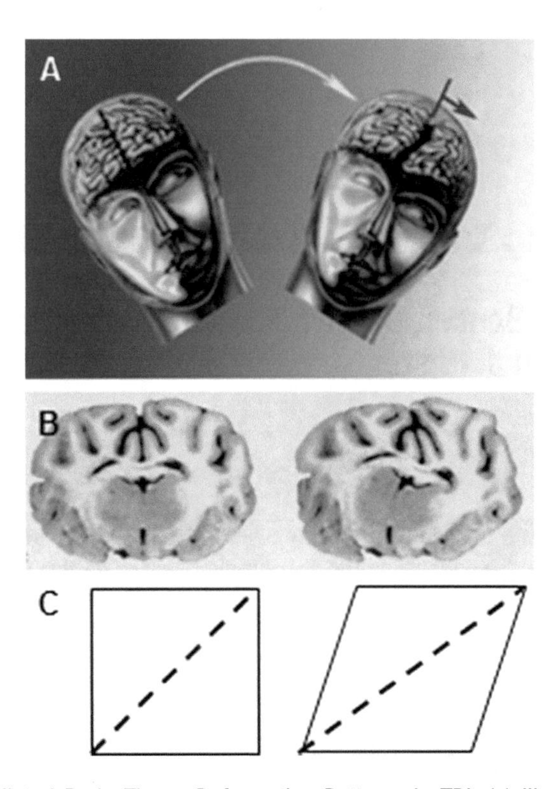

Fig. 1 Predicted Brain Tissue Deformation Patterns in TBI. (**a**) Illustrates how rotational motions can deform brain tissue. (**b**) Shows mock-up of a pig brain deforming in shear due to rotational motion (from [2]). (**c**) Illustrates the concept of shear deformation of a block of brain tissue, translating into a uniaxial stretch of neuronal fibers along the diagonal

Biomechanical studies have shown that rapid rotational motion of the head results in large intracranial shear deformation of the brain with respect to the comparatively rigid skull [3–5]. This macroscopic shear deformation can translate microscopically to rapid stretching of neurons and their axons, Fig. 1 [6, 7].

Understanding the relationship between mechanical loading and the evolution of structural and functional alterations of neural cells is essential, yet can be particularly challenging using animal models. Biomedical engineers have been instrumental in developing in vitro models to study TBI, which offer several advantages including (1) controlled biomechanical inputs designed to mimic loading conditions relevant for clinical TBI, (2) defined neural cellular populations to systematically build complexity, (3) ease of access for repeated measures, and (4) the potential to study cellular responses in real time. Collectively, this high degree of experimental control has resulted in robust platforms that enable the exploration of biological mechanisms involved in the progression of neural cellular injury. Over at least the past three decades, numerous in vitro injury models have been developed that apply mechanical perturbations to cultured neural cells [6–13]. These models replicate many of the morphological and ultrastructural changes observed in vivo and have had significant impact in the field by identifying many important post-injury pathobiological processes [1, 12, 14–22].

Due to the biomechanical etiology of TBI, when utilizing reduced models for scientific investigation it is desirable to apply injury parameters that are scaled from the physical loading experienced by the human brain in TBI. The soft brain tissue reacts slower to the rotation of the head, often causing large intracerebral shear deformation with respect to the comparatively rigid skull. Macroscopic shear deformation of brain tissue translates microscopically to rapid longitudinal or uniaxial stretching along the diagonal of a representative cube, Fig. 1b. Axon fibers aligned along the diagonal will be stretched while perpendicular fibers will be compressed. It is believed that axons are mostly damaged by the stretching forces, whereas compression forces would simply release damaging forces and cause the axon to undulate. A uniaxial stretch to isolated axons would model tensile loading in aligned axonal tracts, believed to be a proximal cause of diffuse axonal injury, the "hallmark" pathology of closed-head TBI.

This chapter presents detailed background and methodology pertaining to two established in vitro models used in the field: (1) "stretch" injury to two-dimensional (2-D) cultures that may be biaxial [23–25] or uniaxial [6, 13, 26, 27], and (2) simple shear deformation applied to three-dimensional (3-D) cell-containing matrices [10, 17]. Most in vitro TBI models, similar to the systems described in this protocol, use a custom fabricated device, although there is at least one commercially available stretch system (Cell Injury Controller; Custom Design & Fabrication, Inc., Sandston, VA). While custom-built systems are not generally commercially available, the developing labs are often open to collaboration and assistance.

1.1 2-D In Vitro Stretch Injury Models

The 2-D stretch injury paradigm uses a rapid pressure-pulse to stretch an elastic silicone membrane on which neural cells are cultured. The deformation can be either biaxial or uniaxial, applied to either neuronal cultures including both soma and axons or to isolated axonal projections. Stretch injury models were first designed to apply a rapid pressure-pulse and cause *biaxial* deformation of the membrane, Fig. 2. To control the pressure pulse, high speed solenoid valves are used to inject compressed air into a chamber that contains the neuronal culture. Aspects of the pressure pulse can be controlled including: (1) peak pressure, which is proportional to the stretch of the axons, (2) pressure rise time, representing the rate of stretch (speed of injury), and (3) pressure pulse duration that defines the impulse of the stretch injury, Fig. 2F. Through calibration, the stretch of the membrane can be related to

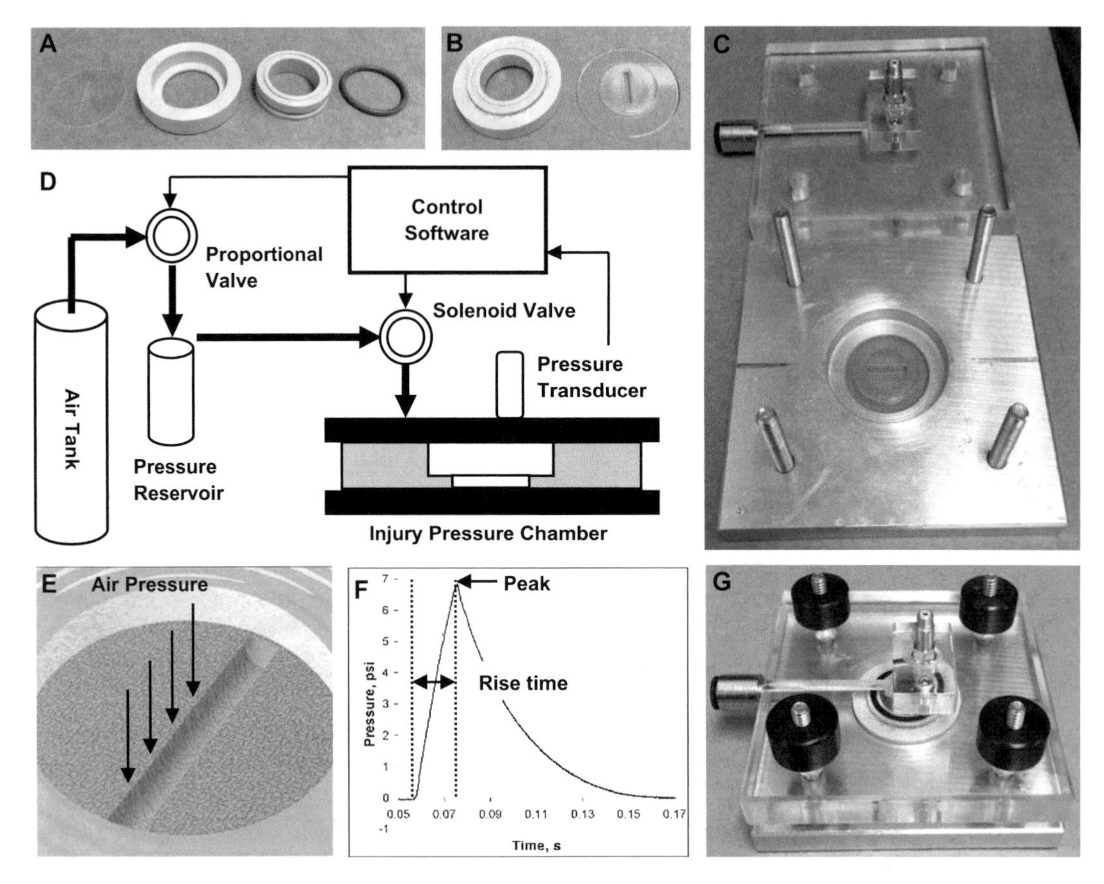

Fig. 2 The In Vitro Stretch Injury System. The stretch injury system uses a pressure-pulse to stretch-injure cultured axons on a deformable substrate. (**a**) Shows the components of the single well design (in order from left to right: Aclar® underside mask, bottom well piece, top well piece, well O-ring). (**b**) Shows the assembled well (left) and 3D-printed deformation mask (right). (**c**) Shows the top and bottom pieces of the injury chamber, while **g** shows the fully assembled injury chamber with culture well. Injuries are produced using an air pulse generating system, illustrated in **d**. Bold lines denote the flow of air through the system. (**e**) Illustrates the concept of uniaxial stretching of axons along a cell-free gap. Deformation is constrained by the deformation mask. (**f**) Shows a sample pressure pulse with relevant parameters, peak pressure amplitude and pressure rise time, labeled (see arrows)

the pressure pulse [6, 13]. The relevant strain fields that are generally considered can span a great amount; 10–70% strain with applied strain rates of 10–100 s^{-1} [6, 28, 29]. The amount of strain to induce injury will vary with the mode of deformation with biaxial requiring less deformation than uniaxial.

Since uniaxial (longitudinal) stretching of the axon is a mechanically relevant deformation in axonal injury, two models have been implemented to deliver a uniaxial stretch injury [7, 13]. Here we describe a protocol based on the method developed by Smith and Meaney where a cell-free gap is created across which cultured axons grow [6, 27]. Cells are restricted from adhering to a 1.5–2 mm line across the well by covering the region with a silicone divider.

Neurons are seeded on both sides of the divider and allowed to adhere to the substrate. Upon removal of the divider, developing axons grow across the gap over 9–11 days in vitro (DIV). This isolated axonal region is stretched by placing a complementary injury mask underneath the silicone substrate, only allowing the pressure pulse to deform this region, Fig. 2e.

1.2 3-D In Vitro Shear Deformation Injury Model

This model replicates two important parameters in TBI: (1) mechanical shear which is the dominant form of tissue deformation that occurs in the brain during trauma, Fig. 1 [5, 10] and (2) the 3-D cellular and extracellular matrix (ECM) environmental effects on injury. Cells in the brain exist in a complex environment in which they extend their processes and come in contact with both ECM and other cells in 3-D. This complex arrangement is an important consideration when examining the effects of physical forces on cells, especially in the study of mechanical trauma such as TBI. For instance, the translation of forces from bulk tissue (macro) to cellular (micro) deformation may occur differently in 2-D and 3-D configurations. The complexity of neural cell morphology, which increases in 3-D, affects the complexity of the strain field to which the cell is exposed [30], resulting in simultaneous combinations of compressive, shear, and tensile forces. As bulk deformation is translated through physical coupling, factors such as cell morphology, matrix mechanical properties, and cell-ECM or cell–cell interactions will have important effects on the transfer of strain to cells, and might underlie localized differences in susceptibility to stress.

Another factor affecting the transfer of bulk to cellular deformation is the bioactivity and physical properties of the ECM material. Viscoelasticity and porosity may affect how reliably bulk deformation is transferred to cells in a 3-D matrix, and these properties can be altered in vitro to approximate those of brain tissue. Cell-ECM and cell-cell contacts may also affect force translation. The number and spatial distribution of these interactions can affect the size and location of stresses that develop on and within cells, which may influence the probability of cellular structural failures. Additionally, these models have the advantage of being able to systematically control various cell culture parameters such as cell composition and matrix constituents that enable the elucidation of the roles of specific factors in bulk to cellular deformation and the associated responses.

Here we describe a protocol based on the method created by LaPlaca and Cargill, which uses 3-D neural cellular constructs (Fig. 3) subjected to controlled 3-D shear-strain based loading profiles [10, 30]. A custom-designed device uses a linear actuator to reproducibly deliver a specified strain field to neural cells within a 3-D bioactive ECM based scaffold. The device input approximates

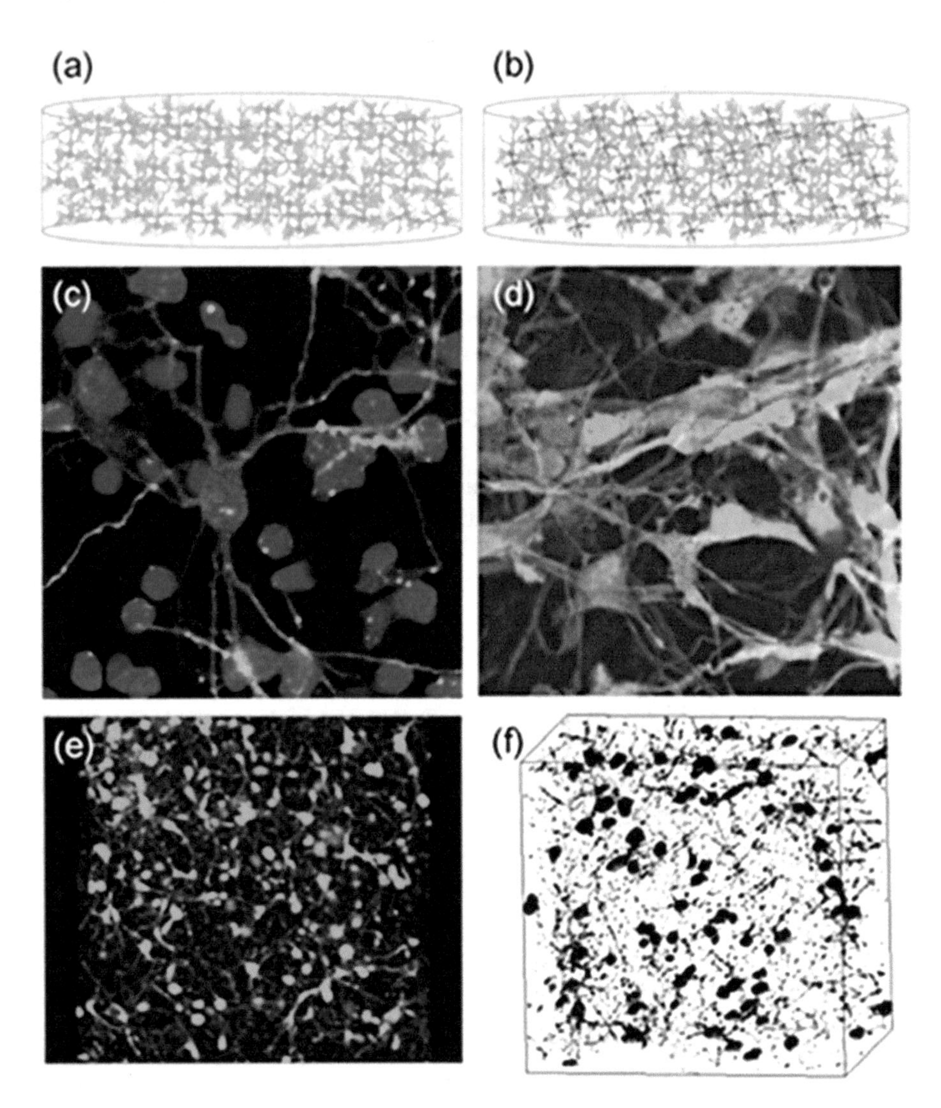

Fig. 3 Concept: Tissue Engineered 3-D Neural Cell Cultures for Neuromechanobiological Investigations In Vitro. Cellular constructs consist of (**a**) neurons or (**b**) neurons mixed with astrocytes, distributed throughout the full thickness of a 3-D matrix/scaffold (>500 μm). (**c**) Neurons in 3-D culture, shown to express neuronal-specific proteins MAP-2 (green) and tau-5 (red) with nuclear marker Hoechst 33258 (blue), assume complex morphologies with 3-D neurite extension. (**d**) These cultures are useful to study neuron–astrocyte interactions in 3-D. (**e**) Live neurons (green) demonstrating survival and network formation throughout the thickness of the 3-D cultures. (**f**) Volumetric rendering of live neurons in 3-D culture. Neurobiological studies of neural cells within 3-D matrices provide enhanced fidelity to in vivo while affording all advantages of traditional (planar) in vitro systems. Panels (**a**), (**b**), (**d**), and (**e**) reproduced with permission from Cullen et al., 2011, Begell House; panel (**c**) reproduced with permission from Irons et al., 2008, IOP Publishing; (and panel (**f**) reproduced from Cullen et al., 2006, Mary Ann Liebert, Inc.)

the 3-D strain fields predicted during closed-head diffuse injury in the human brain. This model has been utilized to study the acute and longer-term responses of 3-D neuronal cultures or 3-D neuronal-astrocytic cocultures to injury.

2 Materials: 2-D In Vitro Stretch Injury Model

- 7.8 mil (199 μm) or 2 mil (51 μm) aclar sheets (Electron Microscopy Sciences).
- Ultrapure water (Milli-Q).
- Sparkleen (Fisher, 04-320-4) or similar labware detergent.
- 70% ethanol (Ricca Chemical, 2546.70-2.5).
- Silicone sheets, 0.005″, 12″ × 12″ (Specialty Manufacturing Inc.).
- Poly-L-lysine (R&D Systems Cultrex, 34-382-0001).
- Leibowitz L-15 Medium (Gibco, 11-415-064).
- Hank's Balanced Salt Solution containing Ca^{2+} and Mg^{2+} (HBSS+, Gibco, 14-025-076).
- Hank's Balanced Salt Solution without Ca^{2+} and Mg^{2+} (HBSS−, Gibco, 14-175-095).
- Phosphate Buffered Saline (PBS, GIbco 10010031).
- Trypsin-EDTA (0.5%) (Gibco, 15-400-054).
- Fetal Bovine Serum (FBS, Gibco, 10-082-139).
- Neurobasal Medium (Gibco, 21-103-049).
- B27 Supplement (Gibco, 17-504-044).
- Penicillin/Streptomycin (P/S, Gibco, 15-140-163).
- GlutaMAX 100× (Gibco, 35-050-061).
- Trypan Blue Solution (0.4%) (Gibco, 15-250-061).
- Dash Number 018 high temperature silicone o-ring (McMaster Carr, 9396K65).
- Fluorescent Beads (Invitrogen, G0100).
- Compressed air tank (AirGas, UN1002).
- Surgical toolset (for instance: small scissors, curved forceps, two fine forceps, straight blade #11 scalpel, surgical spatula, fine angled forceps).

3 Methods: 2-D In Vitro Stretch Injury Model

3.1 Production of Injury Masks

The premise of the design is to unaxially stretch injure isolated tracks of axons (sparing the neuronal somata) to study the direct effect of mechanical stretch on the evolution of axonal pathology [6]. Our axonal injury system utilizes two types of injury masks, one of which is built into the injury well itself on the underside of the silicone membrane (referred to as the "underside mask") and another that is utilized during injuries to create uniaxial stretch (the "injury mask") (Fig. 2). The underside mask is cut from sheets of 7.8 mil (~0.2 mm) Aclar® film using a Cricut Explore One stencil cutter. Details on mask thickness are described in greater detail below in the **Notes 1** and **2**.

Wells can be constructed lacking the underside mask. However this piece serves several important and helpful purposes. First, it constrains the stretch to be uniaxial rather than biaxial. Second, it clearly outlines the injury field, removing possible ambiguity about which cells or neurites were directly deformed. Third, it ensures that cellular elements outside of the injury field remain uninjured (*see* **Notes 1** and **2**).

The injury mask (Fig. 2b) can be 3-D printed. We have used both the Ultimaker 3 thermal deposition printer and Objet 30 Prime polyjet printer. There are not strict requirements concerning the type of print material, only that the mask be rigid enough to not bend significantly when pressure is applied to the injury chamber. A dual extrusion printer is needed, however, as the mask design requires a support material.

3.2 Nonsterile Preparation of Injury Wells for Device Calibration

3.2.1 Overview

The well design allows for rapid assembly by press-fitting two concentric polyetheretherketone (PEEK) rings together, held in place by an O-ring (Fig. 2a, b). The smaller of the PEEK rings contains a groove for the insertion of an O-ring. This O-ring served three purposes: (1) applied pre-stretch to the silicone membrane, (2) held the well together, and (3) acted as a gasket to prevent media leakage. A square of silicone membrane was placed flat over the larger ring, wrinkles in the membrane were straightened and the smaller ring was gently pressed into the larger. The circular geometry ensured that the membrane was pre-stretched equibiaxially and uniformly throughout the well. As previously mentioned, the well design is further modified by including an Aclar® underside mask (Fig. 2a) that sits underneath the silicone sheet and creates a seal with it.

For calibration purposes, the wells do not need to be assembled by the full sterilization procedure. The protocol below outlines a quicker way to prepare injury wells for nonsterile calibration experiments.

3.2.2 Protocol

1. Cut a silicone sheet into 3.5 cm × 3.5 cm^2 (or larger).
2. Place a Dash Number 018 high-temperature silicone O-ring on the small PEEK well piece.
3. Place an Aclar® underside mask within the large PEEK well piece.
4. Dip the large well piece with the Aclar® mask into a beaker of purified reverse osmosis (RO) water, holding the mask in place to prevent it from floating up. Lift up out of the beaker, keeping a pool of water within the well piece atop the mask.

5. Lay a precut square of silicone on top of the large PEEK well piece (*see* **Note 3**).
6. Center the small PEEK well piece on top of the silicone square, then press the well pieces together, ensuring that the silicone placement is more or less centered within the well. The well is now assembled.
7. If desired, trim the excess silicone using a razor.
8. Allow wells to air dry for a minimum of 1 h.

3.3 Calibration: Determining Static Pressure-to-Deformation Relationship

3.3.1 Overview

Each deformation geometry (equibiaxial, uniaxial and width of gap) will have a unique pressure to deformation relationship due to a changing membrane stiffness which is related to how the membrane is constrained. Polystyrene fluorescent beads (Invitrogen) were used to measure the uniaxial deformation of the silicone membrane. A picture of the adherent beads was taken in the undeformed state, that is, 0 psi (Fig. 4a). The chamber pressure was increased incrementally and pictures were taken of the membrane in its deformed state (Fig. 4b, c). All measurements were done using

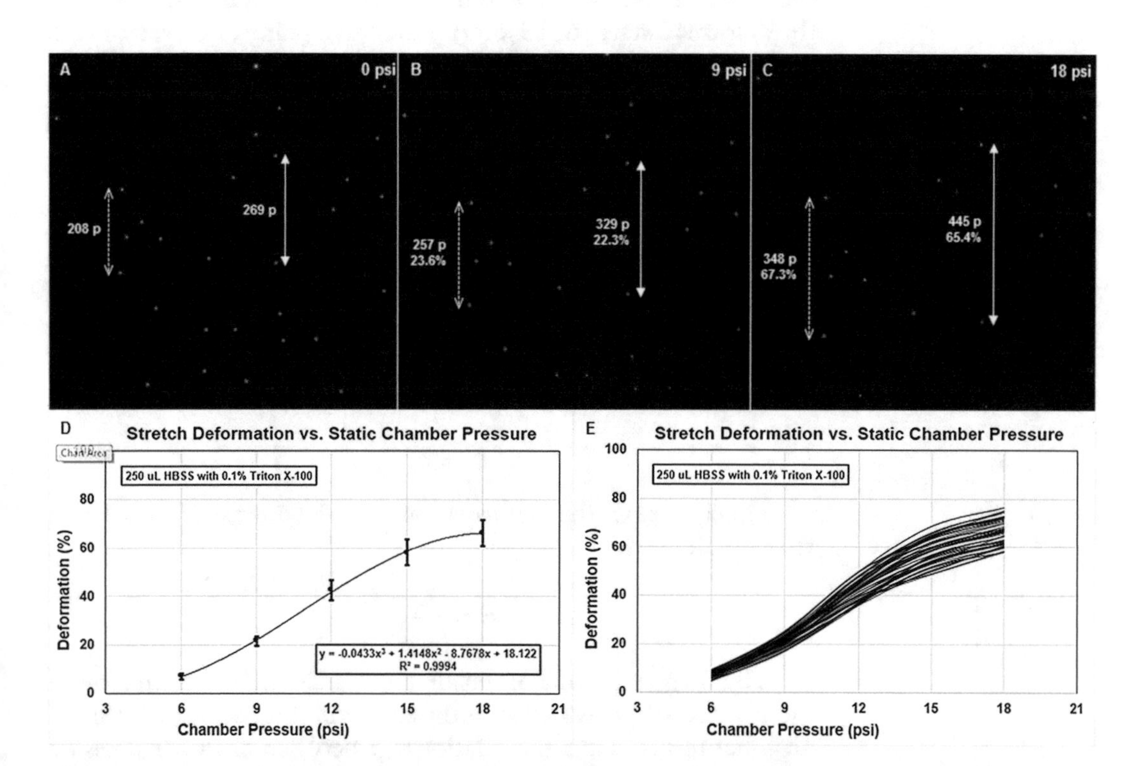

Fig. 4 Determining Relationship between Chamber Pressure and Stretch Deformation. Fluorescent beads are plated on injury wells. (**a–c**) Show representative images of the same field of fluorescent beads at 0 psi, 9 psi, and 18 psi. Arrows denote the distance, measured in image pixels, between two pairs of beads. (**d**) Shows the average percentage of stretch deformation across a range of chamber pressures. (**e**) Shows all measured data points that were used to calculate the average in **d**

deformation masks to create a uniaxial deformation. Strain was calculated using the equation:

$$\text{Strain} = \frac{l_f - l_0}{l_0} \times 100\%$$

Pairs of fluorescent beads were selected for measurement based on two factors: (1) pairs must be aligned within 5° of the stretch direction as to not affect linear measurement and (2) based on the ability to visualize the bead pair throughout the entire sequence of images. The distance between the selected pair of beads was measured in the undeformed state in terms of relative number of pixels (l_0) (Fig. 4d, e). This measurement was repeated for the same pair of beads for pictures of the membrane in its deformed state (l_f).

If desired, the system can also be used to induce biaxial stretch deformation on a cell culture (data not shown). Characterization of biaxial stretch is based on a similar principle using fluorescent beads, as described above. Specifically, three beads are chosen within an imaging field. Pictures of these three points are taken in the undeformed state (0 psi) and at chamber pressure increments in the deformed state [6, 13]. From the three points, two vectors can be defined, $dX^{(1)}$ and $dX^{(2)}$; these vectors represent the membrane stretch. Two vectors are used to define the membrane at each level of chamber pressure: $dx^{(1)}$ and $dx^{(2)}$. $\boldsymbol{F}$ is then defined to be the deformation gradient tensor such that:

$$dx^{(1)} = \boldsymbol{F}dX^{(1)}, \quad dx^{(2)} = \boldsymbol{F}dX^{(2)}$$

From this derivation, $\boldsymbol{F}$ is defined as:

$$\boldsymbol{F} = \left[dx^{(1)}\middle|dx^{(2)}\right]\left[dX^{(1)}\middle|dX^{(2)}\right]^{-1}$$

From this the Cauchy-Green deformation tensor is calculated:

$$\boldsymbol{C} = \boldsymbol{F}^{\mathrm{T}}\boldsymbol{F},$$

Finally, using the deformation, the Lagrangian finite strain tensor is derived:

$$\boldsymbol{E} = \frac{(\boldsymbol{C} - \boldsymbol{I})}{2}$$

The diagonal terms of tensor $\boldsymbol{E}$ provides strain information in the x, y directions, as well as information on shear stresses in the off diagonal terms. Thus the relationship between chamber pressure and biaxial stretch is established.

3.3.2 Protocol

1. Construct injury wells in nonsterile manner as described above.
2. Once fully dried, dilute fluorescent beads to 10^3 beads/mL from stock solution and plate on the injury wells.

3. Rock the wells side-to-side and back-to-front several times to ensure even distribution of the beads.
4. Incubate wells for a minimum of 15 min at room temperature, protected from light.
5. Aspirate fluorescent beads and gently rinse once with PBS, HBSS, or water.
6. Aspirate rinse and allow wells to dry for a minimum of 1 h, protected from light.
7. Once wells are dry, add 200 μL of HBSS with 0.1% Triton X-100 and secure a well inside the injury chamber (*see* **Note 4**).
8. Place injury chamber on an inverted microscope stage.
9. Connect the chamber to a tank containing compressed air with a regulator that allows for pressures up to around 20 psi to be applied to the chamber.
10. Using a 20× or 40× objective lens, focus on the center of the well's injury field (*see* **Note 5**).
11. Capture images of the same field of fluorescent beads at a range of pressure (e.g. 0, 6, 9, 12, 15, and 18 psi).
12. When all images have been captured, well can be rinsed once with PBS, HBSS, or water and allowed to dry.
13. Repeat procedure for desired number of injury wells.
14. For each well's series of images, choose pairs of beads that are aligned with the direction of uniaxial stretch and measure the distance (in pixels) between the beads in each pair, at each psi (Fig. 4a–c).
15. Calculate the percent change in distance for each bead pair relative to the initial 0 psi distance; this represents the percent stretch at each static pressure.
16. Plot the average deformation as a function of chamber pressure; determine the equation of the best fit trend line (Fig. 4d; *see* **Notes 6** and 7).

3.4 Primary Cortical Neuronal Cell Culture Procedures

3.4.1 Sterile Preparation of Injury Wells for Cell Culture

Preparation of the injury wells requires a full day, or the procedure can be split across 2 or more days.

1. Soak the well pieces, Aclar® underside masks, and well O-rings in purified RO water with Sparkleen or similar laboratory detergent; scrub the well pieces with a toothbrush or glassware brush.
2. Rinse the well pieces, Aclar® underside masks, and O-rings thoroughly with RO water 3×.
3. Soak all well components in 70% ethanol and agitate by stirring or shaking; rinse 3× with RO water (*see* **Note 8**).
4. Sonicate all well components in RO water for 15 min, replacing the water after the sonication.

5. While sonicating, cut 3.5 cm × 3.5 cm^2 from a silicone sheet; soak the silicone squares in 70% ethanol (*see* **Note 9**).
6. Upon completion of sonication, assemble the wells; start by placing a Dash Number 018 high-temperature silicone O-ring on the small PEEK well piece.
7. Place an Aclar® underside mask within the large PEEK well piece.
8. Dip the large well piece with the Aclar® mask into a beaker of RO water, holding the mask in place to prevent it from floating up. Lift up out of the beaker, keeping a pool of water within the well piece atop the mask.
9. Lay a precut square of silicone on top of the large PEEK well piece.
10. Center the small PEEK well piece on top of the silicone square, then press the well pieces together, ensuring that the silicone placement is more or less centered within the well (*see* **Note 10**).
11. Trim the excess silicone using a razor.
12. Sonicate completed wells in fresh RO water for 15 min.
13. Transfer the wells to an autoclave-safe container, submerge the wells in RO water, and autoclave for 30 min.
14. Drain the water from the container and allow wells to air dry for at least 1 h in a sterilized biosafety cabinet for a minimum of 1 h (*see* **Note 11**).
15. Add 750 μL of poly-L-lysine solution to each injury well and incubate at 37 °C for at least 1 h.
16. Briefly rinse wells for 5 min with 1 mL of sterile culture-grade water and allow to air dry for at least 1 h; at this point the wells are ready for plating (*see* **Note 12**).
17. If a cell-free gap is desired for an axon-only region, cut 2 mm-wide silicone strips, rinse the strips in 70% ethanol followed by RO water, and autoclave dry for 1 h.
18. Allow the strips to dry, if necessary.
19. Lay the silicone strips across the center of the injury wells using forceps (*see* **Note 13**). The strips should stick well to the silicone well bottom.

3.4.2 Cortical Dissection and Cell Culture

Protocol

1. Prepare 50 mL of Cortical Media by combining 48.9 mL Neurobasal Media (4 °C), 1 mL B27 Supplement, and 100 μL GlutaMAX (0.4 mM).
2. Prepare 2 mL of Cortical Media + Penicillin/Streptomycin (Cortical Media + P/S) by removing 1.98 mL Cortical

Medium to a separate 15 mL tube and adding 20 μL penicillin/streptomycin (final concentration of 1%).

3. Prepare HBSS+/20% FBS/DNAse I solution by combining 1.9 mL HBSS containing calcium and magnesium, 0.5 mL FBS, and 0.1 mL DNAse I (100 units) in a 15 mL tube.
4. Obtain E17 or E18 rat embryos, maintain on ice, and remove embryos from placental sacs. Remove brains from the embryos. Remove meninges and isolate cortices from the rest of the brain, placing cortical fragments in 4 mL L-15 media kept on ice (*see* **Note 14**).
5. Once all cortices are removed, use the spear end of the surgical spatula to break apart the large cortical pieces using a mortar and pestle motion.
6. Allow cortical pieces to settle to bottom of tube and carefully remove the L-15 media.
7. Gently add 4 mL room temperature HBSS without calcium and magnesium to the tube. Allow the cortical pieces to settle to the bottom.
8. Remove the HBSS without calcium and magnesium and gently add 2.5 mL of 37 °C trypsin-EDTA solution (0.05%) diluted in HBSS without calcium and magnesium. Mix (do not vortex) and place at 37 °C for 20 min.
9. After 20 min add 2.5 mL of HBSS+/20% FBS/DNAse I at 37 °C to inactivate the trypsin-EDTA solution.
10. Triterate the cortical pieces using a glass pipette 20 times, then incubate the tube at 37 °C for 20 min (*see* **Note 15**).
11. Spin the tube at low speed (800 rpm) for 5 min.
12. Remove the supernatant, leaving the pellet undisturbed.
13. Resuspend the pellet in 2 mL of Cortical Media with penicillin/streptomycin by triterating 20 times using a glass pipette (*see* **Note 16**).
14. Dilute a portion of the final cell suspension 1:10 in Trypan Blue solution (100 μL cell suspension to 900 μL Trypan Blue). Count cell density with a hemocytometer.
15. Add 900 μL of 37 °C Cortical Media to each well.
16. Dilute a portion of the final cell suspension in a new tube such that each well will receive 100 μL of cell solution containing 175,000 cells per well (roughly 750 cells/mm^2; *see* **Notes 17** and **18**).
17. Slowly add 100 μL of cell dilution directly to the center of each well, and do not disturb the wells for 10 min.
18. Very gently mix the wells by rocking them from side to side, and front to back.

19. Place wells in 37 °C incubator.
20. If wells are to contain a cell-free gap, at least 2 h after plating, gently remove the cell-free gap silicone strip using autoclaved or ethanol-sterilized fine tip forceps.
21. Feed the wells every 2–3 days by replacing half the well volume (*see* **Notes 19** and **20**).

3.5 Basic Injury Device Operation

3.5.1 Hardware

The injury device delivers a precise air pressure pulse to the injury chamber, resulting in uniaxial stretch of the silicone membrane, and the axons or cells on the silicone, above the gap area of the deformation masks (Fig. 2e). In our stretch injury model, the control system needs to be able to control two important parameters, the pressure amplitude within the injury chamber and the duration of pressure release into the injury chamber. The control system must be able to precisely and consistently adjust these parameters to allow for stretch injuries across a diverse range of injury intensities and rates. Further, it is necessary to be able to measure these parameters in real time to know the exact attributes of injury being induced.

The degree of pressure increase in the injury chamber correlates to the magnitude of stretch injury inflicted upon the cells or axons within the injury zone. This is the basis for the static pressure-deformation characterization above; higher chamber pressures produce greater strain. Varying the chamber input pressure affects the rate at which the intra-chamber pressure rises. Modulating the chamber pressure could be controlled manually in a simplistic system using a pressure regulator setup, though the tradeoff would be greatly reduced precision especially with respect to the duration of the pressure pulse. We devised an automatic pressure control system using a Parker Hannifin VSO®-EP Miniature Electronic Pressure Control Unit. This component converts a tunable electrical control signal into a proportional pneumatic output with excellent precision and reproducibility. This unit regulates the pressure within a small reserve air tank, which acts as the reservoir of pressure to be released into the injury chamber (see. Fig. 2d).

The duration of the pressure pulse within the injury chamber correlates to the rise time and duration of the stretch injury. This parameter is dependent on the valve open time, that is, the length of time the pressure is allowed to be released into the chamber. The particular challenges for an in vitro stretch injury model hinge on achieving the fast rise times observed in traumatic brain injury, on the order of milliseconds. To achieve rapid pressure rise times, we utilize a solenoid valve, the Parker Hannifin Series 9 Miniature Calibrant Valve, which enables high speed, very low leak rate, and high flow control of air pressure into the injury chamber. The valve is situated in between the reserve air tank, which stores the pressure to be released into the chamber, and the chamber itself. Opening of the valve releases the pressure pulse into the injury chamber. This valve should be three-way, so that the pressure pulse is vented out of the chamber upon valve closure.

The exact dynamics of the resulting injury are dependent on both the input pressure and valve open time. In general, higher input pressures result in greater degrees of strain and quicker injury rise times; longer valve open times result in longer pressure pulses and greater degrees of strain. These parameters are interdependent. For example, two pressure pulses with the same input pressure but different valve open times will produce two different injury profiles. The faster pulse (shorter valve open time) will also display a reduced pressure amplitude relative to the longer pulse. If a specific type of injury is desired, the device operator must first experimentally determine the necessary pulse parameters more or less by trial and error. Figure 5 shows representative pressure pulses across a range of strain amplitudes and strain rates, up to 60% strain and 90/s strain rate.

There are also other factors that can alter pulse dynamics from one device setup to the next. Namely these are the length and diameter of the tubing used to connect the reserve tank to the valve, and the valve to the injury chamber. For a single well system like the one described here, the air volume within the chamber is small. Therefore, changes to the tubing connecting the valve to the chamber can have a relatively large impact on the effective chamber air volume. A larger chamber volume requires greater input pressure to reach the same level of chamber pressure and therefore silicone stretch, and vice versa.

The tubing connecting the reserve tank to the valve can also play a role in modulating pressure pulse dynamics. Air flow from the reserve tank to the valve to the chamber is constrained by the smallest diameter orifice within the system. In our setup, this is the orifice within the solenoid valve itself, at 0.06″. If smaller diameter tubing or a smaller flow control orifice is placed directly behind the valve, then this will result in slower air flow velocity into the injury chamber. This serves to attenuate the pressure pulse amplitude.

Lastly, the system must be able to measure the pressure pulse delivered to the injury chamber in real time. An advantage of the silicone-based stretch injury model is that, since stretch deformation is related to chamber pressure, the deformation during a pressure pulse can be inferred from pressure amplitude measurements. The main requirement for measurement is a pressure transducer that is small enough to be mounted on the chamber and capable of measuring at high speeds. We have used both PCB transducers, such as the Model 113B28, and Measurement Specialties EPX miniature pressure sensor. Both sensor types are able to operate at kilohertz rates, readily capturing pressure changes taking place over several milliseconds.

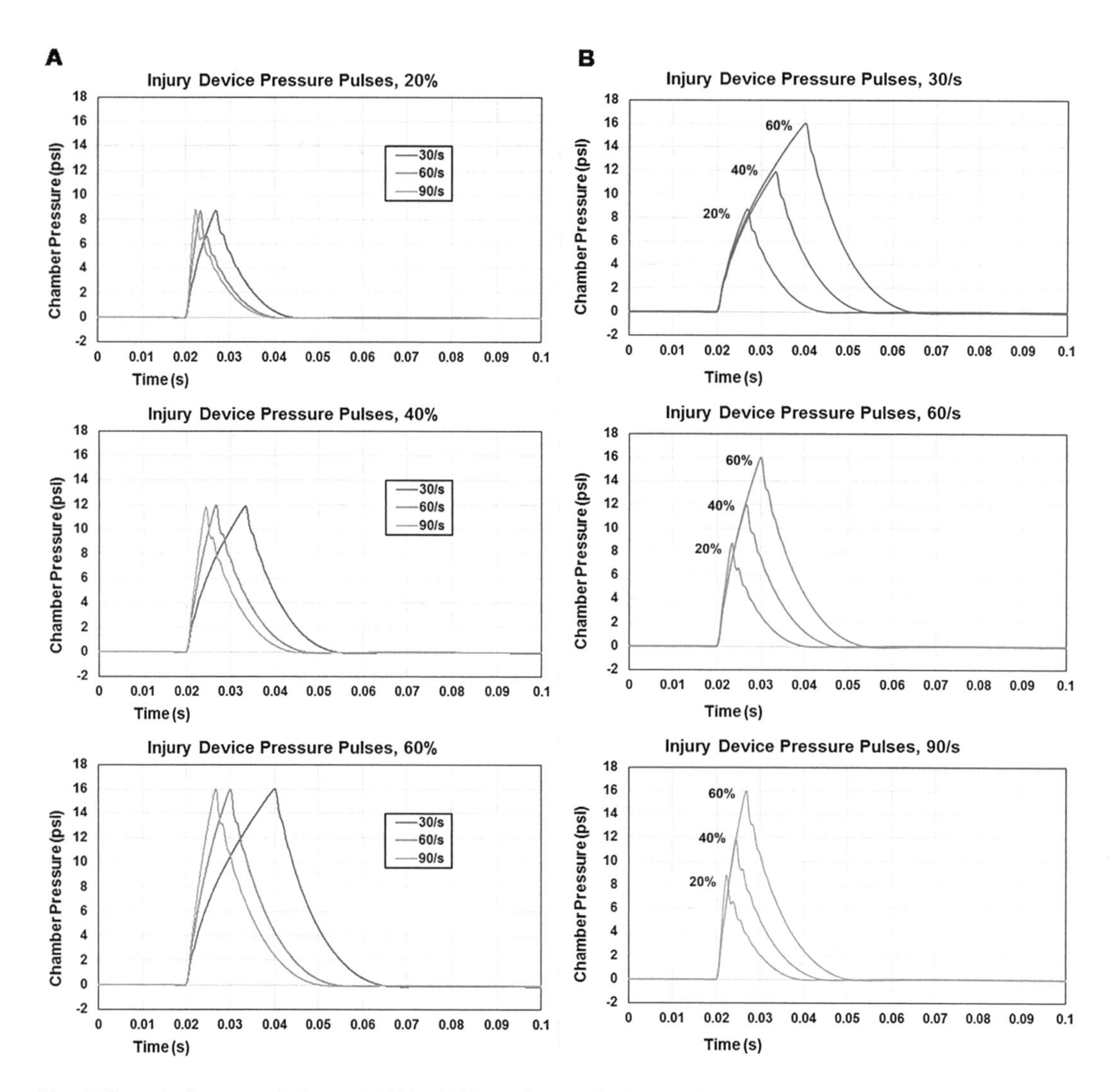

Fig. 5 Sample Pressure Pulses at 20%, 40%, and 60% Strain Amplitude, and 30/s, 60/s, and 90/s Strain Rates. Column **a** shows the individual pressure pulses arranged into separate plots by strain amplitude. Column **b** shows the same individual pressure pulses but arranged into separate plots based on strain rate

3.5.2 *Software*

A custom software program is required to operate the device components discussed above. As an example we will describe the LabView program we use to operate our injury device (Fig. 6). The pressure transducer, solenoid valve, and pressure control unit communicate with LabView via a multifunction DAQ device (we use a National Instruments USB-6211).

As described above, the two relevant parameters to control are the reserve tank pressure and valve open time. Our LabView program contains input boxes for modulating these two parameters (Fig. 6a). The voltage applied to the pressure control unit alters the pressure pumped into the reserve tank. In order for the software to accurately display the chamber pressure, an input needs to be

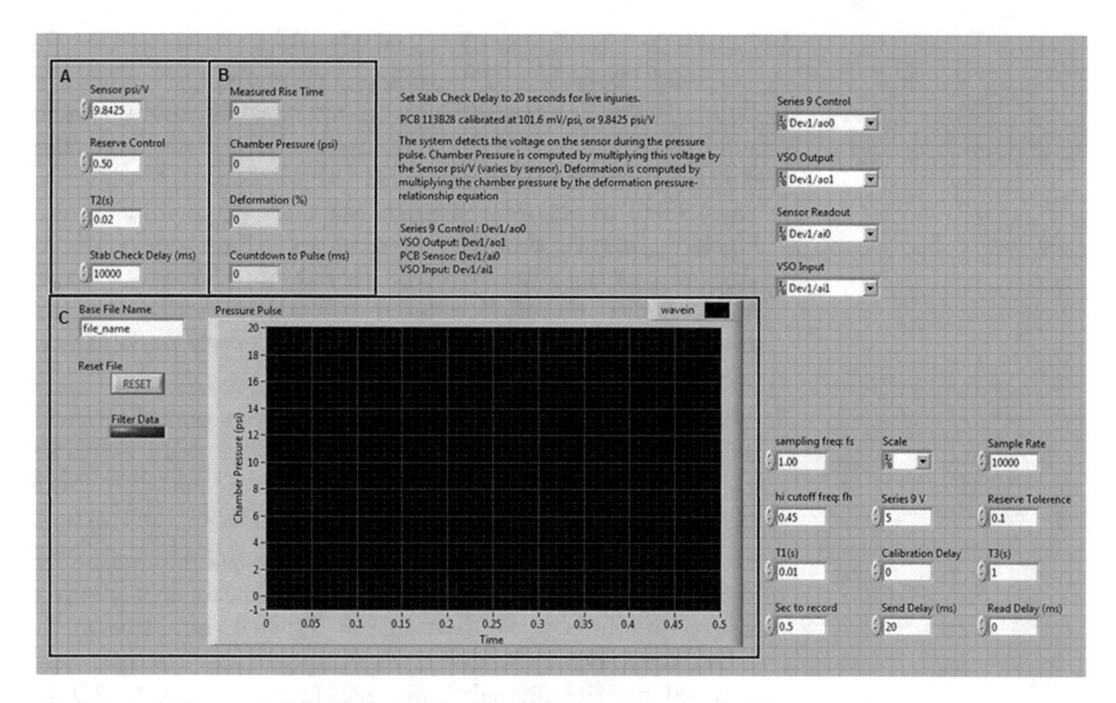

Fig. 6 Sample Custom LabView Program for Controlling Injury Device. (**a**) Shows the device parameters that can be modulated manually, in particular the input pressure (labeled as "Voltage Control") and valve open time (labeled as "T2(s)"). (**b**) Shows the automated outputs based on the pressure detected by the pressure transducer at the injury chamber. (**c**) Shows the graph produced from the sensor's measurements. This data is automatically saved

created to account for the pressure transducer calibration, commonly reported as mV/psi or similar units. It is helpful to see the pressure pulse graphed and logged automatically (Fig. 6c). LabView extracts the rise time based on the recorded data points.

Based on the chamber pressure-deformation relationship determined during the characterization steps, the program can also be designed to report the calculated silicone deformation based on the detected chamber pressure (Fig. 6b).

4 Notes: 2-D In Vitro Stretch Injury Model

1. NOTE 1: We noticed that, quite frequently, the silicone could stick to the injury mask surface. This posed the risk of additional strain during the removal of the culture well from the injury chamber. Additionally, microscopic imperfections in the mask surface, which results especially from thermal deposition prints, could permit unknown and unintended strains to occur in the "uninjured" regions of the culture well. The more rigid Aclar® (relative to the silicone sheet) prevents either of these from occurring, providing much more confidence that the injury is being restricted to the desired area.

2. NOTE 2: The choice of Aclar® film thickness for underside well mask depends on the specific experimental application. We designed this injury setup with certain types of experiments in mind, such that axons or cell bodies could be imaged simultaneously with the injury on an inverted microscope. Specifically, we needed to accommodate an objective lens with a 1 mm working distance; thus we were required to limit the combined thickness of both masks to less than 1 mm. Thinner Aclar® sheets could be used if required; it is only necessary that the masks create a rigid surface that remains largely undeformed when the chamber is pressurized.
3. NOTE 3: The silicone and Aclar® will form a seal when pressured together. Optimally, this seal will be free of any air pockets. Pressing the two together with water in the middle reduces the number and size of air pockets creating a more uniform seal.
4. NOTE 4: It is necessary to perform these calibrations under the same conditions as during an injury experiments. During injury experiments, fluid volume in the well should be minimized to allow for efficient gas exchange. In a well without any cells on the silicone substrate, adding a detergent to HBSS decreases the surface tension of the fluid thereby allowing a smaller volume to spread throughout the entire silicone surface.
5. NOTE 5: This can be achieved by increasing the chamber pressure and identifying the lowest point in the uniaxial curvature of the silicone.
6. NOTE 6: Depending on the pressure range chosen, the trend line may be roughly linear or a third degree (cubic) polynomial. This equation will be plugged into the software program that runs the injury device, as described below in the Injury Device Operation section.
7. NOTE 7: Biaxial calibration, for the most part, follows the same series of steps as far as data collection is concerned, with one major difference. Under biaxial conditions, the silicone will stretch easier, meaning it takes a lower chamber pressure to achieve the same percentage of stretch compared to uniaxial conditions. Consequently, other than the differing mathematical analysis described above, the only other protocol deviation is that the chamber pressure range over which the device is characterized will be narrower, usually only up to about 4 psi maximum. At higher pressures the silicone is likely to blow out.
8. NOTE 8: Ethanol seems to help completely remove any detergent residue on the pieces.
9. NOTE 9: If you notice smudges on the silicone squares, they can be removed by rubbing the sheets in between your fingers with ethanol. We have also noticed that nitrile gloves can sometimes have subtle residues on them which transfer onto

the silicone. Washing one's gloves with ethanol several times can also help eliminate residues and prevent them from transferring to the silicone sheet.

10. NOTE 10: The silicone and Aclar® will form a seal when pressured together. Optimally, this seal will be free of any air pockets. Pressing the two together with water in the middle reduces the number and size of air pockets creating a more uniform seal.
11. NOTE 11: In addition to autoclaving, well pieces can be surface treated using plasma cleaning. Plasma cleaning improves poly-L-lysine coating and cell attachment later on. For plasma cleaning, the autoclaved and dried wells should be placed inside a closed container. Oxygen plasma clean the container for 1 min. After this, the wells are ready for poly-L-lysine coating.
12. NOTE 12: If using wells that have been plasma cleaned, perform three minimum 30-min rinses with sterile culture-grade water. Culture-grade water can be purchased or prepared by autoclaving bottles of RO water.
13. NOTE 13: The two silicone surfaces will form a seal. At plating, cells will loosely adhere to the top of the silicone strips. Upon removal of the strips, a cell-free gap is left underneath allowing for the formation of an axon-only region. With proper plating density, axons will typically traverse the middle of the gap by 9–11 days in vitro. Silicone strips should not be plasma cleaned. If both the strips and wells are plasma cleaned, they will form a very strong seal upon coming into contact with one another.
14. NOTE 14: A typical E17 rat brain can provide at least 6–8 million cortical cells using this protocol, in our experience. E18 brains can provide up to around 10 million cortical cells. The difference is likely due to a combination of slightly increased size and also easier dissection (the meninges is substantially easier to remove) leading to lower cell loss during processing.
15. NOTE 15: DNAse I requires the presence of calcium ions to function. However, trypsin is usually diluted in a calcium-free solution and administered in conjunction with the calcium chelator EDTA, which aids in disrupting cell–cell interactions and dissociating the cells. For this reason, we do not mix the DNAse I with the trypsin-EDTA but instead do sequential treatments. First, cell–cell interactions are weakened with trypsin-EDTA, cells are then dissociated by triteration, and finally any free DNA that results from damaged cells is degraded by DNAse I following dissociation.
16. NOTE 16: Larger cell chunks and other cellular debris can be isolated from the cell suspension by several methods. The

suspension can be passed through a filter. Alternatively, larger pieces of debris can be allowed to sink to the bottom of the tube, the remainder of the fluid moved to a new tube, and the original tube disposed of. This can also be performed in between steps 10 and 11 prior to centrifugation.

17. NOTE 17: Example: Cell solution counted to be approximately 5000 cells/μL

    12 injury wells to plate at 175,000 cells/well.

    175,000 cells/well × 15 wells (*see* NOTE below) = 2,625,000 cells total needed.

    100 μL volume to add per well × 15 wells (*see* NOTE below) = 1500 μL.

18. NOTE 18: When diluting the cell plating suspension, it is advisable to add and account for more volume than is needed as the suspension has a tendency to bubble. In the above example, three extra wells worth of cells and media volume was included.

    2,625,000 cells/5000 cells/ μL = 525 μL

    Add 525 μL final cell suspension to 975 μL Cortical Media and add 100 μL of cell dilution to each well. Discard excess.

19. NOTE 19: To account for evaporation of the media, we assume 50 μL of media is lost each day. So, for example, if it has been 2 days since last feeding, 400 μL would be removed and 500 μL added back in order to maintain a 1 mL volume.

20. NOTE 20: If axons fail to grow across the cell free gap but the cells appear healthy, higher plating densities can be used. We have used plating densities up to 1500 cells/mm^2. If the density is too low, axons will not traverse the cell-free gap. Appropriate plating densities may differ across labs and setups.

5 Materials: 3-D In Vitro Injury Model

5.1 Materials for 3-D Constructs

- Surgical tools (large scissors, small scissors, one set of curved forceps, two sets of fine forceps, one set of fine angled forceps, two micro-scalpel blades, one surgical spatula).
- Hanks Balanced Salt Solution (HBSS, Gibco).
- Compressed CO_2 air tank.
- Guillotine.
- L-15 media (Gibco).
- B27 Supplement (Invitrogen).
- 0.25% tryspin + 1 mM ethylenediaminetetraacetic acid (EDTA, Invitrogen).

- DMEM/F12 media (Gibco).
- Fetal bovine serum (FBS, Atlanta Biologicals).
- DNase I (Roche).
- Magnesium sulfate ($MgSO_4$, Sigma).
- Matrigel (Corning).
- Trypan Blue.
- Neurobasal media (Gibco).
- G5 (100× concentrate, Gibco).
- Glutamax (100× concentrate, Gibco).
- Poly-L-lysine (PLL; Sigma).
- 70% ethanol.

5.2 Materials for Cell Shearing Device for In Vitro Injury

- Custom-designed Cell Shearing Device (Kinetic Software Atlanta) including:
  - Linear voice coil actuator (BEI Kimco; San Marcos, CA).
  - Custom-fabricated digital proportional-integral-derivative (PID) controller (25-kHz sampling rate, 16-bit sampling resolution).
  - Optical position sensor (OPS) (RGH-34, 400 nm resolution; Renishaw, New Mills, UK).
  - Custom-built polycarbonate cell chambers.
- Silicone adhesive.
- Plain glass microscope no. 1½ coverslip or slide (Fisherbrand, 25 × 75 × 1 mm).
- Sylgard 184 (Dow Corning; Midland, MI).
- Polycarbonate cylindrical cutouts (inner diameter roughly 15 mm).
- Scotch tape.
- Razor blade.
- 70% ethanol.

6 Methods: 3-D In Vitro Injury Model

6.1 Introduction to 3-D CSD

All strain fields in the brain are 3-D, yet most in vitro models of TBI use planar (2-D) cultures, which fail to replicate the complex 3-D strain fields of in vivo injury. To address this gap, LaPlaca and colleagues developed and validated the first in vitro model of TBI that employed custom-built 3-D neural cellular constructs subjected to controlled 3-D shear-strain based loading profiles [10]. The custom-designed device is referred to as the 3-D cell shearing device (CSD), and is capable or reproducibly delivering a specified strain field to neural cells within a 3-D bioactive ECM based scaffold. The device input approximates the 3-D strain fields predicted during closed-head diffuse injury in the human brain. In addition to matching the tissue-level biomechanical inputs implicated in the etiology of clinical TBI, engineering neural constructs in vitro offers other advantages for investigating specific mechanisms of injury, including spatial and temporal control of cellular composition, matrix/scaffold properties, mechanical environment, and exogenous factors. Thus, a 3-D culture system more faithfully mimics the complex environment of the brain than 2-D cultures, and permits the application of strain fields to model forces seen in human injury. These 3-D constructs and 3-D CSD permit the investigation of complex responses that occur after traumatic loading and may provide a physiologically-relevant testbed to evaluate the efficacy of potential therapeutic agents.

6.2 Production of CSD Injury Chambers

1. Outline the bottom of the rectuangular polycarbonate well (cell reservoir) with silicon adhesive and attach precleaned, plain glass microscope coverslips or slides and let cure.
2. On the top plates, wrap approximately 30 cm of scotch tape around the cylindrical protrusions (*see* **Note 1**).
3. After the glass slide is sealed to the bottom of the reservoir (e.g., the next day), fill the chambers with PDMS (Sylgard 184 and curing agent mixed 10:1). Let the PDMS sit until the bubbles settle, and apply the top plate so that the two cylinders are centered in the reservoir. Allow this to cure for 24–48 h.
4. After the PDMS has cured sufficiently, slowly remove the top plate (*see* **Note 2**). Use a razor blade or a similar sharp object to cut away excess PDMS at the base of the chamber (on the glass), and remove the tape from the cylinders (*see* **Note 3**).
5. Test the cell chamber on the CSD to ensure that the top plate moves effectively. If it does not, additional tape may be added in order to further widen the holes.
6. To clean the chambers after use, soak in 70% ethanol for at least 4–6 h. Chambers are reusable after cleaning.

6.3 Primary Astrocyte and Neuron Cell Culture Procedures

6.3.1 Overview

We will describe the procedure for generating neuron-astrocyte 3-D cocultures in a bioactive ECM scaffold, though single cell types or other mixes of cells may be used. Embryonic day 17 (E17) Sprague-Dawley rats are used to harvest neurons, while postnatal day 1–2 (P1–2) Sprague-Dawley rat pups are used to harvest astrocytes. The astrocyte harvest should occur several weeks prior to generation of the 3-D constructs, so the astrocytes have a chance to be passaged 2–3 times before use in order to acquire a more mature phenotype.

6.3.2 Neuronal Harvest Protocol

All surgical tools should be sterilized by autoclave prior to dissection. The dissection is performed within a laminar flow hood.

1. Fill 4–6100 mm Petri-dishes with 15 mL each of sterile HBSS and place on ice.
2. Sacrifice the mother Sprague-Dawley rat with E17 pups via prolonged CO_2 and subsequent guillotine. Lay carcass ventral side up and rinse abdomen thoroughly with 70% ethanol. Cut beginning at the lower abdomen extending rostrally to expose the uterus. Remove the uterus and place in a sterile petri-dish with ice-cold sterile HBSS. Remove each fetus from the amniotic sacs and transfer to a new petri-dish containing ice-cold HBSS.
3. Decapitate each fetus and remove the brains by cutting and peeling back the top portion of the skull. Transfer the brains to a new petri-dish containing ice-cold HBSS.
4. To isolate the cortices, remove the hindbrain and perform a midsagittal cut. Remove the midbrain and place hemispheres with lateral surfaces facing down. Remove the olfactory bulbs and turn over the hemispheres such that the medial surfaces face down. Detach the meninges and again turn the hemispheres over such that lateral surfaces face down. Remove the cortical region from the remaining midbrain and cut the hippocampal region from the cortex (*see* **Note 4**).
5. Transfer the cortical regions to a 15 mL centrifuge tube of HBSS (4–6 cortices per tube) and place on ice (*see* **Note 5**).

6.3.3 Astrocyte Harvest Protocol

All surgical tools should be sterilized by autoclave prior to dissection. The dissection is performed within a laminar flow hood.

1. Sacrifice P1–P2 rat pups by placing each pup in the finger of a disposable glove and covering with ice for 5–10 min before dipping in ethanol and decapitating with scissors.
2. Remove the brain and proceed with the cortical isolation as described above for the neuronal harvest.
3. Mince cortices with micro-scalpel blades, then transfer the tissue to a 15 mL centrifuge tube with 900 μL 0.25% trypsin + 1 mM EDTA per brain and place in a 37 °C bath for 5–7 min.
4. After the tissue is clumped together, neutralize the trypsin with 1 mL DMEM/F12 media supplemented with 10% FBS, and also add 400 μL DNase solution (1.5 mg/mL DNase I, 2.5 mg/mL $MgSO_4$, in HBSS). Triturate with a fire-polished pipette until no clumps are visible.
5. Centrifuge at 3500 rpm for 3 min, then remove the supernatant and resuspend the pellet in plating media (DMEM/F12 + 10% FBS).

6. Plate one brain per T75 flask, with a final volume of 10 mL plating media, and place the flask in a tissue culture incubator (37 °C, 5% CO_2, 95% RH).
7. Completely change media about once a week.
8. Passage cells as when they become about 90% confluent. To do so, remove the media and add 5 mL 0.25% trypsin-EDTA. Return the flask to the incubator for 5–10 min, as astrocytes detach from the flask. Once lose, deactivate the trypsin with 1 mL DMEM/F12 + 10% FBS. Transfer the suspension to a 15 mL centrifuge tube and centrifuge at 3500 rpm for 3 min. Remove the supernatant and resuspend the pellet in plating media. Transfer the desired ratio of cells into a new T75 flask (i.e., 1:4, 1:10) with 10 mL plating media, and return flask (s) to incubator. For all steps, keep the tissue on ice when not in the centrifuge or 37 °C bath.

6.4 3-D Coculture Procedures

3-D cultures consist of neural cells embedded within a 3-D bioactive extracellular matrix-based scaffold. We will describe the procedure for generating neuron-astrocyte cocultures, though single cell types or other mixes of cells may be used.

6.4.1 Chamber Preparation

1. Make sure the injury chambers are clean (i.e., soaked in ethanol and rinsed with deionized water) and autoclaved.
2. Pretreat the chambers with 500 μL 0.5 mg/mL PLL and place in incubator (37 °C, 5% CO_2, 95% RH) for at least 4 h. Aspirate the PLL and let the excess evaporate by leaving the lid open under a laminar flow hood for ~2 min.
3. Precoat with 500 μL 0.5 mg/mL Matrigel and place the chambers in an incubator (37 °C, 5% CO_2, 95% RH) for at least 4 h.

6.4.2 Neuronal Dissociation

Tissue and reagents (except trypsin before and during tissue addition) should be kept on ice at all times.

1. Acquire cortices from E17 Sprague-Dawley rats as described above.
2. Mince the cortices with micro-scalpel blades, then transfer the tissue from 2 to 3 brains to a 15 mL centrifuge tube with 5 mL 0.25 tryspin + EDTA and place in a 37 °C bath.
3. After 5–7 min check the tissue for clumping and manually agitate. After the tissue is clumped together, carefully remove the trypsin and gently rinse with 1 mL cold DMEM/F12 + 10% FBS.
4. Remove the serum and gently rinse twice with cold HBSS.

5. After removing the rinse HBSS, add 1.8 mL HBSS + 400 μL DNase solution. Triturate with a fire-polished pipette until no clumps are visible.
6. Centrifuge at 3500 rpm for 3 min, then remove the supernatant and resuspend the pellet in 2 mL neuronal-astrocyte media (Neurobasal + B27 + glutamax + G5).
7. Count the cells in order to dilute them to the desired density. Ensure the cell solution is well-mixed (gently triturate 3–5 times) and remove a small sample to dilute and count (*see* **Notes 6** and 7). Dilute with neuron-astrocyte media to desired density (i.e., 5 million cells/mL).

6.4.3 Plating

1. Place Eppendorf tubes with frozen 500 μL 15 mg/mL Matrigel on ice to begin to thaw to a slush consistency.
2. Acquire astrocytes as if preparing for passaging (see 4.3.3, though do not transfer to a new flask after resuspending pellet), count cells, and dilute with neuron-astrocyte media to desired density (i.e., 5 million cells/mL).
3. Mix the neurons and astrocytes in the desired ratio (i.e., 1:1). From this mixture, take 500 μL of cells and use a cold pipet tip to gently (to avoid bubbles) yet thoroughly mix with the thawed Matrigel tube, bringing the final Matrigel concentration to 7.5 mg/mL (*see* **Note 8**).
4. Bring the injury chambers from the incubator and aspirate the Matrigel precoat from wells. Quickly add the Matrigel/gel suspension at the desired thickness for the construct (i.e., 200 μL) and immediately return the chamber to the incubator to gel.
5. After about 1 h, add 1 mL neuronal-astrocyte media to each well.
6. At 24 h post-plating, feed cells by removing half of the old media and replacing with fresh. Feed cells in this manner every 2–3 days.

Cultures should exhibit minimal clustering, meaning there should be few neurons demonstrating soma-soma contact (assessed visually). If astrocytes are included in the constructs, after several days constructs should look like they are contracting and may develop thicker outer rings as astrocytes remodel the ECM scaffold. Unless experiment-specific, cultures should be used at 14 DIV or later to permit neuronal maturation.

6.5 Injuring 3-D Cultures with Cell Shearing Device

6.5.1 Overview

The 3-D CSD consists of a (1) the device itself—a linear actuator coupled to a closed-loop control system (Fig. 7a), and (2) reusable cell chambers (Fig. 7b). The system was designed by LaPlaca, Cargill, and Cullen, with the final device(s) fabricated by Kinetic Software Atlanta. The cell chamber is comprised of the cell reservoir

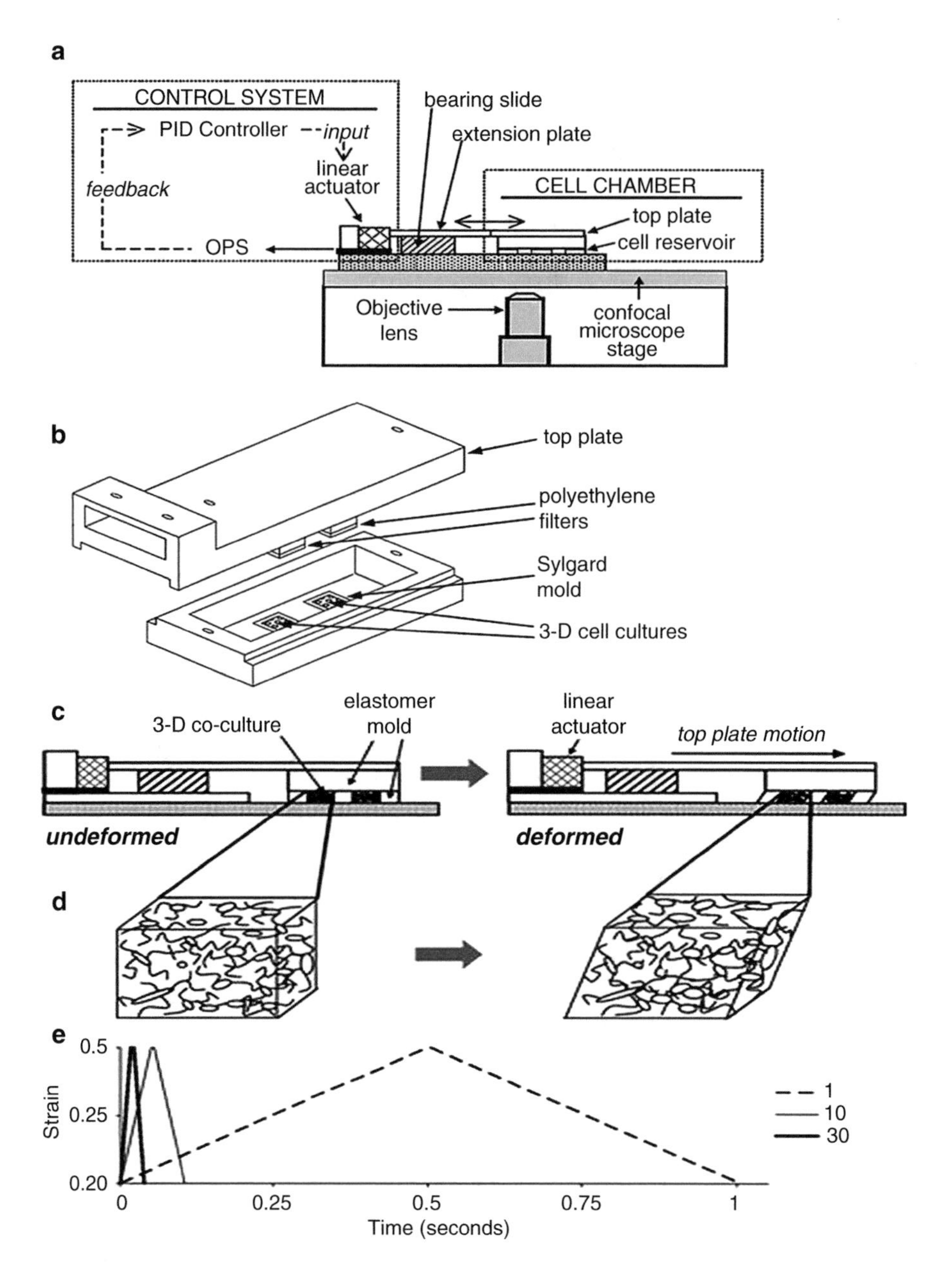

Fig. 7 3-D Cell Shearing Device (3-D CSD) Components. (**a**) A schematic representation of the 3-D CSD. The device can be mounted on a confocal microscope to obtain 3-D images before, during, and after mechanical deformation. A closed-loop control system consisting of a proportional-integral-derivative (PID) control with positional feedback from an optical position sensor (OPS) governs a linear actuator, inducing motion of the cell chamber top plate (not to scale). (**b**) The cell chamber consists of a top plate with polycarbonate protrusions to interface with the 3-D cell cultures. The top plate is mounted above the cell reservoirs and connected to the linear actuator to impart high rate deformation. (**c**) The 3-D CSD top plate motion is driven by a linear actuator. Neuronal–astrocytic cocultures in 3-D were plated throughout the thickness of a bioactive matrix and were laterally constrained by an elastomer mold. (**d**) Shear deformation of the elastomer mold and cell-embedded matrices was induced through horizontal displacement of the cell chamber top plate, which was coupled to the linear actuator. (**e**) Input to the system was a symmetrical trapezoidal input to a constant displacement

and the top plate. The cell reservoir is a rectangular polycarbonate well with a glass floor (no. 1 ½ coverslip or plain glass slide). Parallel shoulders were machined into the upper surface of the two long sides of the polycarbonate walls to provide proper alignment and constrain motion of the top plate to a single axis. The 3-D cultures (two per reservoir) are confined within an elastomer mold composed of Sylgard 184 and cast to the desired thickness. The cultures are contained by cylindrical cutouts (inner diameter of roughly 15 mm). Throughout experimentation, the mold adheres firmly to the bottom glass coverslip without any movement. The top plate is composed of a polycarbonate body. The cylindrical protrusions act as struts to transfer the transverse motion of the top plate to the cell cultures, which are situated, on the chamber base. The top plate is attached to the linear actuator by a polycarbonate extension plate that is mounted to a linear bearing system (Fig. 7).

This device generates a dynamic shear strain by the linear motion of the top plate with respect to the cell reservoir via a linear voice coil actuator coupled to a custom-fabricated digital proportional-integral-derivative (PID) controller with closed-loop motion control feedback from an optical position sensor (OPS). The linear actuator produces a rigid body translation of the top plate assembly. The input signal is the reference input to the PID controller and is compared internally to the actual displacement of the top plate provided by the OPS. The closed loop controller produces a self-correcting signal to minimize the error between the actual and desired position. PID control reduces steady-state error between the input signal and the motion of the top plate, eliminates possible oscillations in the response, and increases sensitivity of the system.

The device is controlled using a custom stimulation control program in LabVIEW. The device may be used on a bench top or mounted above a confocal microscope to image the deformation.

6.5.2 Application

Two parameters that control the severity of injury are the shear strain magnitude and the strain rate. Both can be controlled by a closed-loop actuator/controller system that develops shear strains of up to 0.5 (shear angle up to 45°) at rates from 1 to 50 s^{-1} in order to simulate the spatial and temporal strain patterns associated with inertial TBI. In our system we set several parameters to provide a trapezoidal input: the movement of the top-plate (in millimeters; corresponding to shear strain magnitude) and the desired rise-time (in milliseconds; corresponding to strain rate),

Fig. 7 (continued) (corresponding to 0.50 bulk shear strain) at strain rates of 1, 10 or 30 s^{-1} (corresponding to rise times of 500, 50, or 16.7 ms, respectively; hold time was held constant at 5 ms). Panels (**a**) and (**b**) reprinted with permission from LaPlaca et al., 2005, Elsevier; panels (**c–e**) reprinted with permission from Cullen et al., 2007, Elsevier

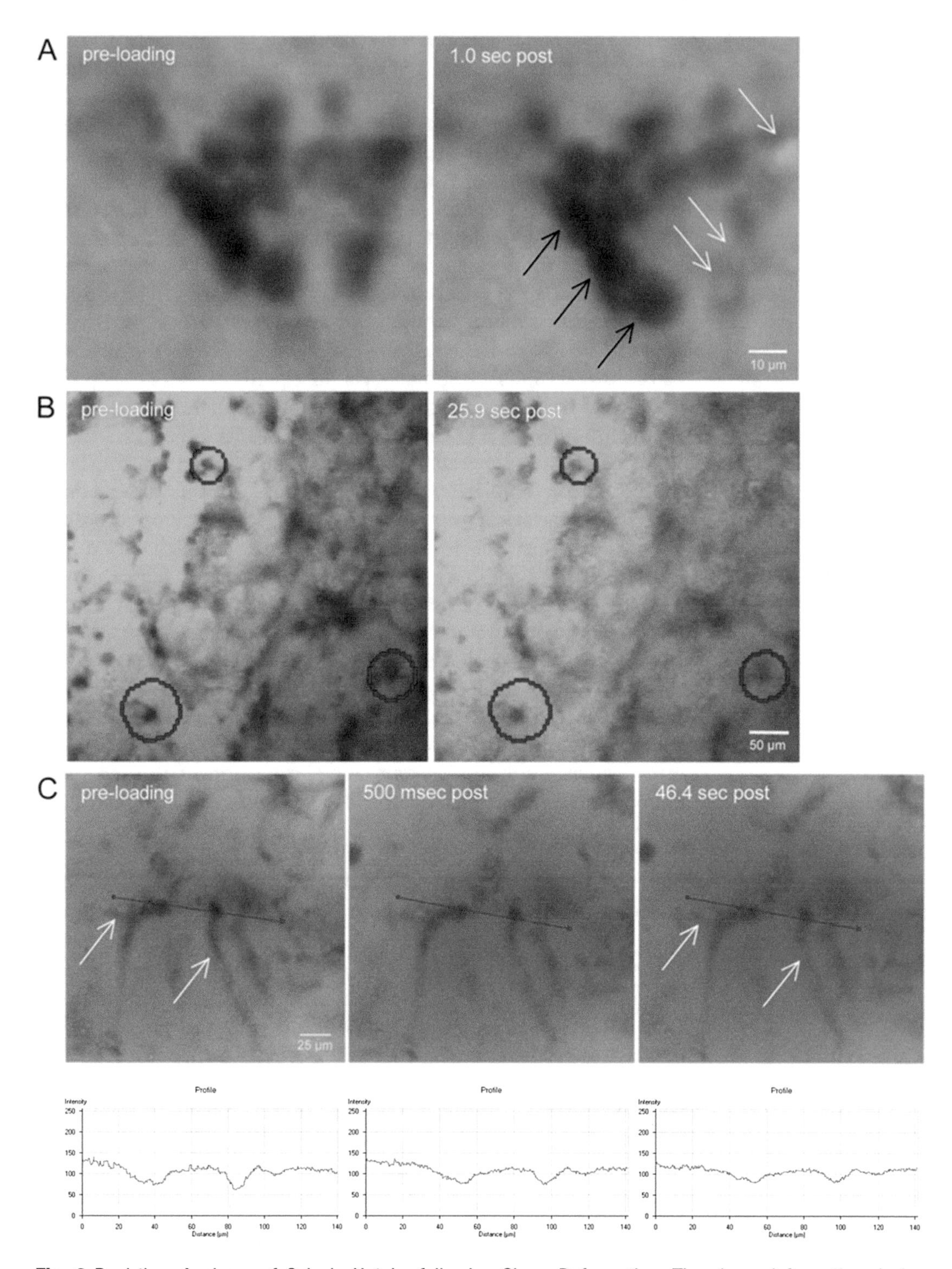

Fig. 8 Real-time Analyses of Calcein Uptake following Shear Deformation. The shear deformation device permits real-time imaging across the 3-D cultures before, during, and after loading. (**a**) Calcein diffused throughout the extracellular space prior to loading (resulting in relatively high fluorescence background during imaging). The 3-D cultures were dynamically deformed at 0.50 shear strain (10 or 30 s^{-1} strain rate).

hold-time, and ramp-down time. The firing of the linear actuator will deflect the top-plate horizontally while the chamber base is held stationary. Static control cultures experience all steps with the exception of device stimulation, thus controlling for the potential effects of media removal, top-plate application, and any temperature fluctuations.

A variety of outcome measures can be used to assess cellular responses after injury. One consequence of traumatic loading is transient micro- or nano-tears in the plasma membranes of cells. Cells that have become mechanoporated but then reseal can be labeled by adding small, nonpermeable molecules (such as calcein or Lucifer Yellow) to the media before an during injury. Because the CSD can be mounted on a confocal microscope, real-time imaging of the permeability event is possible (Fig. 8). Permeability can also be viewed post- injury (Fig. 9). We have shown that the faster the strain rate, the greater extent of membrane permeability, both in terms of the density of permeabilized cells and intensity of dye uptake. Cell death can also be assessed in these cultures. Using a standard viability/cytotoxicity assay (Live/Dead Viability/Cytotoxicity Kit, Molecular Probes), we have shown that at high strain rates, significant cell death and neural network disruption is evident (Fig. 10). The number of cells per construct is also amenable for standard biochemical assays. Moreover, the constructs may be fixed for immunocytochemistry across the full thickness, or the constructs may be embedded and sectioned using protocols similar to those applied for nervous system tissue.

7 Notes: 3-D In Vitro Injury Model

1. NOTE 1: This extra layer of tape makes the cylindrical holes wider than the dimensions of the protrusions (absent the tape), which facilitates the movement of the top plate during strain application.
2. NOTE 2: Be careful—do not remove PDMS mold as well.
3. NOTE 3: This step is not always necessary—if the tape is still in good condition it can be reused to make more chambers.
4. NOTE 4: Hippocampal regions may also be saved.

Fig. 8 (continued) Preloading, there were clear margins delineating neural somata and processes, indicating intact plasmalemma. However, immediately post-insult, there was a subset of cells that became calcein+ (white arrows in **a**), whereas some cells excluded calcein (black arrows in **a**). (**b**) In other cases, a widespread loss of definition was apparent in neurites as well as somata. Calcein uptake was tracked on a per-cell basis throughout loading (**c**). Within 500 ms post-insult, intracellular calcein intensity increased. This process continued throughout the first minute, resulting in processes gradually blending with the background (white arrows in **c**). (Figure reproduced from Cullen et al., 2011, Mary Ann Liebert, Inc.)

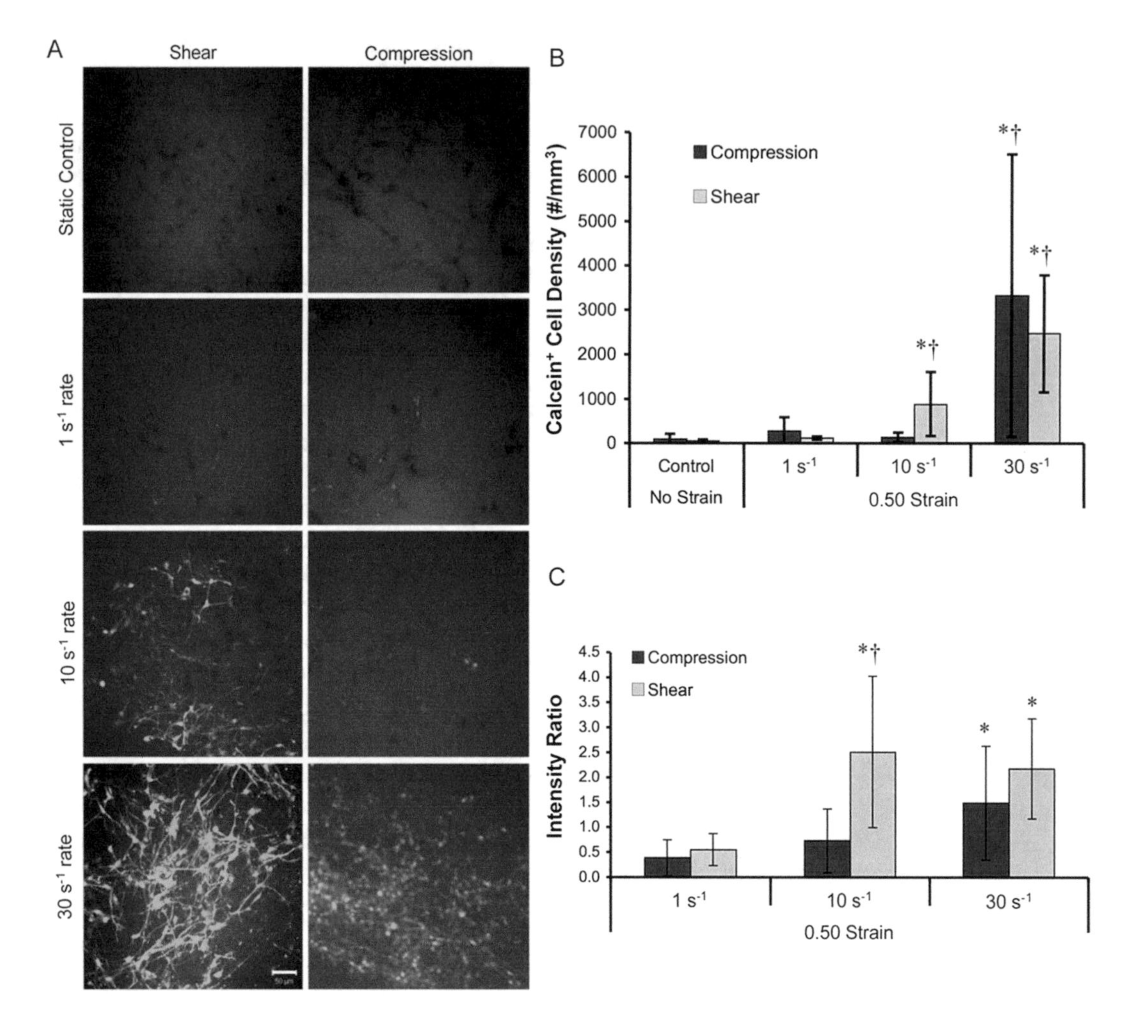

Fig. 9 Alterations in Acute Membrane Permeability in Shear versus Compression. (**a**) Representative confocal reconstructions of 3-D cultures following static control conditions or mechanical loading (0.50 strain at 1, 10, or 30 s^{-1} strain rate). (**b**) Calcein was added to the extracellular space prior to loading and enters permeable cells during or immediately following loading (reconstructions from 50-μm-thick z-stacks taken from cultures post-rinse; scale bar = 50 μm). Cell density of permeabilized (calcein+) cells. The 3-D cell density and percentage of calcein+ cells increased as a function of strain rate, and was highest at the maximum strain rate for both compressive and shear loading ($^*p < 0.05$ versus static control; $^\dagger p < 0.05$ versus lower strain rates). (**c**) The degree of cell permeability following variable rate shear or compressive (0.50 strain) loading. There was a significant increase in the mean intensity of calcein+ cells versus quasi-static loading for both shear and compression at 30 s^{-1} and shear only at 10 s^{-1} ($^*p < 0.05$). Also, at 10 s^{-1} loading, there was enhanced calcein uptake following shear versus compression ($^\dagger p < 0.05$). Note that the 3-D compression injury model is not described in this chapter, and the reader is encouraged to review the original publication, Cullen et al., 2011, for details. Data are mean ± standard deviation (Figure reproduced from Cullen et al., 2011, Mary Ann Liebert, Inc.)

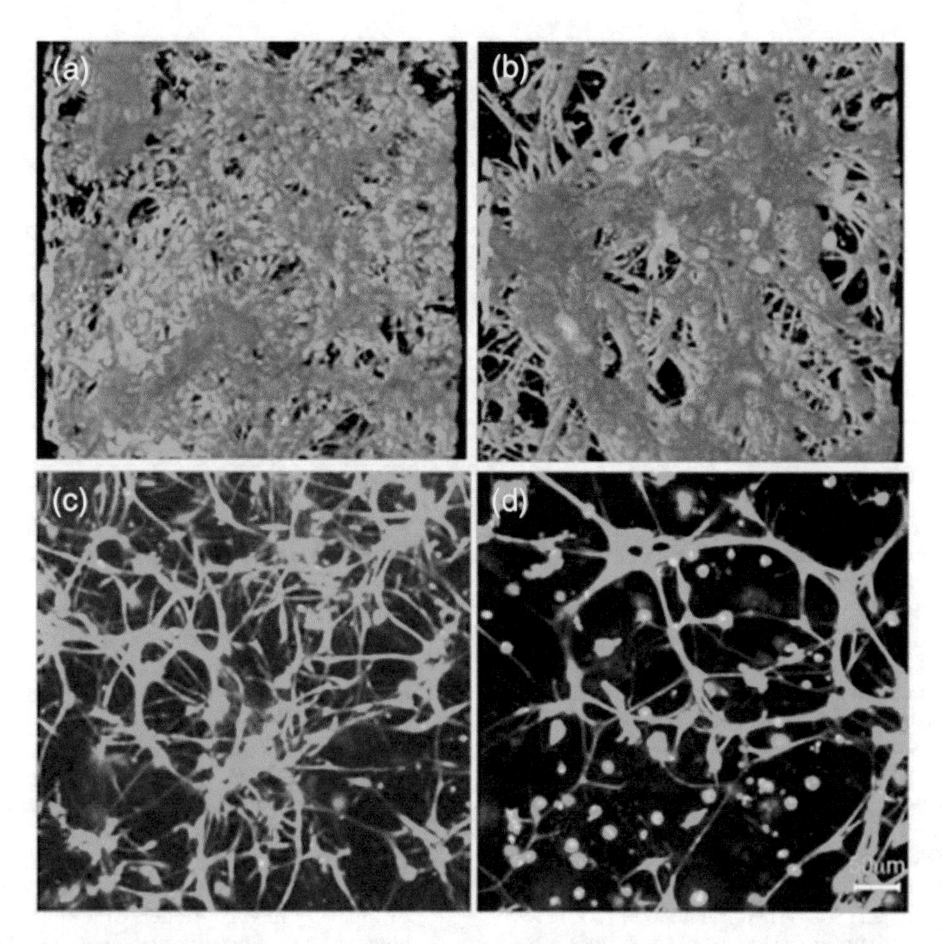

Fig. 10 3-D Neural Cocultures as an In Vitro Model of TBI. (**a**, **b**) Full thickness volumetric rendering and (**c**, **d**) optical slices showing live cells (green) and dead cells (red) in 3-D neuronal-astrocytic cocultures. High strain rate shear deformation (0.50 strain, 30 s^{-1}strain rate) causes significant cell death and neural network degradation at 48 h post-injury (**b**, **d**) compared to uninjured controls (**a**, **c**). Panels (**a**) and (**b**) reproduced with permission from Cullen et al., 2011, Begell House; panels (**c**) and (**d**) reprinted with permission from Cullen et al., 2007, Elsevier

5. NOTE 5: If not immediately proceeding with dissociation, cortices may be stored for 1–2 days. To store, rinse twice with cold HBSS (removing small debris). Add 2 mL L-15 supplemented with 2% B27. Wrap the tube in aluminum foil and place the tube on its side at 4 °C, ensuring that all cortices are submerged.
6. NOTE 6: Typically use a 1:40 dilution consisting of 10 μL cell solution + 190 μL HBSS and 200 μL trypan blue (let the diluted cell solution sit for 30 s prior to trypan blue addition). Ensure the diluted cell solution is well-mixed (gently triturate) and transfer 10 μL to each chamber of a hemocytometer. Count cells on hemocytometer to determine the total cell number and percent viability via trypan blue exclusion.

7. NOTE 7: Total cell yield should be 3–5 million cells/cortical hemisphere. Do not proceed if the yield is substantially less than this amount, or if neuronal viability is less than 90%.
8. NOTE 8: It is critical that the Matrigel remain cold throughout the plating procedure. If the Matrigel is too liquid at the time of plating, cells will sink to the bottom and not be distributed throughout the construct. Furthermore, warm temperatures cause the Matrigel to gel, so exposure to warmth must be avoided prior to placing Matrigel and cells into incubator. Handle the Matrigel with pipet tips that have been stored in a freezer.

References

1. Smith DH, Meaney DF (2000) Axonal damage in traumatic brain injury. Neuroscientist 6 (6):483–495
2. Cullen DK, Harris JP, Browne KD, Wolf JA, Duda JE, Meaney DF, Margulies SS, Smith DH (2016) A porcine model of traumatic brain injury via head rotational acceleration. Methods Mol Biol 1462:289–324. https://doi.org/10.1007/978-1-4939-3816-2_17
3. Margulies SS, Thibault LE, Gennarelli TA (1990) Physical model simulations of brain injury in the primate. J Biomech 23 (8):823–836
4. Meaney DF, Smith DH, Shreiber DI, Bain AC, Miller RT, Ross DT, Gennarelli TA (1995) Biomechanical analysis of experimental diffuse axonal injury. J Neurotrauma 12(4):689–694
5. Meaney DF, Thibault KL, Gennarelli TA, Thibault LE (1993) Experimental investigation of the relationship between head kinematics and intracranial tissue deformation. In: ASME Bioengineering Conference, Breckenridge, Colorado. ASME, pp 8–11
6. Magou GC, Guo Y, Choudhury M, Chen L, Hususan N, Masotti S, Pfister BJ (2011) Engineering a high throughput axon injury system. J Neurotrauma 28(11):2203–2218. https://doi.org/10.1089/neu.2010.1596
7. Smith DH, Wolf JA, Lusardi TA, Lee VM, Meaney DF (1999) High tolerance and delayed elastic response of cultured axons to dynamic stretch injury. J Neurosci 19 (11):4263–4269
8. Cargill RS 2nd, Thibault LE (1996) Acute alterations in [Ca2+]i in NG108-15 cells subjected to high strain rate deformation and chemical hypoxia: an in vitro model for neural trauma. J Neurotrauma 13(7):395–407
9. Ellis EF, McKinney JS, Willoughby KA, Liang S, Povlishock JT (1995) A new model for rapid stretch-induced injury of cells in culture: characterization of the model using astrocytes. J Neurotrauma 12(3):325–339
10. LaPlaca MC, Cullen DK, McLoughlin JJ, Cargill RS 2nd (2005) High rate shear strain of three-dimensional neural cell cultures: a new in vitro traumatic brain injury model. J Biomech 38(5):1093–1105
11. LaPlaca MC, Thibault LE (1997) An in vitro traumatic injury model to examine the response of neurons to a hydrodynamically-induced deformation. Ann Biomed Eng 25 (4):665–677
12. Morrison B 3rd, Elkin BS, Dolle JP, Yarmush ML (2011) In vitro models of traumatic brain injury. Annu Rev Biomed Eng 13:91–126. https://doi.org/10.1146/annurev-bioeng-071910-124706
13. Pfister BJ, Weihs TP, Betenbaugh M, Bao G (2003) An in vitro uniaxial stretch model for axonal injury. Ann Biomed Eng 31 (5):589–598
14. Cohen AS, Pfister BJ, Schwarzbach E, Sean Grady M, Goforth PB, Satin LS (2007) Injury-induced alterations in CNS electrophysiology. Prog Brain Res 161:143–169
15. LaPlaca MC, Simon CM, Prado GR, Cullen DK (2007) CNS injury biomechanics and experimental models. Prog Brain Res 161:13–26. https://doi.org/10.1016/S0079-6123(06)61002-9
16. Morrison B 3rd, Saatman KE, Meaney DF, McIntosh TK (1998) In vitro central nervous system models of mechanically induced trauma: a review. J Neurotrauma 15 (11):911–928
17. Cullen DK, Vernekar VN, LaPlaca MC (2011) Trauma-induced plasmalemma disruptions in three-dimensional neural cultures are dependent on strain modality and rate. J

Neurotrauma 28(11):2219–2233. https://doi.org/10.1089/neu.2011.1841

18. Iwata A, Stys PK, Wolf JA, Chen X-H, Taylor AG, Meaney DF, Smith DH (2004) Traumatic axonal injury induces proteolytic cleavage of the voltage-gated sodium channels modulated by tetrodotoxin and protease inhibitors. J Neurosci 24:4605–4613
19. Magou GC, Pfister BJ, Berlin JR (2015) Effect of acute stretch injury on action potential and network activity of rat neocortical neurons in culture. Brain Res 1624:525–535. https://doi.org/10.1016/j.brainres.2015.07.056
20. Pfister B, Oyler G, Betenbaugh M, Bao G (2004) The effects of BclXL and Bax overexpression on stretch-injury induced neural cell death. Mech Chem Biosyst 1(4):233–243
21. von Reyn CR, Mott RE, Siman R, Smith DH, Meaney DF (2012) Mechanisms of calpain mediated proteolysis of voltage gated sodium channel alpha-subunits following in vitro dynamic stretch injury. J Neurochem 121 (5):793–805. https://doi.org/10.1111/j.1471-4159.2012.07735.x
22. Abdul-Muneer PM, Conte AA, Haldar D, Long M, Patel RK, Santhakumar V, Overall CM, Pfister BJ (2017) Traumatic brain injury induced matrix metalloproteinase2 cleaves CXCL12alpha (stromal cell derived factor 1alpha) and causes neurodegeneration. Brain Behav Immun 59:190–199. https://doi.org/10.1016/j.bbi.2016.09.002
23. Kao CQ, Goforth PB, Ellis EF, Satin LS (2004) Potentiation of GABA(A) currents after mechanical injury of cortical neurons. J Neurotrauma 21(3):259–270
24. Rzigalinski BA, Weber JT, Willoughby KA, Ellis EF (1998) Intracellular free calcium dynamics in stretch-injured astrocytes. J Neurochem 70(6):2377–2385
25. Sherman SA, Phillips JK, Costa JT, Cho FS, Oungoulian SR, Finan JD (2016) Stretch injury of human induced pluripotent stem cell derived neurons in a 96 well format. Sci Rep 6:34097. https://doi.org/10.1038/srep34097
26. Lusardi TA, Rangan J, Sun D, Smith DH, Meaney DF (2004) A device to study the initiation and propagation of calcium transients in cultured neurons after mechanical stretch. Ann Biomed Eng 32(11):1546–1558
27. Wolf JA, Stys PK, Lusardi T, Meaney D, Smith DH (2001) Traumatic axonal injury induces calcium influx modulated by tetrodotoxin-sensitive sodium channels. J Neurosci 21 (6):1923–1930
28. Geddes DM, Cargill RS 2nd, LaPlaca MC (2003) Mechanical stretch to neurons results in a strain rate and magnitude-dependent increase in plasma membrane permeability. J Neurotrauma 20(10):1039–1049
29. Yuen TJ, Browne KD, Iwata A, Smith DH (2009) Sodium channelopathy induced by mild axonal trauma worsens outcome after a repeat injury. J Neurosci Res 87 (16):3620–3625. https://doi.org/10.1002/jnr.22161
30. Cullen DK, LaPlaca MC (2006) Neuronal response to high rate shear deformation depends on heterogeneity of the local strain field. J Neurotrauma 23(9):1304–1319. https://doi.org/10.1089/neu.2006.23.1304

Chapter 4

Drosophila Model to Study Chronic Traumatic Encephalopathy

Rojahne Azwoir and Liam Chen

Abstract

Chronic Traumatic Encephalopathy (CTE) is an established neurodegenerative disease that is closely associated with exposure to repetitive mild traumatic brain injury (mTBI). The mechanisms responsible for its complex pathological changes remain largely elusive, despite a recent consensus to define the neuropathological criteria. Here, we provide details of a novel protocol to develop a model of CTE in fruit fly *Drosophila melanogaster* in an attempt to identify the key genes and pathways that lead to the characteristic hyperphosphorylated tau accumulation and neuronal death in the brain. The advantage of this protocol is that adjustable-strength impacts are delivered directly to the fly head to inflict mild closed injury, subjecting the head to rapid acceleration and deceleration. The less labor- and cost-intensive animal care, short life span, and extensive genetic tools make the fruit fly ideal to study CTE pathogenesis and make it possible to perform large-scale, genome-wide forward genetic and pharmacological screens.

Key words Chronic traumatic encephalopathy, Mild traumatic brain injury, Concussion, *Drosophila melanogaster*, Animal model, Neurodegeneration

1 Introduction

Chronic Traumatic Encephalopathy (CTE) has recently been recognized as a distinct neurodegenerative disorder, separate from other tauopathies such as Alzheimer's disease [1]. Unlike Alzheimer's disease and other common tauopathies whose most important risk factors are advancing age and a family history of dementia, CTE, as indicated by its name, implies a close association with a history of brain trauma, most likely seen in contact sports athletes, such as boxers and football players, as well as in military veterans [2, 3]. One of the most pressing issues is that a definitive diagnosis cannot be made without evidence of degeneration of brain tissue and tau aggregation indications that can only be seen after death. The symptoms of CTE are usually indistinguishable until months and often times years after repeated trauma has occurred. Patients may present symptoms and signs such as cognitive deficits, mood

Amit K. Srivastava and Charles S. Cox, Jr. (eds.), *Pre-Clinical and Clinical Methods in Brain Trauma Research*, Neuromethods, vol. 139, https://doi.org/10.1007/978-1-4939-8564-7_4,

and behavior changes, and movement dysfunction, which overlap significantly with Alzheimer's disease, frontotemporal dementia, Lewy body dementia, and Parkinson's disease [4]. In contrast, post-mortem examinations of brain tissue reveal a distinct pattern of hyperphosphorylated tau accumulation surrounding small blood vessels at the depths of the cortical sulci, a pathognomonic feature not observed in the other degenerative conditions [5].

As yet very little is known about the pathogenesis leading to disease manifestation. This is in large part due to the lack of a faithful animal model. Animal models that faithfully model CTE eatures, including neurophysiological alterations, neuropathological hallmarks, and neurobehavioral deficits, are essential for uncovering disease mechanisms and for developing diagnostic and therapeutic targets. It is understandable that no animal model of a human disease is perfect at mimicking all clinically relevant endpoints. However, we believe that a robust CTE model should satisfy the following three requirements: (1) the impact must be directly applied to a head that has an intact scalp and skull protection; (2) the head should not be immobilized during impact exposure so that rapid acceleration–deceleration and rotational and linear head movements are allowed; and (3) the experimental design should include both single and repetitive regimes, and the impact consequences should be mild in nature, without inflicting visible damage, such as tissue edema, contusion, or frank hemorrhage. Only recently have rodent models been generated [3, 6]. These model organisms have the disadvantages of cost-intensive care and a relatively long life span, which are not well-suited for neurodegenerative disease studies.

Compared to mammalian counterparts, invertebrate animals such as *Drosophila melanogaster* are an excellent alternative, with their cost-effective maintenance, extensive tools for dissecting genetic determinants, and relatively short life span [7]. Remarkably, fly and human brains share evolutionarily conserved molecular and cellular pathways, as well as anatomical similarities [8–10]. Two ingenious *Drosophila* models to study traumatic brain injury have been reported previously [11, 12]. The first "High Impact Trauma" (HIT) device designed by Katzenberger and colleagues contained free-moving flies in a plastic vial that was tied to the free end of a metal spring [11, 13]. When the plastic vial was tilted upright and released, it hit a polyurethane pad and imparted trauma to the flies as they bounced to the vial wall and rebounded. In contrast, Barekat and colleagues designed a different delivery method using the Omni Bead Ruptor-24 homogenizer platform [12]. Flies were incapacitated with CO_2 and placed in a 2 ml screwcap tube that was secured to the homogenizer and subjected to preprogrammed shaking conditions. One benefit of using the tissue homogenizer system is that the experimenter could modulate the intensity of injury, duration of injury, and number of injury bouts. However, both regimes suffer the same drawback: primary injuries to the head are randomly inflicted in terms of impact location and strength. In addition, both methods resulted in considerable mortality, caused by inevitable collateral damage to other parts of the body and internal organs. Here, we outline a novel protocol to induce mTBI in fruit flies. The apparatus consists of a gas-propelled ballistic impactor. Compared to the existing *Drosophila* models, this protocol has the unique advantage of delivering measurable impact, directed only at the free-moving fly head, thus allowing for the accurate control of various factors, such as impact severity, the time interval between impacts, and the total number of impacts sustained.

2 Materials

Material	Company	Catalog number	Comments
Aerosol Barrier	USA Scientific	1120-8810	Used as an impactor
200 μl Pipette Tip	USA Scientific	1111-0706	Used as a fly head holder
1000 μl Pipette Tip	USA Scientific	1122-1830	Used as a connector
1 ml Tuberculin Syringe	Becton Dickinson	309628	
Clear Tubing [1/8 in. (3 mm) I.D.]	Genesee Scientific	59-124C	
60 mm Petri Dishes	Fisher Scientific	FB0875713A	Used as a tracking arena
Flow Regulator	Genesee Scientific	59-122WC	
Standard Clamp Holder/ Stand	EISCO Scientific	CH0688	
Fine Brush	Genesee Scientific	59-204	
Fly CO_2 pad	Genesee Scientific	59-114	
Sylgard Silicone Elastomer	Dow Corning	4019862	
CCD Camera	Microsoft	HD-5000	
Ctrax Walking Fly Tracker	Caltech	Ctrax 0.2.11	
MATLAB Image Processing Toolbox	MATLAB	R2015b	
Nutri-Fly BF *Drosophila* media	Genesee Scientific	66-112	

3 Methods

1. Assembling the Strike Device (Fig. 1a).
   - (a) To make the impactor portion of the apparatus, unwrap and remove the plunger from a 1-ml polycarbonate tuberculin syringe.
   - (b) Cut the barrel of the syringe at the 1-ml mark.
   - (c) Remove the aerosol barrier (3-mm long and 4-mm in diameter) from a 200-μl pipette tip. It will be used as the impactor.
   - (d) Pack the impactor into the syringe barrel with the flat side facing the nozzle. Tap gently to move the impactor sliding towards the nozzle (*see* **Note 1**).
   - (e) Attach the syringe barrel to one end of a 3-ft-long clear tubing.
   - (f) Connect the other end of tubing to a carbon dioxide flow regulator. The regulator must have an on/off toggle switch which has good control of the gas flow rate.
   - (g) Clamp the barrel vertically to an iron stand such that the impactor will fall by gravity to the bottom of the barrel.

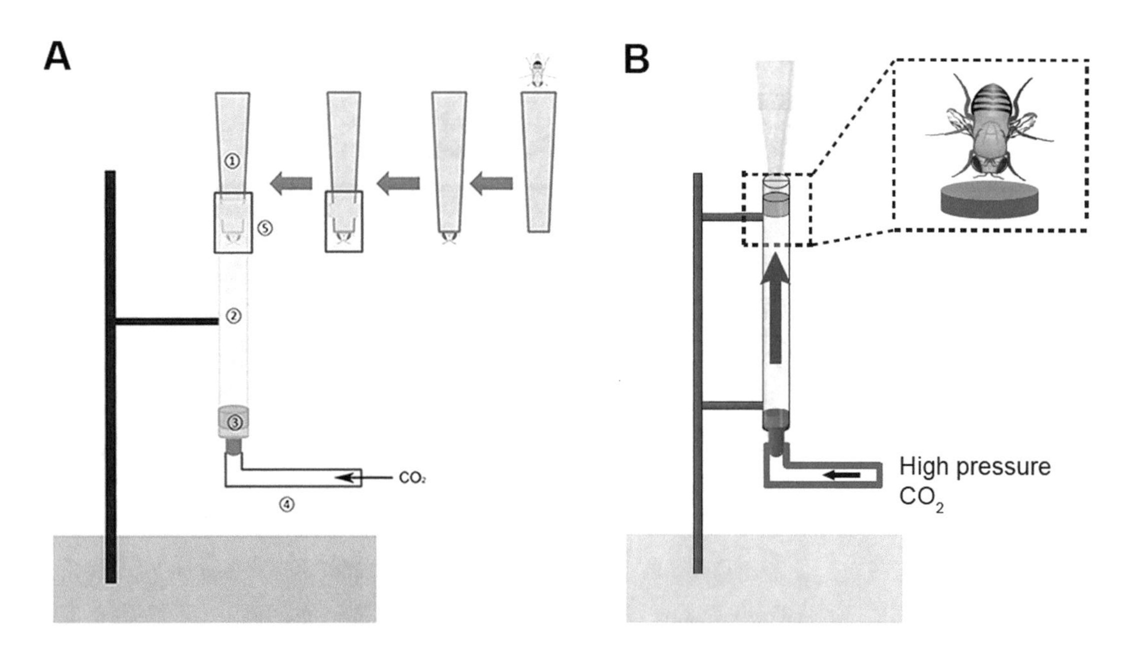

Fig. 1 Diagram of the strike device. (**a**) The device is made from materials that are readily available in a *Drosophila* lab: (1) 200 μl pipette tip, (2) 1 ml tuberculin syringe, (3) impactor, (4) connecting tube hooked to a fly anesthesia station, and (5) connector to tighten the tip onto the syringe. (**b**) High-pressure CO_2 would drive the compactor upwards to inflict direct hit to the fly head

(h) Next, make the fly holder. Start with a 200-μl pipette tip. Cut the pipette tip 4 mm from the small end to make a 0.8 mm diameter opening (*see* **Note 2**).

(i) Lastly, make the connector to tighten the fly holder to the impactor device.

- Cut a 1-ml pipette tip 44 mm from the tip opening.
- Next, cut a 6-mm-length segment from the needle cover of a 1-ml syringe.
- Push the needle cover segment into the tip segment to finish the assembly.

2. Operation of the Strike Device (Fig. 1b).

(a) Anesthetize a single 2-day-old adult female fly using carbon dioxide. From the CO_2 pad, use a fine brush to gently transfer the fly into the holder. Then, gently tap the holder until the head protrudes from the tip (*see* **Notes 3** and **4**).

(b) Once the fly is loaded, tighten the holder to the syringe barrel using the connector such that the fly head faces downward.

(c) Next, set the gas pressure to 100 kilo-Pascals and adjust the flow as needed (**Note 5**).

(d) Toggle the switch on to send a burst of gas that moves the impactor to strike the fly once and only once (**Notes 6** and 7).

(e) Detach the fly holder and dump the fly back out onto the CO_2 pad.

(f) Brush the fly into an empty vial until it recovers. Keep one fly per vial—the recovery takes only a few minutes (**Note 8**).

(g) Repeat the process to inflict impacts to the rest flies in the experimental group (**Note 9**).

(h) Process two groups, one experimental and one sham group. Treat shams the same as flies in the experimental group except no aerosol barrier in the impactor tube.

3. Video-Assisted Movement Tracking

(a) First, fill a 6-cm dish with transparent silicon elastomer to make the tracking arena.

(b) Smoothing the gel out in the dish so there is an even 3-mm gap to the lid once the silicon hardens (**Note 10**).

(c) Anesthetize four flies from the same treatment group and place them in the arena. Let the flies acclimate for an hour to the arena, which should be under full light and kept at 22 °C.

(d) Use a charge-coupled device camera positioned above the arena to record the flies' activity for 5 min.

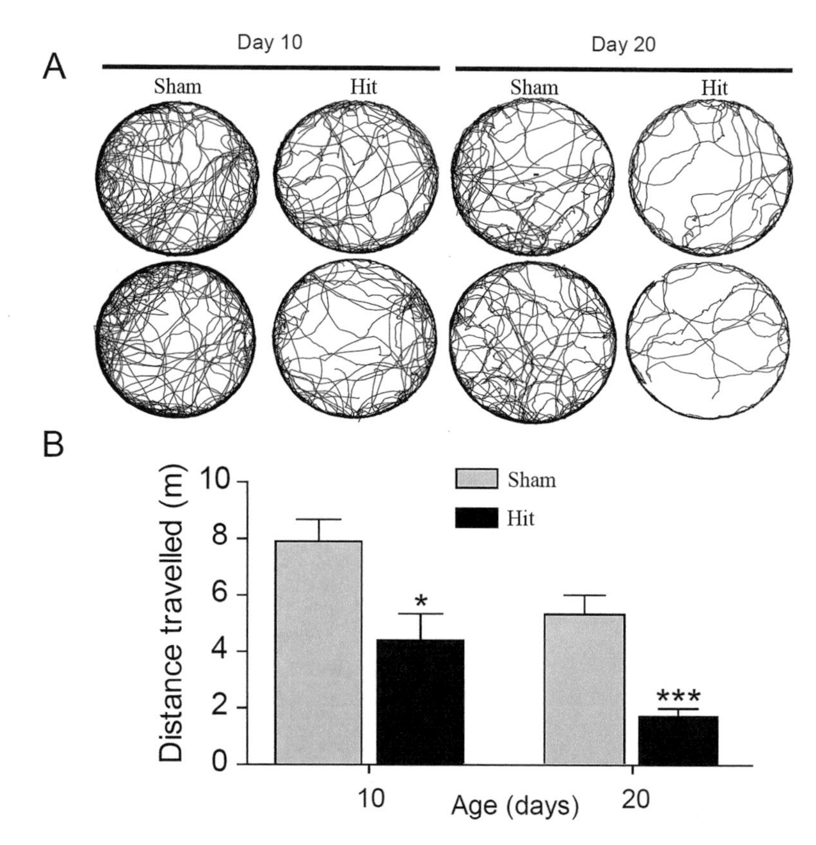

Fig. 2 Video-assisted movement tracking. Repeated strikes over 5 days resulted in significantly impaired locomotor activity. Representative walking tracks recorded at 10 days and 20 days post-treatment over 5 min for four sham and treated flies are shown in **a**. Statistic analysis results are shown in **b**. Error bars, SEM. $^{*}p < 0.05$, $^{***}p < 0.01$, Mann-Whitney *U*-test with a Bonferroni correction. Total flies: treated group $n = 100$, sham group $n = 129$

(e) After the recording, anesthetize the flies in the arena and return them to a new vial.

(f) Later, analyze the recordings using the freely available C-trax software [14] (*see* **Note 11**). This software generates tracking data which can be exported in a programmable language compatible format, such as the Matlab format [15] (Fig. 2a).

(g) From the data, calculate the distance traveled per frame, the mean walking distance for each fly and the average distance travelled per fly (Fig. 2b, *see* **Notes 12** and **13**).

4. Life span analysis.

(a) To evaluate long-term effects, a repetitive traumatic brain injury protocol of five strikes over 5 days (one strike per day) was applied.

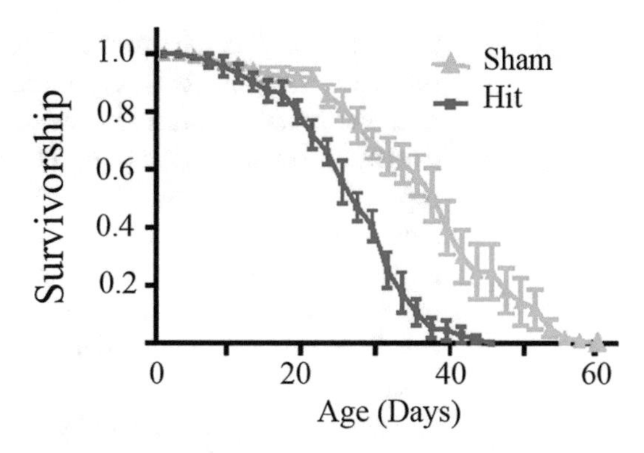

Fig. 3 Life span analysis. Representative life span curves of female adult flies from sham and treated group aged at 25 °C. Treated group has significantly shortened life span compared to that of the sham group

(b) We recommend Nutri-Fly Bloomington Formulation that consists of cornmeal, syrup, yeast in an agar base (http://flystocks.bio.indiana.edu/Fly_Work/media-recipes/bloomfood.htm) to make *Drosophila* food (**Note 14**).

(c) Follow standard environmental conditions for maintenance to keep the flies in an incubator at 25 °C with a 12:12 h light/dark cycle and 60% relative humidity.

(d) During the aging period, transfer flies onto new vials containing fresh food every 2 days (**Note 15**).

(e) During each vial transfer, record the age, count the dead flies in the old vial. Continue to repeat transfers until the last survivor is dead (**Note 16**).

(f) The survivorship curve displays the probability that an individual survives to a given age and is typically calculated using a Kaplan-Meier approach (Fig. 3).

4 Notes

The protocol described here is distinguished from other available methods that inflict traumatic brain injury in flies [11, 12]. Both the Bead Ruptor model and the HIT device inflict injury with random impact location and strength. In both methods, variable numbers of flies contained in either a standard plastic vial or a 2 ml screwcap tube contact the container wall and rebound multiple times during each strike or injury bout. Since the flies are freely movable inside the container, the primary injuries inflicted are considerably different for individual flies, as different parts of their bodies and/or heads are injured with different forces. Thus, it is not surprising that, using those models, the flies suffered from intestinal barrier dysfunction. This was directly correlated with the mortality rate in both regimens [12, 13], strongly indicating that non-neuronal alterations were responsible for trauma-induced death in both systems. In contrast, this protocol is capable of repeatedly delivering direct head impacts to the same location. The body is protected from any direct exposure, eliminating the potential caveats that death might be caused by damage to other parts of the body or to internal organs. In addition, the flow rate of the CO_2 gas used to propel the weight can be regulated to achieve controlled, adjustable strength.

Most importantly, the fly head is not constrained at the time of strike, allowing for a very rapid acceleration–deceleration process that mimics the most common form of mTBI that occurs in the human population [16]. Although it would be ideal to install an automatic on/off switch to control the CO_2 flow regulator, the current manual switch does not appear to affect the reproducibility of the impactor force because the design of our device allows the force to be quickly exerted once the impactor strikes the head.

The following notes address the potential pitfalls that may occur along the specific procedures, and we have provided steps to avoid these dangers. In summary, this protocol provides a reliable regime to better mimic mTBI, introducing a new platform to model CTE. Like other *Drosophila* models that have proven valuable at deciphering human neurodegenerative disorders, it is anticipated that the ongoing detailed characterization of the model in terms of neuronal loss, tau hyperphosphorylation, TDP-43 proteinopathy, and neuroinflammatory response to the inflicted mTBI will generate important mechanistic insights into CTE disease processes and will help to answer some of the critical questions in the field.

1. Note 1. The aerosol barrier has a smooth side and rough side. It should be oriented with the smooth side facing the nozzle in the syringe barrel.
2. Note 2. The opening should not be larger than 0.8 mm in diameter so that a fly head will fit through this hole but the thorax will not.
3. Note 3. If the proboscis is exposed, use a blunted 1-ml syringe needle to gently tuck it back inside the tip.
4. Note 4. It is critical to keep the fly body, especially the mouthpart, inside the holder. Otherwise the fly may die from damage to the internal organs or from not being able to ingest food.
5. Note 5. To establish a chronic traumatic encephalopathy model, the effectiveness of a single closed-head injury made by the impacting apparatus at different speeds was analyzed using 2-day-old Canton-S females [17]. The gas pressure was held at 100 kilo-Pascals. Flies exposed to a single strike at the highest tested flow rate exhibited minimal external defects.

6. Note 6. Despite no clear evidence for external damage, injuries at a flow rate of 15 l/min were acutely lethal, resulting in less than a 10% survival within 24 h [17]. Survival increased with lower flow rates. 100% survival was achieved at 5 l/min or less, so 5 l/min was selected as the standard for generating the model.
7. Note 7. In case when a higher flow rate is needed, an optional toothpick can be applied inside the fly holder to prevent the fly from moving.
8. Note 8. Flies gradually recovered their mobility within 4 min. Recovery for the shams was about a minute quicker.
9. Note 9. Processing one group of four-six flies takes about 20–30 min.
10. Note 10. Leave a 3-mm gap between the silicon and the lid so that the flies can walk freely in the gap, but not take flight.
11. Note 11. C-trax is freely available from http://ctrax.sourceforge.net.
12. Note 12. For statistical significance, expect to test a minimal of 100 flies per treatment group.
13. Note 13. During the 2 days after the impact, locomotion slowly restored to normal. This evidence for gradual recovery is in line with CTE recovery in humans.
14. Note 14. Food should be allowed to solidify and evaporate for 12–24 h prior to storage. Because the nutrient environment can substantially impact longevity, consistency in cooking processes is essential for comparison between experiments.
15. Note 15. This step will ensure that the feeding environment is not disrupted by the presence of larvae. This transfer should be completed without CO_2 anesthesia, which can induce acute mortality, particularly in older flies.
16. Note 16. Be aware that as the flies age, some flies may lie on their back and appear dead due to their inactiveness. Tap on the side of the vials to determine if there are leg movements to make sure these flies are still alive.
17. Note 17. Following this procedure, other methods like standard histology, immunohistochemical and immunofluorescent stains can be performed in order to answer additional questions about mTBI-induced neurodegeneration and tau hyperphosphorylation.

References

1. McKee AC, Cairns NJ, Dickson DW, Folkerth RD, Keene CD, Litvan I, Perl DP, Stein TD, Vonsattel JP, Stewart W, Tripodis Y, Crary JF, Bieniek KF, Dams-O'Connor K, Alvarez VE, Gordon WA, Group TC (2016) The first NINDS/NIBIB consensus meeting to define neuropathological criteria for the diagnosis of chronic traumatic encephalopathy. Acta

Neuropathol 131(1):75–86. https://doi.org/10.1007/s00401-015-1515-z
2. Omalu BI, DeKosky ST, Minster RL, Kamboh MI, Hamilton RL, Wecht CH (2005) Chronic traumatic encephalopathy in a National Football League player. Neurosurgery 57 (1):128–134. discussion 128–134
3. Goldstein LE, Fisher AM, Tagge CA, Zhang XL, Velisek L, Sullivan JA, Upreti C, Kracht JM, Ericsson M, Wojnarowicz MW, Goletiani CJ, Maglakelidze GM, Casey N, Moncaster JA, Minaeva O, Moir RD, Nowinski CJ, Stern RA, Cantu RC, Geiling J, Blusztajn JK, Wolozin BL, Ikezu T, Stein TD, Budson AE, Kowall NW, Chargin D, Sharon A, Saman S, Hall GF, Moss WC, Cleveland RO, Tanzi RE, Stanton PK, McKee AC (2012) Chronic traumatic encephalopathy in blast-exposed military veterans and a blast neurotrauma mouse model. Sci Transl Med 4(134):134ra160. https://doi.org/10.1126/scitranslmed.3003716
4. Mez J, Stern RA, McKee AC (2013) Chronic traumatic encephalopathy: where are we and where are we going? Curr Neurol Neurosci Rep 13(12):407. https://doi.org/10.1007/s11910-013-0407-7
5. McKee AC, Stern RA, Nowinski CJ, Stein TD, Alvarez VE, Daneshvar DH, Lee HS, Wojtowicz SM, Hall G, Baugh CM, Riley DO, Kubilus CA, Cormier KA, Jacobs MA, Martin BR, Abraham CR, Ikezu T, Reichard RR, Wolozin BL, Budson AE, Goldstein LE, Kowall NW, Cantu RC (2013) The spectrum of disease in chronic traumatic encephalopathy. Brain 136 (Pt 1):43–64. https://doi.org/10.1093/brain/aws307
6. Petraglia AL, Plog BA, Dayawansa S, Chen M, Dashnaw ML, Czerniecka K, Walker CT, Viterise T, Hyrien O, Iliff JJ, Deane R, Nedergaard M, Huang JH (2014) The spectrum of neurobehavioral sequelae after repetitive mild traumatic brain injury: a novel mouse model of chronic traumatic encephalopathy. J Neurotrauma 31(13):1211–1224. https://doi.org/10.1089/neu.2013.3255
7. Hirth F (2010) Drosophila melanogaster In the study of human neurodegeneration. CNS Neurol Disord Drug Targets 9(4):504–523
8. Littleton JT, Ganetzky B (2000) Ion channels and synaptic organization: analysis of the Drosophila genome. Neuron 26(1):35–43
9. Appel LF, Prout M, Abu-Shumays R, Hammonds A, Garbe JC, Fristrom D, Fristrom J (1993) The Drosophila stubble-stubbloid gene encodes an apparent transmembrane serine protease required for epithelial morphogenesis. Proc Natl Acad Sci U S A 90 (11):4937–4941
10. Piyankarage SC, Featherstone DE, Shippy SA (2012) Nanoliter hemolymph sampling and analysis of individual adult Drosophila melanogaster. Anal Chem 84(10):4460–4466. https://doi.org/10.1021/ac3002319
11. Katzenberger RJ, Loewen CA, Wassarman DR, Petersen AJ, Ganetzky B, Wassarman DA (2013) A Drosophila model of closed head traumatic brain injury. Proc Natl Acad Sci U S A 110(44):E4152–E4159. https://doi.org/10.1073/pnas.1316895110
12. Barekat A, Gonzalez A, Mauntz RE, Kotzebue RW, Molina B, El-Mecharrafie N, Conner CJ, Garza S, Melkani GC, Joiner WJ, Lipinski MM, Finley KD, Ratliff EP (2016) Using Drosophila as an integrated model to study mild repetitive traumatic brain injury. Sci Rep 6:25252. https://doi.org/10.1038/srep25252
13. Katzenberger RJ, Loewen CA, Bockstruck RT, Woods MA, Ganetzky B, Wassarman DA (2015) A method to inflict closed head traumatic brain injury in Drosophila. J Vis Exp 100:e52905. https://doi.org/10.3791/52905
14. Branson K, Robie AA, Bender J, Perona P, Dickinson MH (2009) High-throughput ethomics in large groups of Drosophila. Nat Methods 6(6):451–457. https://doi.org/10.1038/nmeth.1328
15. Straw AD, Dickinson MH (2009) Motmot, an open-source toolkit for realtime video acquisition and analysis. Source Code Biol Med 4:5. https://doi.org/10.1186/1751-0473-4-5
16. Theadom A, Parmar P, Jones K, Barker-Collo S, Starkey NJ, McPherson KM, Ameratunga S, Feigin VL, Group BR (2015) Frequency and impact of recurrent traumatic brain injury in a population-based sample. J Neurotrauma 32 (10):674–681. https://doi.org/10.1089/neu.2014.3579
17. Sun M, Chen LL (2017) A novel method to model chronic traumatic encephalopathy in Drosophila. J Vis Exp 125. https://doi.org/10.3791/55602

Chapter 5

Controlled Cortical Impact for Modeling Traumatic Brain Injury in Animals

Nicole Osier and C. Edward Dixon

Abstract

Controlled cortical impact (CCI) is one of the most common experimental models of traumatic brain injury (TBI). The original CCI device used a pneumatic piston to induce TBI in ferrets; CCI has since been adapted for use in other species of test animals and electromagnetically driven devices are now commercially available. Whether a pneumatic or electromagnetic device is used, the reproducible injury parameters can be adjusted to achieve a desired injury severity in a given species of test animal. The three main goals of this chapter are to: (1) give an overview of the CCI model, including its historical development, key features, and device options; (2) provide a detailed protocol including materials and methods; and (3) discuss considerations relevant to execution, troubleshooting, and dissemination.

Key words Traumatic brain injury (TBI), Experimental brain trauma, Controlled cortical impact (CCI), Preclinical research, Animal models, Cortical contusion

1 Introduction to the Controlled Cortical Impact Model

1.1 Historical Development

Animal models have been used to study TBI for over a century [1–4]. Substantial additions to and improvements in available injury induction methods have occurred over the past several decades, including: expanding injury induction methods, improving model consistency/reproducibility, and enhancing control over confounding variables [2, 5–14]. During the past 30 years, these efforts have included the development, extension, and refinement of the controlled cortical impact (CCI) model. J.W. Lighthall and his colleagues developed the initial ferret model of pneumatic CCI in the late 1980s and refined it during the 1990s [15, 16]. CCI's utility, control, and reproducibility gave rise to efforts to scale the model for use in other species of laboratory animals. In the early 1990s, C.E. Dixon and colleagues modified the model for use in rats [17]. In the mid-1990s as the availability and use of transgenic mice grew, D.H. Smith and colleagues scaled the model for use in mice [18]. Later, the model was further modified for use in swine

Amit K. Srivastava and Charles S. Cox, Jr. (eds.), *Pre-Clinical and Clinical Methods in Brain Trauma Research*, Neuromethods, vol. 139, https://doi.org/10.1007/978-1-4939-8564-7_5, © Springer Science+Business Media, LLC, part of Springer Nature 2018

[19, 20] and nonhuman primates [21], allowing researchers to test the effects of CCI on animals with larger, more human-like brains. In recent years, there has been a reappearance of CCI studies using ferrets [22, 23]. Researchers may now choose from either pneumatic or electromagnetic CCI devices [24–27]. The purpose of this chapter is to provide an overview of the CCI model, and provide a detailed protocol including materials, methods, and notes relevant to execution, troubleshooting, and dissemination.

1.2 Overview of CCI's Key Features

There are several key features of CCI that contribute to its rigorous nature and widespread use. Overall, CCI allows researchers to control a variety of injury induction parameters including: the tip size/geometry, craniotomy size, velocity at which the tip moves, impact depth, impact angle, and dwell time. Notably, the quantitative control afforded by the CCI model is maintained within a wide range of contact velocities. This control contributes to the model being adapted for use in several species of test animals [15, 17–21], and extended to numerous research applications (e.g. closed- and open-skull TBI; repetitive TBIs) [28–33]. CCI has also been induced in combination with other forms of injury, such as concussive injury [34], hemorrhagic shock [35, 36], polytrauma [37], or blast-induced TBI [38].

1.3 Subtypes of the CCI Model

1.3.1 Pneumatic CCI

The original device used in ferrets had a gas-driven piston; pneumatic devices remain popular today [39–43]. An example of a pneumatic CCI device is depicted in Fig. 1. The piston holds the tip, and has a maximum stroke length of typically 50 mm; together the piston and tip deliver an impact of desired depth and dwell time. A stereotaxic frame facilitates adjustment of the tip to be either perpendicular or angled with respect to the injury site. The impact speed is monitored by a velocity sensor. Many devices have the option of interchangeable commercially available or custom tips allowing researchers to adjust the size, geometry, and composition to best meet the experimental goals. Notably, beyond CCI, another pneumatically driven methods for inducing brain trauma has become available; the Closed-Head Impact Model of Engineered Rotational Acceleration (CHIMERA) method is a surgery-free method of inducing TBI that has been found to result in diffuse axonal injury and microgliosis [44].

1.3.2 Electromagnetic CCI

Electromagnetically driven CCI devices have also become available and gained popularity; an example of an electromagnetic CCI device is depicted below (Fig. 2) [24–27]. As with pneumatic CCI, electromagnetic CCI devices afford similar control over injury location via use of a stereotactic frame and often offer interchangeable tips. In addition to use of standard tips available from the manufacturer of the CCI device, in-house modifications to tips have been reported in the literature; one electromagnetic CCI

Fig. 1 Pneumatic CCI device. A pneumatic CCI device is depicted along with a stereotaxic frame. Not pictured is the compressed gas cylinder used to drive the piston

study covered a standard commercial tip with vulcanized rubber to mimic sports-related traumas caused by lacrosse balls [32]. In comparison to the pneumatic CCI devices, the lighter weight of the electromechanical CCI devices affords increased portability. Notably cylinders of compressed gases are not needed to drive the piston, however they may be required for anesthesia delivery.

Electromagnetic CCI has been found to produce graded injury [24, 25]. More recently a study used high speed videography to evaluate the performance of an electromagnetic CCI device [45]. While some horizontal movement of the tip was detected, overall variation in the impact depth and velocity were found to be modest and the device was found to operate consistently when tested four times over a 14 month period [45]. When five machines of the same model were compared, the velocity of three were found to slightly exceed the set value, highlighting the importance of publishing details surrounding the device's functioning and resulting neuropathology to promote comparisons across devices and laboratories [45]. Current commercial suppliers of both electromagnetic and pneumatic CCI devices are provided below (Table 1).

Fig. 2 Electromagnetic CCI device. An electromagnetic CCI device is depicted along with a stereotaxic frame. No compressed gas cylinder is required for moving the piston, as an electromagnetic actuator is used

Table 1
CCI device suppliers and models by category (electromagnetic vs. pneumatic)

Category	Supplier	Model name/number
Electromagnetic	Hatteras Instruments Leica	Pinpoint PCI3000 Precision Cortical Impactor™ Impact One™
Pneumatic	AmScien Instruments Precision Systems & Instrumentation, LLC	The Pneumatic (Cortical) Impact Device AMS 201™ Head Impactor TBI-0310™

2 Materials

2.1 Test Animal

CCI devices have been used to induce TBI in a variety of species, including: rats [17], mice [18], ferrets [15, 22], swine [19, 20, 46–49], and nonhuman primates [21]. For the purposes of this chapter, we will provide a protocol specific to adult rats, the most frequently used test animal in our laboratory, as is most common in our laboratory.

1. Adult, male Sprague Dawley rats (Charles River Labs, Raleigh, VA, USA).

2.2 Anesthesia Induction Equipment

1. Anesthetic gas of choice (in our laboratory, we use isoflurane).
2. Anesthesia chamber.
3. Compressed gas cylinders (N_2O and O_2).
4. Gas mixer.
5. Gas scavenging system.
6. Cannula for intubation (or nose cone).
7. Laryngoscope to assist with intubation (if needed).

2.3 Thermo-regulation and Physiological Monitoring Supplies

1. Homeothermic heating system.
2. Rectal temperature probe with associated readout equipment/software.
3. Blood oxygenation monitoring system with associated readout equipment/software.

2.4 Surgical Supplies and Equipment

1. Autoclave or other sterilizer.
2. Surgical instruments (e.g. scalpel, scissors, clamps, rongeurs, forceps, periosteal elevator).
3. Sterile surgical drape.
4. Animal hair trimmers (cordless preferred).
5. Betadine.
6. Cotton-tipped applicators.
7. Gauze.
8. Suture kit.

2.5 Medications

1. Topical analgesic.
2. Test agent (if evaluating the effects of a drug on TBI outcomes).

2.6 CCI Induction Equipment

1. Pneumatic (or electromagnetic) CCI device.
2. Tip of desired size, shape, and composition.
3. Stereotaxic frame compatible with CCI device.
4. Cylinder of compressed N_2 (for pneumatic CCI only).
5. Pneumatic drill with drill bits (for open-skull CCI only).
6. Cylinder of compressed air for drill (for open-skull CCI only).

3 Methods

3.1 Animal Care, Housing, and Husbandry

As with all animal research, national and institutional guidelines for use of laboratory animals must be followed, along with all additional requirements associated with certain protected species. All planned experimental activities must be approved by the

Institutional Animal Care and Use Committee (IACUC) and the associated institution(s). When planning for and writing IACUC protocols, researchers should consider the choice of test animal, anesthesia regimen, monitoring for pain/distress, sacrifice method, and criteria for exclusion from the study and/or premature euthanasia.

Housing and husbandry controls affect rigor in all animal research, including CCI studies. Standardization of care across all test animals and comparison groups is especially important. The following is a brief summary of the housing and husbandry used in our rat CCI studies. All rats are allowed to acclimate in the housing room for a period of 7 days upon arrival at the facility; this allows for adjustment to the new environment and recovery from any stress associated with the shipping and handling processes. Throughout the acclimation and study period, all rats are maintained on a constant 12 h light/12 h dark cycle. Moreover, due to the demonstrated neuroprotective effects of environmental enrichment [50–52], rats are kept in group cages with only standard nesting supplies (i.e. no toys or other potentially enriching items). All test animals are given ad libitum (i.e. free) access to food and water and are regularly attended to by trained husbandry and veterinary personnel.

3.2 Anesthesia Induction

In most laboratories, including ours, test animals used in CCI as well as other experimental TBI studies are routinely anesthetized as part of the effort to minimize pain and suffering. Anesthesia is typically induced prior to craniectomy and maintained throughout the surgery. Depending on the laboratory protocol and proficiency of the surgeon, anesthesia is discontinued either immediately pre- or post-closure of the surgical site. Occasionally, anesthesia is discontinued post-craniectomy, but pre-impact [53], or omitted altogether [32]. In our laboratory, when performing CCI on rats, we anesthetize animals using 4.0% isoflurane (in a 2:1 mixture of N_2O: O_2) and then lower the dose to 2.0% throughout the surgery, only increasing/decreasing the dose if the rat shows signs of regaining consciousness (emergence of a paw pinch reflex) or being overly sedated (loss of spontaneous respiration).

3.3 Detailed Protocol for Open-Skull CCI in Rats

1. Examine the CCI device and its maintenance logs to ensure that it is in proper working order and the maintenance schedule has been adhered to (*see* **Note 1**).
2. Place the rat into the chamber and induce anesthesia using isoflurane (4.0%) in a 2:1 mixture of N_2O:O_2.
3. Ensure adequate anesthesia is achieved by use of the toe pinch test or a similar method.
4. Intubate the test animal (*see* **Note 2**).

5. Place the intubated rat into the stereotaxic frame and stabilize the head using an incisor bar and bilateral ear bars.
6. Place all physiologic monitoring equipment needed for data collection and maintaining body temperature, such as a temperature probe, pulse oximeter, etc.
7. Reduce the isoflurane to the maintenance dose (2.0%); if signs of regaining consciousness are noted (e.g. whisker movement), gradually increase the dose.
8. Shave the rat's head over the surgical site using hair trimmers.
9. Apply a sterile drape over the rat to maintain aseptic technique, exposing only the surgical site.
10. Apply antiseptic solution (e.g. betadine) to the scalp using sterile gauze.
11. Use a sterile scalpel to make a 20 mm long midline incision on the rat's scalp.
12. Use sterile microdissecting forceps, periosteal elevator, and a cotton-tipped applicator to expose the skull by carefully reflecting the skin, fascia, and muscle.
13. Use the pneumatic drill to create a circular bone window (i.e. craniectomy), 7 cm in diameter (approximately 1 mm larger than the tip diameter), and centered between the lambda/bregma (anterior-to-posterior) and coronal ridge/ sagittal suture (medial-to-lateral) (*see* **Note 3**).
14. Use microdissecting forceps to gently lift and remove the bone flap without disrupting the dura mater; discard the bone flap.
15. If the craniectomy is not large enough for unobstructed tip clearance, enlarge it with sterile rongeurs.
16. Extend the shaft of the CCI device manually to verify correct positioning of the test animal within the stereotaxic frame such that the impactor tip is centered within the bone window (*see* **Note 4**).
17. Gently zero the impact tip to the cortical surface, ensuring the piston is statically pressured and in the full-stroke position.
18. Carefully withdraw the impactor tip and set the piston assembly to 2.8 mm, or the desired depth of injury specified in your protocol.
19. Use the remote to actuate the device.
20. Use sutures or another method (e.g. staples; glue) to close the surgical site and apply lidocaine or a similar topical drug to minimize pain.
21. Discontinue the anesthesia and extubate the rat.
22. Loosen the ear bars, remove the teeth from the incisor bar, and lift the animal from the stereotaxic frame.

23. Depending on the experimental goals, monitor any post-operative outcomes of interest such as the righting reflex.
24. Supervise the rat while it recovers from anesthesia to the point that spontaneous locomotion returns.
25. Return the rat to its cage and resume standard housing and husbandry.
26. Continue to monitor the rat for pain and administer analgesic per institutional and federal guidelines for the treatment of laboratory animals and in accordance with your IACUC protocol.

4 Notes

1. **Note 1**:Test fire the device to verify the piston fires freely. Compare the velocity sensor to the device's speed setting to verify consistency.
2. **Note 2**: Use a laryngoscope if necessary to facilitate cannula insertion.
3. **Note 3**: Exercise care to standardize the size and shape of the craniectomies across test animals.
4. **Note 4**: When verifying the position of the impactor tip, exercise care to not disrupt the dura.

When a well-designed CCI protocol is followed by trained technicians, high-quality, reproducible data results. The first step in the design of a CCI protocol is a thorough review of the literature, especially if you are new to using the model or are making modifications to the protocol used in your laboratory. Pilot testing can help to identify issues with the specific injury parameters chosen. Likewise, pilot testing can reveal need to modify the protocol or retrain personnel. Pilot testing may also lead to the identification of confounding variables for targeted control in a larger study. Several key considerations for conducting CCI studies will be described below, including choosing an anesthesia regimen (Sect. 4.1.1.), deciding on a craniectomy protocol (Sect. 4.1.2), selecting an impactor tip (Sect. 4.1.3), additional troubleshooting considerations (Sect. 4.2), as well as reporting data in accordance with CDEs (Sect. 4.3).

4.1 Experimental Design

4.1.1 Choosing an Anesthesia Regimen

Almost always, CCI studies anesthetize both the CCI injured and sham control animals, though there is variability with respect to the choice of drug, regimen, and duration. Typically, anesthesia is induced using a higher dose, then maintained at a lower dose once the test animal is in the stereotaxic frame, and discontinued either right before or right after closure of the surgical site (as described briefly in Sect. 4.1.2). Occasionally, anesthesia is used when creating the bone window but then discontinued prior to injury induction itself [53] or omitted altogether and another strategy (e.g. restraint) employed [32]. Either way, alterations in a standard anesthesia regimen are usually made to minimize the potential confounding effects of the anesthetic agents (i.e. neuroprotection, neural suppression) on study outcomes.

Indeed, common anesthetic agents (e.g. isoflurane [54], halothane [55], and ketamine [56]) have demonstrated neuroprotective effects. One study found that compared to fentanyl, isoflurane-treated animals exhibited less hippocampal damage and behavioral deficits. Thus, careful selection of anesthesia is important, even when anesthetized sham animals are used to control for neuroprotective effects. A related consideration is the half-life of the chosen agent, as a long-acting anesthetic may interfere with collection of acute variables of interest [17]. Another strategy is to try to standardize the duration of anesthesia exposure, which can be accomplished through organization of the surgical suite, proficiency in the procedure, and documenting the anesthesia start- and stop-times.

While performing the surgery, under- or over-sedation should be avoided by monitoring the test animal and increasing or decreasing the dose accordingly. Monitoring efforts can include looking for a response to touching the whiskers and/or firm toe pinch. In non-intubated animals, respiratory rate can also be used when monitoring sedation level. In intubated animals, researchers should look for evidence the animal is fighting (i.e. bucking) the ventilator, which may indicate the dose is too low or the tube is displaced and blocking the flow of anesthesia. Often repositioning the cannula results in proper sedation. Less frequently, a kink, leak, or crack in the tubing could be contributing to the problem, highlighting the importance of personal protective equipment and environmental and engineering controls (e.g. a gas scavenging system; electronic monitors of individual personnel's exposure levels).

4.1.2 Deciding on a Craniectomy Protocol

If an open-skull CCI model is used, the choice of how to make the craniectomy is paramount. Most commonly, a drill (either pneumatic or electric) is used for the craniectomy procedure. However, the use of a drill is associated with heat production, which could confound results. To minimize the consequences of heat production, the surgical site can be periodically irrigated with sterile saline via a syringe during the drilling process. Alternatively, a handheld trephine may be used. One study empirically compared the methods for inducing craniectomy and found that a handheld trephine resulted in smaller lesions, less inflammation, and better performance on behavioral tasks compared to animals craniectomized with a drill [57]. However, a manual trephine still resulted in pathophysiological changes and behavioral deficits compared to naïve animals who received anesthesia only [57]. In addition to the method used to make the bone window, the location is also important. Midline craniectomy is typically associated with more bleeding than parasagittal craniectomy [17, 52, 58]. Bilateral craniectomies can be made to facilitate lateral movement of brain tissue and/or produce multiple contusions in distinct locations [59, 60].

After careful selection of whether or not to perform a craniectomy, how many to make, and where, proficiency in the procedure is needed to minimize confounding effects associated with inconsistency across test animals. Records should be kept documenting whether the craniectomy was clean or if there was evidence of dura breach, tissue herniation, or bleeding present. Any mortalities in the sham group should also be noted as they may provide insights into a need for personnel retraining on the craniectomy protocol or the presence of an underlying issue in the colony (e.g. sickness). The details surrounding the craniectomy procedure and any issue that arises should be reported consistent with established CDEs (described in more detail in Sect. 4.3).

4.1.3 Selecting an Impactor Tip

Researchers can choose from commercially available CCI tips with a variety of diameters, geometries and compositions, depending on the test animal and research question. Today, most CCI tips are either round (the choice in the early models [15, 16] as well as many contemporary studies [37, 61, 62]) or the flat-beveled alternative [18, 63–70]. One study empirically evaluated the effect of tip geometry on outcomes and found, compared to round alternatives, flat beveled tips resulted in more extensive cortical hemorrhaging and neuronal loss [71]. Another notable finding was that flat tips produced more rapid neurodegenerative changes than round tips; for researchers interested in studying acute neurodegeneration in the minutes-to-hours after TBI, flat tips may be the better choice [71].

4.2 Other Troubleshooting Considerations

While careful study planning and pilot work can help to minimize issues arising, sometimes unexpected problems occur that require troubleshooting, such as adjusting the cannula placement in cases of under-sedation. Unexpected variability that was intended to be controlled as part of the experimental design may occur and require attention. For example, if exogenous heating sources (e.g. heating pads, heating lamps) are insufficient to maintain the temperatures of test animals during surgery, the variability may confound study outcomes, especially if hypothermic temperatures are reached [21, 72]. In such an instance, new heating devices can be pursued and/or a new temperature monitoring system can be employed. If there is no injury effect between CCI and sham-injured animals, possible troubleshooting includes: improving the cleanness and consistency of sham animals, increasing the injury depth, and/or using high speed videography to evaluate the performance of the CCI device and repair the machine as needed.

4.3 Reporting CCI Data Including Common Data Elements (CDEs)

To maximize the quality of the experimental TBI knowledge base, and promote the ability for published findings to be replicated by independent research teams, CCI researchers should publish following the guidelines for reporting CDEs as proposed by the National Institute of Neurological Diseases and Stroke (NINDS). Indeed, the NINDS guidelines should be considered both during study planning and manuscript development [73, 74]. In addition to the generic CDEs for all preclinical TBI studies, there are a set of CDEs specific to CCI. In all experimental TBI studies, details should be provided concerning the test animal used (e.g. type, strain, supplier weight, sex(es)), animal care (e.g. pre-injury housing), and outcome assessment methods and timing (e.g. righting reflex time; body weight change; memory retention tests, histopathology). In CCI studies, specifically, researchers should report details regarding whether a craniectomy was performed (and if so, the procedure used), the tip (shape; rigidity; angle), and the injury parameters (e.g. velocity; depth setting; dwell time).

5 Conclusion

CCI continues to be a widely used model of TBI, nearly 30 years after the initial ferret paper was published [15]. Since the original development of a pneumatic model, electromagnetic devices have become available. CCI has been extended to several research applications including studying CHI, repeated injuries, and/or the combined effects of two neurologic insults via use of combination models. This chapter introduced the historical development and use of the CCI model and provided a detailed protocol for a rat CCI study, including all materials and methods. In addition, discussion was included regarding how to make decisions related to experimental design, execution, troubleshooting, and dissemination.

Acknowledgements

The authors would like to acknowledge the following federal funding sources which have supported this chapter: NIH-NINDS grant R01-NS079061; and Department of Veterans Affairs grant VAI01RX001127. We would also like to thank Marilyn K. Farmer for her continued editorial support and Michael D. Farmer for his assistance with the figures.

References

1. Kramer SP (1896) A contribution to the theory of cerebral concussion. Anim Surg. 23:163–173
2. Rinder L, Olsson Y (1968) Studies on vascular permeability changes in experimental brain concussion. I. Distribution of circulating fluorescent indicators in brain and cervical cord after sudden mechanical loading of the brain. Acta Neuropathol. 11(3):183–200
3. Denny-Brown D, Russell W (1941) Experimental concussion. Brain.:93–184
4. Lindgren S, Rinder L (1965) Experimental studies in head injury. I. Some factors influencing results of model experiments. Biophysik. 2 (5):320–329
5. Gennarelli TA, Thibault LE, Adams JH, Graham DI, Thompson CJ, Marcincin RP (1982) Diffuse axonal injury and traumatic coma in the primate. Ann Neurol. 12(6):564–574
6. Govons SR, Govons RB, VanHuss WD, Heusner WW (1972) Brain concussion in the rat. Exp Neurol. 34(1):121–128
7. Nilsson B, Pontén U, Voigt G (1977) Experimental head injury in the rat. Part 1: Mechanics, pathophysiology, and morphology in an impact acceleration trauma model. J Neurosurg. 47(2):241–251
8. Ommaya AK, Geller A, Parsons LC (1971) The effect of experimental head injury on one-trial learning in rats. Int J Neurosci. 1(6):371–378
9. Ommaya AK, Gennarelli TA (1974) Cerebral concussion and traumatic unconsciousness. Correlation of experimental and clinical observations of blunt head injuries. Brain. 97 (4):633–654
10. Sullivan HG, Martinez J, Becker DP, Miller JD, Griffith R, Wist AO (1976) Fluid-percussion model of mechanical brain injury in the cat. J Neurosurg. 45(5):521–534
11. Parkinson D, West M, Pathiraja T (1978) Concussion: Comparison of humans and rats. Neurosurgery. 3(2):176–180
12. Onyszchuk G, Al-Hafez B, He Y-Y, Bilgen M, Berman NEJ, Brooks WM (2007) A mouse model of sensorimotor controlled cortical impact: characterization using longitudinal magnetic resonance imaging, behavioral assessments and histology. J Neurosci Methods. 160 (2):187–196
13. Koliatsos VE, Cernak I, Xu L, Song Y, Savonenko A, Crain BJ et al (2011) A mouse model of blast injury to brain: initial pathological, neuropathological, and behavioral characterization. J Neuropathol Exp Neurol. 70 (5):399–416
14. Fritz HG, Walter B, Holzmayr M, Brodhun M, Patt S, Bauer R (2005) A pig model with secondary increase of intracranial pressure after severe traumatic brain injury and temporary blood loss. J Neurotrauma. 22(7):807–821
15. Lighthall JW (1988) Controlled cortical impact: a new experimental brain injury model. J Neurotrauma. 5(1):1–15
16. Lighthall JW, Goshgarian HG, Pinderski CR (1990) Characterization of axonal injury produced by controlled cortical impact. J Neurotrauma. 7(2):65–76
17. Dixon C, Clifton G, Lighthall J, Yaghmai A, Hayes R, Dixon CE, Clifton GL, Lighthall JW, Yaghmai AAHR et al (1991) A controlled cortical impact model of traumatic brain injury in the rat. J Neurosci Methods. 39(3):253–262
18. Smith D, Soares H, Pierce J, Perlman K, Saatman K, Meaney D et al (1995) A model of parasagittal controlled cortical impact in the mouse: cognitive and histopathologic effects. J Neurotrauma. 12(2):169–178
19. Manley GT, Rosenthal G, Lam M, Morabito D, Yan D, Derugin N et al (2006) Controlled cortical impact in swine: pathophysiology and biomechanics. J Neurotrauma. 23(2):128–139
20. Kilbaugh TJ, Bhandare S, Lorom DH, Saraswati M, Robertson CL, Margulies SS (2011) Cyclosporin A preserves mitochondrial function after traumatic brain injury in the immature rat and piglet. J Neurotrauma. 28 (5):763–774
21. King C, Robinson T, Dixon CE, Rao GR, Larnard D, Nemoto CEM (2010) Brain temperature profiles during epidural cooling with the ChillerPad in a monkey model of traumatic brain injury. J Neurotrauma. 27 (10):1895–1903
22. Schwerin SC, Hutchinson EB, Radomski KL, Ngalula KP, Pierpaoli CM, Juliano SL (2017) Establishing the ferret as a gyrencephalic animal model of traumatic brain injury: optimization of controlled cortical impact procedures. J Neurosci Methods. 285:82–96
23. Hutchinson EB, Schwerin SC, Radomski KL, Irfanoglu MO, Juliano SL, Pierpaoli CM (2016) Quantitative MRI and DTI abnormalities during the acute period following CCI in the Ferret. Shock. 46(3 Suppl 1):167–176
24. Brody DL, Mac Donald C, Kessens CC, Yuede C, Parsadanian M, Spinner M et al (2007) Electromagnetic controlled cortical impact device for precise, graded experimental

traumatic brain injury. J Neurotrauma. 24 (4):657–673

25. Washington PM, Forcelli PA, Wilkins T, Zapple DN, Parsadanian M, Burns MP (2012) The effect of injury severity on behavior: a phenotypic study of cognitive and emotional deficits after mild, moderate, and severe controlled cortical impact injury in mice. J Neurotrauma. 29(13):2283–2296
26. Xiong L-L, Hu Y, Zhang P, Zhang Z, Li L-H, Gao G-D et al (2017) Neural stem cell transplantation promotes functional recovery from traumatic brain injury via brain derived neurotrophic factor-mediated neuroplasticity. Mol Neurobiol
27. Hill JL, Kobori N, Zhao J, Rozas NS, Hylin MJ, Moore AN et al (2016) Traumatic brain injury decreases AMP-activated protein kinase activity and pharmacological enhancement of its activity improves cognitive outcome. J Neurochem. 139(1):106–119
28. Donovan V, Kim C, Anugerah AK, Coats JS, Oyoyo U, Pardo AC et al (2014) Repeated mild traumatic brain injury results in long-term white-matter disruption. J Cereb Blood Flow Metab. 34(4):715–723
29. Woertgen C, Rothoerl RD, Brawanski A (2001) Neuron-specific enolase serum levels after controlled cortical impact injury in the rat. J Neurotrauma [Internet] 18 (5):569–573. [cited 2012 Nov 12]; Available from: http://www.ncbi.nlm.nih.gov/pubmed/11393260
30. Klemenhagen KC, O'Brien SP, Brody DL (2013) Repetitive concussive traumatic brain injury interacts with post-injury foot shock stress to worsen social and depression-like behavior in mice. PLoS One. 8(9):e74510
31. Shitaka Y, Tran HT, Bennett RE, Sanchez L, Levy MA, Dikranian K et al (2011) Repetitive closed-skull traumatic brain injury in mice causes persistent multifocal axonal injury and microglial reactivity. J Neuropathol Exp Neurol [Internet] 70(7):551–567. [cited 2013 Mar 31]; Available from: http://www.pubmedcentral.nih.gov/articlerender.fcgi?artid=3118973&tool=pmcentrez&rendertype=abstract
32. Petraglia AL, Plog BA, Dayawansa S, Chen M, Dashnaw ML, Czerniecka K et al (2014) The spectrum of neurobehavioral sequelae after repetitive mild traumatic brain injury: a novel mouse model of chronic traumatic encephalopathy. J Neurotrauma. 31(13):1211–1224
33. Jamnia N, Urban JH, Stutzmann GE, Chiren SG, Reisenbigler E, Marr R et al (2017) A clinically relevant closed-head model of single and repeat concussive injury in the adult rat using a controlled cortical impact device. J Neurotrauma. 34(7):1351–1363
34. Dapul HR, Park J, Zhang J, Lee C, DanEshmand A, Lok J et al (2013) Concussive injury before or after controlled cortical impact exacerbates histopathology and functional outcome in a mixed traumatic brain injury model in mice. J Neurotrauma [Internet] 30 (5):382–391. [cited 2013 May 25]; Available from: http://www.ncbi.nlm.nih.gov/pubmed/23153355
35. Shein SL, Shellington DK, Exo JL, Jackson TC, Wisniewski SR, Jackson EK et al (2014) Hemorrhagic shock shifts the serum cytokine profile from pro- to anti-inflammatory after experimental traumatic brain injury in mice. J Neurotrauma. 31(16):1386–1395
36. Jin G, DeMoya MA, Duggan M, Knightly T, Mejaddam AY, Hwabejire J et al (2012) Traumatic brain injury and hemorrhagic shock: evaluation of different resuscitation strategies in a large animal model of combined insults. Shock. 38(1):49–56
37. Mirzayan MJ, Probst C, Samii M, Krettek C, Gharabaghi A, Pape HC et al (2012) Histopathological features of the brain, liver, kidney and spleen following an innovative polytrauma model of the mouse. Exp Toxicol Pathol. 64 (3):133–139
38. Ko J, Hemphill MA, Gabrieli D, Wu L, Yelleswarapu V, Lawrence G et al (2016) Smartphone-enabled optofluidic exosome diagnostic for concussion recovery. Sci Rep 6:31215
39. Febinger HY, Thomasy HE, Pavlova MN, Ringgold KM, Barf PR, George AM et al (2015) Time-dependent effects of CX3CR1 in a mouse model of mild traumatic brain injury. J Neuroinflammation. 12:154
40. Osier ND, Pham L, Pugh BJ, Puccio A, Ren D, Conley YP et al (2017) Brain injury results in lower levels of melatonin receptors subtypes MT1 and MT2. Neurosci Lett. 650:18–24
41. Songarj P, Luh C, Staib-Lasarzik I, Engelhard K, Moosmann B, Thal SC (2015) The antioxidative, non-psychoactive tricyclic phenothiazine reduces brain damage after experimental traumatic brain injury in mice. Neurosci Lett. 584:253–258
42. Dong T, Zhang Q, Hamblin MR, Wu MX (2015) Low-level light in combination with metabolic modulators for effective therapy of injured brain. J Cereb Blood Flow Metab. 35 (9):1435–1444
43. Song S, Kong X, Acosta S, Sava V, Borlongan C, Sanchez-Ramos J (2016) Granulocyte colony-stimulating factor promotes

behavioral recovery in a mouse model of traumatic brain injury. J Neurosci Res. 94 (5):409–423
44. Namjoshi DR, Cheng WH, McInnes KA, Martens KM, Carr M, Wilkinson A et al (2014) Merging pathology with biomechanics using CHIMERA (Closed-Head Impact Model of Engineered Rotational Acceleration): a novel, surgery-free model of traumatic brain injury. Mol Neurodegener Engl 9:55
45. Kim Y, Fu AH, Tucker LB, Liu J, McCabe JT (2018) Characterization of controlled cortical impact devices by high-speed image analysis. J Neurosci Res 96(4):501–511
46. Duhaime AC, Margulies SS, Durham SR, O'Rourke MM, Golden JA, Marwaha S et al (2000) Maturation-dependent response of the piglet brain to scaled cortical impact. J Neurosurg. 93(3):455–462
47. Sindelar B, Bailes J, Sherman S, Finan J, Stone J, Lee J et al (2017) Effect of internal jugular vein compression on intracranial hemorrhage in a porcine controlled cortical impact model. J Neurotrauma. 34(8):1703–1709
48. Pareja JCM, Keeley K, Duhaime A-C, Dodge CP (2016) Modeling pediatric brain trauma: piglet model of controlled cortical impact. Methods Mol Biol. 1462:345–356
49. Hawryluk GWJ, Phan N, Ferguson AR, Morabito D, Derugin N, Stewart CL et al (2016) Brain tissue oxygen tensiòn and its response to physiological manipulations: influence of distance from injury site in a swine model of traumatic brain injury. J Neurosurg. 125(5):1217–1228
50. Bondi CO, Klitsch KC, Leary JB, Kline AE (2014) Environmental enrichment as a viable neurorehabilitation strategy for experimental traumatic brain injury. J Neurotrauma. 31 (10):873–888
51. Cheng JP, Shaw KE, Monaco CM, Hoffman AN, Sozda CN, Olsen AS et al (2012) A relatively brief exposure to environmental enrichment after experimental traumatic brain injury confers long-term cognitive benefits. J Neurotrauma. 29(17):2684–2688
52. Shin SS, Bales JW, Yan HQ, Kline AE, Wagner AK, Lyons-Weiler J et al (2013) The effect of environmental enrichment on substantia nigra gene expression after traumatic brain injury in rats. J Neurotrauma. 30(4):259–270
53. Adelson PD, Fellows-Mayle W, Kochanek PM, Dixon CE (2013) Morris water maze function and histologic characterization of two age-at-injury experimental models of controlled cortical impact in the immature rat. Childs Nerv Syst. 29(1):43–53
54. Statler KD, Alexander H, Vagni V, Holubkov R, Dixon CE, Clark R et al (2006) Isoflurane exerts neuroprotective actions at or near the time of severe traumatic brain injury. Brain Res. 1076(1):216–224
55. McPherson RW, Kirsch JR, Salzman SK, Traystman RJ (1994) The neurobiology of central nervous system trauma. Oxford University Press, New York, pp 12–27
56. McDonald JW, Roeser NF, Silverstein FS, Johnston MV (1989) Quantitative assessment of neuroprotection against NMDA-induced brain injury. Exp Neurol. 106(3):289–296
57. Cole JT, Yarnell A, Kean WS, Gold E, Lewis B, Ren M et al (2011) Craniotomy: true sham for traumatic brain injury, or a sham of a sham? J Neurotrauma. 28(3):359–369
58. Shin SS, Bray ER, Dixon CE (2012) Effects of nicotine administration on striatal dopamine signaling after traumatic brain injury in rats. J Neurotrauma. 29(5):843–850
59. Meaney DF, Ross DT, Winkelstein BA, Brasko J, Goldstein D, Bilston LB et al (1994) Modification of the cortical impact model to produce axonal injury in the rat cerebral cortex. J Neurotrauma. 11(5):599–612
60. He J, Evans C-O, Hoffman SW, Oyesiku NM, Stein DG (2004) Progesterone and allopregnanolone reduce inflammatory cytokines after traumatic brain injury. Exp Neurol. 189 (2):404–412
61. Mirzayan MJ, Klinge PM, Ude S, Hotop A, Samii M, Brinker T et al (2008) Modified calcium accumulation after controlled cortical impact under cyclosporin A treatment: a 45Ca autoradiographic study. Neurol Res. 30 (5):476–479
62. Eslami M, Ghanbari E, Sayyah M, Etemadi F, Choopani S, Soleimani M et al (2015) Traumatic brain injury accelerates kindling epileptogenesis in rats. Neurol Res 38(3):269–274
63. Dennis AM, Haselkorn ML, Vagni VA, Garman RH, Janesko-Feldman K, Bayir H et al (2009) Hemorrhagic shock after experimental traumatic brain injury in mice: effect on neuronal death. J Neurotrauma. 26(6):889–899
64. Fox GB, Fan L, LeVasseur RA, Faden AI (1998) Sustained sensory/motor and cognitive deficits with neuronal apoptosis following controlled cortical impact brain injury in the mouse. J Neurotrauma. 15(8):599–614
65. Sandhir R, Berman NEJ (2010) Age-dependent response of CCAAT/enhancer binding proteins following traumatic brain injury in mice. Neurochem Int. 56 (1):188–193

66. Hemerka JN, Wu X, Dixon CE, Garman RH, Exo JL, Shellington DK et al (2012) Severe brief pressure-controlled hemorrhagic shock after traumatic brain injury exacerbates functional deficits and long-term neuropathological damage in mice. J Neurotrauma. 29 (12):2192–2208
67. Monaco CM, Mattiola VV, Folweiler KA, Tay JK, Yelleswarapu NK, Curatolo LM et al (2013) Environmental enrichment promotes robust functional and histological benefits in female rats after controlled cortical impact injury. Exp Neurol. 247:410–418
68. Griesbach GS, Hovda DA, Gomez-Pinilla F, Sutton RL (2008) Voluntary exercise or amphetamine treatment, but not the combination, increases hippocampal brain-derived neurotrophic factor and synapsin I following cortical contusion injury in rats. Neuroscience. 154(2):530–540
69. Thompson SN, Gibson TR, Thompson BM, Deng Y, Hall ED (2006) Relationship of calpain-mediated proteolysis to the expression of axonal and synaptic plasticity markers following traumatic brain injury in mice. Exp Neurol. 201(1):253–265
70. Whalen MJ, Clark RSB, Dixon CE, Robichaud P, Marion DW, Vagni V et al (1999) Reduction of cognitive and motor deficits after traumatic brain injury in mice deficient in poly(ADP-ribose) polymerase. J Cereb Blood Flow Metab. 19(8):835–842
71. Pleasant JM, Carlson SW, Mao H, Scheff SW, Yang KH, Saatman KE (2011) Rate of neurodegeneration in the mouse controlled cortical impact model is influenced by impactor tip shape: implications for mechanistic and therapeutic studies. J Neurotrauma. 28 (11):2245–2262
72. Lee JH, Wei L, Gu X, Wei Z, Dix TA, Yu SP (2014) Therapeutic effects of pharmacologically induced hypothermia against traumatic brain injury in mice. J Neurotrauma. 31 (16):1417–1430
73. RIGOR. Improving the quality of NINDS-supported pre-clinical and clinical research through rigorous study design and transparent reporting [Internet]. Available from: www.ninds.nih.gov/funding/transparency_in_reporting_guidance.pdf
74. Smith DH, Hicks RR, Johnson VE, Bergstrom DA, Cummings DM, Noble LJ et al (2015) Pre-clinical traumatic brain injury common data elements: toward a common language across laboratories. J Neurotrauma 32 (22):1725–1735

Chapter 6

Fluid Percussion Model of Traumatic Brain Injury

Rachel K. Rowe, Daniel Griffiths, and Jonathan Lifshitz

Abstract

Research models of traumatic brain injury (TBI) hold significant validity towards the human condition, with each model replicating a subset of clinical features and symptoms. After 30 years of characterization and implementation, fluid percussion injury (FPI) is firmly recognized as a clinically relevant model of TBI and the hallmarks of TBI in man can be faithfully reproduced. Variations in the surgical procedure provide the ability to induce focal, diffuse, or mixed focal and diffuse brain injury in various laboratory species. Being fully scalable, fluid percussion can induce mild, moderate, or severe brain injury in subjects of either sex, at any age. This chapter outlines the procedures for FPI in adult male rats and mice. With these procedures, it becomes possible to generate brain-injured laboratory animals for studies of injury-induced pathophysiology and behavioral deficits, for which rational therapeutic interventions can be evaluated.

Key words Fluid percussion, Rat, Mouse, Brain injury, Concussion, Trauma, Diffuse, Focal, Righting reflex, Fencing response

1 Introduction

Midline fluid percussion permits the study of experimental traumatic brain injury (TBI) in a model that is reproducible, clinically relevant, and scalable between species and injury severities. Brain injury is induced by a rapid (~20 ms) fluid pulse through a craniectomy onto the intact dura that follows the inner curvature of the skull and creates an elastic decompression of the brain [1, 2]. While fluid percussion injury (FPI) necessitates breaching the cranial vault, the skull is sealed to the injury device, recreating a closed system, which approximates a closed head injury with decompressive craniectomy. The mechanical forces disrupt cell membranes, blood vessels, and neuronal processes. By increasing the angle from which the pendulum hammer falls, greater pressures can be generated to travel through the fluid-filled cylinder and

Electronic supplementary material: The online version of this chapter (https://doi.org/10.1007/978-1-4939-8564-7_6) contains supplementary material, which is available to authorized users.

Amit K. Srivastava and Charles S. Cox, Jr. (eds.), *Pre-Clinical and Clinical Methods in Brain Trauma Research*, Neuromethods, vol. 139, https://doi.org/10.1007/978-1-4939-8564-7_6, © Springer Science+Business Media, LLC, part of Springer Nature 2018

impact the brain. At a moderate level of injury, 20–25% of animals die as a result of the injury within the acute posttraumatic period (15 min), generally from respiratory failure and pulmonary edema. This is a normal and desired feature of TBI models, as it reflects human TBI.

In laboratories worldwide, subtle variations in surgical and injury procedures reproduce the spectrum of brain injuries found in the human population. Primarily, the location of injury site determines the major features of the injury, where a midline impact location induces a diffuse injury and a lateral impact location induces a focal injury with a diffuse component [3–5]. Fluid percussion injury reproduces the acute reflex suppression, functional deficits, and histopathology evident after TBI in man [6–8]. The model continues to be implemented to evaluate pathophysiological mechanisms underlying histological and behavioral deficits, and therapeutic interventions to mitigate degeneration and promote the recovery of function [7, 8].

2 Materials

2.1 Animals

Fluid percussion brain injury has been successfully performed on various species, including cats, rabbits, pigs, rats, and mice. The adaptation of fluid percussion to rats [6, 9, 10] was followed by its implementation in mice [11]. The procedures outlined in this chapter focus on midline fluid percussion in 8-week-old adult male C57BL/6 mice (approximately 20–30 g) and adult male Sprague-Dawley rats (approximately 300–400 g). To maximize the success of brain injury, examine all animals for any signs of ill health (e.g., rough coat, bleeding or dirty eyes, runny or bleeding nose, and scratched around eyes or nose area). Weigh all animals prior to surgery in order to track injury-induced weight loss.

2.2 Equipment

2.2.1 Injury Device

- Fluid percussion injury device (Fig. 1).
- Custom Design and Fabrication.
- Virginia Commonwealth University.
- http://www.radiology.vcu.edu/research/customdesign.html.
- Product information including assembly manual, operation manual, and product brochure are provided on the manufacturer's website.
- Recording oscilloscope (recommended: Tektronix, Model 1001B).
- Industrial Velcro to secure the device to the bench to prevent movement.
- High-vacuum grease (e.g., Fisher Scientific, #14-635-5D).

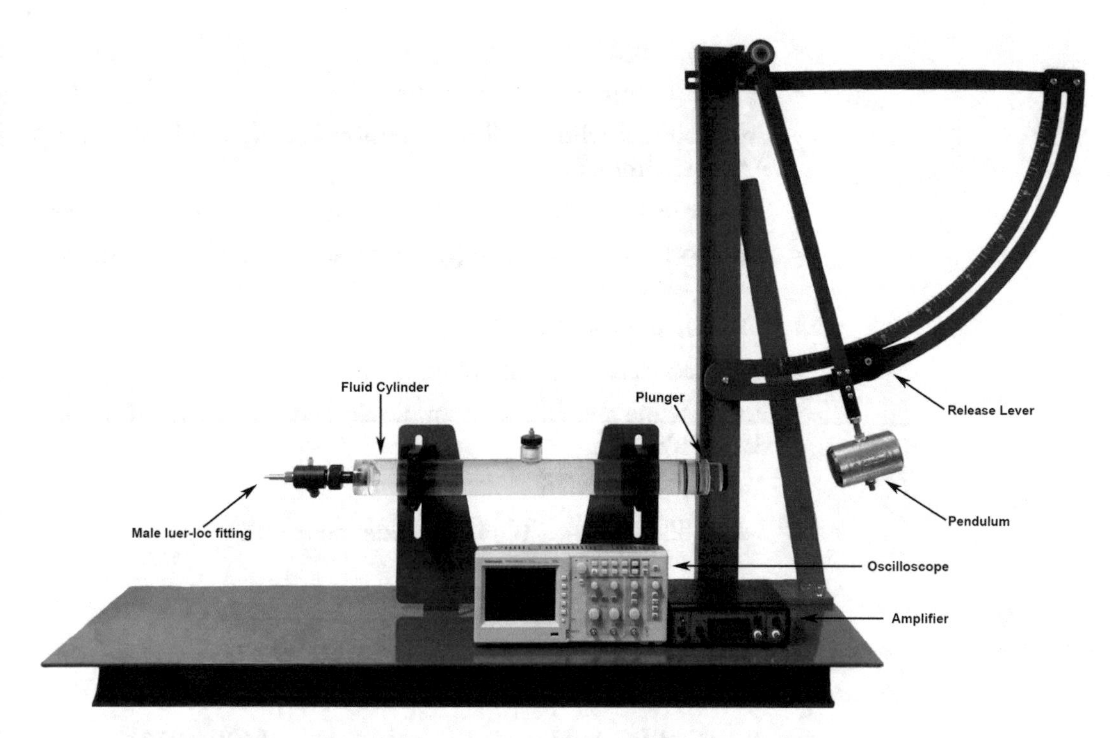

Fig. 1 Fluid percussion injury device. Injury is induced by a 20 ms fluid pulse delivered onto the intact dura via a craniectomy and surgically implanted injury hub. The fluid pulse is generated by the pressure wave produced when the weighted end of the pendulum arm strikes the plunger of a fluid-filled cylinder. The force of the pulse is detected by a transducer and the signal is amplified before being sent to the oscilloscope which outputs the millivolts. The millivolts can then be converted to atmospheres of pressure

- Dishwashing solution to clean fluid cylinder.
- Jet Dry finishing rinse to minimize air bubbles in the cylinder upon filling.

2.2.2 Anesthesia

- Vaporizer for delivery of inhaled anesthesia.
- Tubing/petcocks.
- Induction chamber.
- Isoflurane.
- Oxygen and regulator.
- Rodent nose cone for inhaled anesthetic that is compatible with the stereotaxic frame.

2.2.3 Surgical Supplies

- Gauze sponges.
- Cotton tip applicators.
- Heating pad (recommended: Deltaphase isothermal heating pad, BrainTree Scientific, #39DP).
- 20 gauge needles (recommended: 1″ length).

- 1-mL syringes.
- $\geq$ 10-mL syringes, Luer-lock tip.
- Small animal trimmer for fur removal (e.g., Wahl, Mini Arco Animal Trimmer).
- Ophthalmic ointment to prevent drying of eyes during surgery.
- 4% chlorhexidine solution (or Betadine scrub) for preparation of the incision.
- 70% ethanol (or alcohol pads).
- Cyanoacrylate (e.g. Super Glue).
- Perm Reline and Repair resin, liquid and powder (All for Dentist, #H00327).
- Antibiotic Ointment.
- Saline-filled syringe, blunted needle bent 90°.

2.2.4 Surgical Instruments

- Small animal stereotaxic frame.
- Scalpel handle and blade.
- Delicate bone scraper (Fine Science Tools, #10075-16).
- Wedelstaedt Chisels ¾ DE (Henry Schein, #600-4972).
- Bull Dog clips (Fine Science Tools, #18050-28, #18051-28).
- Needle holder and scissors.

Mouse Surgical Instruments

- Trephine (3.0 mm) (Machine Shop, Arizona State University, Tempe, AZ) contact Rachel Rowe, rkro222@email.arizona.edu.
- Weed whacker line for cranial disc (1.7 mm diameter).
- Side-grasping forceps (7 × 7) (Henry Schein, #6-124XL).
- 3M Vetbond tissue adhesive (Henry Schein, #700-3449).

Rat Surgical Instruments

- Dremel tool with engraving cutter #106.
- Trephine: 4.7 mm-adult (Miltex, #26-140).
- Fingernail drill with 5/64″ drill bit (Miltex, #33-232).
- Stainless steel skull screws (2–56 × 3/16″) (Small parts Inc., #MX-0256-03B-25).

2.2.5 Injury Hub (Fig. 2)

- 1 ½″ needle (20 gauge) (Becton Dickinson, #305176).
- Syringe (1 cc).
- Razor blades.
- Tissue forceps (Henry Schein, #6-114).

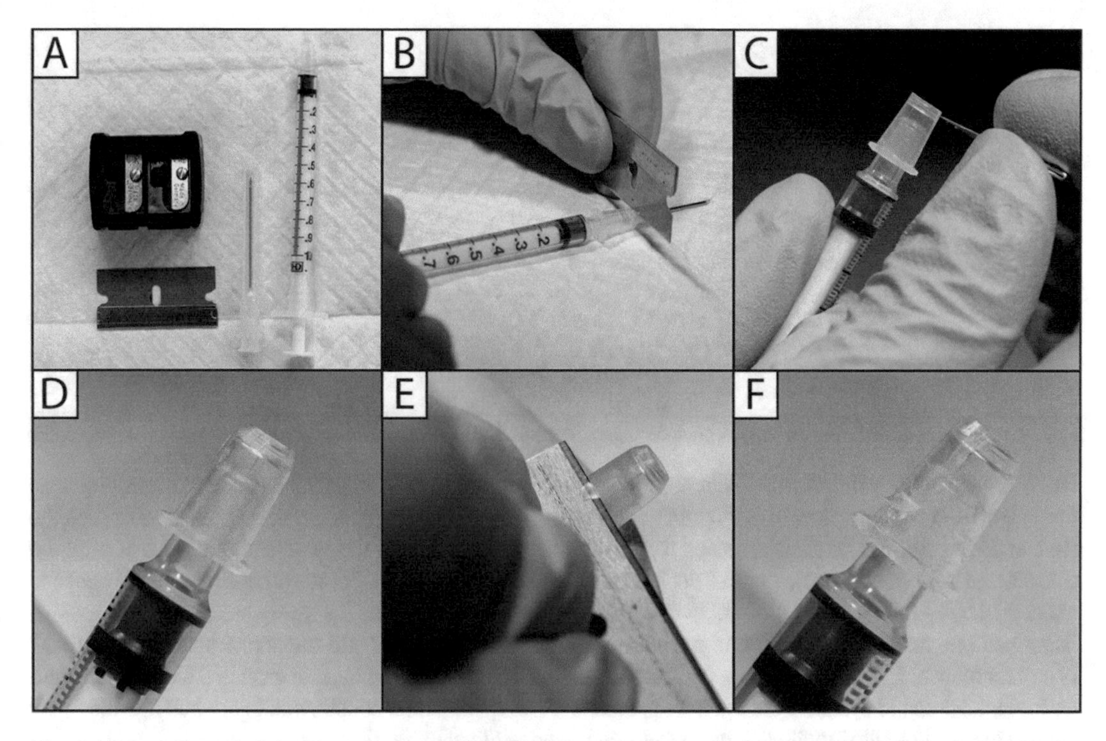

Fig. 2 Injury hub materials (**a**) and construction. Attach a 20 gauge needle to a 1 cc syringe and insert the needle into a laboratory bench pad to prevent the needle from becoming projected after it is cut (**b**). Use a razor blade to cut off the tip of the needle (**b**) and check and refine to make level (**c**). For the rat, the injury hub is beveled using a cosmetic pencil sharpener (**d**). Using a razor blade, score the exterior of the hub making burrs at even intervals around the hub (**e**). When finished, the cut end should be flat and even, and parallel to the Luer-Loc plane (**f**)

Mouse Injury Hub

- Luer-loc extension tubing (Baxter, #2C5643).

Rat Injury Hub

- Cosmetic pencil sharpener.

3 Methods

3.1 Record Keeping

A standard surgery sheet should be used to record information pertaining to the surgical procedure, injury, and both immediate and long-term postoperative care (see Supplementary Material). Postoperative observation and treatment of each animal should be recorded and include notes about the general condition of the animal and any supportive care the animal received (e.g. saline injections).

3.2 Preoperative Preparation

Appropriate personal protective equipment should be worn: clean lab coat or scrubs, gloves, face mask, hair covering, and protective

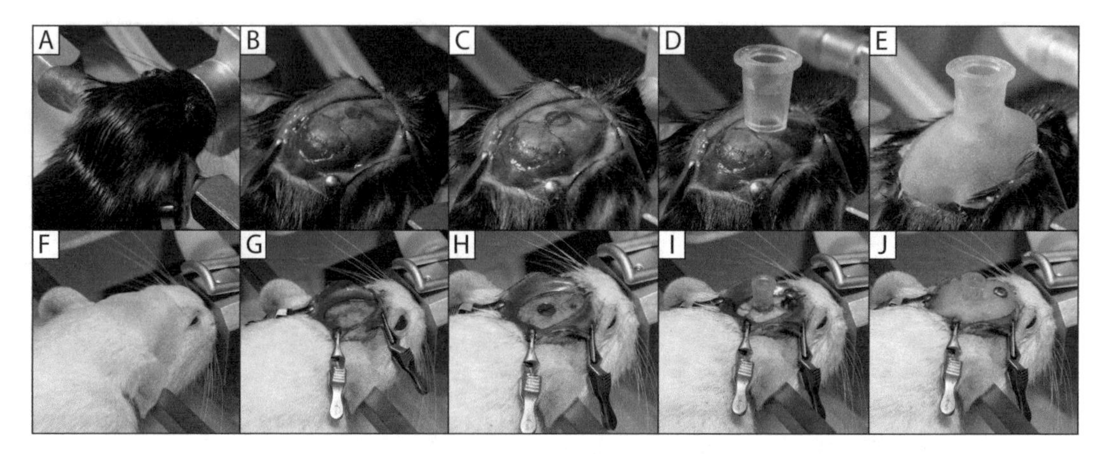

Fig. 3 Cranial surgery for midline FPI in the mouse (**a–e**) and rat (**f–i**). The animal is secured in a stereotaxic frame with a continuous flow of isoflourane via a nose cone (**a**, **f**). A midline incision is made to expose the skull and the overlying fascia is removed (**b**, **g**). For the mouse, Vetbond tissue adhesive is used to secure a disc shaved from weed whacker line at the location of the craniectomy (**b**). For the rat, a Dremel tool is used to make pilot holes for the screw placement and for securing the trephine pin. The screw hole is expanded with a finger nail drill and 5/64″ drill bit and a stainless steel screw is secured into the screw hole (**h**). A trephine (3 mm for mouse, 4.7 mm for rat) is used to create a cranial disc that is removed to expose the underlying dura (**c**, **h**). Small drops of cyanoacrylate gel are placed on the outside of the constructed injury hub and the injury hub is covered in methyl methacrylate cement and filled with saline (**e**, **j**)

eyewear. Assess the animal for signs of pain, distress, or disease and record this information on the data sheet (e.g., abnormal posture, movement, poor grooming, and evidence of porphyrin accumulation on eyes, nose, or fur).

3.3 Administer Anesthesia and Secure in Head Holder

- Anesthetize the animal with 5% isoflurane for 5 min in an induction chamber.
- Shave or remove hair from scalp, as appropriate.
- Secure animal in a stereotaxic frame equipped with a nose cone for continuous inhalation of isoflurane (2.0–2.5%) (Fig. 3a, f). The back of the front incisors should be flush with the bite bar, without tension applied to the teeth. If you observe mouth breathing, check the positioning of the teeth over the bite bar and/or reposition the nose cone to allow for normal respiration.
- Apply ophthalmic ointment to the eyes to keep them moist during the surgery.
- Prepare the surgical area with 70% alcohol and betadine solution (antiseptic).
- Monitor anesthesia by observing muscle relaxation, in addition to assessing the toe pinch reflex. Animals under appropriate anesthesia will have a steady respiration rate.

3.4 Cranial Surgery for Hub Placement

- Make a midline sagittal incision extending from between the eyes, to the base of the skull, just past the ears. To avoid excessive bleeding, avoid cutting the muscle at the base of the skull (Fig. 3b, g).
- Expose the skull and scrape the fascia from the skull using a delicate bone scraper, cotton swabs, and gauze. Clear away temporal muscle as necessary. If greater exposure is needed, stretch the skin by applying pressure with the fingers.
- Attach Bull Dog clips to the edges of the incision (two anterior, two posterior) to expose the surgical site. When the Bull Dog clips fall down, the weight will hold the incision open (Fig. 3b, g).

3.4.1 Cranial Surgery for Hub Placement-Mouse

- Shave weed whacker line with a razor blade as thin as possible to make a circular disc that is an equal thickness on all sides. Disc should be level when placed on the skull.
- Pick up the disc with side-grasping forceps. Dip the cranial disc into a drop of Vetbond tissue adhesive placed on a nonabsorbent surface.
- Place the disc at the location of the craniectomy (midway between bregma and lamda on the sagittal suture for mFPI). To drop the disc, release the forceps and use a wooden applicator stick to properly position the disc. Once in position, use a Kimwipe tissue to wick away any excess Vetbond (Fig. 3b). Allow the Vetbond to fully dry before beginning to trephine.
- Place the 3.0 mm trephine over the disc and perform the craniectomy by continually turning and spinning the trephine without disrupting the underlying dura. Keep trephine clean by using a toothbrush to remove bone debris from the trephine. Apply saline to moisten the bone and aid in trephination. As needed, angle the trephine to evenly cut around the craniectomy.
- Frequently check the progress of the craniectomy by applying mild pressure to the center of the craniectomy. As the skull thins, the craniectomy will be able to move independently of the skull. The craniectomy is complete when the bone can move freely in all directions.

  Under magnification, remove the bone piece working around the circumference using the wedelstaedt and scalpel, or two wedelstaedt instruments without disrupting the dura (Fig. 3c). When the bone has been removed, gently clear any blood from the craniectomy site.

3.4.2 Cranial Surgery for Hub Placement-Rat

- Mark the locations on the skull for the screw hole(s) and craniectomy center. The skull screw is used to secure the injury hub in place.
  - For midline FPI, position one screw hole 1 mm lateral to bregma and 1 mm rostral to the coronal suture on the right side.
  - For lateral FPI, position the one screw hole 1 mm lateral to bregma and 1 mm rostral to the coronal suture on the ipsilateral side to the craniectomy; position the second hole midway between bregma and lambda and 1 mm lateral to the central suture contralateral to the craniectomy.
- Drill pilot holes at both markings using the Dremel tool and burr bit.
- Expand the screw hole with a finger nail drill and 5/64″ drill bit (Fig. 3h).
- Place the centering pin inside the 4.7 mm diameter trephine. Anchor the centering pin in the pilot hole at the craniectomy center.
- Continually turn and spin the trephine to make a craniectomy without disrupting the underlying dura. Keep trephine clean by using a toothbrush to remove bone debris from the trephine teeth. Apply saline to moisten the bone and aid in trephination. As needed, angle the trephine to evenly cut around the craniectomy.
- Frequently check the progress of the craniectomy by applying mild pressure to the center of the craniectomy. As the skull thins, the craniectomy will be able to move independently of the skull. The craniectomy is complete when the bone can move freely in all directions.
- Remove the bone piece working around the circumference using the wedelstaedt and scalpel, or two wedelstaedt instruments without disrupting the dura. When the bone has been removed, gently clear any blood from the craniectomy site (Fig. 3h).
- Secure a stainless steel screw in the skull screw hole. Hold the screw with forceps and advance the screw with a screwdriver (Fig. 3i).

3.5 Injury Hub

3.5.1 Injury Hub Construction

- Attach a 22 gauge, 1 ½″ needle to a 1 cc syringe. Place the needle into a laboratory bench pad (Fig. 2a, b).
- Cut the female Luer-Loc hub from the needle using a razor blade (Fig. 2b). The cut is made parallel to the Luer-loc with an outer diameter of ~4.7 mm for the rat, and ~3.0 mm for the mouse.

- Inspect the cut edge of the injury hub and trim to size and level as necessary (Fig. 2c).
- For the rat, bevel the cut edge of the injury hub with a cosmetic pencil sharpener (Fig. 2d).
- Shave thin burrs around the injury hub starting at the Luer-Loc edge in the direction of the cut edge using a razor blade (Fig. 2e, f).

3.5.2 Injury Hub Placement

- Hold the hub in tissue forceps (behind the teeth). Apply small drops of cyanoacrylate gel on the outside of the hub, just above the cut end.
- For the mouse, using magnification, position the hub over the craniectomy. The injury hub should fit outside the craniectomy (Fig. 3d). For the rat, the injury hub fits inside the craniectomy (Fig. 3i).
- Using a wooden applicator stick (cut a sharp angle) gently scrape the cyanoacrylate gel down the injury hub onto the skull. Apply more cyanoacrylate gel if needed to the junction between the injury hub and the skull to firmly adhere the injury hub to the skull in addition to creating a seal.
- After the cyanoacrylate gel dries, cover the injury hub (and screw) in methyl methacrylate cement (Fig. 3e, j). Apply the methyl methacrylate cement from a 1 cc syringe when it is thick enough to hold shape.
- When the methacrylate cement has dried, fill the injury hub with saline.
- Place a suture at both the anterior and posterior edges of the incision.
- Remove the animal from the stereotaxic frame and anesthesia. Place the animal in a recovery cage on a heating pad until the animal is awake and alert. Monitor animals for outward signs of pain or distress.

3.6 Injury

Before using the injury device, check that when the weighted pendulum arm is hanging in a neutral position (at 0°) that it is flush and centered on the foam pad at the end of the plunger. Adjust as needed. Drop the pendulum hammer several times to prime the device.

- Reanesthetize the animal after an approximately 60 min recovery period from surgery.
- Visually inspect inside the injury hub for debris, blood or dried dental acrylic. Clean out the injury hub using a small cotton tip applicator or irrigate with saline if necessary.

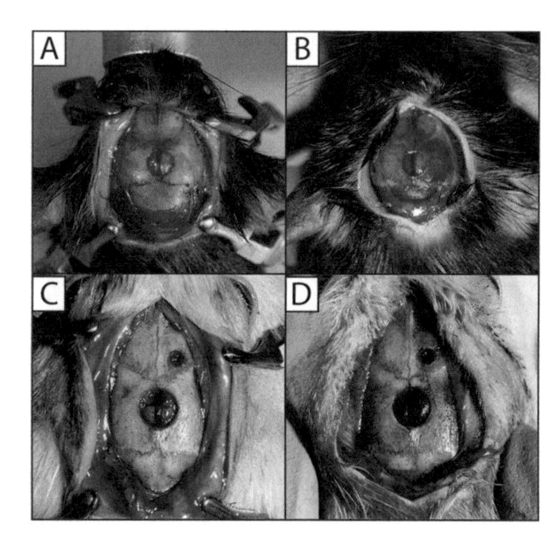

Fig. 4 Craniectomy before (**a**, **c**) and after mFPI (**b**, **d**). After the injury induction the craniectomy site and skull should be observed and notes should be taken. Immediately following injury it is common to have hematoma in both the mouse (**b**) and rat (**d**). Check for herniation of the dura, there should be uniform herniation which is indicative of an intact dura (**b**, **d**). Uneven or protruding herniation is characteristic of a dura breach and the animal should be removed from the study as a surgical failure

- Fill the injury hub with sterile saline until a bead of fluid is formed by surface tension. Remove any air bubbles from inside the hub.
- To avoid air between the hub and device, press the plunger so that a drop of fluid is produced at the end of the injury device. Connect the female Luer-Loc injury hub on the animal to the male Luer-Loc fitting on the injury device. Create continuity between the fluid of the cylinder and the fluid in the injury hub.
- Check the animal for a toe pinch response. Once a normal breathing pattern returns (1–2 breaths per second) and the animal has a positive toe pinch response, release the pendulum to injure the animal. Secure the pendulum after it strikes the plunger and return it to the catch.
- Immediately after the injury, start a timer to measure the duration of the suppression of the righting reflex.
- Remove the injury hub by pressing on the bridge of the nose for leverage (Fig. 4b, d).
- Observe and record the duration and extent of apnea or seizure. Note the condition/appearance of the surgical site and brain tissue beneath the injury site and record brain herniation and hemorrhage (Fig. 4b, d). If the dura is breached, the animal should be euthanized and not included in the study.

- Control bleeding if necessary. Leave the craniectomy open. Close the wound (i.e., suture or staple) and apply topical lidocaine and antibiotic ointment.
- Place the animal in a supine position on a heating pad. The time elapsed until the animal spontaneously rights is recorded as the righting reflex time.
- Once the animal has righted, place it in a designated recovery area equipped with a heating pad.
- When the animal regains normal ambulatory behavior, it can be returned to its home cage.

3.7 Postoperative Care

3.7.1 Postoperative Evaluations

- Following injury, animals should be visually monitored for continued recovery every 10 min post-injury (for the first hour). Within 15–20 min after injury, surviving animals should be alert. Within 1 h after injury, animals should be ambulatory. Brain-injured and uninjured control animals typically show no outward effects once they have recovered from anesthesia, and resume normal eating, drinking, and grooming patterns. Typically animals return to sleep, as the injury occurs during their sleep cycle.
- Postoperative evaluations should be done daily (for a minimum of 3 days). Follow the Postoperative Evaluation Sheet to record the external examination, physical examination, suture site, and a pain evaluation. Typically, animals require no special supportive care after surgery. This injury does not produce overt signs of postoperative pain, and does not call for pain monitoring or drugs to manage pain. Caution should be taken in administering such compounds, as they can influence outcome (for review see [12]).

3.7.2 Postoperative Weight

- Weigh animals daily. Record weights on the evaluation sheet.
- Animals can lose up to 20% of their body weight after surgery and injury. It is beneficial to prophylactically provide mash (chow + water) and/or place normal chow on the floor of the cage to facilitate weight gain.
- If by the second day post-injury, there is continued weight loss, the animals will likely require fluid injections (0.9% sterile saline) to prevent dehydration. Consult a local veterinarian for advice.
- Weight loss exceeding 20% of pre-injury body weight indicates significant injuries that require intensive postoperative care or euthanasia.

4 Notes

- The plunger impact pad on the fluid cylinder should be replaced every 8–12 months. Information and instructions for the setup, cleaning, and maintenance of the FPI device can be found in the FPI Operation Manual: http://www.radiology.vcu.edu/docs/FPIOperationManual.pdf.
- If the surgery site continues to bleed when the skull is removed, lightly remove blood with gauze. Adding saline can create hydrostatic pressure that will reduce bleeding. If the site continues to bleed, control the bleeding with Gelfoam. Excessive wiping or dabbing at the craniectomy site will prevent blood clotting and worsen the bleed.
- When using an inhaled anesthetic, it is recommended that all procedures are performed in a well-ventilated area, on a downdraft or similar table, or in a type II biosafety cabinet to minimize anesthesia exposure to the surgeon (current OSHA recommendation for halogenated gasses is <2 ppm).
- If the disc comes off while trephining during a mouse surgery, clean excess dried glue from the area and apply a new disc using Vetbond. However, if the bone can move independently of the skull in an area, use a small dot of superglue to attach a new disc. Vetbond will run and may touch the surface of the dura compromising the surgery.
- When constructing the injury hub for mice, to confirm the proper diameter you can place the trephine through the hub and confirm a tight fit.
- For the injury, rats should be held in your left hand lying on their right side. Attach the rat directly to the device.
- During the cranial surgery, the dura can be compromised by the trephination or removal of the bone. When the injury is induced, pressure from the fluid pulse will cause the dura to tear and the brain will herniate through the craniectomy. If the dura is compromised, the injury becomes inconsistent and should be classified as a technical failure. A dura breach will extend the opening of the blood brain barrier and displace neural tissue. Animals with a dura breach should be excluded from any study.
- When placing the injury hub over the craniectomy, cyanoacrylate gel can spread on to the dura. If the cyanoacrylate is not thoroughly dry forming a seal, the methyl methacrylate can also spread under the injury hub and onto the dura. These substances on the dura will change mechanical properties and alter the injury. Visual inspection is necessary to identify cyanoacrylate gel or methyl methacrylate on the dura, as well as any other obstruction over the injury site, such as a blood clot.

- Air bubbles in the FPI device can prevent an accurate measurement of the injury magnitude. When air is present in the device, the oscilloscope reading will have many jagged peaks instead of a smooth curve with one peak. The syringe ports can be used to remove any air that enters the device. One way for air to become trapped in the fluid cylinder is after cleaning of the cylinder. This can be minimized by rinsing with a spot remover solution for an automatic dishwasher (e.g., Jet Dry).
- Air bubbles can also enter the device during the impact. To prevent air bubbles it is necessary to use two 10 mL syringes during preparation of the device between rat injuries. After injury, remove the rat from the male Luer-loc fitting on the end of the device. Next, a 10 mL syringe with a female Luer-loc fitting should be attached to the device. Pull up on the syringe to remove fluid contaminated with blood or air from the device. Lastly, a second 10 mL syringe containing clean deionized water should be attached to the device. Pull up on the syringe to remove any air bubbles from the device. Check that when the weighted pendulum arm is hanging in a neutral position (at 0°) that it is flush and centered on the foam pad at the end of the plunger. Adjust as needed. Drop the pendulum hammer several times to prime the device. Between every rat injury a syringe should be attached and "dirty" water removed, then a second "clean" syringe should be attached to prime the device.
- It is important to make sure the Luer-loc extension tubing for mouse injury is free of air bubbles before each mouse injury. Hold the end of the tubing higher than the connection point on the device. Lightly tapping the extender tubing will force air bubbles to the end of the tubing where they can easily be removed. Between each mouse injury, remove all air bubbles from the tubing. Check that when the weighted pendulum arm is hanging in a neutral position (at 0°) that it is flush and centered on the foam pad at the end of the plunger. Adjust as needed. Drop the pendulum hammer several times to prime the device.
- The procedures presented in this book chapter for mFPI have been adapted by our group for postnatal day (PND)-17 and PND-35 rats. For PND-17 rats, the 3.0 mm mouse trephine is used for the crainectomy. We have designed and manufactured a 4.0 mm trephine for the PND-35 rats to accommodate the size difference of their skull compared to PND-17 and adult rats. Scaling the trephine sizes based on skull size is an essential consideration for all experimental brain injury models. PND-17 and PND-35 rats often experience a cessation of breathing as denoted by absence of chest movement associated with breathing. This apnea duration is significantly longer than

apnea noted in adult rats or adult mice. PND-17 rats also experience injury-induced seizures defined as one or more observed characteristic movements: unnatural body contortions, vigorous tail whipping or spinning, or spontaneous overt muscle contraction/relaxation.

References

1. Dixon CE, Lighthall JW, Anderson TE (1988) Physiologic, histopathologic, and cineradiographic characterization of a new fluid-percussion model of experimental brain injury in the rat. J Neurotrauma 5(2):91–104
2. Thibault LE, Meaney DF, Anderson BJ, Marmarou A (1992) Biomechanical aspects of a fluid percussion model of brain injury. J Neurotrauma 9(4):311–322
3. Floyd CL, Golden KM, Black RT, Hamm RJ, Lyeth BG (2002) Craniectomy position affects Morris water maze performance and hippocampal cell loss after parasagittal fluid percussion. J Neurotrauma 19(3):303–316. https://doi.org/10.1089/08977150275359487\3
4. Vink R, Mullins PG, Temple MD, Bao W, Faden AI (2001) Small shifts in craniotomy position in the lateral fluid percussion injury model are associated with differential lesion development. J Neurotrauma 18(8):839–847. https://doi.org/10.1089/089771501316919201
5. Iwamoto Y, Yamaki T, Murakami N, Umeda M, Tanaka C, Higuchi T, Aoki I, Naruse S, Ueda S (1997) Investigation of morphological change of lateral and midline fluid percussion injury in rats, using magnetic resonance imaging. Neurosurgery 40(1):163–167
6. Dixon CE, Lyeth BG, Povlishock JT, Findling RL, Hamm RJ, Marmarou A, Young HF, Hayes RL (1987) A fluid percussion model of experimental brain injury in the rat. J Neurosurg 67(1):110–119
7. Thompson HJ, Lifshitz J, Marklund N, Grady MS, Graham DI, Hovda DA, McIntosh TK (2005) Lateral fluid percussion brain injury: a 15-year review and evaluation. J Neurotrauma 22(1):42–75. https://doi.org/10.1089/neu.2005.22.42
8. Lifshitz J, Rowe RK, Griffiths DR, Evilsizor MN, Thomas TC, Adelson PD, McIntosh TK (2016) Clinical relevance of midline fluid percussion brain injury: acute deficits, chronic morbidities and the utility of biomarkers. Brain Inj 30(11):1293–1301. https://doi.org/10.1080/20699052.2016.1193628
9. McIntosh TK, Noble L, Andrews B, Faden AI (1987) Traumatic brain injury in the rat: characterization of a midline fluid-percussion model. Cent Nerv Syst Trauma 4(2):119–134
10. McIntosh TK, Vink R, Noble L, Yamakami I, Fernyak S, Soares H, Faden AL (1989) Traumatic brain injury in the rat: characterization of a lateral fluid-percussion model. Neuroscience 28(1):233–244
11. Carbonell WS, Maris DO, McCall T, Grady MS (1998) Adaptation of the fluid percussion injury model to the mouse. J Neurotrauma 15(3):217–229
12. Rowe RK, Harrison JL, Thomas TC, Pauly JR, Adelson PD, Lifshitz J (2013) Using anesthetics and analgesics in experimental traumatic brain injury. Lab Anim 42(8):286–291

Chapter 7

Development of a Rodent Model of Closed Head Injury: The Maryland Model

Erik Hayman, Kaspar Kaledjian, Vladimir Gerzanich, and J. Marc Simard

Abstract

Brain injury due to closed frontal head impact is a common mechanism in civilian traumatic brain injury (TBI). Researchers have developed a variety of models of traumatic brain injury in rodents, using both open and closed methods of injury. However, these models fail to reproduce the frontal impact of force commonly found in human TBI, result in significant focal injury such as skull fractures or focal contusions, and, in certain cases, carry an unacceptably high mortality. The Maryland TBI model provides an alternative rodent model to address these shortcomings. Here, we describe the rationale for the development of the Maryland TBI model. We then provide a detailed procedural overview of the model. We then summarize relevant pathological findings in the model. Finally, we compare the model to other existing closed head injury models in rodents, both with regard to advantages and limitations of the model.

Key words Traumatic brain injury, Diffuse axonal injury, Rodent, Closed head injury, Frontal impact

1 Introduction: Frontal Impact, Diffuse Axonal Injury, and TBI

Traumatic brain injury (TBI) refers to a heterogenous group of brain injuries and brain injury mechanisms that affect more than 1.5 million Americans annually, accounting for nearly 50,000 deaths and significant costs associated with long-term disability in survivors [1–3]. Although advances in surgical interventions, critical care, rehabilitation, and trauma system organization have undoubtedly led to incrementally improved outcomes following traumatic brain injury, the sometimes suboptimal results of even aggressive intervention mandate a better understanding of the disease. The need for a better pathophysiologic understanding of TBI is especially important given the numerous recent failures of surgical [4, 5], medical [6, 7], and pharmacologic [8] interventions to improve outcome in TBI despite their ostensibly solid pathophysiologic grounding. Animal models present an invaluable resource both for understanding TBI at a pathophysiologic level and for improving TBI therapies.

Amit K. Srivastava and Charles S. Cox, Jr. (eds.), *Pre-Clinical and Clinical Methods in Brain Trauma Research*, Neuromethods, vol. 139, https://doi.org/10.1007/978-1-4939-8564-7_7,

The heterogenous nature of TBI, ranging from mild concussion to diffuse brain injury to focal lesions such as traumatic contusions, necessitates a variety of models. Nevertheless, all animal models of TBI ought to strive to fulfill several criterion, namely (a) reproduction of mechanistic forces at play during actual injury, (b) production of pathology similar to that of human TBI, (c) consistent reproducibility of injury with well-defined parameters between experimental subjects and experimental groups, and (d) practicality with regard to cost and technical ease. Although a number of rodent models of TBI, such as the lateral fluid percussion model [9] and the controlled cortical impact model [10], fulfill these latter two requirements, these models rely on highly artificial experimental injury, direct brain deformation via a craniotomy window, to impart mechanical injury to the brain.

A significant proportion of human TBI, by contrast, results from sudden acceleration-deceleration as well as rotational forces and subsequent shear stresses imparted during impact, as occurs during high-speed motor vehicle collision or sports injury. These shear stresses induce a widespread disruption of axons throughout the brain, a form of pathologic injury characterized as diffuse axonal injury [11]. Initial theories of DAI posited a strictly mechanical basis for axonal disruption, suggesting that DAI represents a form of primary brain injury [12]. However, both in vitro and in vivo evidence suggests that shear stress initiates a secondary biochemical cascade culminating in axonal destruction via diverse mechanisms, such as pathologic activation of mechanosensitive channels [13], progressive instability of cytoskeletal elements [14], and abnormal activation of proteases such as calpain [15]. The delay between initiation of mechanical stress and actual disruption of axons suggests that DAI is amenable to treatment, although no such therapies for DAI exist clinically, motivating further research.

Given the importance of shear stress mechanisms in human TBI, a number of animal models of DAI exist. The earliest models employed primates [16] subjected to head acceleration without impact, allowing for creation of relatively reproducible injury which, importantly, models human TBI both mechanistically and pathologically. However, pragmatic, financial, and ethical concerns associated with primate experiments severely limit the use of this model. For this reason, researchers have developed a number of rodent models of impact-acceleration to create diffuse brain injury and DAI. The earliest model of acceleration injury in rats, the so-called Marmarou impact-acceleration model, impacts the dorsal aspect of the skull with a weight dropped from height to induce diffuse injury [17]. Given that the brain, and especially its white matter, demonstrates anisotropy with regard to strain [18], other rodent models incorporating various combinations of linear and rotational acceleration attempt to understand their role in diffuse brain injury [19–23].

Frontal impact, such as that experienced during a head-on car collision or football injury, consists of rapid changes with regard to linear acceleration in the direction of impact as well lateral rotation of the head. These forces differ significantly from the dorsal-ventral translational acceleration of the Marmarou model [17] or the rotational energy of the Genarelli model and its derivatives [21]. Moreover, the direct application of energy to the cranial vault risks skull deformation with potential for confounding of injury via skull fracturing with brain deformation [24]. These twin concerns motivated the development of an impact-acceleration model involving application of anterior-posterior linear acceleration in combination with a sagittal rotational acceleration to mimic the forces experienced during a frontal impact TBI. Importantly, application of force to an area of the skull remote from the cranial vault removes potential confounding due to cranial vault deformation, providing a pure model of diffuse brain injury due to acceleration and rotation. This article describes both the original Maryland frontal impact model [25] in the rat in procedural detail; moreover, a formal comparison is made to other existing models of impact-acceleration, both with regard to strengths and limitations.

2 Materials and Methods

2.1 Frontal Impact Device

The frontal impact device consists of three components: a 500 g steel ball (Small Parts, Inc., Miami), a pair of sloping rails in a hockey-shaped configuration, and a force coupling mechanism consisting of a pair of thin brass rods protruding from a solid plastic cylinder (Fig. 1a). The pair of sloping rails, formed from 19 mm-diameter tubular steel electrical conduit joined with a 3 cm gap via nuts and bolts, are bent to form a 66 degree angle with ground as depicted in the figure. A wooden collecting chamber sits at the bottom of the device and serves to hold the coupling mechanism. Rolling the ball from any given height allows the experimenter to impart a reproducible and calculable quantity of energy to the coupling arm via impact, with a vertical height of 2.1 m used for severe TBI and a height of 0.25 m used for mild TBI.

The coupling mechanism (Fig. 1b) consists of three components, a 25 × 70 mm delrin acetal cylinder (Small Parts, Inc.) and two 3.2 × 60 mm brass rods (type C330 ASTM B135; Small Parts, Inc.). Two holes, drilled 6 mm from the center of the acetal cylinder to a depth of 25 mm, serve to receive these brass rods; of note, the rods are secured in the holes such that there is a 3 mm offset between the lengths of the two rods. The rods themselves protrude ~3 cm from the surface of the acetal rod, in order to accommodate the rodent's snout without facial injury during frontal impact, while

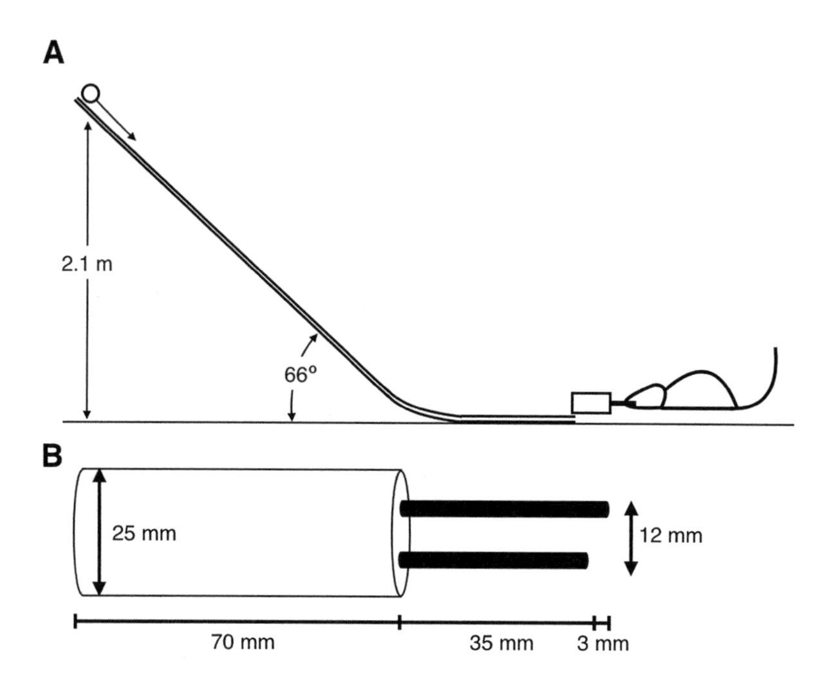

Fig. 1 (**a**) Diagram depicting overview of the Maryland model frontal impact model. A ~500 g metal ball is allowed to roll down a 66° incline from a height of 2.1 m until it strikes a coupling device at the bottom of the incline. The coupling device serves to impart force to the malar processes of an anesthetized rat, producing the frontal impact injury. (**b**) Schematic depicting the coupling mechanism, consisting of an acetal cylinder with two protruding metal rods. Drawings are not to scale

the offset between rods provides a rotational component to the injury. Of note, the exposed surfaces of these rods have a small V-shaped groove.

2.2 Animal Preparations

All procedures described have received approval from the University of Maryland Institutional Animal Care and Use Committee; procedures may require modification to conform to local institutional approval. For frontal impact TBI, we employ male Long-Evans rats weighing between 250–275 g (Harlan, Indianapolis, IN), although strain, weight, and gender conceivably may be varied as experimental purposes dictate. Animals are allowed a minimum of 1 week to acclimate following shipment. All animals are housed in our center's animal facility until time of injury. On the evening prior to injury, animals are fasted until time of injury.

2.3 Frontal Impact Injury

Prior to injury, all animals undergo anesthesia with a single intraperitoneal dose of ketamine (60 mg/kg) and xylazine (7.5 mg/kg). Loss of pedal reflex to toe pinch confirms depth of anesthesia. Rats are spontaneously allowed to breathe room air throughout the surgical procedure. We maintain core temperature at 37° using a

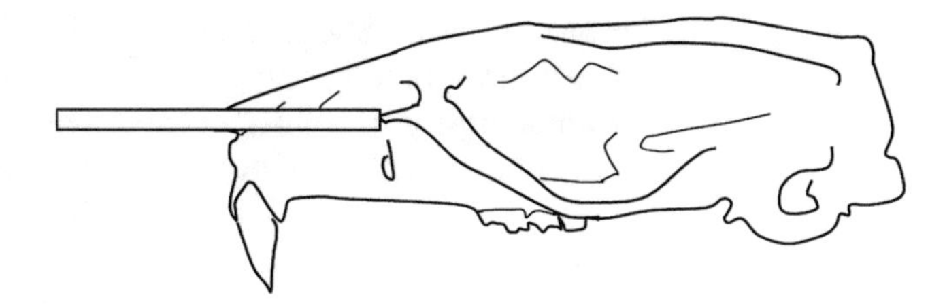

Fig. 2 Diagram demonstrating correct placement of the rods on the malar process of a rat's skull

heating pad regulated by a rectal thermal probe (Harvard Apparatus, Holliston, MA). Oxygen saturation throughout the injury procedure is monitored using a hind limb pulse oximeter (Mouse Ox; STARR Life Sciences Corp, Oakmont, PA).

After anesthesia induction, the infraorbital areas of the anesthetized rat are shaved and cleansed with a combination of betadine and alcohol. Surgical sites are then anesthetized with 1% lidocaine injected in subcutaneous manner. Bilateral infraorbital incisions are then performed sharply with a scalpel down the underlying to malar processes, with any remaining muscular tissue remove using blunt dissection. Hemostasis is obtained with thermal cautery. Following exposure, the malar processes are then situated within the V-shaped grooves on the coupling mechanism's rods (Fig. 2). The coupling mechanism is then secured to the rodent's head using rubber bands. When performed correctly, the rat's nasal passages remain clear, allowing the rat to continue spontaneous breathing. The rat is then placed prone, with the coupling mechanism situated at the end of the rails. The rat is gently secured in place using adhesive tape across its thorax.

To induce injury, the metal ball is released from a predetermined height down the rails. After gaining momentum, the ball strikes the coupling mechanism, imparting both translational and rotational force to the skull via the rat's malar processes. Following injury, the coupling mechanism is removed and the infraorbital incisions sutured closed. For sham injury, an identical surgical procedure and coupling mechanism placement is performed, but the rat is not subjected to a ball strike. After injury, rats are monitored until recovery from anesthesia and then returned to their cages for further experimentation.

3 The Maryland Model and Other Closed Head Injury Models

3.1 The Maryland Model

The Maryland model of TBI provides a simple, reproducible model of diffuse brain injury due to frontal impact. The experimental model possesses a number of significant advantages. The injury apparatus is readily constructed from inexpensive, commercially available parts. The surgical procedure is simple and well within the capabilities of a laboratory with rudimentary rodent surgical technique. Injury mortality even at severe levels of TBI is low, with minimal disruption of normal physiology following injury, avoiding confounding from factors such as hypoxia or hypotension [25]. Finally, coupling the injury force to the malar processes, a structure relatively remote from the cranial vault, decreases the risk of inadvertent skull fracture or direct brain injury due to deformation of the cranial vault.

Diffuse brain injury in human injuries is thought to result from shear stresses placed on the brain by rapid changes acceleration or rotation, rather than skull deformation or impact of the brain against the skull (coup-contrecoup injury). The Maryland model represents a relatively specific model of this process. On gross pathological review following injury, animals do not demonstrate evidence of significant frontal contusions or other signs of coup injury; although cerebellar subarachnoid hemorrhage, thought to represent a form of countercoup injury, is a relatively common feature in this model, it does not represent the dominant pathology. Histologic evaluation demonstrates multifocal petechial hemorrhages predominating in the frontal and parietal lobe white matter, as well as the corpus callosum, deep nuclei, and brain stem after severe frontal impact. Of note this pattern of petechial hemorrhage mirrors diffuse axonal injury in human TBI [11]. Immunohistochemical study of these sections conforms to the gross histology. Beta-amyloid precursor protein upregulation, an established marker of DAI, demonstrates increased expression with the neuronal perikarya [26] as an atypical beaded appearance within damaged axons [27, 28] following DAI. Following frontal impact in the Maryland model, both of these pathological B-APP features coincide with areas of gross histologic injury, including the brainstem, hippocampus, and cerebral white matter, with two important exceptions: the thalamus demonstrates significant perikaryal up-regulation without significant axonal injury while the basal ganglia demonstrates axonal injury without neuronal up-regulation. More recently, study of white matter tracts following the Maryland frontal impact model using MRI diffusion tensor imaging corroborates these histologic findings of significant white matter injury, particularly in the fimbria and the splenium of the corpus callosum [29].

3.2 The Marmarou Model

The Marmarou model is the oldest model of diffuse rodent brain injury [17, 30]. The model employs brass weights dropped from prespecified heights through a plexiglass tube to impact a rat or mouse's skull centrally between the lambda and bregma. A metal disc directly affixed to the skull using cement protects the skull from the weight drop, reducing both the effects of skull deformation and the incidence of skull fractures. The rat rests on a deep foam bed at time of injury. The primary mode of injury observed in the Marmarou model is diffuse axonal and microvascular injury in the corpus callosum, internal capsule, and long tracts of the brainstem due to significant dorsal-ventral brain acceleration (900 G) at impact. Due to its long history, low cost, and simplicity, the Marmarou model remains a popular model of diffuse brain injury.

Despite its advantages the Marmarou model suffers from a number of significant drawbacks. First and foremost, skull protection via a metal disc does not entirely eliminate skull deformation, which occurs to a depth of 0.3 mm underneath the metal disc [17]. Furthermore, severe injury in the Marmarou model produces skull fractures with an incidence of 12.5% despite the protection afforded by the disc. Thus, especially at severe levels of injury, the Marmarou model may not represent a pure model of impact injury, as injury to skull deformation/fracture with direct brain compression serves to confound conclusions regarding acceleration-induced injury. Neuronal injury primarily occurs in the cortex underneath the disc, suggesting a local component to the diffuse injury [30]. High peri-procedural mortality represents another significant drawback of the Marmarou model, with a mortality of nearly 60% following severe injury due to apnea [30]. Aside from the increased resources in the form of experimental subjects required to compensate for this high mortality, the significant apnea and associated hypoxia induced by the injury serve to confound significantly experimental conclusions. Finally, the possibility of a second impact due to rebounding of the weight as well as lateralization of the weight as it falls within the tube serve to reduce reproducibility of the injury. None of these issues are insurmountable. The undesirable injury mechanisms such as skull deformation and fracture may be reduced with use of a less severe injury [17]. Mechanical ventilation reduces mortality and hypoxia due to apnea. Modern variations of the model replacing the weight drop with a computer-controlled impactor minimize intraprocedural variability [31, 32]. However, steps such as mechanical ventilation and specialized impactors significantly increase the complexity and cost of the model, eliminating some of its chief advantages.

3.3 Other Rodent Models of Diffuse Brain Injury

Aside from the Maryland and the Marmarou models of diffuse brain injury, a number of other models of rodent diffuse TBI exist. Although diffuse brain injury models for higher mammals, such as pigs and primates, exist, they lie outside the scope of our discussion. Aside from addressing the aforementioned practical issues arising from the Marmarou model, these models address another key limitation of the Marmarou model, namely confinement of acceleration to a single plane without any rotational component. This concern arises from early studies by Genarelli demonstrating the importance of plane of rotation to DAI-induced coma in primates, with coma chiefly attributable to coronal rotation [33]. Several models attempt to translate these primate studies into rodents. One study employing rapid coronal skull rotation in rats demonstrated evidence of diffuse injury, both grossly and histologically, in the brain white matter, particularly of the brain stem, with fairly high rates of mortality (17%) [22], consistent with Genarelli's findings in primates. One key limitation of this model, however, is possible confounding due to cervical spinal cord injury, with histologic evidence of upper cervical cord injury a common histologic feature. A more recent report describes an alternative model of rapid coronal rotational head injury generated by laterally striking a constrained helmet with an impactor [21, 23]. Although this model succeeds with regard to a number of criteria such as minimizing injury due to brain deformation, low mortality, and reproducibility of impact, a key limitation noted by the authors is the absence of histological evidence for DAI, specifically axonal bulb swelling, following injury, limiting extension of this model to human injuries.

Although pure coronal rotation represents a common mode of experimental injury, at least one model has explored pure sagittal rotation as an injury mechanism [34]. In this model, an impactor anteriorly strikes a bar affixed above a rat's head by means of a skull cap, causing the head to rotate rearwards. An affixed accelerometer measures rotation. Like coronal models of acceleration injury, this model yields both functional and histological findings consistent with DAI with limited macroscopic injury.

A common concern in most of the aforementioned injury models is direct brain injury from skull deformation. Although the Maryland model bypasses this issue by impacting the malar processes, rather than the cranial vault directly, one intriguing alternative is an air cannon induced injury developed in mice. Exposure of mice to a high-pressure air cannon results in rapid rotational acceleration of their heads, leading to diffuse functional deficits [35]. Aside from the advantages of a noncontact injury mechanism, the use of an air cannon precludes the need for surgical intervention and its confounding effects. This model, however, does possess a number of disadvantages. Although rotational injury represents the predominant injury mechanism, the unconstrained nature of the head motion theoretically adds significant variability between injuries; furthermore, attempts to quantify head rotation with a gyroscope significantly increased procedural mortality, precluding standardization of injury. Moreover, although the absence of a measurable blast wave and protection of the thorax during cannon exposure reduces systemic off-target effects, the possibility of confounding, particularly due to aural injury, represents an important limitation of this model.

Given that injuries sustained by humans are unlikely to consist of movements within a single plane or around a single axis, more modern models attempt to reproduce the more complex movements that occur during human diffuse brain injury, such as the combination of lateral rotation and anterior-posterior linear acceleration observed with the Maryland model. One such model employs a defined combination of lateral translation and coronal rotation to induce widespread histologic DAI and long-lasting behavioral deficits in rats [20]. Similar, the recent Closed-Head Impact Model of Engineered Rotational Acceleration (CHIMERA) combines dorsal-ventral acceleration with sagittal acceleration in lightly anesthetized mice [32]. This latter model has a number of advantages including the absence of surgical intervention, histologic pattern of injury similar to human diffuse injury, and production of long-lasting behavioral deficits. Although both of these combined acceleration-rotation models provide satisfactory models of human diffuse injury both from a histologic and behavioral perspective, they suffer from two key disadvantages. Both models require relatively specialized equipment, limiting their widespread applicability. Perhaps more importantly, the combinations of forces described do not model common modes of injury in human beings, unlike the Maryland model, which models the combinations of forces typically experienced during frontal impact.

4 Conclusions

Diffuse brain injury represents a clinical problem refractory to clinical intervention beyond supportive care, mandating further experimental study. The Maryland model has a number of advantages over other models of rodent TBI. The combination of linear translation and sagittal rotation accurately models a common mode of human injury, namely frontal impact. The widespread white matter injury with relative brainstem sparing reproduces human histologic findings without suffering from high mortality or confounding due to physiologic alteration. The absence of specialized impactors or other equipment significantly reduces costs associated with the model. Finally, use of the malar processes, rather than the cranium, as the point of energy transfer prevents unintended brain deformation which serves as a source of confounding in other models. While other models incorporate a number of these features, the Maryland model combines them all, providing it with a number of advantages in the study of diffuse brain injury.

References

1. Coronado VG, Xu L, Basavaraju SV, LC MG, Wald MM, Faul MD, Guzman BR, Hemphill JD, Centers for Disease C, Prevention (2011) Surveillance for traumatic brain injury-related deaths—United States, 1997-2007. MMWR Surveill Summ 60(5):1–32

2. Thurman DJ, Alverson C, Dunn KA, Guerrero J, Sniezek JE (1999) Traumatic brain injury in the United States: a public health perspective. J Head Trauma Rehabil 14 (6):602–615
3. Langlois JA, Rutland-Brown W, Wald MM (2006) The epidemiology and impact of traumatic brain injury: a brief overview. J Head Trauma Rehabil 21(5):375–378
4. Cooper DJ, Rosenfeld JV, Murray L, Arabi YM, Davies AR, D'Urso P, Kossmann T, Ponsford J, Seppelt I, Reilly P, Wolfe R, Investigators DT, Australian, New Zealand Intensive Care Society Clinical Trials G (2011) Decompressive craniectomy in diffuse traumatic brain injury. N Engl J Med 364(16):1493–1502. https://doi.org/10.1056/NEJMoa1102077
5. Hutchinson PJ, Kolias AG, Timofeev IS, Corteen EA, Czosnyka M, Timothy J, Anderson I, Bulters DO, Belli A, Eynon CA, Wadley J, Mendelow AD, Mitchell PM, Wilson MH, Critchley G, Sahuquillo J, Unterberg A, Servadei F, Teasdale GM, Pickard JD, Menon DK, Murray GD, Kirkpatrick PJ, Collaborators RET (2016) Trial of decompressive craniectomy for traumatic intracranial hypertension. N Engl J Med 375(12):1119–1130. https://doi.org/10.1056/NEJMoa1605215
6. Chesnut RM, Temkin N, Carney N, Dikmen S, Rondina C, Videtta W, Petroni G, Lujan S, Pridgeon J, Barber J, Machamer J, Chaddock K, Celix JM, Cherner M, Hendrix T, Global Neurotrauma Research G (2012) A trial of intracranial-pressure monitoring in traumatic brain injury. N Engl J Med 367(26):2471–2481. https://doi.org/10.1056/NEJMoa1207363
7. Andrews PJ, Sinclair HL, Rodriguez A, Harris BA, Battison CG, Rhodes JK, Murray GD, Eurotherm Trial C (2015) Hypothermia for intracranial hypertension after traumatic brain injury. N Engl J Med 373(25):2403–2412. https://doi.org/10.1056/NEJMoa1507581
8. Wright DW, Yeatts SD, Silbergleit R, Palesch YY, Hertzberg VS, Frankel M, Goldstein FC, Caveney AF, Howlett-Smith H, Bengelink EM, Manley GT, Merck LH, Janis LS, Barsan WG, Investigators N (2014) Very early administration of progesterone for acute traumatic brain injury. N Engl J Med 371 (26):2457–2466. https://doi.org/10.1056/NEJMoa1404304
9. Dixon CE, Lighthall JW, Anderson TE (1988) Physiologic, histopathologic, and cineradiographic characterization of a new fluid-percussion model of experimental brain injury in the rat. J Neurotrauma 5(2):91–104. https://doi.org/10.1089/neu.1988.5.91
10. Dixon CE, Clifton GL, Lighthall JW, Yaghmai AA, Hayes RL (1991) A controlled cortical impact model of traumatic brain injury in the rat. J Neurosci Methods 39(3):253–262
11. Adams JH, Doyle D, Ford I, Gennarelli TA, Graham DI, McLellan DR (1989) Diffuse axonal injury in head injury: definition, diagnosis and grading. Histopathology 15(1):49–59
12. Adams JH, Graham DI, Murray LS, Scott G (1982) Diffuse axonal injury due to nonmissile head injury in humans: an analysis of 45 cases. Ann Neurol 12(6):557–563. https://doi.org/10.1002/ana.410120610
13. Wolf JA, Stys PK, Lusardi T, Meaney D, Smith DH (2001) Traumatic axonal injury induces calcium influx modulated by tetrodotoxin-sensitive sodium channels. J Neurosci 21 (6):1923–1930
14. Tang-Schomer MD, Patel AR, Baas PW, Smith DH (2010) Mechanical breaking of microtubules in axons during dynamic stretch injury underlies delayed elasticity, microtubule disassembly, and axon degeneration. FASEB J 24 (5):1401–1410. https://doi.org/10.1096/fj.09-142844
15. McGinn MJ, Kelley BJ, Akinyi L, Oli MW, Liu MC, Hayes RL, Wang KK, Povlishock JT (2009) Biochemical, structural, and biomarker evidence for calpain-mediated cytoskeletal change after diffuse brain injury uncomplicated by contusion. J Neuropathol Exp Neurol 68 (3):241–249. https://doi.org/10.1097/NEN.0b013e3181996bfe
16. Gennarelli TA, Adams JH, Graham DI (1981) Acceleration induced head injury in the monkey. I. The model, its mechanical and physiological correlates. Acta Neuropathol Suppl 7:23–25
17. Marmarou A, Foda MA, van den Brink W, Campbell J, Kita H, Demetriadou K (1994) A new model of diffuse brain injury in rats. Part I: Pathophysiology and biomechanics. J Neurosurg 80(2):291–300. https://doi.org/10.3171/jns.1994.80.2.0291
18. Prange MT, Meaney DF, Margulies SS (2000) Defining brain mechanical properties: effects of region, direction, and species. Stapp Car Crash J 44:205–213
19. Wang HC, Duan ZX, Wu FF, Xie L, Zhang H, Ma YB (2010) A new rat model for diffuse axonal injury using a combination of linear acceleration and angular acceleration. J Neurotrauma 27(4):707–719. https://doi.org/10.1089/neu.2009.1071
20. Li XY, Li J, Feng DF, Gu L (2010) Diffuse axonal injury induced by simultaneous moderate linear and angular head accelerations in rats.

Neuroscience 169(1):357–369. https://doi.org/10.1016/j.neuroscience.2010.04.075

21. Ellingson BM, Fijalkowski RJ, Pintar FA, Yoganandan N, Gennarelli TA (2005) New mechanism for inducing closed head injury in the rat. Biomed Sci Instrum 41:86–91
22. Xiao-Sheng H, Sheng-Yu Y, Xiang Z, Zhou F, Jian-ning Z (2000) Diffuse axonal injury due to lateral head rotation in a rat model. J Neurosurg 93(4):626–633. https://doi.org/10.3171/jns.2000.93.4.0626
23. Fijalkowski RJ, Stemper BD, Pintar FA, Yoganandan N, Crowe MJ, Gennarelli TA (2007) New rat model for diffuse brain injury using coronal plane angular acceleration. J Neurotrauma 24(8):1387–1398. https://doi.org/10.1089/neu.2007.0268
24. De Mulder G, Van Rossem K, Van Reempts J, Borgers M, Verlooy J (2000) Validation of a closed head injury model for use in long-term studies. Acta Neurochir Suppl 76:409–413
25. Kilbourne M, Kuehn R, Tosun C, Caridi J, Keledjian K, Bochicchio G, Scalea T, Gerzanich V, Simard JM (2009) Novel model of frontal impact closed head injury in the rat. J Neurotrauma 26(12):2233–2243. https://doi.org/10.1089/neu.2009.0968
26. Itoh T, Satou T, Nishida S, Tsubaki M, Hashimoto S, Ito H (2009) Expression of amyloid precursor protein after rat traumatic brain injury. Neurol Res 31(1):103–109. https://doi.org/10.1179/016164108X323771
27. Blumbergs PC, Scott G, Manavis J, Wainwright H, Simpson DA, McLean AJ (1995) Topography of axonal injury as defined by amyloid precursor protein and the sector scoring method in mild and severe closed head injury. J Neurotrauma 12(4):565–572. https://doi.org/10.1089/neu.1995.12.565
28. Stone JR, Singleton RH, Povlishock JT (2000) Antibodies to the C-terminus of the beta-amyloid precursor protein (APP): a site specific marker for the detection of traumatic axonal injury. Brain Res 871(2):288–302
29. Herrera JJ, Bockhorst K, Kondraganti S, Stertz L, Quevedo J, Narayana PA (2017) Acute white matter tract damage after frontal mild traumatic brain injury. J Neurotrauma 34(2):291–299. https://doi.org/10.1089/neu.2016.4407
30. Foda MA, Marmarou A (1994) A new model of diffuse brain injury in rats. Part II: Morphological characterization. J Neurosurg 80(2):301–313. https://doi.org/10.3171/jns.1994.80.2.0301
31. Cernak I, Vink R, Zapple DN, Cruz MI, Ahmed F, Chang T, Fricke ST, Faden AI (2004) The pathobiology of moderate diffuse traumatic brain injury as identified using a new experimental model of injury in rats. Neurobiol Dis 17(1):29–43. https://doi.org/10.1016/j.nbd.2004.05.011
32. Namjoshi DR, Cheng WH, McInnes KA, Martens KM, Carr M, Wilkinson A, Fan J, Robert J, Hayat A, Cripton PA, Wellington CL (2014) Merging pathology with biomechanics using CHIMERA (Closed-Head Impact Model of Engineered Rotational Acceleration): a novel, surgery-free model of traumatic brain injury. Mol Neurodegener 9:55. https://doi.org/10.1186/1750-1326-9-55
33. Gennarelli TA, Thibault LE, Adams JH, Graham DI, Thompson CJ, Marcincin RP (1982) Diffuse axonal injury and traumatic coma in the primate. Ann Neurol 12(6):564–574. https://doi.org/10.1002/ana.410120611
34. Davidsson J, Risling M (2011) A new model to produce sagittal plane rotational induced diffuse axonal injuries. Front Neurol 2:41. https://doi.org/10.3389/fneur.2011.00041
35. Sabbagh JJ, Fontaine SN, Shelton LB, Blair LJ, Hunt JB Jr, Zhang B, Gutmann JM, Lee DC, Lloyd JD, Dickey CA (2016) Noncontact rotational head injury produces transient cognitive deficits but lasting neuropathological changes. J Neurotrauma 33(19):1751–1760. https://doi.org/10.1089/neu.2015.4288

Chapter 8

Rodent Model of Primary Blast-Induced Traumatic Brain Injury: Guidelines to Blast Methodology

Venkatasivasai Sujith Sajja, Peethambaram Arun, Stephen A. Van Albert, and Joseph B. Long

Abstract

Neuropsychological symptoms in warfighters after exposures to blast have triggered considerable research interest in the pathophysiological manifestations of blast-induced traumatic brain injury (bTBI). Preclinical research models of blast are attractive tools to understand the prognosis of behavioral changes, identify relevant biomarkers and characterize the neurobiological underpinnings of blast injury. However, the lack of standardization among preclinical bTBI studies has led to numerous inconsistencies in the data. Inadequate characterization of blast simulators, incomplete understanding and interpretation of blast physics, improper use of animal restraining techniques, and misapplication of biomechanical loading conditions in animal research have led to laboratory results that all-too-often bear little resemblance and relevance to injuries sustained by warfighters. Another major challenge for the bTBI research community is inadequate reporting of methodological conditions such as total pressure, static pressure, positive pressure duration, negative pressure duration, and impulse, to name a few, which has also contributed to ambiguous and sometimes conflicting research outcomes. This report focuses on the requirements for standardization of rodent experimental blast exposure conditions, blast simulator characterization, and guidelines on the dissemination of blast injury research methodology.

Key words Blast, Primary blast, Characterization, Blast simulator, Guidelines, Traumatic brain injury

1 Introduction

Early research primarily associated blast and explosions with extremity injuries to warfighters caused by projectiles and shrapnel [1–4]. Although the first reported clinical evidence of blast-related concussion dates to the early 1940s [4], traumatic brain injury (TBI) has been primarily studied in the context of automobile crashes or head impacts (e.g. falls, contact sports) resulting in a concussive or subconcussive event. However, the biomechanics of this type of impact injury are characteristically very different from bTBI [5]. Effects of blast on the brain garnered increased attention during the late 1990s to early 2000s primarily due to the

Amit K. Srivastava and Charles S. Cox, Jr. (eds.), *Pre-Clinical and Clinical Methods in Brain Trauma Research*, Neuromethods, vol. 139, https://doi.org/10.1007/978-1-4939-8564-7_8, © Springer Science+Business Media, LLC, part of Springer Nature 2018

preponderance of closed head TBI from warfighters returning from Operation Iraqi Freedom (OIF) and Operation Enduring Freedom (OEF) [5–8]. Increased use of improvised explosive devices (IEDs) in modern asymmetric warfare coupled with civilian terrorist attacks has continued to be major drivers of bTBI research.

Blast injury can be divided into four major categories: (1) primary blast injury, in which injury results solely from the blast pressure wave; (2) secondary blast injury, in which explosion-driven shrapnel or projectiles contribute to injuries; (3) tertiary blast injury, in which injury results from impact, including impact with debris launched by the explosion; and (4) quaternary blast injury, in which injuries result from heat, radiation, and other modalities that are not encompassed by primary, secondary, or tertiary injuries [9]. While close proximity to blast can result in a combination of primary, secondary, and tertiary injuries, primary bTBI typically results at distances greater than the effective projectile range [10]. Thus, proximity to blast plays a critical role determining the extent to which bTBI is accompanied by injury to other major organs of the body, which can be multifactorial and interactive with bTBI. In most cases, bTBI resulting in secondary, tertiary, and quaternary injuries present physical manifestations of trauma for treatment and diagnosis, while primary mild bTBI does not, often times resulting in an occult injury [9].

Over the years, body armor (e.g. helmet and chest protection vest) has been developed to protect against penetrating and blunt trauma to the head and upper body. However, at present there is no body armor or helmet standard for protection against primary blast injury. It is also important to note that, according to statistics assembled by the Department of Defense, the breakdown of TBIs reported worldwide from 2000 to 2017 (Q2) reveals that the majority (82.3%) of cases were categorized as mild TBI, while the least number of injuries were severe and penetrating brain injuries (1.1% and 1.4% respectively). These figures highlight the importance of preventive strategies, diagnostic markers and treatments targeting mTBI [6]. Several theories have been proposed to account for the biomechanical mechanisms of primary bTBI. The leading hypotheses include: (1) direct loading of primary blast pressure on the brain, (2) coup and contrecoup injury due to head acceleration, (3) skull flexure due to blast loading, and (4) vascular transmission. Although these competing theories remain to be validated, recent studies have demonstrated that intercranial pressure of the brain closely parallels the ambient peak static pressure profile and associated skull flexure due to the shock front in rodent models [10, 12].

bTBI resulting from primary blast has been a major challenge to diagnose and treat primarily due to delayed onset of symptoms and no physical manifestation of injury, particularly with mild forms of bTBI (mbTBI). Thus far, there are no FDA-approved diagnostic modalities to identify moderate to mild bTBI. To understand injury mechanisms, develop diagnostic modalities and biomarkers of bTBI, rodent models have been widely used in the laboratory [12]. With well-conserved neuroanatomical features and neural structures, rodents (mice and rats) have generally been the most widely used experimental models for high-throughput research in neuroscience. A large number of neurodegenerative diseases and mental health disorders such as Alzheimer's disease & dementia, Parkinson's disease, autism, depression, and addiction have been widely studied using rodent models [14, 15]. Characterizing TBI resulting from exposures to blast or explosions is a relatively new consideration in the field of neuroscience. For this purpose, laboratory-scale blast simulators are typically used to mimic free-field-like primary blast injury with a waveform that resembles a Friedlander wave. A Friedlander waveform is an idealized blast wave with a near-instantaneous rise (on the order of nanoseconds) from ambient pressure to a sharp peak followed by exponential decay resulting in positive pressure duration on the order of milliseconds (0.2–6 millisecond (ms) for an improvised explosive device (IED)-like blast (Fig. 1). The relative importance of blast wave parameters such as impulse (the area of positive pressure duration) and strain (i.e. loading) rates that translates to injuries produced in rodents is unknown. The influences of these parameters, species difference, and scaling parameters on the brain injury are currently undefined.

Flow conditions of blast waves have been thoroughly studied over the past 75 years by the blast physics community. Information and techniques to measure and understand different components of blast shown in Fig. 1 have been nicely summarized by Dewey [16]. Despite the insights of the blast physics community, a considerable amount of blast-related biomedical research has been conducted improperly, using blast simulators without careful characterization of flow conditions, leading to artifacts and exaggerated injury outcomes that bear no resemblance to primary blast injury [17]. This mischaracterization is principally due to the inherent focus of research biologists to simulate biological injury outcomes as opposed to using appropriate mechanical forces to understand outcomes of the injury (effect–cause vs. cause–effect relationship). While the majority of neurodisorders are understood to be the result of aberrant biological processes (e.g., prion protein aggregation in Alzheimer's disease, Parkinson's disease and multiple sclerosis due to genetic predisposition and environmental factors), traumatic brain injury is the result of mechanical forces in the form of blunt impact or blast overpressure. Thus, proper mechanical characterization of the loading conditions of the shock tube is essential to define injury outcomes of bTBI. *The discussion here is confined to the context of idealized "free-field"-like primary blast; primary complex blast is not discussed due to limited research and characterization at the laboratory scale.*

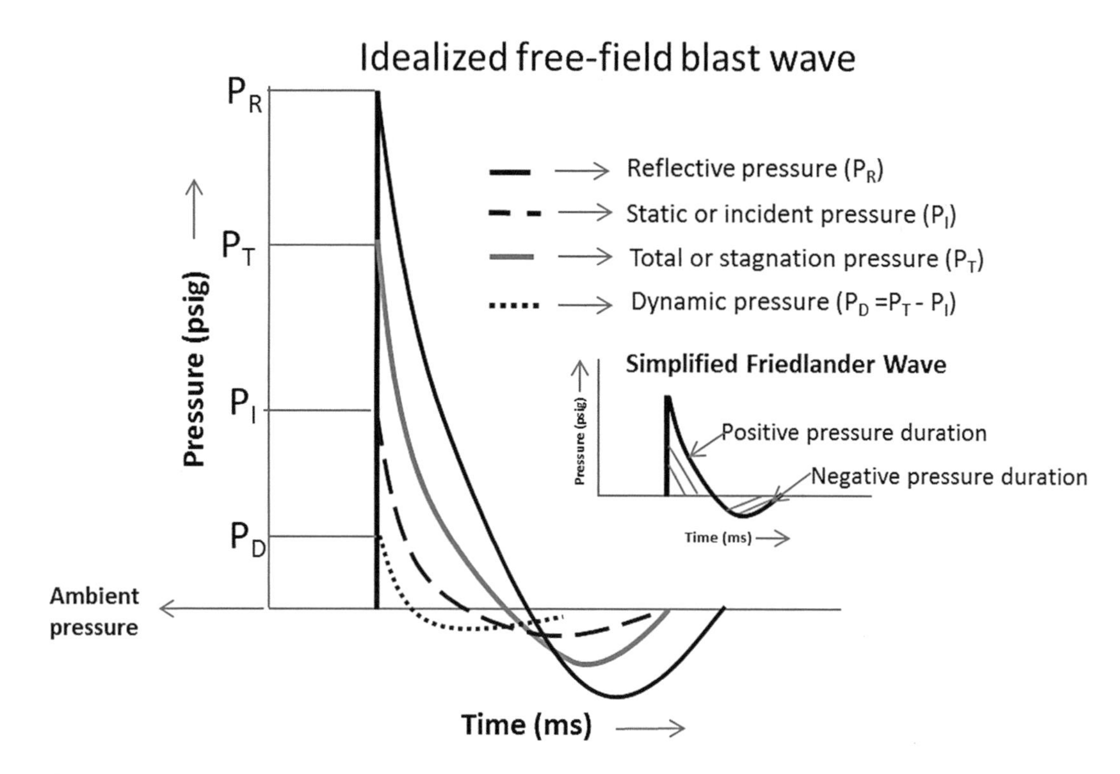

Fig. 1 Idealized wave form depicting the relationship of reflective pressure (P_R), stagnation or total pressure (P_T), static or incident pressure (P_I), and dynamic pressure (P_D) components with positive and negative pressure durations. Each of the components of the larger idealized wave can be represented as a simplified Friedlander-like wave as shown in the inset. Depicting one waveform alone is not adequate to convey the actual exposure conditions of the blast. Simple stated, a stagnation or a total pressure is recorded at 180° (isentropic) to the flow of a planar transmitted shock wave brought to complete rest (i.e. stagnated), while a reflective peak is the highest blast peak and largest loading force that occurs when there is any form of impedance (non-isentropic) to the shock wave propagation. Static pressure is measured 90° to the flow of the shock wave. The static (i.e. crushing) pressure is the pressure typically reported in bTBI research. Dynamic pressure is too difficult to measure directly in a blast. Rather, dynamic pressure is inferred as the difference between the total and static pressures. Dynamic pressure gives rise to the blast wind during the blast wave. (Refer to Dewey 2016 for further description of flow conditions and measurement methods [13]). Note: depicted pressure profiles and their durations are not scaled to ideal conditions derived from Rankine–Hugoniot equations

1.1 Current Issues or Limitations with Characterization of Primary Blast Injury

1.1.1 Explosive Charge Detonation

Few groups have studied primary blast by detonating explosives in the free-field using a rodent model [18, 19]. Ideally, this would be the research method of choice to study blast injury. However, explosive-charge field work requires facilities and trained personnel that are beyond the reach of most research programs. Other groups have attempted to use explosives inside a blast simulator (shock tube) to mimic free-field blast using a variety of charges (e.g., 2,4,6-trinitrotoluene (TNT) or 1,3,5-Trinitro-1,3,5-triazinane) [20]. The first study on bTBI in a rodent model, conducted by a group in National University of Singapore using TNT blast exposure albeit in an enclosed space to mimic blast in bunker, demonstrated that blast can cause neuropathological changes [18]. In addition to reproducibility challenges, conducting an experiment on a blast range or with explosives to generate free-field blast exposure has many regulatory and logistics requirements, including costs, planning associated with live explosive charges, certification of personnel to use live explosives, animal transportation and monitoring, all of which pose considerable challenges. Other potentially important limitations are repeatability of blast exposure; change in size, shape, and type of explosive resulting in varied exposure levels; inconsistencies of blast waveforms; and impulse of waveforms. In addition, the stand-off distance, ground height of explosive to animal location, and environmental factors (altitude, temperature, and humidity) can change the exposure intensity. Given these challenges, laboratory simulators have generally become the preferred choice among researchers to simulate primary blast with high throughput and repeatability.

1.1.2 Laboratory Blast Simulators

Since 1899, constant diameter shock tubes have been used to understand the physics of supersonic and aerodynamic flow of gas at different pressures (either using an explosive or a calibrated membrane to generate a supersonic flow) [21]. Research in the early 1940s extensively studied blast effects on the human body, concluding that ears and air-filled organs are most vulnerable to blast injury [22, 23]. Studies were primarily focused on evaluating the effects of blast on ear and lungs until early 1980s. With the emergence of mTBI in Operation Iraqi Freedom (OIF) and Operation Enduring Freedom (OEF), attention was redirected to understanding effects of blast on the brain. Due to the lack of better technology and understanding, conventional constant diameter shock tubes were used to study brain injury from primary blast in rodent models across various research laboratories, including the authors' [13, 24–27]. However, constant diameter shock tubes characteristically impart high-dynamic forces compared to free-field blast, yielding artifactual flow conditions quite different from those encountered by warfighters. Other limitations with constant diameter blast simulators include longer impulse durations (>8 ms positive pressure duration), plateaued peak pressures, and the absence of a negative phase of the shock wave (Fig. 2).

In many cases, laboratory studies have been conducted using end muzzle exposures outside of the shock tube, which expose experimental subjects to greatly exaggerated end-jet effects that bear little resemblance to the static and dynamic forces associated with free-field blast [28–30]. Upon emergence from the tube, shock waves suddenly expand, causing formation of rarefaction waves associated with complex and artifactual high-flow gradients such that slight changes in position impart large changes in pressure conditions and unrealistic pressure profiles [16] (Fig. 3). In addition, a number of researchers have used cylindrical blast

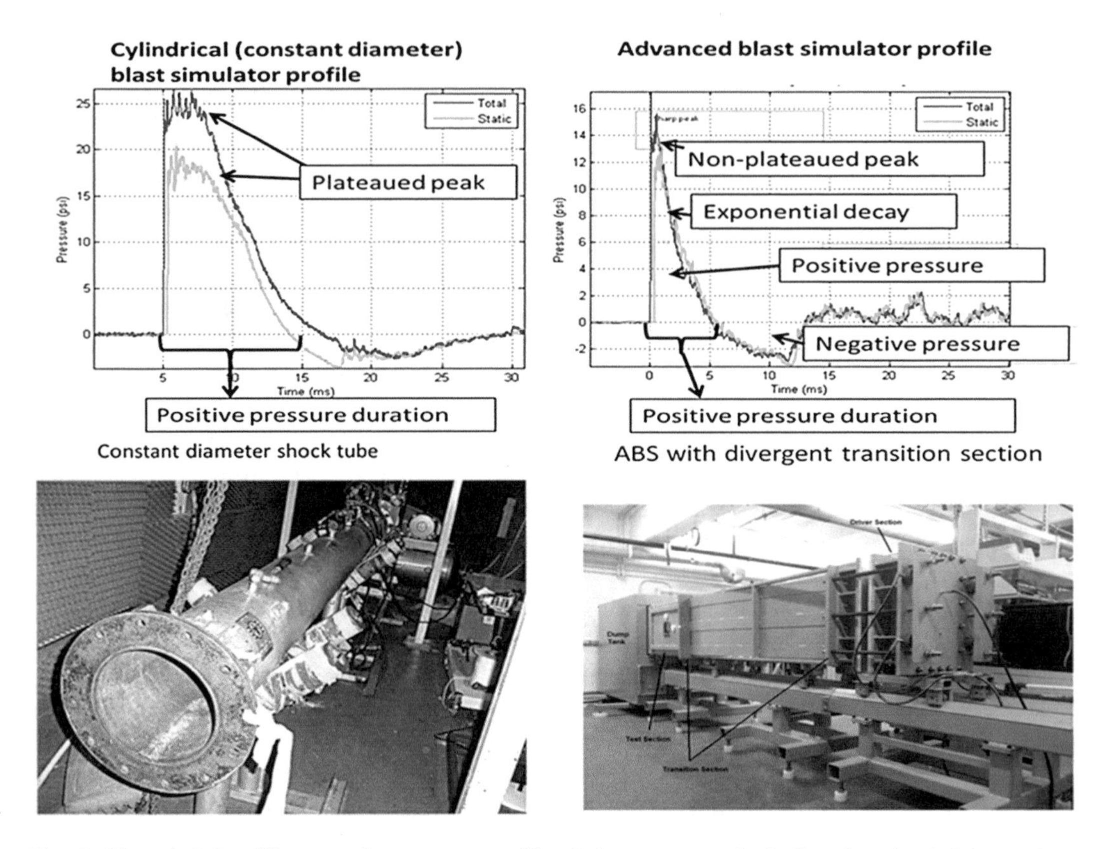

Fig. 2 Characteristic differences in pressure profiles between a constant diameter shock tube and an Advanced Blast Simulator (ABS): a plateaued peak, absence of fidelic negative phase, and longer positive pressure duration is observed in constant diameter shock tube pressure profile vs. a free-field-like characteristic profile produced in the ABS

simulators with very small diameters (<6 in.) resulting in constriction of the blast waves, leading to artifactual flow conditions and reflections. In other cases, laser-induced shock waves have been applied; however, these shockwaves have impulse durations of a few microseconds and are categorically very different from shock/blast waves that have impulses lasting in a milliseconds timescale [31]. These laser models have been shown to produce focal injuries but do so with loading conditions that are likely to be invalid. Proponents point to similar impulse (area under the positive pressure curve) as blast waves; however, the increased tissue loading characteristics of laser-induced shock waves have not been validated. In general, experimental artifacts associated with these invalidated approaches very likely will cause irrelevant and overwhelming brain injuries in rodent models that are not representative of primary blast injury. These experimental artifacts occur predominantly in models where the animals: (1) are exposed to blast at or outside the mouth of blast simulators, (2) create high obstruction of occlusion of the blast simulator, (3) are improperly

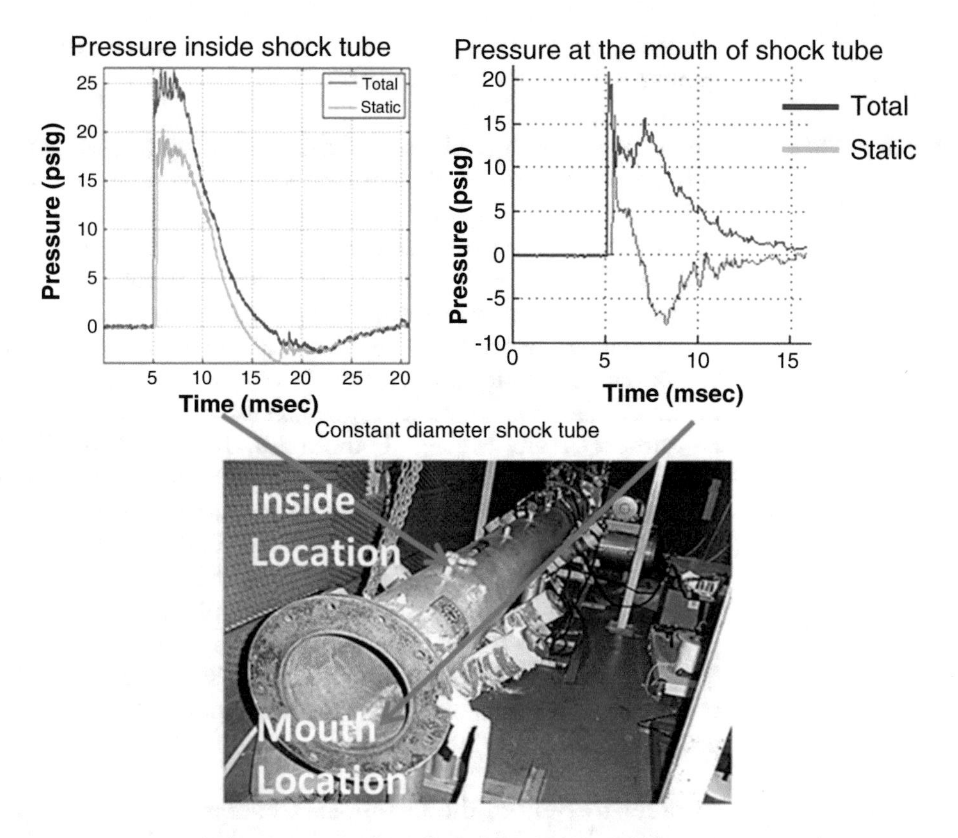

Fig. 3 Pressure inside the shock tube of the constant diameter has a different pressure profile when compared to the pressure profiles at the mouth of the shock tube that leads to the end jet effects. In this case, the static pressure profile has a shorter positive pressure duration coupled with an abrupt decay and exaggerated negative phase. Commonly mistaken for a Friedlander-like profile, when this pressure trace is shown as the sole representation of the blast, the true nature of the exposure is not embodied. When taken together with the total pressure (blue trace), a characteristic increased dynamic pressure is observed which imparts high accelerative forces on the rodents body (whipping motion when restrained) leading to spurious injury outcomes

restrained in cylindrical blast simulators or (4) are subjected to invalid injury exposures. Recently developed divergent cross-section blast simulators present a state–of-the-art technology to simulate free-field-like primary blast to study bTBI and have been employed at multiple research centers [32–34]. The advanced blast simulator (ABS™) has a divergent transition cross-section that generates blast waves closely mimicking free-field blast (Friedlander wave-form) that is described in Sect. 3 below.

2 Materials

Rats and/or mice are the rodent model of choice for primary bTBI research; however, some groups have used non-rodent models such as swine and ferrets. The materials discussed here are applied primarily to rodent models, although not exclusively. Similarly, the following materials and methods described pertain to ABS or divergent cross-section blast simulators.

2.1 Primary bTBI

2.1.1 Membranes

There are various membranes of choice that can be readily used to generate a range of blast intensities. Acetate membranes are widely used to generate shockwaves across various laboratories. Acetate membranes can be obtained in various thicknesses from Grafix Plastics Inc. (Maple Heights, OH) which is useful for producing varied overpressure exposures. A drawback is that shards of acetate membrane are generated after blast and can act as projectiles that can potentially injury the animals. In our laboratory, the currently preferred choice of material for higher blast intensities (>13 psi) is Valmex® from Mehler Texnologies, Inc. (Martinsville, VA); acetate membranes are used to generate lower intensity exposures (≤13 psi)[32]. Aluminum membranes have been used; however, due to the labor-intensive machining required to score these membranes for consistent bursting, they are not currently in favor [33]. Though widely used, Mylar membrane material is compromised by the manner in which it ruptures. Mylar tends to "petal" outward when it ruptures which disrupts the flow. A long transition section is required for the shock wave to achieve proper planarity.

2.1.2 Compressed Gas

In order to rupture the membranes, compressed air or helium is typically delivered from compressed gas cylinders [32–35]. In some cases, other research laboratories have used air compressors to rupture the membranes of choice as discussed in Sect. 2.1.1 [13]. Helium will produce sharper rise times, higher peak pressures, and shorter positive phase durations compared to air for the same membranes, but is significantly more expensive.

2.1.3 Instrumentation

It is ideal to gather the critical static (Ps) and dynamic (Pd) pressures (biomechanical loading) of the blast wave on the experimental subject, which are fully recorded by the combination of static (aka incident or side-on pressure) and total (aka stagnation or head-on) pressure gauges. *In case both the pressures cannot be obtained for each blast due to size limitations of the blast simulator, it is essential to characterize the blast simulator with two sensor gauges to obtain total and static pressures (Sect. 4.1)*. Once calibration profiles of animals and holders have been completed (Sect. 4.1), presenting only static (incident) pressure is acceptable to define loading conditions on animals.

The pressure sensor data from these gauges should be acquired at a minimum sampling rate of 250,000 Hz, although higher sampling rates are recommended to achieve more accurate pressure profiles, especially for static and reflective pressures. High-frequency pressure sensors from Endevco (Irvine, CA) or PCB

Piezotronics (Depew, NY) [16] are widely used to acquire blast exposure data. A variety of commercial data acquisition systems are currently available, to name a few: TDAS PRO (Diversified Technical Systems, Inc., Seal Beach, CA), LabVIEW (National instruments, Austin, TX, USA), TMX-18 data acquisition system (AstroNova, Inc., West Warwick, RI). It is important to not apply hardware or software filter settings while collecting data in order to capture fast rise times, peak overpressure, rapid decay and reflections of the blast. There are several probes available to study intracranial and hemodynamic pressures: authors prefer the use of Mikro-Tip® pressure catheters from Millar Inc. (Houston, TX, USA) due to the size and geometry for small animal testing. An alternative choice is fiber optics probes from FISO Inc. (Quebec, QC, Canada). Always check the response frequency for any gauges used in blast work. In addition, care must be exercised when using company-provided amplifiers or conditioners employing filtering, which will compromise the signal response (e.g. Millar). Calibrating these types of sensors using relevant static pressures is critical to maintain data integrity.

2.1.4 High-Speed Video Camera

A variety high-speed cameras are commercially available which can record at >20,000 frames per second (fps). Video records help to characterize forces during the exposure and understand any artifacts associated from blast wind. A frame-wise recording of blast exposure is demonstrated in Fig. 4.

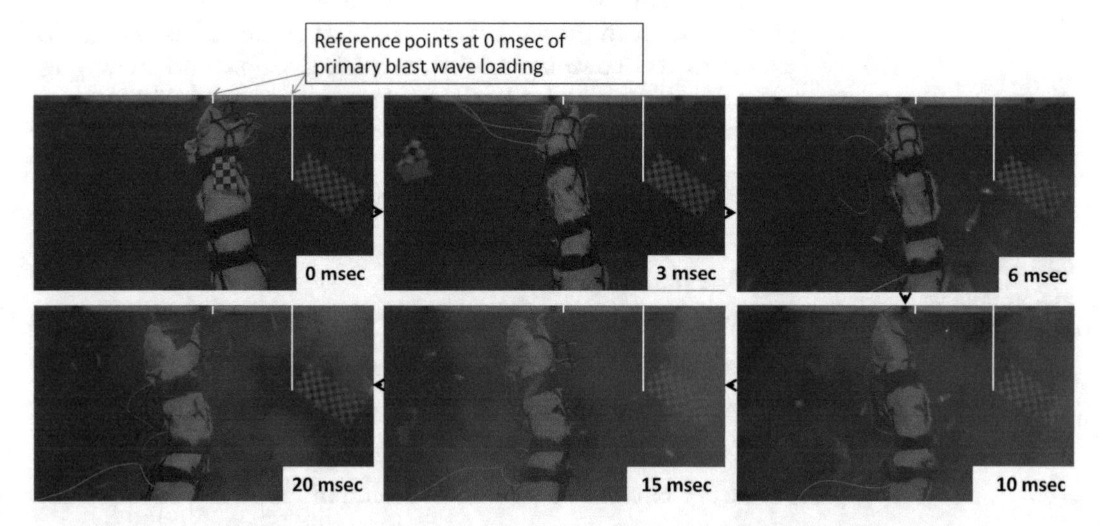

Fig. 4 Frame-wise recording of untethered animal motion (0–20 ms) following primary blast loading (see Sect. 4.1.2). The animal is in parasagittal plane which presents the maximum surface area to the primary blast. As shown, minimal gross motion is associated with primary blast wave with 15 psi static pressure (speed of shockwave: 1.37 mach)

3 Methods

3.1 Animal Preparation

3.1.1 Approvals

All animal experiments should be conducted in accordance with the Animal Welfare Act and other federal statutes and regulations relating to animals and experiments involving animals, and should adhere to principles stated in the Guide for the Care and Use of Laboratory Animals (NRC Publication 2011 edition) using an Institutional Animal Care and Use Committee-approved protocol. Male Sprague Dawley rats, 8–9 weeks old that weighed 270–290 g (Charles River Laboratories, Wilmington, MA) were housed at 20–22 °C (12 h light/dark cycle) with free access to food and water ad libitum.

3.1.2 Blast Exposure

In our laboratory, rats are anesthetized with 4% isoflurane in 2 ft^3/min of air for 6 min. Other injectable anesthetics can be used, but researchers are cautioned to understand the neuroprotective effects of each anesthetic. Animals are then restrained in a custom made sling at 9.5 ft from the membrane in a custom-built Advanced Blast Simulator (ORA Inc., Marion, NC). The ABS is 21.5 ft long and comprises a 2 ft long compression chamber, a 9.5 ft divergent transition section, and a 4 ft long test section coupled to an end wave eliminator (6 ft long) (Fig. 5). Compressed air is used to rupture the acetate (<13 psi blast) or Valmex® (>13 psi) membranes. Piezoresistive gauges (Endevco, Irvine, CA) incorporated into a low profile aluminum holder are used to record the static (i.e. side-on pressure) and total pressure to which each animal is exposed using Astro-med TMX-18 acquisition system at 800,000 Hz sampling rate. The mach speed of the shock wave (calculated using the distance between the gauges) is then used in the Rakine-Hugoniot equations to determine reflective, total, dynamic and static pressures. The high-speed video (Fig. 4) is recorded at 25,000 fps using a Phantom v1212 camera (Vision Research Inc., Wayne, NJ).

3.2 Guidelines to Minimum Common Data Elements to Disseminate Methodology of Blast Research

1. Dimensions of blast tube or explosive type.
2. Location of the animal.
3. Orientation of the animal.
4. Weight of the animal.
5. Sampling rate of blast pressure profile recording.
6. Sensor locations of blast profile recording.
7. Pressure profiles of blast.
8. Representative total, static, and impulse of the pressure.
9. Positive and negative pressure duration and impulse.
10. Mach number and rise time of the blast wave.
11. Instrumentation used for pressure data collection.

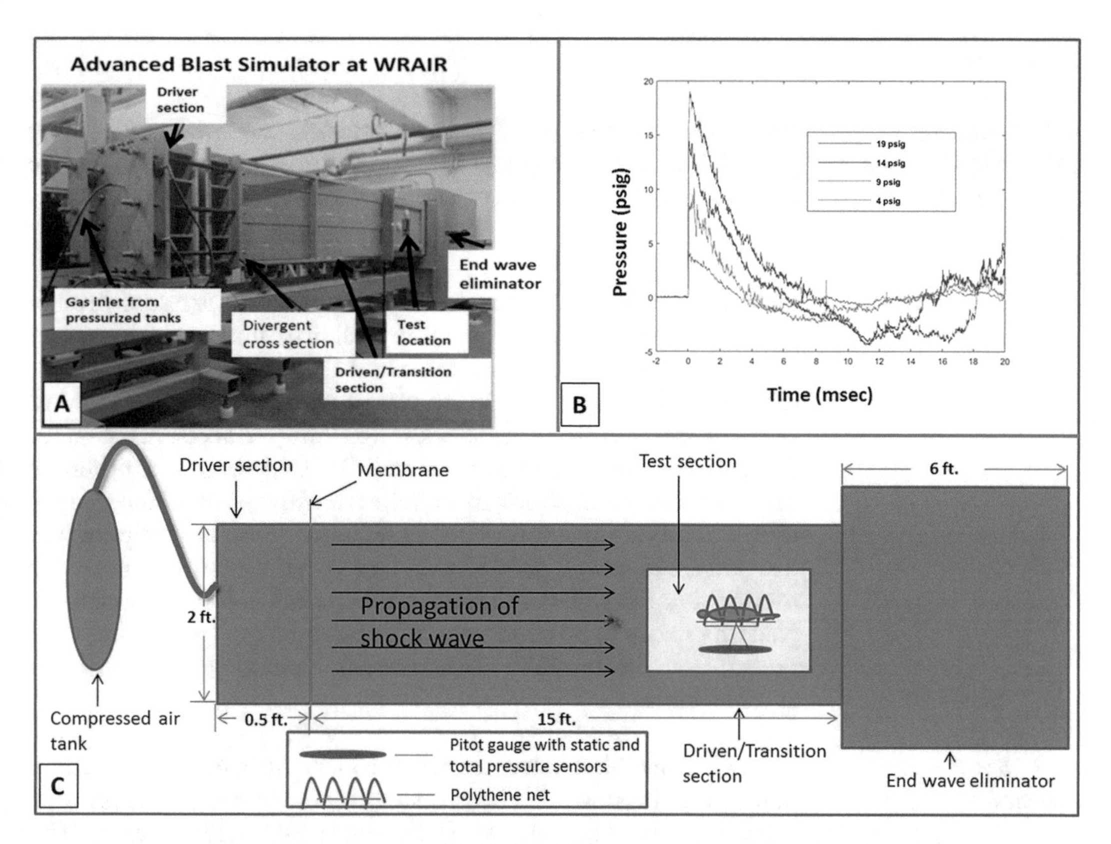

Fig. 5 (**a**) Advanced Blast Simulator (ABS) located at Walter Army Institute of Research. (**b**) A range of static pressures generated using the ABS (only static pressures are shown for clarity in illustration). (**c**) Side of view of the ABS schematic showing the propagation of wave, the rodent subject exposed to blast in frontal orientation and an end wave eliminator that eliminates rarefaction waves from propagating back to the subject

(a) Sensor type.
(b) Data capturing program (e.g., TMX-18 from Astroview Inc).
(c) Sampling rate.
(d) Filtering of data (not recommended).

12. Righting time reflex (as applicable).

In order to be qualified for a well-calibrated animal model in addition to disseminating above data elements in manuscripts, it is essential to follow Sect. 3. This adherence would not only standardize primary blast injury research, but also help to compare injury outcomes across the laboratories, which is not achievable in the current way of research dissemination.

4 Notes

4.1 Guidelines to Characterize Blast Simulators for Animal Research

Proper characterization of the injury model is essential to effectively study bTBI. Prior to animal experimentation, steps 1–3 should be followed to validate the blast simulator or shock tube.

4.1.1 Step 1: Pressure Characterization

Characterize the blast simulator by mapping the pressure profiles across the animal test section using blast-quality pressure gauges that can specifically measure static and total pressure loading conditions. Two gauges should be placed a known distance apart to calculate the velocity of the shock wave front (see Sects. 2.1.2–2.1.4 for instrumentation and parameters). These pressures can then be validated using the shock front velocity with the Rankine–Hugoniot equations for appropriate gaseous conditions (temperature, humidity, and driver gas) [32]. In cases where pressure sensors are mounted in the walls of the shock tube, a validation for blast wave planarity needs to be conducted with pitot gauges (measuring total pressure) at the location of the animals (including any platforms and restraints).

4.1.2 Step 2: Acceleration

True IED-like blast waves actually produce little dynamic pressure that creates blast wind. The gross motion imparted by the primary blast wave is minimal, especially on streamlined rodents. The momentum and acceleration of the rodents needs to be characterized using a high-speed camera capable of at least 20,000 frames per second. Tracking the linear and rotational acceleration of the animal's head along with pressure gauges can help characterize the loading conditions that are relevant to blast exposure (see work from Sawyer TW, 2016) [35]. The acceleration should be characterized with video whenever the position of the animal is changed with respect to blast exposure (frontal, transverse, or parasagittal plane). Exaggerated blast wind and dynamic forces impart higher acceleration that is not a characteristic of pure blast waves. These aberrant conditions are usually seen in cases where the animals are blasted: at the mouth of the driven section, too close to the membrane, in tubes with long duration positive phase (>7–8 ms), in constant diameter shock tubes with plateaued pressure traces, while obstructing a significant portion of the shock tube, too close to the walls or poorly retrained. Animals should not be placed on the walls of shock tubes due to the drag and planarity disruption caused by shock tube wall boundary effects.

4.1.3 Step 3: Intracranial Pressure (ICP)

Several laboratories have independently validated pressure loading of brain (ICP) from primary blast (not to be confused with the much slower intracranial pressure changes associated with brain swelling/edema) on rodents and observed that blast loading on brain (ICP) is a function of static pressure (see Sect. 2.1.3 for instrumentation) [11, 36, 37]. It is important to understand the loading conditions to relate the pressure to biological changes in brain and develop standardization across the research communities. One of the main drawbacks in the bTBI research field is the lack of standardization and guidelines to include minimal data points required in the methods of blast exposure. The guidelines for methodology dissemination in manuscripts are discussed in Sect. 3.2.

Note: Characteristically, these recommendations should be applied to auditory and vision research associated with primary blast exposure.

4.2 Pathology Associated with Primary Blast Injury

Information on clinical pathology associated with bTBI with no other comorbidities is rare. Recent studies indicate that chronic traumatic encephalopathy (CTE) and tau protein-linked neurodegenerative disorder in autopsy specimens from deceased athletes may extend to victims of blast exposure [38, 39]. Specifically, phosphorylated Tau (pTau) protein neuropathology with perivascular neurofibrillary degeneration which is recognized as a distinct feature of CTE has been observed postmortem in the brains of blast exposed victims [38, 39]. Accordingly, preclinical studies also showed tauopathy in the brain after blast exposure [40, 41]. A recent clinical report has demonstrated that unique astrogliosis patterns (astroglial scarring) were observed in surface and layer 1 of the cortex, while traditional impact TBI did not demonstrate similar pathology [42]. Several magnetic resonance imaging studies have demonstrated that pathology associated with bTBI is unique and may not have similar mechanisms seen with impact TBI [43, 44]. In accordance with clinical reports, several preclinical studies have demonstrated astrogliosis as one of the common pathological outcomes following bTBI across various research laboratories following bTBI [35, 41, 45] (Fig. 6). The majority of bTBI research is conducted at pressures (>14 psi) which also results in damage to lung [46]. Oxidative stress [47] and neuroinflammation [41] have been widely demonstrated after blast in rodent models using divergent cross-section blast simulators.

4.3 Conclusions

Characterizing the blast injury model is absolutely essential before the research studies are conducted in brain injury. Determining minimum data points to report along with research findings will provide a basis for inter-laboratory comparison. We have presented a list of data used to characterize blast systems as well as **the best practices to conduct blast TBI research** which have been developed in conjunction with blast physicists to provide a basis for collaboration and validation.

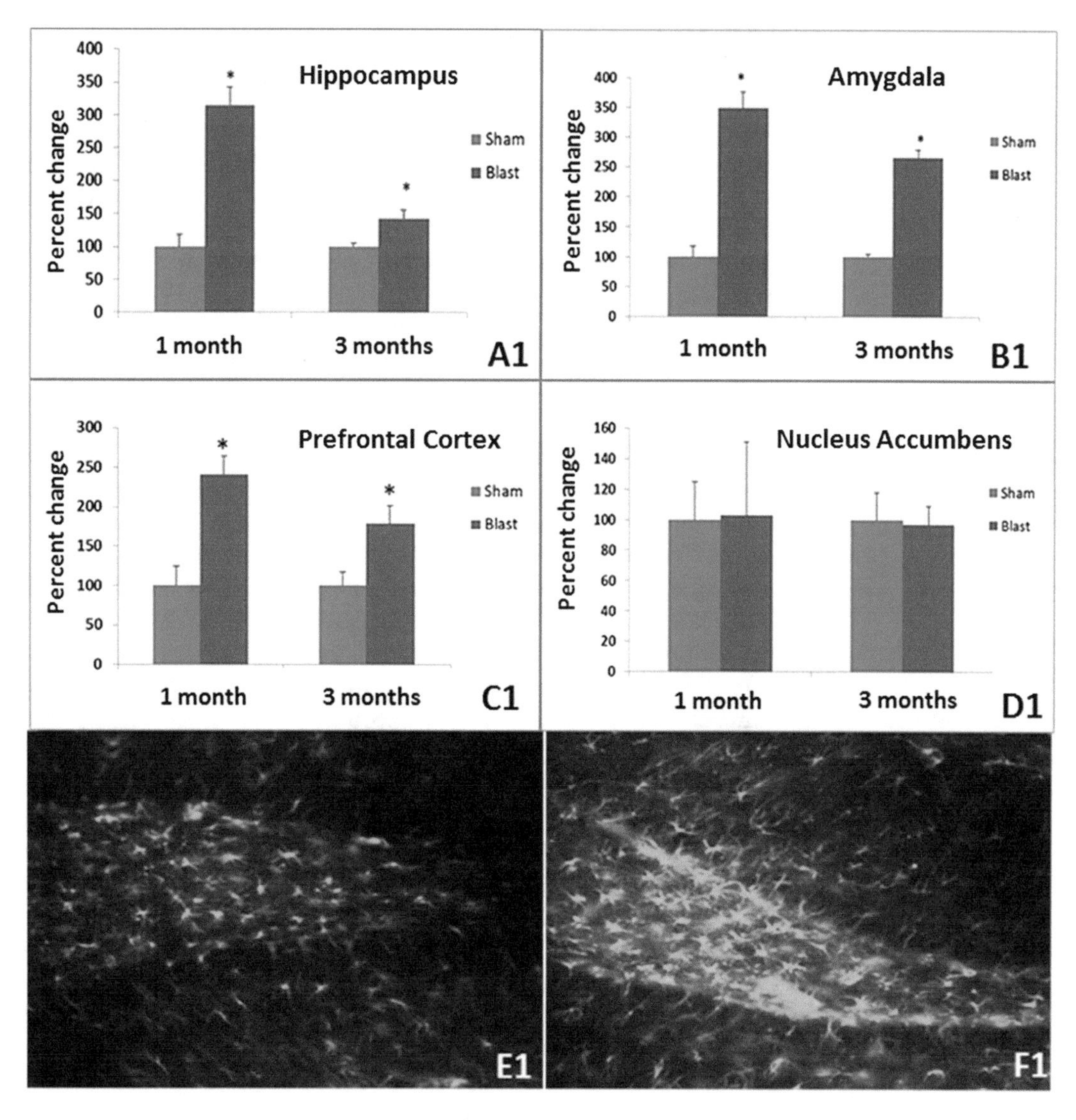

Fig. 6 Activation of astroglia at 1 and 3 months following single blast exposure in hippocampus (**a1**), amygdala (**b1**), prefrontal cortex (**c1**), and nucleus accumbens (**d1**). Representative images dentate gyrus in sham (**e1**) and blast exposure animal at 17 psi (**f1**). Adopted from Sajja et al., Sci Rep. 2016 [40]

Disclaimer

The contents, opinions, and assertions contained herein are the private views of the authors and are not to be construed as official or reflecting the views of the Department of the Army or the Department of Defense.

References

1. Adams RW (1951) Small caliber missile blast wounds of the hand; mechanism and early management. Am J Surg 82:219–226
2. Beal SL, Blaisdell FW (1989) Traumatic hemipelvectomy: a catastrophic injury. J Trauma 29:1346–1351
3. Connolly M, Ibrahim ZR, Johnson ON 3rd (2016) Changing paradigms in lower extremity reconstruction in war-related injuries. Mil Med Res 3:9
4. Morrissey EJ (1944) Head and blast injuries. Cal West Med 61:196–199
5. Chapman JC, Diaz-Arrastia R (2014) Military traumatic brain injury: a review. Alzheimers Dement 10:S97–S104
6. DoD Worldwide Numbers for TBI (2017) DVBIC statistics
7. Bhattacharjee Y (2008) Shell shock revisited: solving the puzzle of blast trauma. Science 319:406–408
8. Hoge CW et al (2008) Mild traumatic brain injury in U.S. soldiers returning from Iraq. JAMA 358:453–463
9. Center for Disease Control (2013) Explosions and blast injuries: a primer for clinicians
10. Carr W et al (2016) Repeated low-level blast exposure: a descriptive human subjects study. Mil Med 181:28S–39S
11. Dal Cengio Leonardi A et al (2012) Head orientation affects the intracranial pressure response resulting from shock wave loading in the rat. J Biomech 45:2595–2602
12. Moss WC, King MJ, Blackman EG (2009) Skull flexure from blast waves: a mechanism for brain injury with implications for helmet design. Phys Rev Lett 103:108702
13. Sajja VS et al (2018) Neurolipids and microRNA changes in blood following blast traumatic brain injury in mice: an exploratory study. J Neurotrauma 35:353–361
14. Ellenbroek B, Youn J (2016) Rodent models in neuroscience research: is it a rat race? Dis Model Mech 9:1079–1087
15. McGraw CM, Ward CS, Samaco RC (2017) Genetic rodent models of brain disorders: perspectives on experimental approaches and therapeutic strategies. Am J Med Genet C Semin Med Genet 175:368–379
16. Dewey JM (2016) Measurement of the physical properties of blast waves in experimental methods of shock wave research. Springer International Publishing, Switzerland, pp 53–85
17. Needham CE et al (2015) Blast testing issues and TBI: experimental models that lead to wrong conclusions. Front Neurol 6:72
18. Kaur C et al (1995) The response of neurons and microglia to blast injury in the rat brain. Neuropathol Appl Neurobiol 21:369–377
19. Woods AS et al (2013) Gangliosides and ceramides change in a mouse model of blast induced traumatic brain injury. ACS Chem Neurosci 4:594–600
20. Säljö A et al (2000) Blast exposure causes redistribution of phosphorylated neurofilament subunits in neurons of the adult rat brain. J Neurotrauma 17:719–726
21. Igra O (2016) Shock tubes in experimental methods of shock wave research. Springer International Publishing, Switzerland, pp 3–52
22. Fearnley GR (1945) Blast injury to the lungs. Br Med J 1:474–477
23. Henry GA (1945) Blast injuries of the ear. Laryngoscope 55:663–672
24. Elsayed NM (1997) Toxicology of blast overpressure. Toxicology 121:1–15
25. Cernak I et al (2001) Ultrastructural and functional characteristics of blast injury-induced neurotrauma. J Trauma 50:695–706
26. Rodriguez O et al (2016) Manganese-enhanced magnetic resonance imaging as a diagnostic and dispositional tool after mild-moderate blast traumatic brain injury. J Neurotrauma 33:662–671
27. Haghighi F et al (2015) Neuronal DNA methylation profiling of blast-related traumatic brain injury. J Neurotrauma 32:1200–1209
28. Bricker-Anthony C, Rex TS (2015) Neurodegeneration and vision loss after mild blunt trauma in the C57Bl/6 and DBA/2J mouse. PLoS One 10:e0131921
29. Kabu S et al (2015) Blast-associated shock waves result in increased brain vascular leakage and elevated ROS levels in a rat model of traumatic brain injury. PLoS One 10:e0127971
30. Toklu HZ et al (2015) The functional and structural changes in the basilar artery due to overpressure blast injury. J Cereb Blood Flow Metab 35:1950–1956
31. Miyazaki H et al (2015) Thoracic shock wave injury causes behavioral abnormalities in mice. Acta Neurochir 157:2111–21120
32. Ritzel DV et al (2018) Acceleration from short duration blast. Shock Waves 28:101–114
33. Alphonse VD et al (2014) Membrane characteristics for biological blast overpressure testing

using blast simulators. Biomed Sci Instrum 50:248–253

34. Sawyer TW et al (2016) High-fidelity simulation of primary blast: direct effects on the head. J Neurotrauma 33:1181–1193
35. Skotak M et al ((2013)) Rat injury model under controlled field-relevant primary blast conditions: acute response to a wide range of peak overpressures. J Neurotrauma 3:1147–1160
36. Chavko M et al (2011) Relationship between orientation to a blast and pressure wave propagation inside the rat brain. J Neurosci Methods 195:61–66
37. Meabon JS et al (2016) Repetitive blast exposure in mice and combat veterans causes persistent cerebellar dysfunction. Sci Transl Med 8:321ra6
38. McKee AC, Robinson ME (2014) Military-related traumatic brain injury and neurodegeneration. Alzheimers Dement 10:S242–S253
39. Arun P et al (2015) Acute decrease in alkaline phosphatase after brain injury: a potential mechanism for tauopathy. Neurosci Lett 609:152–158
40. Sajja VS et al (2015) Enduring deficits in memory and neuronal pathology after blast-induced traumatic brain injury. Sci Rep 5:15075
41. Shively SB et al (2016) Characterisation of interface astroglial scarring in the human brain after blast exposure: a post-mortem case series. Lancet Neurol 15:944–953
42. Hayes JP et al (2015) The nature of white matter abnormalities in blast-related mild traumatic brain injury. Neuroimage Clin 8:148–156
43. Sorg SF et al (2014) White matter integrity in veterans with mild traumatic brain injury: associations with executive function and loss of consciousness. J Head Trauma Rehabil 29:21–32
44. Hubbard WB et al (2017) Distinguishing the unique Neuropathological profile of blast polytrauma. Oxidative Med Cell Longev 2017:5175249
45. Abdul-Muneer PM, Chandra N, Haorah J (2015) Interactions of oxidative stress and neurovascular inflammation in the pathogenesis of traumatic brain injury. Mol Neurobiol 51:966–979
46. VS Sajja, et al., (2017) Pulmonary injuries and systemic disruptions in rats in response to repeated daily exposures to blast overpressure. Military Health System Research Symposium (MHSRS). Kissimmee FL, 27–30 Aug
47. Chandra N, Sundaramurthy A, Gupta RK (2017) Validation of laboratory animal and surrogate human models in primary blast injury studies. Mil Med 182:S105–S113

Chapter 9

Cognitive and Motor Function Assessments in Rodent Models of Traumatic Brain Injury

Danielle Scott and Kathryn E. Saatman

Abstract

Cognitive and motor dysfunction is common in people who have experienced a traumatic brain injury (TBI). These deficits can include memory loss, learning impairment, dizziness, difficulty with balance, and loss of fine motor control and coordination. Cognitive function and vestibulomotor tasks have been widely used in clinically relevant rodent models of experimental TBI to study the relationship of neurobehavioral dysfunction to injury severity, secondary injury mechanisms, or putative therapeutic interventions. Here we describe paradigms for the novel object recognition task, a test of memory, and beam walking and rotarod tasks, tests of coordinated motor function. Key advantages and disadvantages are presented, and potential problems and adaptations of these behavioral tests are discussed.

Key words Traumatic brain injury, Novel object recognition, Beam walking, Rotarod

1 Introduction

Traumatic brain injury (TBI) is the leading cause of death and disability for people under the age of 45 years [1]. It is estimated that about 3.8 million people sustain a TBI every year in the United States alone [1]. Symptoms of TBI include nausea, dizziness, headaches, memory loss, motor coordination and balance deficits, vision impairment, anxiety, irritability, and depression, leading to a severely decreased quality of life [2]. The type and extent of pathophysiology resulting from a TBI depends on the magnitude of acceleration, deceleration, and impact forces associated with the traumatic insult. Aspects of TBI-induced cellular damage include cell death, gliosis, axonal injury, membrane depolarization, disruption of vasculature, and ischemia [3]. Collectively, cell damage and dysfunction across multiple brain regions manifest as deficits in neurobehavior. Neurologic function assessments in animal models of TBI provide researchers a way to evaluate the health status of subjects, determine the initial effect of TBI on aspects of motor and cognitive function, and monitor functional recovery after TBI

Amit K. Srivastava and Charles S. Cox, Jr. (eds.), *Pre-Clinical and Clinical Methods in Brain Trauma Research*, Neuromethods, vol. 139, https://doi.org/10.1007/978-1-4939-8564-7_9,

[4]. The combination of an assessment of learning or memory, using the novel object recognition test or Morris water maze, with a test for motor coordination and balance, such as the beam walking task or the rotarod task, has been used extensively in rodent models of TBI [5, 6]. This chapter offers detailed protocols for the novel object recognition, beam walking, and rotarod tasks while highlighting their advantages, modifications, and the common issues that can arise when utilizing these neurological tests.

2 Novel Object Recognition

The novel object recognition (NOR) test was first described for rats by Ennaceur and Delacour in 1988 [7]. It has now become a widely used model for the evaluation of recognition memory in many animal models of human disease and injury where cognition is impaired. This task exploits the innate and spontaneous tendency of rodents to explore novel items, contrasting the amount of time that rodents spend exploring familiar and novel objects [8, 9]. Animals with TBI exhibit impaired recognition memory. Failure to recognize a familiar object results in a decreased propensity for exploring a novel object over the familiar one.

2.1 Animals and Injury

The novel object task has been evaluated in both rats [7, 10, 11] and mice [12–14]. NOR has been applied across multiple TBI models including fluid percussion [15], controlled cortical impact (CCI) [16, 17], closed head injury [2], weight drop [12], and blast [13] at mild, moderate, and severe levels. The NOR protocol for mice presented in this chapter is based on a paradigm implemented by Tsenter and colleagues [18] and modified by Schoch and associates [19].

2.2 Materials

1. Sheet to record notes and times.
2. Two stopwatches.
3. One timer.
4. One 10.5″ × 19″ × 8″ Plexiglas rat cage or box per mouse.
5. Two unique pairs of identical objects (e.g. mugs, animal figurines, light bulbs).

2.3 Working Protocol

2.3.1 Habituation/Acclimation

1. One day before surgery/injury/treatment, place each mouse into an empty cage for 1 h.
2. Each mouse should be acclimated to a cage dedicated to that mouse for the duration of the experiment (all trials/time points). During acclimation, all mice can be habituated to their cages simultaneously, as long as each cage is placed in

the same spot as that used for testing and mice cannot see each other during evaluation.

3. Take note of the surroundings of each cage and make sure any visual cues remain the same and in the same orientation throughout evaluation.
4. Return mice to home cages.
5. Each cage can be wiped out with a fresh, dry paper towel to remove waste, but do not clean the cage with cleaner throughout the duration of testing.

2.3.2 Testing

1. Testing consists of one 5 min "pretrial" with two identical objects followed 4 h later by a 5 min "NOR" trial with one of the previously used objects and one novel object.
2. Make sure the cage surroundings are the same as they were during acclimation. Perform each test trial on only one mouse at a time.
3. Place the two identical objects in opposite corners of the cage (cattycorner to each other), about 2 in. from the walls of the cage to allow space for the mouse to walk around the objects.
4. Place the test mouse in the center of the cage, equidistant from the two objects, and start the timer to count up to 5 min.
5. Use one stopwatch to time the mouse's exploration of object "A" and a second to time the exploration of object "B." A mouse is considered to be exploring if its nose is positioned toward the object at a distance of 2 cm or less. Sitting on top of the object or using the object to get to a higher position does not count as exploring.
6. If total exploration time after 5 min is less than 10 s, continue the trial until the subject has explored for at least 10 s. Record the duration of additional exploration trial time.
7. Return the mouse to its home cage.
8. Record exploration times.
9. Wipe out cage with a dry paper towel only. Wipe down each object with a disinfectant cleaner between each trial/mouse and dry thoroughly.
10. Four hours later, place one of the initial objects (familiar object) in one corner as in step 3, with a second, novel object in the opposite corner.
11. Follow steps 4–9.

2.4 Notes

2.4.1 Advantages

The NOR task offers many advantages over other memory tests for assessing recognition memory. It is quick to administer and does not require any subject training other than arena habituation [20]. It relies primarily on a rodent's natural preference for novelty, and thus incorporates no stressful conditions [8]. NOR also requires no external motivation, reward or punishment, and can be performed easily and often due to its low cost [20]. This test is easily modified, and can be configured to measure working memory, long-term memory, attention, anxiety, preference for novelty, or therapeutic efficacy [9]. NOR is also an excellent test to evaluate improvement of cognitive function longitudinally with repeated testing over several timepoints. As long as distinct novel objects are used at each timepoint, the innate tendency to explore novelty will not be affected by the previous trials [2, 9].

2.4.2 Disadvantages

Although the NOR task has been widely utilized within neurobiological studies of memory, variations across NOR protocols, such as those described in Sect. 2.4.4, complicate comparisons across studies of memory [20]. A meta-analysis of a large subset of published reports inferring the role of the rodent hippocampus in object memory using the NOR task revealed many differences in the characteristics of objects, arena type, duration of inter-session delay, trial duration, and minimum exploration requirement [20]. Because the NOR test is quite adaptable and easy to set up without a machine-made apparatus, comparison of results across studies should be made with caution, appreciating the potential for differences in implementation of the test.

The NOR task has a limited range of potential scores, with naïve mice achieving a 70–75% recognition index (RI) and a complete loss of recognition memory yielding a 50% RI (see Sect. 2.4.4). Since in many models of TBI injury results in a RI of 50%, it can be difficult to detect worsening of recognition memory due to genetic alterations or pharmacological interventions using NOR.

2.4.3 Potential Problems

Many protocols in the literature fail to mention troubleshooting or characterization of the NOR task prior to final testing. It is very important to establish that all objects to be used in a study are equally interesting on initial exposure. This is done by testing objects in all combinations of pairs on naïve (control) mice. Any object with systematically higher or lower exploration times than the others should be eliminated to prevent a confound in interpreting responses after injury [20]. Additionally, animals that exhibit low total exploration times, such as only a few seconds, may receive abnormally high or low recognition indices due to inadequate sampling. If this problem arises, it is acceptable to continue the trial until the subject has explored for a set minimum exploration time. Here we suggest 10 s, but this cutoff differs across protocols and species [2, 11, 20]. The use of automated tracking during video recorded NOR trials (see Sect. 2.4.4) requires vigilance to avoid introduction of artifacts. Unless the camera is positioned in such a way that it can see all sides of both objects, it can be difficult to track exploration of an animal when it goes behind, over, or under an object. In this case, it may be preferential to do manual scoring or to choose objects that are small enough to see around but difficult for the animal to climb on or underneath.

2.4.4 Adaptations

The NOR task has numerous adaptable aspects that allow it to be modified to fit a variety of studies. For example, the first phase of the typical NOR protocol—habituation—can be performed prior to injury [2], following injury [12], or on repeated days [16]. Habituation before testing eliminates possible confounds such as motor deficits or anxiety deficits that could affect object exploration when the animals are not accustomed to the testing arena. Although some studies do not perform a separate arena habituation stage at all [17], a 5 min familiarization trial with the arena containing the objects may not be adequate to separate environment and object exploration behaviors. Longer habituation intervals or repeated habituation trials, such as described by Cai and colleagues who employed 4 consecutive days of habituation in an empty cage followed by 2 consecutive days in the arena with two identical objects [21], can ensure more complete environmental habituation, but increase the time and labor involved in conducting the entire evaluation.

The relative difficulty of, and type of memory studied by, the NOR test can be adjusted by increasing or decreasing the delay between the familiarization and test phases [20]. A shorter inter-session delay (i.e. 1 h) can be used to measure short-term memory, and can also adjust the sensitivity of the test by making it easier [14]. A longer intersession delay (i.e. 24 h) can be used to measure long-term memory and can increase the sensitivity of the test by making it more difficult [15]. For example, Prins et al. showed that after a 1 h inter-trial delay, rats with sham injury, single mild TBI, and repeated mild TBI could all differentiate between familiar and novel objects, but after a 24 h inter-trial delay, the repeatedly injured group could not [22].

Not only can the phases of the test be modified, but so too can the reported measures. There are two common measures used to assess exploration performance. The recognition index is determined by dividing the exploration time of the novel object by the total object exploration time. A value above 0.5 is indicative of a preference for the novel object, whereas a value below 0.5 suggests preference for the familiar object. A recognition index of 0.5 represents chance performance. Madathil et al. showed that animals with TBI exhibit on average a recognition index around 0.5, whereas control mice (C57BL/6) had mean indices of about 0.75 [3]. Alternatively, the discrimination ratio is calculated by the difference in exploration time between the novel and familiar objects divided by the total object exploration time. This ratio ranges from −1 to +1, with negative scores indicating familiarity preference and positive scores signifying novelty preference [20]. However, Pérez-García and colleagues chose to report raw exploration times in seconds [23], and Lee et al. measured number of explorations rather than exploration times, reporting the results as a ratio of the novel object explorations over the familiar object explorations [24].

The NOR task can be manually assessed or automated, allowing for data to be collected more quickly. Tracking software can be used with video recording, permitting multiple animals to be tested at once. This also allows for verification of results by manual timing using the videos at any time after testing, as well as later qualitative assessment of behavior [14]. Silvers et al. took automation of NOR even further, using an infrared photocell grid to measure total locomotor activity, number of explorations, and object exploration times [10].

3 Beam Walking Test

The beam walking task was the first motor test utilized in rodent models of TBI [6]. This test evaluates motor balance and coordination by measuring the latency and/or number of foot faults of an animal crossing a narrow beam. Animals with brain injuries have a tendency to have a longer latency to complete the task, with more footslips, than sham or naïve animals [6].

3.1 Animals and Injury

The beam walking task has been evaluated in both rats [25–27] and mice [2, 15, 16]. Beam walking has been used in fluid percussion [15], controlled cortical impact [16, 26, 27], and weight drop [25] models of experimental TBI at the mild, moderate, and severe level. The beam walking protocol for mice presented in this chapter is based on Tsenter and colleagues' Neurological Severity Score (NSS) [18], modified by Pleasant and associates [28].

3.2 Materials

1. Sheet to record notes and times.
2. Stopwatch or timer.
3. Two ring stands.
4. Two C-clamps.
5. Four 60 cm long Plexiglas beams of different widths (3.0, 2.0, 1.0, and 0.5 cm).
6. One 60 cm long rod (0.5 cm diameter).

3.3 Working Protocol

Apparatus Setup

1. Attach clamps to ring stands, placed 60 cm apart on a clean testing surface.
2. Clamp each beam to the stands so that the beams are level and 47 cm above the testing surface.

3. Clean beams with disinfectant thoroughly between mice.

Training/Acclimation

1. Animals are trained on the beam walking task 1 day before injury.
2. Use only the 3 cm-wide beam for training [2].
3. Place the mouse at one end of the beam and start the timer. During training, all mice should be able to cross the beam with ease.
4. Allow each mouse 30 s to cross the beam.
5. Once the mouse has reached the other side, return it to its home cage.
6. Thoroughly clean the beam between each mouse.

Testing

1. Testing consists of one trial on each of the four beams in order of decreasing size followed by one trial on the circular rod.
2. Place the mouse at one end of the beam or rod and start the timer.
3. Allow the mouse 30 s to cross each beam.
4. If the mouse inverts beneath the beam or rod, the mouse is righted and allowed to continue until 30 s have elapsed.
5. If the mouse falls off the beam or rod, the trial is terminated.
6. Record score for each mouse based on 14 point scale described below.
7. After each trial, return mouse to home cage and clean beam or rod.
8. Perform trial with remaining mice in the cohort before beginning testing on the next beam or rod, allowing mice to rest between trials.

Neurologic Severity Rating Scale

Each mouse can receive a maximum of 14 points. Each beam has a maximum of 3 points. A mouse receives 3 points if it successfully crosses the beams with normal forelimb and hindlimb position, 2 points if it successfully crosses the beam but exhibits one or more footfall, 1 point for crossing despite inverting beneath the beam, and 0 points if it fails to cross or falls off the beam. The rod has a maximum of 2 points. A mouse receives 2 points for successfully crossing the rod, 1 point for crossing despite inverting more than three times, and 0 points for failing to cross the rod or falling off the rod.

3.4 Notes

3.4.1 Advantages

The beam walking task is one of the easier vestibulomotor tasks to utilize due to its simplicity and low cost. All parts can be manufactured in the lab, and each animal only takes about 5 min to perform the entire set of tasks. Compared to the rotarod task, the beam walking task is a better measurement for motor coordination, whereas with rotarod, it can be difficult to determine whether a rodent's performance is due to motor coordination or to endurance. Beam walking measures can incorporate how quickly the task is performed as well as the quality or characteristics of movement, whereas rotarod only measures whether the task was performed [29]. The beam walking test can be administered at multiple time points to assess recovery of function as performance on this test is less affected by motor learning than on the rotarod. The beam walking test is particularly effective in brain injury models because it is sensitive to most injury severities, injury lateralization, and pharmacologic manipulation in both mice and rats [6].

3.4.2 Disadvantages

While this test is cheap and easy to utilize, its measures are only semiquantitative. Therefore the beam walking task can have low inter-rater reliability. If the test is evaluated manually at the time of the animal's performance, a foot slip can be missed and experimenters can vary in scoring. The beam walking task, while sensitive to moderate and severe injury severity, is not particularly sensitive to mild TBI. With mild TBI, an injury effect is often not detected [30]. Another disadvantage is that this test is very sensitive to animal size, as size can affect balance in this test.

3.4.3 Potential Problems

A lack of motivation to walk on the beams can confound assessment of motor impairment. If a mouse sits in one place or turns back toward the starting point, it is acceptable to gently nudge the animal with a rod or pen until it begins to move toward the opposite side. To aide in motivating the animal to cross the beam, the subject's home cage can be placed at the end of the beams [26, 27]. Alternatively, a bright light above the starting point can be employed as an aversive stimulus and a dark box can be placed at the end of the beams as an escape [15, 31]. Nesting material can be placed in the escape box to reduce the animal's anxiety [15, 31]. White noise can be implemented as an aversive stimulus to motivate mice to cross the beam as well. In this case, an escape box should be placed at the end of the beam and the white noise should be turned off whenever the mouse is in the box, to associate the inside of the box with silence [5].

Issues with standardizing what counts as a foot fall (e.g. dragging foot during entire step cycle vs. slipping for a brief moment) can be solved by video-taping the trials to allow for later confirmation of scoring [15, 25, 27].

3.4.4 Adaptations

The beam walking task consists of many components that can be modified to be more relevant to different studies. Rather than using a series of increasingly difficult (more narrow) beams, some experimenters choose to evaluate animals on only one beam over several trials and/or time points to measure recovery over time. Ouyang and colleagues used a 1 m long beam that was only 6 mm wide, performing three trials a day, at 1, 3, 7, 9, 11, 14, and 16 days post-injury [15]. Bolton Hall and associates modified their protocol by only using the smallest beam and the rod in order to enhance sensitivity for detection of deficits after repeated mild TBI [2]. Some use a single tapered beam, which incorporates the concept of increasing difficulty in a one-beam test. Simon-O'Brien and collegues [27], for example, utilized a 165 cm long, tapered beam for four trials a day at 2, 5, and 7 days post-injury. In this study, performance was scored by calculating the percentage of hindlimb faults as the number of foot faults divided by the total number of steps and multiplied by 100.

There are several ways to evaluate the performance of animals in the beam walking test. Like the protocol described above, the beam walking test can be used as a modified NNS with a point scaling system. Schoch and colleagues [19] utilized a 14 point scale, like Pleasant et al. [28]. Madathil et al. [16] increased the sensitivity of the NSS by expanding Pleasant's scale [28] to a scale with a maximum of 18 points, due to the increased manifestation of motor deficits with their severe model of TBI. In this modified scale, with a maximum of 4 points per beam, 3 points were given to mice with either a forelimb or hindlimb fault, whereas only 2 points were rewarded if the subject had both a forelimb and a hindlimb fault. In this way, smaller differences in movement can be detected and accounted for in the scoring. An adaptation to the scoring method of a point scale is that performance on the beam walking task can be evaluated by measuring latency to cross the beam [31, 32] or percent foot slips from the total number of steps taken [27, 33], so as to allow the assessment to be more quantitative and therefore more accurate.

4 Rotarod

The rotarod test was first developed by Dunham and Miya in 1957 [34], and was modified by Hamm et al. to evaluate deficits following injury [30]. This test is used to evaluate coordinated motor function and is particularly sensitive and reliable in the assessment of therapeutic intervention and neurodegenerative disease models [5, 35]. Brain-injured animals tend to fall off the rotating rod earlier than uninjured animals.

4.1 Animals and Injury

The rotarod task has been evaluated in both rats [30, 36, 35] and mice [33, 37, 38]. Rotarod has been used in fluid percussion [30, 38], weight drop [39], impact acceleration [40], closed head injury [41], and controlled cortical impact (CCI) [33] models of experimental traumatic brain injury at the mild, moderate, and severe levels [6, 26, 42]. The rat rotarod protocol presented in this chapter is modified from Rozas et al. [36].

4.2 Materials

1. Rotamex-5 apparatus or equivalent rotarod apparatus (e.g. Panlab or San Diego Instruments).
2. Standard PC computer for system software.

Rat/Mouse Rotarod Apparatus

The Rotamex-5 rotarod apparatus and software (Columbus Instruments, Columbus, Ohio) consists of a metal frame with a motorized, rotating solid axle, 7.0 cm in diameter for rats or 3.0 cm in diameter for mice, covered with a rubberized material to aide in gripping. The axle sits 44.5 cm from the bottom of the box. Large Plexiglas disks separate each of four lanes (9.5 cm wide) to allow for more than one animal to be tested at a time. The rotational speed of the axle is controlled by system software that requires a standard PC, with a speed range of 0–99.9 rpm and accelerations ranging from 0.1 rpm to 20 rpm per second. Infrared beams are used to detect when a subject has fallen from the rotarod or spins around on the rod for one full rotation without attempting to walk. When the subject has fallen, the system logs this as the end of the experiment for that subject. The system records the total time running on the rotarod and the current rotational speed at the time of the subject's fall.

4.3 Working Protocol

Training/Acclimation

1. Animals are trained on the rotarod, individually or up to four animals per trial, for a total of ten trials over 2 consecutive days prior to or after injury, with five trials a day, before testing trials begin.
2. Set the rotarod to increase in velocity by 2 rotations per minute (rpm) every 18 s, with a maximum speed of 30 rpm. Start the rotarod rotating at 9 rpm and place the rat(s) on the rotarod facing the direction opposing the rotation.
3. Begin the trial and allow the animal(s) to remain on the device for a maximum of 300 s per trial or until all animals have fallen.
4. The latency for each animal to fall off the rotarod and the speed at the time of the subject's fall is recorded by the software.
5. Remove the animal(s) and place back into home cage until next trial.

6. Allow a resting period of at least 10 min between trials for each animal.
7. Clean the entire apparatus with a cleaning solution after every trial.

Testing

1. Testing consists of three trials over one day. Testing can be repeated on multiple days after injury. If animals are to be tested at several chronic timepoints after TBI, both days of training can be repeated at each time point, with five trials per day, on the two days before testing is done.
2. Set the rotarod to start rotating at 5 rpm and to increase the speed by 2 rpm every 18 s until it reaches a maximum speed of 30 rpm.
3. Start the rotarod and place the animal(s) on the rotarod facing the direction opposing rotation of the rod.
4. Begin the trial and allow the animal(s) to remain on the rotarod for 300 s per trial or until all animals have fallen.
5. Replace the subject(s) back into home cage.
6. Allow a resting period of at least 10 min between trials for each animal.
7. Clean the entire apparatus with a cleaning solution after every trial.

4.4 Notes

4.4.1 Advantages

The rotarod task measures features of motor dysfunction that are not assessed by many other vestibular tasks in TBI animal models. In a comparative study of performances on the beam balancing task, the beam walking task, and the rotarod task in rats with mild or moderate TBI, the rotarod task was able to detect motor impairment induced by both mild and moderate levels of TBI, whereas the other tasks were only sensitive to dysfunction induced by moderate TBI [30]. Rotarod is even sensitive enough to distinguish effects of single and repeated mild TBI [43]. In addition, Hamm et al. showed that the rotarod task continued to detect motor deficits after brain-injured rats had recovered normal function in beam walking and beam balancing tests, concluding that the rotarod measures different or additional aspects of motor deficit not assessed by beam balance or beam walking [30]. Due to the component of learning involved in the rotarod task, TBI-induced dysfunction can be detected on the rotarod at timepoints up to 11 weeks post-injury in rats with moderate TBI [44]. Furthermore, vestibulomotor tests such as rotarod can detect the benefit of therapeutic interventions in clinically relevant models of TBI [5]. The rotarod test can be easily modified to better meet the requirements of the researchers and to be more relevant to a particular study, as described in the Adaptations section (Sect. 4.4.4).

4.4.2 Disadvantages

The rotarod task incorporates a component of learning across trials that can confound conclusions about motor deficits or motor improvement over time in models of TBI. As a result, it can be difficult to determine whether improvement in rotarod performance is due to inherent learning on the rotarod task or recovery of motor function or therapeutic efficacy [45]. A way to attenuate this issue is to utilize longer training periods in order to reach a stable baseline. O'Connor et al. showed that following 5 days of pre-injury training with two trials a day, test performance of sham rats was stable, while rats with TBI showed improvement over time after their initial injury deficit [45]. However, Yang et al. showed that after 5 days of pre-injury and 5 days post-injury training, with three trials each day, mice with mild TBI, as well as shams, exhibited increased rotarod latencies during testing [46]. This suggests the importance of optimizing the amount of rotarod training prior to using the test for a specific study, as the species, genotype or specific rotarod parameters could affect the magnitude of inherent learning in this task.

4.4.3 Potential Problems

In some cases, animals may learn that it is safe to jump down to the platform below rather than remaining on the rotating axle. This can cause a subject's performance on the rotarod to appear worse because its latencies are cut short. The use of aversion, for example, by incorporating an electrical stimulus in the floor of the rotarod, is one way to attenuate this problem [36].

Another issue arises when subjects become accustomed to a certain speed and are comfortable moving around on the axle. If there is space on either side of the animal in the testing lane, it may decide to turn around and face the wrong direction on the axle, causing the animal to fall more quickly to the floor, again shortening latencies on the rotarod. However, using a rotarod apparatus with adjustable Plexiglas disks to customize the lane width can prevent the subject from turning around.

4.4.4 Adaptations

The rotarod task consists of many components that can be modified to be more relevant to any individual study. Shiotsuki and colleagues chose to modify the rotarod task in an attempt to evaluate motor learning skills selectively, rather than maximal gait performance, by having a nonaccelerating, wider axle with a hard surface which was difficult for mice to grip [47]. Their results showed that daily latencies on test day were four times those of the first training day of with the modified protocol, whereas latencies only increased 1.5-fold with the typical accelerating protocol. This is also shown by O'Connor et al. and Yan et al. [45, 48]. Hamm et al. used an assembly of rods, arranged in a cylindrical fashion, rather than a solid axle, so as to increase test sensitivity by incorporating grip strength, which also decreases with TBI [30]. Altering the number of trials and the amount of resting time between trials is just one more adaptation to the rotarod test [38].

The speed and type of rotation can also affect performance on the rotarod test. Im et al. utilized both a constant velocity rotarod task and a constant acceleration paradigm over several timepoints to study motor skill development after neonatal brain injury [37]. However, when maximum sensitivity is required to detect small changes, such as when measuring treatment-derived recovery, an incremental increase in rotational speed may be more appropriate than constant acceleration [45, 48]. On the other hand, the constant acceleration rotarod may be more suitable in studies evaluating the magnitude or extent of impairment [35].

Acclimation of study animals to the rotarod prior to testing can be helpful in teaching the animals to remain on the rod [49]. For example, Shiotsuki and colleagues habituated the animals to stay on the stationary rod for 3 min prior to testing [47]. Alternatively, a short stationary period of 10 s can be incorporated into the beginning of each testing trial [30].

Animals are shown to improve throughout rotarod testing due to the test's component of learning [30]. Incorporation of pretraining into the protocol to determine a stable baseline can remove the need for inclusion of a control (uninjured) group and can also eliminate the confounding variable of intrinsic learning of rodents on the rotarod. Rustay et al. found that once performance stabilizes with training, basal ability remains consistent over repeated sessions, whereas no training over repeated sessions could lead to longer latencies during testing due to learning, rather than drug treatment or other intervention [50]. Thus, when using rotarod in TBI studies, establishing a baseline performance through pretraining can allow better isolation of posttraumatic motoric improvements from inherent learning.

Data reporting and analysis can be done in several ways for the rotarod test. When testing animals over multiple trials, the most common method is to report the group mean for latency to fall and/or final speed, calculated from daily means for all the trials per day for each animal [30, 37, 42, 47]. Because trial latencies even within a given animal can be variable (See Potential Problems, Sect. 4.4.3), trial extremes are sometimes omitted prior to calculation of daily means. For example, Srodulski and colleagues tested rats on the rotarod a total of four times per day over 5 days, but chose to discard each animal's shortest performance and calculated the average of the other three [49]. Scherbel et al. removed both the shortest *and* longest of four trials [51]. Baseline performance achieved through pretraining can be incorporated into data analysis. Rather than reporting the average latency to fall, Scherbel et al. calculated postinjury latencies for each animal as a percentage of their respective pre-injury baseline [51]. This can be useful when experimental groups have differing initial abilities due to factors such as sex, body weight, or genotype.

Acknowledgements

Supported, in part, by NIH R01 NS072302 and Kentucky Spinal Cord and Head Injury Research Trust grants 14-12A and 14-13A.

References

1. Langlois JA, Rutland-Brown W, Wald MM (2006) The epidemiology and impact of traumatic brain injury: a brief overview. J Head Trauma Rehabil 21(5):375–378
2. Bolton Hall AN, Joseph B, Brelsfoard JM, Saatman KE (2016) Repeated closed head injury in mice results in sustained motor and memory deficits and chronic cellular changes. PLoS One 11(7):e0159442. https://doi.org/10.1371/journal.pone.0159442
3. Madathil SK, Saatman KE (2015) IGF-1/IGF-R signaling in traumatic brain injury: impact on cell survival, neurogenesis, and behavioral outcome. In: Kobeissy FH (ed) Brain neurotrauma: molecular, neuropsychological, and rehabilitation aspects. Frontiers in Neuroengineering, Boca Raton, FL
4. Bales JW, Macfarlane K, Edward Dixon C (2012) Neurobehavioral assessments of traumatic brain injury. In: Chen J, Xu XM, Xu ZC, Zhang JH (eds) Animal models of acute neurological injuries II, vol vol II. Humana Springer Science+Business Media, LLC, New York, pp 377–384. https://doi.org/10.1007/978-1-61779-782-8_34
5. Bales JW, Macfarlane K, Edward Dixon C (2012) Vestibular assessments following traumatic brain injury. In: Chen J, Xu XM, Xu ZC, Zhang JH (eds) Animal models of acute neurological injuries II, vol vol II. Humana Springer Science+Business Media, LLC, New York, pp 385–396. https://doi.org/10.1007/978-1-61779-782-8_33
6. Fujimoto ST, Longhi L, Saatman KE, Conte V, Stocchetti N, McIntosh TK (2004) Motor and cognitive function evaluation following experimental traumatic brain injury. Neurosci Biobehav Rev 28(4):365–378. https://doi.org/10.1016/j.neubiorev.2004.06.002
7. Ennaceur A, Delacour J (1988) A new one-trial test for neurobiological studies of memory in rats. 1: behavioral data. Behav Brain Res 31 (1):47–59
8. Grayson B, Leger M, Piercy C, Adamson L, Harte M, Neill JC (2015) Assessment of disease-related cognitive impairments using the novel object recognition (NOR) task in rodents. Behav Brain Res 285:176–193. https://doi.org/10.1016/j.bbr.2014.10.025
9. Antunes M, Biala G (2012) The novel object recognition memory: neurobiology, test procedure, and its modifications. Cogn Process 13 (2):93–110. https://doi.org/10.1007/s10339-011-0430-z
10. Silvers JM, Harrod SB, Mactutus CF, Booze RM (2007) Automation of the novel object recognition task for use in adolescent rats. J Neurosci Methods 166(1):99–103. https://doi.org/10.1016/j.jneumeth.2007.06.032
11. Goulart BK, de Lima MN, de Farias CB, Reolon GK, Almeida VR, Quevedo J, Kapczinski F, Schroder N, Roesler R (2010) Ketamine impairs recognition memory consolidation and prevents learning-induced increase in hippocampal brain-derived neurotrophic factor levels. Neuroscience 167(4):969–973. https://doi.org/10.1016/j.neuroscience.2010.03.032
12. Rachmany L, Tweedie D, Rubovitch V, Yu QS, Li Y, Wang JY, Pick CG, Greig NH (2013) Cognitive impairments accompanying rodent mild traumatic brain injury involve p53-dependent neuronal cell death and are ameliorated by the tetrahydrobenzothiazole PFT-alpha. PLoS One 8(11):e79837. https://doi.org/10.1371/journal.pone.0079837
13. Rubovitch V, Zilberstein Y, Chapman J, Schreiber S, Pick CG (2017) Restoring GM1 ganglioside expression ameliorates axonal outgrowth inhibition and cognitive impairments induced by blast traumatic brain injury. Sci Rep 7:41269. https://doi.org/10.1038/srep41269
14. Muradashvili N, Benton RL, Saatman KE, Tyagi SC, Lominadze D (2015) Ablation of matrix metalloproteinase-9 gene decreases cerebrovascular permeability and fibrinogen deposition post traumatic brain injury in mice. Metab Brain Dis 30(2):411–426. https://doi.org/10.1007/s11011-014-9550-3
15. Ouyang W, Yan Q, Zhang Y, Fan Z (2017) Moderate injury in motor-sensory cortex causes behavioral deficits accompanied by electrophysiological changes in mice adulthood. PLoS One 12(2):e0171976. https://doi.org/10.1371/journal.pone.0171976
16. Madathil SK, Carlson SW, Brelsfoard JM, Ye P, D'Ercole AJ, Saatman KE (2013) Astrocyte-specific overexpression of insulin-like growth

Factor-1 protects hippocampal neurons and reduces behavioral deficits following traumatic brain injury in mice. PLoS One 8(6):e67204. https://doi.org/10.1371/journal.pone.0067204

17. Haus DL, Lopez-Velazquez L, Gold EM, Cunningham KM, Perez H, Anderson AJ, Cummings BJ (2016) Transplantation of human neural stem cells restores cognition in an immunodeficient rodent model of traumatic brain injury. Exp Neurol 281:1–16. https://doi.org/10.1016/j.expneurol.2016.04.008
18. Tsenter J, Beni-Adani L, Assaf Y, Alexandrovich AG, Trembovler V, Shohami E (2008) Dynamic changes in the recovery after traumatic brain injury in mice: effect of injury severity on T2-weighted MRI abnormalities, and motor and cognitive functions. J Neurotrauma 25(4):324–333. https://doi.org/10.1089/neu.2007.0452
19. Schoch KM, Evans HN, Brelsfoard JM, Madathil SK, Takano J, Saido TC, Saatman KE (2012) Calpastatin overexpression limits calpain-mediated proteolysis and behavioral deficits following traumatic brain injury. Exp Neurol 236(2):371–382. https://doi.org/10.1016/j.expneurol.2012.04.022
20. Cohen SJ, Stackman RW Jr (2015) Assessing rodent hippocampal involvement in the novel object recognition task. A review. Behav Brain Res 285:105–117. https://doi.org/10.1016/j.bbr.2014.08.002
21. Cai L, Gibbs RB, Johnson DA (2012) Recognition of novel objects and their location in rats with selective cholinergic lesion of the medial septum. Neurosci Lett 506(2):261–265. https://doi.org/10.1016/j.neulet.2011.11.019
22. Prins ML, Hales A, Reger M, Giza CC, Hovda DA (2010) Repeat traumatic brain injury in the juvenile rat is associated with increased axonal injury and cognitive impairments. Dev Neurosci 32(5-6):510–518. https://doi.org/10.1159/000316800
23. Perez-Garcia G, Guzman-Quevedo O, Da Silva Aragao R, Bolanos-Jimenez F (2016) Early malnutrition results in long-lasting impairments in pattern-separation for overlapping novel object and novel location memories and reduced hippocampal neurogenesis. Sci Rep 6:21275. https://doi.org/10.1038/srep21275
24. Lee YA, Kim YJ, Goto Y (2016) Cognitive and affective alterations by prenatal and postnatal stress interaction. Physiol Behav 165:146–153. https://doi.org/10.1016/j.physbeh.2016.07.014
25. Gao Y, Li J, Wu L, Zhou C, Wang Q, Li X, Zhou M, Wang H (2016) Tetrahydrocurcumin provides neuroprotection in rats after traumatic brain injury: autophagy and the PI3K/AKT pathways as a potential mechanism. J Surg Res 206(1):67–76. https://doi.org/10.1016/j.jss.2016.07.014
26. Yu YW, Hsieh TH, Chen KY, Wu JC, Hoffer BJ, Greig NH, Li Y, Lai JH, Chang CF, Lin JW, Chen YH, Yang LY, Chiang YH (2016) Glucose-dependent insulinotropic polypeptide ameliorates mild traumatic brain injury-induced cognitive and sensorimotor deficits and neuroinflammation in rats. J Neurotrauma 33(22):2044–2054. https://doi.org/10.1089/neu.2015.4229
27. Simon-O'Brien E, Gauthier D, Riban V, Verleye M (2016) Etifoxine improves sensorimotor deficits and reduces glial activation, neuronal degeneration, and neuroinflammation in a rat model of traumatic brain injury. J Neuroinflammation 13(1):203. https://doi.org/10.1186/s12974-016-0687-3
28. Pleasant JM, Carlson SW, Mao H, Scheff SW, Yang KH, Saatman KE (2011) Rate of neurodegeneration in the mouse controlled cortical impact model is influenced by impactor tip shape: implications for mechanistic and therapeutic studies. J Neurotrauma 28(11):2245–2262. https://doi.org/10.1089/neu.2010.1499
29. Stanley JL, Lincoln RJ, Brown TA, McDonald LM, Dawson GR, Reynolds DS (2005) The mouse beam walking assay offers improved sensitivity over the mouse rotarod in determining motor coordination deficits induced by benzodiazepines. J Psychopharmacol 19(3):221–227. https://doi.org/10.1177/0269881105051524
30. Hamm RJ, Pike BR, O'Dell DM, Lyeth BG, Jenkins LW (1994) The rotarod test: an evaluation of its effectiveness in assessing motor deficits following traumatic brain injury. J Neurotrauma 11(2):187–196. https://doi.org/10.1089/neu.1994.11.187
31. Luong TN, Carlisle HJ, Southwell A, Patterson PH (2011) Assessment of motor balance and coordination in mice using the balance beam. J Vis Exp 49. https://doi.org/10.3791/2376
32. Statler KD, Alexander H, Vagni V, Holubkov R, Dixon CE, Clark RS, Jenkins L, Kochanek PM (2006) Isoflurane exerts neuroprotective actions at or near the time of severe traumatic brain injury. Brain Res 1076(1):216–224. https://doi.org/10.1016/j.brainres.2005.12.106
33. Desai A, Park T, Barnes J, Kevala K, Chen H, Kim HY (2016) Reduced acute neuroinflammation and improved functional recovery after traumatic brain injury by alpha-linolenic acid

supplementation in mice. J Neuroinflammation 13(1):253. https://doi.org/10.1186/s12974-016-0714-4

34. Dunham NW, Miya TS (1957) A note on a simple apparatus for detecting neurological deficit in rats and mice. J Am Pharm Assoc Am Pharm Assoc 46(3):208–209

35. Monville C, Torres EM, Dunnett SB (2006) Comparison of incremental and accelerating protocols of the rotarod test for the assessment of motor deficits in the 6-OHDA model. J Neurosci Methods 158(2):219–223. https://doi.org/10.1016/j.jneumeth.2006.06.001

36. Rozas G, Guerra MJ, Labandeira-Garcia JL (1997) An automated rotarod method for quantitative drug-free evaluation of overall motor deficits in rat models of parkinsonism. Brain Res Brain Res Protoc 2(1):75–84

37. Im SH, Yu JH, Park ES, Lee JE, Kim HO, Park KI, Kim GW, Park CI, Cho SR (2010) Induction of striatal neurogenesis enhances functional recovery in an adult animal model of neonatal hypoxic-ischemic brain injury. Neuroscience 169(1):259–268. https://doi.org/10.1016/j.neuroscience.2010.04.038

38. Yu W, Parakramaweera R, Teng S, Gowda M, Sharad Y, Thakker-Varia S, Alder J, Sesti F (2016) Oxidation of KCNB1 potassium channels causes neurotoxicity and cognitive impairment in a mouse model of traumatic brain injury. J Neurosci 36(43):11084–11096. https://doi.org/10.1523/JNEUROSCI.2273-16.2016

39. Mannix R, Berglass J, Berkner J, Moleus P, Qiu J, Andrews N, Gunner G, Berglass L, Jantzie LL, Robinson S, Meehan WP 3rd (2014) Chronic gliosis and behavioral deficits in mice following repetitive mild traumatic brain injury. J Neurosurg 121(6):1342–1350. https://doi.org/10.3171/2014.7.JNS14272

40. Byard RW, Donkin J, Vink R (2017) The forensic implications of amphetamine intoxication in cases of inflicted blunt craniocerebral trauma. J Forensic Sci 63:151–153. https://doi.org/10.1111/1556-4029.13509

41. Dachir S, Shabashov D, Trembovler V, Alexandrovich AG, Benowitz LI, Shohami E (2014) Inosine improves functional recovery after experimental traumatic brain injury. Brain Res 1555:78–88. https://doi.org/10.1016/j.brainres.2014.01.044

42. Lagraoui M, Latoche JR, Cartwright NG, Sukumar G, Dalgard CL, Schaefer BC (2012) Controlled cortical impact and craniotomy induce strikingly similar profiles of inflammatory gene expression, but with distinct kinetics. Front Neurol 3:155. https://doi.org/10.3389/fneur.2012.00155

43. Laurer HL, Bareyre FM, Lee VM, Trojanowski JQ, Longhi L, Hoover R, Saatman KE, Raghupathi R, Hoshino S, Grady MS, McIntosh TK (2001) Mild head injury increasing the brain's vulnerability to a second concussive impact. J Neurosurg 95(5):859–870. https://doi.org/10.3171/jns.2001.95.5.0859

44. Lindner MD, Plone MA, Cain CK, Frydel B, Francis JM, Emerich DF, Sutton RL (1998) Dissociable long-term cognitive deficits after frontal versus sensorimotor cortical contusions. J Neurotrauma 15(3):199–216. https://doi.org/10.1089/neu.1998.15.199

45. O'Connor CA, Cernak I, Johnson F, Vink R (2007) Effects of progesterone on neurologic and morphologic outcome following diffuse traumatic brain injury in rats. Exp Neurol 205(1):145–153. https://doi.org/10.1016/j.expneurol.2007.01.034

46. Yang SH, Gustafson J, Gangidine M, Stepien D, Schuster R, Pritts TA, Goodman MD, Remick DG, Lentsch AB (2013) A murine model of mild traumatic brain injury exhibiting cognitive and motor deficits. J Surg Res 184(2):981–988. https://doi.org/10.1016/j.jss.2013.03.075

47. Shiotsuki H, Yoshimi K, Shimo Y, Funayama M, Takamatsu Y, Ikeda K, Takahashi R, Kitazawa S, Hattori N (2010) A rotarod test for evaluation of motor skill learning. J Neurosci Methods 189(2):180–185. https://doi.org/10.1016/j.jneumeth.2010.03.026

48. Yan EB, Johnstone VP, Alwis DS, Morganti-Kossmann MC, Rajan R (2013) Characterising effects of impact velocity on brain and behaviour in a model of diffuse traumatic axonal injury. Neuroscience 248:17–29. https://doi.org/10.1016/j.neuroscience.2013.05.045

49. Srodulski S, Sharma S, Bachstetter AB, Brelsfoard JM, Pascual C, Xie XS, Saatman KE, Van Eldik LJ, Despa F (2014) Neuroinflammation and neurologic deficits in diabetes linked to brain accumulation of amylin. Mol Neurodegener 9:30. https://doi.org/10.1186/1750-1326-9-30

50. Rustay NR, Wahlsten D, Crabbe JC (2003) Influence of task parameters on rotarod performance and sensitivity to ethanol in mice. Behav Brain Res 141(2):237–249

51. Scherbel U, Raghupathi R, Nakamura M, Saatman KE, Trojanowski JQ, Neugebauer E, Marino MW, McIntosh TK (1999) Differential acute and chronic responses of tumor necrosis factor-deficient mice to experimental brain injury. Proc Natl Acad Sci U S A 96(15):8721–8726

Chapter 10

Pre-Procedural Considerations and Post-Procedural Care for Animal Models with Experimental Traumatic Brain Injury

Mary A. Robinson, Samer M. Jaber, Stacey L. Piotrowski, and Thomas H. Gomez

Abstract

In vivo models of traumatic brain injury (TBI) are powerful means of examining the progression and sequelae after neurologic injury at the cellular and molecular level. Drafting the animal use protocol, acquiring animals, and preparing the animals for the procedure must be considered in advance of the start of the actual study. The provision of supportive care in the post-procedure period cannot be overlooked as a critical component of experimental success. This chapter introduces these topics to the researcher studying TBI in animal models.

Key words Animal welfare, Animal use protocol, Traumatic brain injury, Experimental models, Support, Assessments

1 Introduction

Animal models of TBI allow for the study of the biochemical, cellular, and molecular events after TBI, which often cannot be evaluated in human patients in the clinical setting [1]. Additionally, outcomes of behavior, learning, and memory tests with these models offer insight into the cognitive and behavioral effects of the injury. A thoughtfully chosen animal model combined with careful pre-injury planning and attentive post-procedural care contributes to overall success with these studies. The beginning of this chapter provides information to guide the researcher during the planning and pre-procedural stages of an animal study of TBI. The second section outlines materials and methods for supportive care after the TBI procedure to facilitate animal recovery.

Amit K. Srivastava and Charles S. Cox, Jr. (eds.), *Pre-Clinical and Clinical Methods in Brain Trauma Research*, Neuromethods, vol. 139, https://doi.org/10.1007/978-1-4939-8564-7_10, © Springer Science+Business Media, LLC, part of Springer Nature 2018

2 Pre-procedural Considerations

2.1 Animal Use Protocol and IACUC Approval

Once an appropriate animal model of TBI has been chosen, the next step is to submit an animal use protocol (AUP) for review by the Institutional Animal Care and Use Committee (IACUC). The IACUC is charged with confirming that the protocol is consistent with federal law [2] to the extent to which it applies to the particular project [3] and the provisions in the *Guide for the Care and Use of Laboratory Animals* [4]. A description of the procedure(s) to be performed, species and strain(s) to be used, timeline, justification for the use of and the number of animals requested, plus other information as specified on the individual institution's AUP, should be stated in clear, concise language. Both experimental and humane endpoints should be clearly defined in the AUP. Experimental endpoints are defined as the times at which scientific aims and objectives have been reached [4]. Alternative endpoints, called humane endpoints, are criteria used to remove an animal from the study earlier in order to preempt or end unrelieved pain, discomfort, or distress [5, 6]. Due to the potential for pain and distress with all models of TBI, humane endpoints should be incorporated into the AUP. A literature search using keywords that are specific for the type of procedure may reveal refinements that could mitigate pain and distress, or replacements for the use of live animals for certain aspects of the study. Such a search is required for AUPs involving species that are covered by the Animal Welfare Act [2], such as swine, and is recommended for AUPs that involve non-USDA covered species, such as mice and rats. Guidance on any aspect of the AUP should be sought from experienced researchers, veterinarians, or the IACUC office staff. In all cases, the AUP with complete and accurate descriptions of the studies must be approved by the IACUC before any animal work commences.

2.2 Staff Training

Proper training of the researcher performing the procedure(s) is essential for the success of the study [5]. The principal investigator, laboratory manager, or other personnel in a supervisory capacity should ensure that research staff at all levels are knowledgeable and have the required skill set to perform the tasks associated with the project [4]. The experience level of each staff member should be assessed to identify areas in which training is needed with regard to the particular animal model. All research staff should have basic animal research methodology training in the areas of animal restraint, identification methods, administration of injections, and identification of and monitoring for pain and distress. Often, training in basic animal handling and methodology is available from the animal care resource veterinary staff or training specialist. Those who will conduct the surgical portion of the experiment should have in-depth, hands-on training for the specific type of surgery.

A tiered approach to training allows the animal surgeon to gain familiarity with the procedure through the following steps. Utilizing carcasses as a first step acquaints the researcher with the anatomy of the surgical area and general flow of the procedure. Progressing to performing nonsurvival surgery provides the surgeon-in-training with realistic physiological responses to tissue manipulation without the added challenge of survivability. Once some expertise has been gained, the researcher progresses to a survival procedure under supervision of an experienced animal surgeon. A laboratory member who has successfully performed the procedure previously is often the best resource for training new staff in specific surgical methods. Although some basic animal care and use training can be found online (CITI Program, AALAS), electronic resources cannot adequately substitute for hands-on learning.

Both the research team and animal care staff must work together to enhance animal welfare during the course of the project [5]. Communication is critical so that all personnel involved have a similar understanding of the impact the study will have on the animals' health and well-being. Some institutions request that the research group meet with individuals from the veterinary and animal care staff to review study goals, predict animal welfare issues, and discuss other specifics about the study. If special care such as a softened diet or a padded cage must be provided post-procedure, the researchers should make these arrangements with animal care and veterinary personnel before the study begins. In some cases, husbandry care that is usually provided by the animal care staff may instead be handled by the researcher, which should be discussed with the animal care staff or veterinarian in advance [5]. In general, providing information regarding post-procedural care in a written format in addition to any verbal communication is helpful to the animal care and veterinary staff. In situations where both animal care and research staff play a role in providing husbandry care, clearly defined procedures for each group and good documentation of completed duties is important to ensure appropriate support of animals [5]. The research staff should provide contact information for the responsible laboratory member(s) as animal health concerns could be found after working hours or on weekends by animal care staff on duty. Posting contact names and phone numbers in a prominent location near or in the animals' housing room facilitates rapid notification of the research staff of any significant issues.

2.3 Animal Acclimation and Quarantine

Newly arrived animals should be given time to adjust to their new housing environment and caretakers before the TBI procedure [7, 8]. Studies have shown that noise [9], vibration [10, 11], inhalation of particulate matter such as from exhaust fumes [12], changes in housing conditions [13, 14], and temperature [11] alter both animal physiology and behavior. The act of transportation itself was found to increase anxiety behaviors in control rats in a study of blast TBI [15], highlighting the direct effect of transportation stress in these models. The acclimation period allows for the stabilization of the animal's physiologic and behavioral parameters before the introduction of the stress of the procedure. Several factors should be considered when determining the length of the acclimation period, including type and duration of transport, the species, the complexity and clinical consequences of the procedure, and any preexisting health conditions, such as in some genetically modified (GM) models [4]. Although recommendations for the length of the acclimation and stabilization period vary by institution, the minimum is generally considered to be 2–3 days for rodent species and 5–7 days for swine [8, 16, 17]. During this time, animals should be observed for appropriate food and water intake and normal elimination levels.

In addition to animal acclimation, facilities may require the newly arrived animals to be initially quarantined from the general population of animals [18] to protect in-house colonies from the inadvertent introduction of pathogens. The quarantine program may include use of specific personal protective equipment (PPE), restricted access to the animals for research staff, and pathogen testing of animals [4]. The length and nature of the quarantine program is determined by the veterinarian at each institution or facility. Although information from the sending institution or vendor often enables the veterinarian to determine the likely health status of incoming animals, stress from transportation may cause recrudescence of undetected, subclinical infections [4]. Awareness of the facility's quarantine program on the part of the researcher allows for appropriate scheduling of procedures and prevents unanticipated delay to the start of the study while awaiting the completion of the quarantine period [8, 19].

The end of quarantine or facility acclimation may be an appropriate time to begin training phases for certain behavioral tasks that might be required of the animals after the injury procedure. This can include tests that evaluate learning and memory, locomotion, or affective state. Training can also include acclimation to certain restraint devices, such as body slings for swine or restraint tubes or cones for rodents.

2.4 Animal Health/Condition

Rodents used in TBI research are generally obtained from commercial vendors. Reputable sources can be found on the internet, through the animal care department, from collaborators, and in the Laboratory Animal Science Buyer's Guide (www.laboratoryanimalsciencebuyersguide.com). These vendors offer specific-pathogen-free rodents from stocks and strains that routinely undergo genetic monitoring. Using such well-defined animals in TBI research mitigates concerns associated with intersubject variability due to genetic changes [20]. The use of

GM animals, particularly mice, permits exploration of specific molecular pathways in TBI [21]. The general health of GM strains should be assessed prior to injury as undetected health or mobility issues can affect post-injury evaluations [5].

Pre-procedural evaluation of swine prior to experimental TBI is important to rule out confounding spontaneous disease conditions. Central nervous system disease may be infectious, congenital, or secondary to water deprivation or sodium ion intoxication [22, 23]. Nonneurological disease may also impact the ability to use swine as experimental TBI models. Cardiac or respiratory diseases may render swine unacceptable anesthetic candidates [24], and musculoskeletal disorders may interfere with behavioral assays that have outcomes dependent on locomotion [25, 26].

2.5 Fasting/Withholding of Water

Withholding food from certain species of animals prior to anesthetic events is common practice in veterinary medicine to prevent aspiration pneumonia secondary to passive regurgitation of stomach contents [27]. Mice and rats do not require fasting as these species lack the ability to vomit [28]. Fasting adult pigs for experimental TBI procedures is usually not required for longer than 6–8 h. This timeframe is sufficient to empty gastric contents and mitigate post-anesthetic or analgesic-induced vomiting [17]. Although vomiting after head injury may be a common finding in human TBI cases [29], the authors have rarely witnessed this after experimental TBI in swine. Due to the risk of hypoglycemia, pediatric swine may not be fasted at all, especially in the neonatal period when metabolic alterations can be most profound [30]. Presumably, most patients experiencing traumatic brain injury events do so while euglycemic; therefore, research models should approximate those homeostatic conditions as closely as possible. The shifts in metabolic pathways that occur during hypoglycemia in pigs [31] may confound outcomes during cerebral injury [32]. One example is the increase in ketones seen during episodes of hypoglycemia. Ketones may have neuroprotective effects and feeding of ketogenic diets have been shown to decrease lesion volume in rodents after experimental TBI [33] and fasting induced hypoglycemia has been shown to improve outcomes in neonatal rats [34]. More complete reviews of the neuroprotective effects of cerebral ketone metabolism are available in the literature [35–38]. Due to the risk of dehydration and associated hypovolemia [39], withholding of water prior to experimental TBI in any species is not recommended.

2.6 Sample Checklists

Ensuring that all elements of the study are considered and that the researcher is prepared to perform animal procedures requires strong attention to detail. The PREPARE Guidelines Checklist offers an organizational tool to guide the researcher during the planning stage [40]. This checklist is divided into three sections:

(1) formulation of the study, (2) dialogue between scientists and the animal facility, and (3) quality control of the components in the study. The first section introduces points to consider during the development of the study and subsequent creation of the AUP. The second section speaks to communication between the researcher and animal facility personnel as an important facet of the preparation before initiation of the study. The third section delineates broad areas that require a certain amount of management to reduce their intrinsic variability and thus promote a more reliable preclinical model. The PREPARE Guidelines Checklist are considered to be dynamic and subject to evolving to align with published best practices in animal research.

Adhering to a simple pre-procedural checklist will assist in reducing inconsistent practices among laboratory members and aid in making preparation for the procedures a straightforward process. Components of the checklist should include drugs and supplies needed for the procedure as well as items for supportive care of animals after the procedure. A sample checklist is in Fig. 1.

- ❑ Anesthetic (isoflurane, injectable drug, etc.)
- ❑ If gas anesthesia to be used:
    - ❑ Anesthetic machine set up and checked for leaks
    - ❑ Sufficient oxygen supply available
- ❑ Eye lubricant
- ❑ Analgesic (local such as bupivacaine or systemic, if using)
- ❑ Drapes/pads for work surface/table
- ❑ Clippers with sharpened blade (replace as needed) or depilatory cream
- ❑ 70% alcohol
- ❑ Skin disinfectant (chlorhexidine or povidone-iodine scrub)
- ❑ Nonsterile gauze or cotton-tip applicators for surgical prep
- ❑ Sterile syringes/needles
- ❑ Sterile drapes (transparent plastic for rodent, disposable or cloth for swine)
- ❑ Sterile surgeons gloves
- ❑ Sterile gauze or cotton-tip applicators
- ❑ Sterile saline
- ❑ Sterile surgery pack (autoclaved or pre-packaged irradiated)
- ❑ Suture material or skin glue
- ❑ Heat source to support core body temperature during procedure and recovery (warm water circulating pad, far infrared warming pad, BAIR hugger, etc.)
- ❑ Clean and dry cage ready for recovery
- ❑ Soft food or recovery diet available, if needed
- ❑ Water source within easy reach of animal

Fig. 1 Sample pre-procedural checklist

3 Surgical Recovery

3.1 Anesthetic Recovery

Experimental animals should be monitored post-procedure, with careful observation until fully recovered. An animal is considered to be fully recovered if it is able to normally ambulate in its enclosure. If inhalant anesthesia is used, recovery usually occurs rapidly, as the inhalant anesthesia is naturally reversed via normal respirations. If intraperitoneal injectable anesthesia is utilized in rodents, recovery may be prolonged, particularly if repeated doses of the injectable anesthesia are administered. Repeat bolus dosing can also lead to increased mortality in rodents [41]. The use of total intravenous anesthesia in swine during brain injury procedures is possible, and sometimes desirable, to mitigate hypotension caused by isoflurane. This is especially true for prolonged procedures where invasive neuromonitoring is used and intracranial pressure (ICP) or cerebral blood flow are measured outcomes [42, 43].

Recovery time from injectable anesthesia can be shortened with reversal agents. The appropriate reversal agent and dose will be dictated by the anesthesic regimen. For example, atipamezole can be used to reverse the anesthetic effects of alpha-2 adrenergic receptor agonists used in common veterinary anesthetic cocktails (e.g., ketamine-dexmedetomidine and ketamine-xylazine) [44–46]. Anesthetic and reversal regimens should be developed with the aid of veterinary staff and must be listed in the approved AUP. If recovery is prolonged, the animal can be rotated between left-side and right-side recumbency at regular intervals (e.g., every 10–15 min) in order to improve respirations and help shorten recovery time.

3.2 Support

Timely and smooth recovery can also be facilitated through the use of various support mechanisms. Animals should be singly housed during the anesthesia recovery period in order to prevent injury from conspecifics. A recovery surface should be provided, and this surface should be free of standard bedding material (e.g., wood chips or paper squares) as small particles can potentially obstruct respirations or be ingested during the recovery period [47]. An empty cage or a cage with bedding material covered with padding or paper that can be removed once the animal is fully recovered are examples of appropriate recovery cages. Padded stanchions or padded caging may be used for swine after experimental TBI. Pigs should not be returned to enclosures with smooth or wet floors, as this may lead to musculoskeletal injuries due to hyperabduction of the pelvic limbs. Thoracic harnesses that do not put pressure on the neck, such as those designed for dogs, can be used to stabilize piglets while they recover. In the authors' experience, this is useful to prevent injury in piglets <10 kg that are not accustomed to close human contact or being manually restrained. In rare circumstances, a mild sedative (such as low dose acepromazine or dexmedetomidine) may be used to calm dysphoric or disoriented swine that are in danger of harming themselves or staff. During the recovery, swine may vocalize loudly; if this is sustained for more than a brief period, the animal should be administered an appropriate analgesic or sedative as approved on the AUP or recommended by the veterinarian.

Thermal support should be provided as needed during the recovery period. Supplemental heating sources may be used to prevent hypothermia [48] and facilitate anesthetic recovery. Heating sources should provide gentle, constant warmth. Direct or close contact of the animal with the source should be avoided. Preferred heating sources include forced-air warming systems, recirculating water pads, or warming lamps. Recirculating water pads should be insulated and placed under the recovery cage, containing bedding or padding, to prevent direct exposure of the animal to the heat source. Electric heating blankets are not recommended as they can easily cause thermal burns and injury. Warming lamps and forced-air warming systems must be placed at a distance that provides gentle heat at the cage level in order to prevent burns or overheating of animals.

Hypothermia has been used therapeutically to mitigate adverse outcomes after TBI [49] and should be avoided unless the effects of hypothermia are accounted for in the study design. Conversely, hyperthermia has been associated with negative outcomes after TBI [50], thus body temperature should be monitored to prevent unnecessary or excessive warming of the animal.

Intensive care unit (ICU) caging can also be utilized to provide both thermal support and oxygen supplementation. Oxygen supplementation can also be provided via other methods including a facemask or flow-by supply near the recovering animal. Animal models of diffuse axonal injury may experience periods of apnea [51, 52]. Ventilatory support is recommended in survival models in order to maintain proper oxygenation and normocarbia. Ventilatory parameters, such as respiratory rate and tidal volume, should be tailored to maintain an end-tidal carbon dioxide partial pressure of 35–45 mmHg. Generally, animals can be extubated when they regain the gag reflex and the ability to chew and swallow, have a sustained head lift, and are able to maintain SpO_2 above 92%. A laryngoscope and appropriately sized endotracheal tube should be available until the animal is ambulatory and fully recovered in the event re-intubation is required.

Based on the nature and length of the procedure and the incidence of any hemorrhaging, warmed subcutaneous or intraperitoneal fluids (in rodents) or intravenous fluids (in larger animals) may be warranted to support proper hydration and blood volume [47]. While fluid therapy is clearly important in the hypotensive patient for maintenance of adequate cerebral perfusion pressure in the face of rising ICP, it is important to note that any procedural complications resulting in hemorrhage may also require volume replacement. Hemorrhage in combination with TBI can lead to cerebral pathology not seen with TBI or hemorrhage alone, despite normotensive conditions [53].

Prolonged recovery may also occur secondary to hypoglycemia, especially in pediatric subjects as mentioned in the pre-procedural section of this chapter. Monitoring blood glucose levels can be advantageous for some animals where hypoglycemia is a high risk. Supplementation with intravenous dextrose can be used to restore euglycemia. It is also important to note that many anesthetic agents, such as alpha-2 agonists and isoflurane, can affect blood glucose levels [54, 55]. The potential confounding effect of various anesthetics on blood glucose values in TBI models should be considered during the planning phase of these experiments. Hyperglycemia has been shown to correlate with adverse outcomes in TBI patients [56]. Severe injury models may be more prone to this phenomenon than mild to moderate injury models.

3.3 Assessments

Initial assessments of the animals can be performed during the anesthetic recovery period. The likelihood of certain procedural complications, such as skull fracture or neurological abnormalities, is model-dependent. For example, skull fractures are not desirable in models of sports-related TBI [1], thus animal subjects should be examined post-procedure for this complication. In addition, diastasis of the cranial sutures may be criteria for exclusion in some rodent models [57]. Seizure activity can be seen in the acute phase [58] and can be treated with antiepileptic medications (e.g., diazepam or levetiracetam), if not contraindicated by the study design. Seizure activity may go unnoticed or may be masked by the residual effects of anesthetic agents. Posttraumatic epilepsy in animals can develop soon after the traumatic procedure or as part of a long-term syndrome [59].

Surgical incisions should be observed regularly by the researcher to ensure that it remains intact and that hemostasis has been achieved. Once an animal has become ambulatory, it can be considered to be recovered and be returned to the home cage.

4 Short-Term Post-procedural Care (up to 48 h)

4.1 Support

As pigs are very social species, isolation from conspecifics is not recommended as a routine procedure, even after traumatic procedures. Even brief exposure to social isolation, especially in younger piglets, can induce central neuroendocrine and gene expression changes consistent with moderate to severe stress that may affect immune function [60]. Therefore, it is recommended that swine have some auditory and visual contact with conspecifics at a minimum and are cohoused in socially compatible pairs or small groups when possible. After the immediate anesthetic recovery period, rodents should also be group housed to allow for normal conspecific interactions and behavior.

The use of analgesics for pain control is based on the specific TBI model and the measured experimental variables [61]. There is evidence that some forms of analgesics, particularly nonsteroidal anti-inflammatory drugs, may influence experimental outcomes [62, 63]. Some studies have utilized local analgesics, such as bupivicaine, at the incision site [64, 65] or injectable opioids such as buprenorphine [66] to provide post-procedural pain control. Buprenorphine, a partial opioid agonist, is commonly used to mitigate pain after experimental TBI in swine [51, 67]. Nonsteroidal anti-inflammatory medications should be used with caution in some models, as premedication with these agents prior to rotational nonimpact TBI in piglets has been associated with the development of subdural and subarachnoid hematoma and higher mortality prior to experimental endpoint [68]. In the authors' experience, topical analgesics can also be applied at the surgical site to provide localized incisional pain relief. The use or withholding of analgesics in TBI studies should be discussed with veterinary staff. Studies in which analgesics are withheld during a painful or stressful condition are required to be classified as USDA Category E procedures for USDA covered species, such as swine. The withholding of analgesics in this instance requires additional justification in the AUP [2].

Corticosteroids have been evaluated as a treatment in TBI, namely to help reduce inflammatory responses. However, in the face of currently available data on mortality rates and adverse outcomes associated with their use, such as gastric ulceration and infections, corticosteroids are contraindicated [69, 70].

Depending on the clinical condition of the animals after the procedure, nutritional support may be offered. Rodents that may be unable to reach food hoppers and water bottles due to their postprocedural condition will benefit from supplementation with commercially available specialized gel diets that support adequate body condition and hydration [71]. Use of these specialized diets should be discussed with veterinary and husbandry staff and be indicated in the approved AUP.

Gastrointestinal health is important for animal models of TBI to make a full recovery. This includes ensuring proper nutritional support and the prevention of gastric erosion and ulceration. Gastrointestinal protectant agents, such as histamine-2 receptor antagonists, proton pump inhibitors, and sucralfate may help alleviate abdominal discomfort and decrease incidence of gastric erosions and ulceration after TBI. Gastrointestinal stasis has been described in human TBI patients [72] and should be evaluated when possible in animal models. Quality of appetite, abdominal distension, normal sounds on auscultation, and fecal output can be assessed when considering treatment. Treatment with promotility agents may be beneficial in the relief of these signs, if present. Enteral nutrition is preferred for TBI patients and helps not only with gastrointestinal health, but also with immune and metabolic health [73]. Therefore, enteral feeding is recommended for TBI models and should be started as soon as the animal's condition allows.

In addition to the use of intravenous crystalloid fluids as previously mentioned, the use of hyperosmotic agents, such as hypertonic saline or mannitol, may benefit some animals with severely increased ICP [70]. It is important that these treatments are considered during the planning phase of the experiments to adequately address whether their use aligns with the goals of the study. Frequently, a craniotomy/craniectomy is created during TBI model induction. This allows for some expansion of the soft brain parenchyma and may help alleviate severe increases in ICP. Elevation of the head at a 15–30° angle can be performed in sedated or comatose patients to facilitate venous drainage from the head, aiding in reduction of ICP. Any bending of the neck must be avoided if employing this technique. [70].

Nearly 50% of human TBI patients have concomitant injuries, ranging from spinal cord injuries to external injuries [74]. To replicate these conditions, a variety of preclinical rodent models of TBI and concomitant injuries have been utilized [75]. These models usually result in more compromised animals and may require additional personnel training, support, and assessments than those outlined here. For example, studies involving TBI and long bone injuries require perioperative antibiotics to avoid infection and post-procedural analgesia to provide proper pain control [76, 77]. Additional measures needed to provide adequate support for these concomitant models should be discussed with veterinary and husbandry staff.

4.2 Assessments

A nonimpact, rapid inertial model of brain injury in swine has been described (summarized in [51]) and is an established, clinically relevant model of TBI. The apparatus used to rotate heads for these models is attached to the animals via a custom-built bite plate. With proper securing this should pose no problem, but laxity in the attachment can cause disconjugated motion of the pig's mouth versus the apparatus, resulting in dental, lingual, or other orofacial trauma. When using this type of device, a thorough orofacial examination should be performed to rule out injuries unrelated to the intended TBI that may complicate the disease model or impact ability of the animal to humanely reach the intended endpoint.

Neurological examination of swine can be challenging, but gross deficits in ambulation and proprioception, responsiveness to auditory or visual cues, and the presence of nystagmus, head tilt, or anisocoria can be used to help determine the suitability of an individual animal to humanely reach their previously determined study endpoint. In addition to neurological abnormalities from the primary injury disrupting brain tissue, many neurological abnormalities may be due to secondary injury after the trauma event, including increased ICP, excitotoxicity, reactive oxygen species, and inflammation [70].

Pain should be assessed at regular intervals in the immediate post-procedural period, usually dependent on institutional policies and monitoring regimens described in the AUP. Since rodents are a prey species, recognizing and interpreting signs of pain can be challenging for the inexperienced researcher [78]. Well-characterized and widely utilized facial expression recognition scales (grimace scales) are available for both mice [79] and rats [80]. Similar scales have been developed and utilized in swine, but are not as widely utilized in TBI models [81, 82]. Non-species-specific indicators of pain, such as decreased food consumption, decreased activity, decreased fecal and urine production, and decreased body weight, can also be monitored [78]. Headaches are common clinical sequelae to traumatic brain injury in people [83]. Because large animal models are used to mimic the human condition, it should be assumed that headaches are also common in post-TBI animals. Signs of pain in swine may include listlessness, bruxism, vocalization or avoidance of palpation relative to non-surgical areas, and inappetence. Any indications of unexpected pain should be addressed by administration of approved analgesics listed in the AUP or after consultation with veterinary staff.

The surgical site should be monitored at least daily for signs of seroma formation or infection. Signs that may warrant additional veterinary medical attention include dehiscence, redness, swelling, focal pain, and discharge, especially if the exudate is opaque or foul-smelling.

Recordkeeping is extremely important, particularly in the early period after a procedure has been performed. While many researchers keep records in their laboratory, recordkeeping at the cage level for rodents can be a means of communication between researchers, veterinary staff, and husbandry staff. Useful information includes notations of which procedure has been performed and when any analgesics or other treatments have been given. An example of a cage card that can be utilized for rodent post-procedural monitoring is provided in Fig. 2. For large animal species such as swine, individual animal records should be maintained by veterinary or research staff.

Front of card

RODENT POST-OPERATIVE CARE

Surgeon_______________ Surgery date__________

Type of surgery_____________________________

Date	Initials	Antibiotic AM/PM	Analgesic AM/PM	Comments
		/	/	
		/	/	
		/	/	
		/	/	
		/	/	

Research staff must post a work order explaining post-operative care of these animals.

Back of card

Date	Initials	Antibiotic AM/PM	Analgesic AM/PM	Comments
		/	/	
		/	/	
		/	/	
		/	/	
		/	/	
		/	/	

Fig. 2 Sample cage card for post-procedural documentation

5 Long-Term Post-procedural Care

Rodent models of TBI have typically focused on the pathophysiology experimental outcomes for up to a month following injury. Increasingly, studies have focused on pathology or effects of delayed or extended therapeutics for up to a year after TBI [84]. Accordingly, observation and support of injured animals beyond the initial post-operative period is often warranted. Although gross motor or cognitive deficits are not expected beyond the initial recovery period, assessments should still be made regularly between and during scheduled behavioral assays to verify that experimental injuries or normal aging have not produced chronic illness and that animals remain in good health.

Body weights may be obtained at weekly or other fixed intervals to provide an assessment of general health, with rapid or progressive weight loss being an indicator of underlying impairment or disease. Weight assessments may be combined with or substituted by measurement of overall body condition, particularly in cases where loss of body weight may be masked by other changes (e.g., progressive weight gain in free-fed, adult male rats). Scoring systems for body condition have been developed for rodents and other species [85, 86] and can provide a useful method of easily assessing overall health status.

Other clinical signs may also be general indicators of progressive disease and should be evaluated. These include abdominal enlargement, progressive dermatitis, unkempt appearance, hunched posture, and lethargy or decreased response to stimulation [87]. Humane endpoints should be defined in the AUP, along with plans for therapeutic intervention or removal from the study and euthanasia once the endpoints have been met. During long-term care, the surgical site should be monitored for any signs of infection or dehiscence. If nonresorbable sutures or wound clips were used for superficial wound closure, these should be removed 7–14 days after the procedure at the time of wound healing.

References

1. Berkner J, Mannix R, Qiu J (2016) Clinical traumatic brain injury in the preclinical setting. Methods Mol Biol 1462:11–28. https://doi.org/10.1007/978-1-4939-3816-2_2
2. Animal Welfare Act as Amended (2016) 7 USC §2131-2159. vol 7 USC
3. Office of Laboratory Animal Welfare (2015) Public health service policy on humane care and use of laboratory animals. National Institutes of Health, Bethesda
4. Institue of Laboratory Animal Research (2011) Guide for the care and use of laboratory animals, 8th edn. National Academies, Washington. https://doi.org/10.17226/12910
5. Institue of Laboratory Animal Research (2003) Guidelines for the care and use of mammals in neuroscience and behavioral research. National Academies, Washington
6. Office of Laboratory Animal Welfare (2002) Institutional animal care and use guidebook, 2nd edn. National Institutes of Health, Bethesda
7. Brown MJ, Winnicker C (2015) Animal Welfare. In: Fox JG, Anderson LC, Otto GM, Pritchett-Corning KR, Whary MT (eds) Laboratory animal medicine, 3rd edn. Academic, Boston, pp 1653–1672. https://doi.org/10.1016/B978-0-12-409527-4.00039-0
8. Conour LA, Murray KA, Brown MJ (2006) Preparation of animals for research--issues to consider for rodents and rabbits. ILAR J 47 (4):283–293
9. Di G, He L (2013) Behavioral and plasma monoamine responses to high-speed railway noise stress in mice. Noise Health 15(65):217–223. https://doi.org/10.4103/1463-1741.113506
10. Ishitake T (1990) Hemodynamic changes in skin microcirculation induced by vibration stress in the conscious rabbit. Kurume Med J 37(4):235–245
11. Toth LA, Trammell RA, Ilsley-Woods M (2015) Interactions between housing density and ambient temperature in the cage environment: effects on mouse physiology and behavior. J Am Assoc Lab Anim Sci 54(6):708–717

12. Tzamkiozis T, Stoeger T, Cheung K, Ntziachristos L, Sioutas C, Samaras Z (2010) Monitoring the inflammatory potential of exhaust particles from passenger cars in mice. Inhal Toxicol 22(Suppl 2):59–69. https://doi.org/10.3109/08958378.2010.519408
13. Duke JL, Zammit TG, Lawson DM (2001) The effects of routine cage-changing on cardiovascular and behavioral parameters in male Sprague-Dawley rats. Contemp Top Lab Anim Sci 40(1):17–20
14. Febinger HY, George A, Priestley J, Toth LA, Opp MR (2014) Effects of housing condition and cage change on characteristics of sleep in mice. J Am Assoc Lab Anim Sci 53(1):29–37
15. Kamnaksh A, Kovesdi E, Kwon SK, Wingo D, Ahmed F, Grunberg NE, Long J, Agoston DV (2011) Factors affecting blast traumatic brain injury. J Neurotrauma 28(10):2145–2153. https://doi.org/10.1089/neu.2011.1983
16. Capdevila S, Giral M, Ruiz de la Torre JL, Russell RJ, Kramer K (2007) Acclimatization of rats after ground transportation to a new animal facility. Lab Anim 41(2):255–261. https://doi.org/10.1258/002367707780378096
17. Swindle MM, Sistino JJ (2015) Anesthesia, analgesia, and perioperative care. In: Swindle MM, Smith AC (eds) Swine in the laboratory: surgery anesthesia, imaging, and experimental techniques, 3rd edn. CRC, Boca Raton, pp 39–88. https://doi.org/10.1201/b19430-31201/b19430-3
18. Magden ER, Mansfield KG, Simmons JH, Abee CR (2015) Nonhuman primates. In: Fox JG, Anderson LC, Otto GM, Pritchett-Corning KR, Whary MT (eds) Laboratory animal medicine, 3rd edn. Academic, Boston, pp 771–930. https://doi.org/10.1016/B978-0-12-409527-4.00017-1
19. Smith AC, Swindle MM (2006) Preparation of swine for the laboratory. ILAR J 47 (4):358–363
20. Berry ML, Linder CC (2007) Breeding systems: considerations, genetic fundamentals, genetic background, and strain types. In: Fox JG, Barthold SW, Davisson MT, Newcomer CE, Quimby FW, Smith AL (eds) The mouse in biomedical research, vol 1, 2nd edn. Academic, Burlington, pp 53–78. https://doi.org/10.1016/B978-012369454-6/50016-9
21. Osier N, Dixon CE (2016) The controlled cortical impact model of experimental brain trauma: overview, research applications, and protocol. Methods Mol Biol 1462:177–192. https://doi.org/10.1007/978-1-4939-3816-2_11
22. Done S (1995) Diagnosis of central nervous system disorders in the pig. In Pract 17 (7):318–327. https://doi.org/10.1136/inpract.17.7.318
23. Gelberg HB (2010) Neurologic disease in a pig. Vet Pathol 47(3):576–578. https://doi.org/10.1177/0300985810367894
24. Philips BH, Loria KO, Sirivelu MP, Jaber SM, Allen-Worthington KH, Veeder CL, Brice AK (2016) Pathology in practice. J Am Vet Med Assoc 249(7):755–757. https://doi.org/10.2460/javma.249.7.755
25. Friess SH, Ichord RN, Owens K, Ralston J, Rizol R, Overall KL, Smith C, Helfaer MA, Margulies SS (2007) Neurobehavioral functional deficits following closed head injury in the neonatal pig. Exp Neurol 204(1):234–243. https://doi.org/10.1016/j.expneurol.2006.10.010
26. Sullivan S, Friess SH, Ralston J, Smith C, Propert KJ, Rapp PE, Margulies SS (2013) Improved behavior, motor, and cognition assessments in neonatal piglets. J Neurotrauma 30(20):1770–1779. https://doi.org/10.1089/neu.2013.2913
27. Davies JA, Fransson BA, Davis AM, Gilbertsen AM, Gay JM (2015) Incidence of and risk factors for postoperative regurgitation and vomiting in dogs: 244 cases (2000-2012). J Am Vet Med Assoc 246(3):327–335. https://doi.org/10.2460/javma.246.3.327
28. Horn CC, Kimball BA, Wang H, Kaus J, Dienel S, Nagy A, Gathright GR, Yates BJ, Andrews PL (2013) Why can't rodents vomit? A comparative behavioral, anatomical, and physiological study. PLoS One 8(4):e60537. https://doi.org/10.1371/journal.pone.0060537
29. Mishra RK, Munivenkatappa A, Prathyusha V, Shukla DP, Devi BI (2017) Clinical predictors of abnormal head computed tomography scan in patients who are conscious after head injury. J Neurosci Rural Pract 8(1):64–67. https://doi.org/10.4103/0976-3147.193538
30. Goodwin RF (1957) The relationship between the concentration of blood sugar and some vital body functions in the new-born pig. J Physiol 136(1):208–217
31. Swiatek KR, Kipnis DM, Mason G, Chao KL, Cornblath M (1968) Starvation hypoglycemia in newborn pigs. Am J Phys 214(2):400–405
32. Chang YS, Park WS, Ko SY, Kang MJ, Han JM, Lee M, Choi J (1999) Effects of fasting and insulin-induced hypoglycemia on brain cell membrane function and energy metabolism during hypoxia-ischemia in newborn piglets. Brain Res 844(1–2):135–142
33. Prins ML, Hovda DA (2009) The effects of age and ketogenic diet on local cerebral metabolic rates of glucose after controlled cortical impact

injury in rats. J Neurotrauma 26 (7):1083–1093. https://doi.org/10.1089/neu.2008.0769
34. Yager JY, Heitjan DF, Towfighi J, Vannucci RC (1992) Effect of insulin-induced and fasting hypoglycemia on perinatal hypoxic-ischemic brain damage. Pediatr Res 31(2):138–142. https://doi.org/10.1203/00006450-199202000-00009
35. Prins ML (2008) Cerebral metabolic adaptation and ketone metabolism after brain injury. J Cereb Blood Flow Metab 28(1):1–16. https://doi.org/10.1038/sj.jcbfm.9600543
36. Prins ML, Matsumoto JH (2014) The collective therapeutic potential of cerebral ketone metabolism in traumatic brain injury. J Lipid Res 55(12):2450–2457. https://doi.org/10.1194/jlr.R046706
37. White H, Venkatesh B (2011) Clinical review: ketones and brain injury. Crit Care 15(2):219. https://doi.org/10.1186/cc10020
38. White H, Venkatesh K, Venkatesh B (2017) Systematic review of the use of ketones in the management of acute and chronic neurological disorders. J Neurol Neurosci 8:2. https://doi.org/10.21767/2171-6625.1000188
39. Houpt TR, Yang H (1995) Water deprivation, plasma osmolality, blood volume, and thirst in young pigs. Physiol Behav 57(1):49–54
40. Smith AJ, Clutton RE, Lilley E, Hansen KEA, Brattelid T (2017) PREPARE: guidelines for planning animal research and testing. Lab Anim 52:135–141. https://doi.org/10.1177/0023677217724823
41. Jaber SM, Hankenson FC, Heng K, McKinstry-Wu A, Kelz MB, Marx JO (2014) Dose regimens, variability, and complications associated with using repeat-bolus dosing to extend a surgical plane of anesthesia in laboratory mice. J Am Assoc Lab Anim Sci 53 (6):684–691
42. Bruins B, Kilbaugh TJ, Margulies SS, Friess SH (2013) The anesthetic effects on vasopressor modulation of cerebral blood flow in an immature swine model. Anesth Analg 116 (4):838–844. https://doi.org/10.1213/ANE.0b013e3182860fe7
43. Clevenger AC, Kilbaugh T, Margulies SS (2015) Carotid artery blood flow decreases after rapid head rotation in piglets. J Neurotrauma 32(2):120–126. https://doi.org/10.1089/neu.2014.3570
44. Baker NJ, Schofield JC, Caswell MD, McLellan AD (2011) Effects of early atipamezole reversal of medetomidine-ketamine anesthesia in mice. J Am Assoc Lab Anim Sci 50(6):916–920
45. Izer JM, Whitcomb TL, Wilson RP (2014) Atipamezole reverses ketamine-dexmedetomidine anesthesia without altering the antinociceptive effects of butorphanol and buprenorphine in female C57BL/6J mice. J Am Assoc Lab Anim Sci 53(6):675–683
46. Janssen CF, Maiello P, Wright MJ Jr, Kracinovsky KB, Newsome JT (2017) Comparison of atipamezole with yohimbine for antagonism of xylazine in mice anesthetized with ketamine and xylazine. J Am Assoc Lab Anim Sci 56 (2):142–147
47. Hoogstraten-Miller SL, Brown PA (2008) Techniques in aseptic rodent surgery. Curr Protoc Immunol 82:1.12.1–1.12.14. https://doi.org/10.1002/0471142735.im0112s82
48. Caro AC, Hankenson FC, Marx JO (2013) Comparison of thermoregulatory devices used during anesthesia of C57BL/6 mice and correlations between body temperature and physiologic parameters. J Am Assoc Lab Anim Sci 52 (5):577–583
49. Peterson K, Carson S, Carney N (2008) Hypothermia treatment for traumatic brain injury: a systematic review and meta-analysis. J Neurotrauma 25(1):62–71. https://doi.org/10.1089/neu.2007.0424
50. Thompson HJ, Tkacs NC, Saatman KE, Raghupathi R, McIntosh TK (2003) Hyperthermia following traumatic brain injury: a critical evaluation. Neurobiol Dis 12(3):163–173
51. Cullen DK, Harris JP, Browne KD, Wolf JA, Duda JE, Meaney DF, Margulies SS, Smith DH (2016) A porcine model of traumatic brain injury via head rotational acceleration. Methods Mol Biol 1462:289–324. https://doi.org/10.1007/978-1-4939-3816-2_17
52. Hellewell SC, Ziebell JM, Lifshitz J, Morganti-Kossmann MC (2016) Impact acceleration model of diffuse traumatic brain injury. Methods Mol Biol 1462:253–266. https://doi.org/10.1007/978-1-4939-3816-2_15
53. Glass TF, Fabian MJ, Schweitzer JB, Weinberg JA, Proctor KG (1999) Secondary neurologic injury resulting from nonhypotensive hemorrhage combined with mild traumatic brain injury. J Neurotrauma 16(9):771–782. https://doi.org/10.1089/neu.1999.16.771
54. Saha JK, Xia J, Grondin JM, Engle SK, Jakubowski JA (2005) Acute hyperglycemia induced by ketamine/xylazine anesthesia in rats: mechanisms and implications for preclinical models. Exp Biol Med 230(10):777–784
55. Tanaka T, Nabatame H, Tanifuji Y (2005) Insulin secretion and glucose utilization are impaired under general anesthesia with

sevoflurane as well as isoflurane in a concentration-independent manner. J Anesth 19(4):277–281. https://doi.org/10.1007/s00540-005-0341-1

56. Syring RS, Otto CM, Drobatz KJ (2001) Hyperglycemia in dogs and cats with head trauma: 122 cases (1997-1999). J Am Vet Med Assoc 218(7):1124–1129
57. Henninger N, Bouley J, Sikoglu EM, An J, Moore CM, King JA, Bowser R, Freeman MR, Brown RH Jr (2016) Attenuated traumatic axonal injury and improved functional outcome after traumatic brain injury in mice lacking Sarm1. Brain 139(Pt 4):1094–1105. https://doi.org/10.1093/brain/aww001
58. Nilsson P, Ronne-Engstrom E, Flink R, Ungerstedt U, Carlson H, Hillered L (1994) Epileptic seizure activity in the acute phase following cortical impact trauma in rat. Brain Res 637(1–2):227–232
59. Kharatishvili I, Nissinen JP, McIntosh TK, Pitkanen A (2006) A model of posttraumatic epilepsy induced by lateral fluid-percussion brain injury in rats. Neuroscience 140(2):685–697. https://doi.org/10.1016/j.neuroscience.2006.03.012
60. Kanitz E, Puppe B, Tuchscherer M, Heberer M, Viergutz T, Tuchscherer A (2009) A single exposure to social isolation in domestic piglets activates behavioural arousal, neuroendocrine stress hormones, and stress-related gene expression in the brain. Physiol Behav 98(1–2):176–185. https://doi.org/10.1016/j.physbeh.2009.05.007
61. Rowe RK, Harrison JL, Thomas TC, Pauly JR, Adelson PD, Lifshitz J (2013) Using anesthetics and analgesics in experimental traumatic brain injury. Lab Anim 42(8):286–291. https://doi.org/10.1038/laban.257
62. Browne KD, Iwata A, Putt ME, Smith DH (2006) Chronic ibuprofen administration worsens cognitive outcome following traumatic brain injury in rats. Exp Neurol 201(2):301–307. https://doi.org/10.1016/j.expneurol.2006.04.008
63. Thau-Zuchman O, Shohami E, Alexandrovich AG, Trembovler V, Leker RR (2012) The anti-inflammatory drug carprofen improves long-term outcome and induces gliogenesis after traumatic brain injury. J Neurotrauma 29(2):375–384. https://doi.org/10.1089/neu.2010.1673
64. Bolton Hall AN, Joseph B, Brelsfoard JM, Saatman KE (2016) Repeated closed head injury in mice results in sustained motor and memory deficits and chronic cellular changes. PLoS One 11(7):e0159442. https://doi.org/10.1371/journal.pone.0159442
65. Thelin EP, Frostell A, Mulder J, Mitsios N, Damberg P, Aski SN, Risling M, Svensson M, Morganti-Kossmann MC, Bellander BM (2016) Lesion size is exacerbated in hypoxic rats whereas hypoxia-inducible Factor-1 alpha and vascular endothelial growth factor increase in injured normoxic rats: a prospective cohort study of secondary hypoxia in focal traumatic brain injury. Front Neurol 7:23. https://doi.org/10.3389/fneur.2016.00023
66. Cheng JS, Craft R, Yu GQ, Ho K, Wang X, Mohan G, Mangnitsky S, Ponnusamy R, Mucke L (2014) Tau reduction diminishes spatial learning and memory deficits after mild repetitive traumatic brain injury in mice. PLoS One 9(12):e115765. https://doi.org/10.1371/journal.pone.0115765
67. Margulies SS, Kilbaugh T, Sullivan S, Smith C, Propert K, Byro M, Saliga K, Costine BA, Duhaime AC (2015) Establishing a clinically relevant large animal model platform for TBI therapy development: using Cyclosporin A as a case study. Brain Pathol 25(3):289–303. https://doi.org/10.1111/bpa.12247
68. Friess SH, Naim MY, Kilbaugh TJ, Ralston J, Margulies SS (2012) Premedication with meloxicam exacerbates intracranial haemorrhage in an immature swine model of non-impact inertial head injury. Lab Anim 46(2):164–166. https://doi.org/10.1258/la.2011.011084
69. Alderson P, Roberts I (1997) Corticosteroids in acute traumatic brain injury: systematic review of randomised controlled trials. BMJ 314(7098):1855–1859
70. Sande A, West C (2010) Traumatic brain injury: a review of pathophysiology and management. J Vet Emerg Crit Care 20(2):177–190. https://doi.org/10.1111/j.1476-4431.2010.00527.x
71. Crumrine RC, Marder VJ, Taylor GM, Lamanna JC, Tsipis CP, Scuderi P, Petteway SR Jr, Arora V (2011) Intra-arterial administration of recombinant tissue-type plasminogen activator (rt-PA) causes more intracranial bleeding than does intravenous rt-PA in a transient rat middle cerebral artery occlusion model. Exp Transl Stroke Med 3(1):10. https://doi.org/10.1186/2040-7378-3-10
72. Kao CH, ChangLai SP, Chieng PU, Yen TC (1998) Gastric emptying in head-injured patients. Am J Gastroenterol 93(7):1108–1112. https://doi.org/10.1111/j.1572-0241.1998.00338.x
73. Taylor SJ, Fettes SB, Jewkes C, Nelson RJ (1999) Prospective, randomized, controlled trial to determine the effect of early enhanced enteral nutrition on clinical outcome in

mechanically ventilated patients suffering head injury. Crit Care Med 27(11):2525–2531

74. Leitgeb J, Mauritz W, Brazinova A, Majdan M, Wilbacher I (2013) Impact of concomitant injuries on outcomes after traumatic brain injury. Arch Orthop Trauma Surg 133 (5):659–668. https://doi.org/10.1007/s00402-013-1710-0
75. McDonald SJ, Sun M, Agoston DV, Shultz SR (2016) The effect of concomitant peripheral injury on traumatic brain injury pathobiology and outcome. J Neuroinflammation 13(1):90. https://doi.org/10.1186/s12974-016-0555-1
76. Brady RD, Grills BL, Church JE, Walsh NC, McDonald AC, Agoston DV, Sun M, O'Brien TJ, Shultz SR, McDonald SJ (2016) Closed head experimental traumatic brain injury increases size and bone volume of callus in mice with concomitant tibial fracture. Sci Rep 6:34491. https://doi.org/10.1038/srep34491
77. Tsitsilonis S, Seemann R, Misch M, Wichlas F, Haas NP, Schmidt-Bleek K, Kleber C, Schaser KD (2015) The effect of traumatic brain injury on bone healing: an experimental study in a novel in vivo animal model. Injury 46 (4):661–665. https://doi.org/10.1016/j.injury.2015.01.044
78. Mayer J (2007) Use of behavior analysis to recognize pain in small mammals. Lab Anim (N Y) 36(6):43–48
79. Langford DJ, Bailey AL, Chanda ML, Clarke SE, Drummond TE, Echols S, Glick S, Ingrao J, Klassen-Ross T, Lacroix-Fralish ML, Matsumiya L, Sorge RE, Sotocinal SG, Tabaka JM, Wong D, van den Maagdenberg AM, Ferrari MD, Craig KD, Mogil JS (2010) Coding of facial expressions of pain in the laboratory mouse. Nat Methods 7(6):447–449. https://doi.org/10.1038/nmeth.1455
80. Sotocinal SG, Sorge RE, Zaloum A, Tuttle AH, Martin LJ, Wieskopf JS, Mapplebeck JC, Wei P, Zhan S, Zhang S, McDougall JJ, King OD, Mogil JS (2011) The Rat Grimace Scale: a partially automated method for quantifying pain in the laboratory rat via facial expressions. Mol Pain 7:55. https://doi.org/10.1186/1744-8069-7-55
81. Stasiak KL, Maul D, French E, Hellyer PW, VandeWoude S (2003) Species-specific assessment of pain in laboratory animals. Contemp Top Lab Anim Sci 42(4):13–20
82. Viscardi AV, Hunniford M, Lawlis P, Leach M, Turner PV (2017) Development of a piglet grimace scale to evaluate piglet pain using facial expressions following castration and tail docking: a pilot study. Front Vet Sci 4:51. https://doi.org/10.3389/fvets.2017.00051
83. Mayer CL, Huber BR, Peskind E (2013) Traumatic brain injury, neuroinflammation, and post-traumatic headaches. Headache 53 (9):1523–1530. https://doi.org/10.1111/head.12173
84. Xiong Y, Mahmood A, Chopp M (2013) Animal models of traumatic brain injury. Nat Rev Neurosci 14(2):128–142. https://doi.org/10.1038/nrn3407
85. Hickman DL, Swan M (2010) Use of a body condition score technique to assess health status in a rat model of polycystic kidney disease. J Am Assoc Lab Anim Sci 49(2):155–159
86. Ullman-Cullere MH, Foltz CJ (1999) Body condition scoring: a rapid and accurate method for assessing health status in mice. Lab Anim Sci 49(3):319–323
87. Morton DB, Griffiths PH (1985) Guidelines on the recognition of pain, distress and discomfort in experimental animals and an hypothesis for assessment. Vet Rec 116(16):431–436

Chapter 11

Laser Capture Microdissection of Single Cells, Cell Populations, and Brain Regions Affected by Traumatic Brain Injury

Harris A. Weisz, Deborah R. Boone, Stacy L. Sell, and Helen L. Hellmich

Abstract

Since its introduction, laser capture microdissection (LCM) methods have been extensively employed to study cell-specific functions in complex, heterogeneous tissues composed of multiple cell types. Laser capture microdissection is particularly suited to studies of the mammalian brain, which, because of its heterogeneity, presents a major challenge in studies that attempt to correlate region or cell type-specific function with distinct gene expression profiles. We have used LCM to study genomic changes in rat brain after experimental traumatic brain injury (TBI). The use of LCM allows precise measures of TBI-induced changes in gene expression in identified populations of brain cells and in anatomically distinct subregions of the rat hippocampus. We have been able to study gene expression in specific populations of dying and surviving hippocampal neurons after TBI and to detect circadian clock dysfunction in the suprachiasmatic nucleus after TBI. We have also used LCM to study epigenetic changes following TBI, mediated in part by small, noncoding microRNAs in different brain regions. We found strikingly different microRNAs are expressed in laser-captured single neurons compared to laser-captured brain areas from which they originate and manually dissected brain areas, indicating the importance of this technology to the study of TBI-induced changes in specific cell types.

Key words Laser capture microdissection, Traumatic brain injury, Hippocampus, Suprachiasmatic nucleus, microRNA, Region-specific gene expression

1 Introduction

Molecular studies of the injured brain are limited by the lack of effective methods to isolate specific cell types or cell populations. Cell sorting methods are efficient at isolating specific cell types but depend on cell-specific antibodies [1] and can only isolate a few cell populations at a time. Therefore, manual dissection of defined brain regions remains the standard in CNS studies. Here, we describe the use of laser capture microdissection (LCM) techniques that enable precise microdissection of single brain cells or defined regions of the brain for studies of experimental traumatic brain injury.

Amit K. Srivastava and Charles S. Cox, Jr. (eds.), *Pre-Clinical and Clinical Methods in Brain Trauma Research*, Neuromethods, vol. 139, https://doi.org/10.1007/978-1-4939-8564-7_11,

Following a brief overview, we will discuss the specific application of LCM methods used in our laboratory to address critical questions in the field of brain trauma research.

Since its development by Lance Liotta's group at the National Institutes of Health in 1996 [2, 3], LCM has been used in every field of investigation, especially in cancer and brain research, to study the molecular underpinnings of cell-specific or tissue-specific function [4–15]. With the present focus on big data [13], LCM is a powerful tool which is essential for the revolution in personalized medicine [16, 17]. To allow interrogation of cell-specific function, LCM can also be combined with other well-established histological and immunohistochemistry methods to study cell- or tissue-specific function [18–21]. Basically, the original method of LCM involves using an infrared (IR) laser to capture single cells or regions of cells onto a thermoplastic membrane (bonded to a cap) from stained sections of tissues (prepared RNase free). Subsequently, the cells are lysed in appropriate buffers, and nucleic acids (DNA, RNA) or proteins are isolated from the caps for subsequent genomic [22–24] or proteomic analysis [4, 25, 26]. This technique is illustrated in our JoVE video article by Boone et al., (https://www.ncbi.nlm.nih.gov/pubmed/23603738) [27]. Another method of laser capture utilizes an ultraviolet (UV) laser to cut the boundaries of cells or regions of interest and transfer of those cells into a collection tube or capture onto the thermoplastic membrane-coated cap. This LCM technique is illustrated in our second JoVE article by Weisz et al. (https://www.ncbi.nlm.nih.gov/pubmed/28930995) [28].

Although manually microdissected regions of the brain such as the hippocampus have been used for genome-wide transcriptional profiling in the past [29], we now know that cell types in the brain show cell- and brain area-specific patterns of expression [30]. Thus, in our experience, we found that LCM is necessary to study the transcriptional profile of distinct cell types in the brain [31].

Applications of LCM to research on the central nervous system vary widely. Some examples include the use of LCM to determine differences in TBI-induced changes in expression of GABA receptor subunits in the hippocampus and thalamus [32], identification of cell type-specific marker genes in the ventromedial hypothalamus [23], transcriptome profiling of epileptogenesis in the rat amygdala [33], determining how transplanted human bone marrow stem cells alter the expression of Nogo-A (an inhibitor of axonal regeneration) in oligodendrocytes and affect axonal morphology [34], transcriptional profiling of sprouting neurons in the somatosensory cortex of young and aged rats after experimental stroke [35], gene expression profiling of individual hypothalamic neurons in a rat model of obesity [24], microarray analysis of pools of astrocytes in a mouse model of amyotrophic lateral sclerosis [36], studies of intron retention and splicing in hippocampal neurons of wild-type and human apoE knock-in mice [37] and studies of *Bdnf* and trkB expression in rat spinal cord motor neurons [38]. LCM has also been used to recover transplanted human microglial cells and compare their gene expression to host glial cells [39].

There are several LCM systems that have been used in brain research, for instance, the Leica [40], and PALM systems [41]. The methods and discussion of LCM applications from our TBI studies are confined to the two instruments that we use, the Arcturus Pixcell IIe and Arcturus XT LCM systems (ThermoFisher). The Arcturus XT that we use has also been used in other brain studies to investigate the effects of breaching the blood brain barrier [42] and to laser capture retrogradely labeled (with cholera toxin subunit B conjugated to Alexa Fluor 488) phrenic motoneurons from rat spinal cord for qPCR analysis of the effects of TrkB gene therapy in a rat cervical spinal cord injury model [43].

Although some studies of cell death using LCM in the brain had been reported within a few years since the introduction of this technique [44] and LCM studies revealed the existence of large differences in gene expression between subregions [22], our research group were among the first to use LCM to study gene expression in the rodent hippocampus after experimental TBI, and specifically, to show that distinct subregions and adjacent glia expressed different pro-survival and pro-death genes up to one year after injury [45, 46]. These studies demonstrated that we were able to precisely laser capture distinct subregions (CA1–CA3 pyramidal layers, dentate gyrus, and glial cell layers) from the hippocampus and show that the cell populations in each subregion have distinct gene expression profiles that are differentially affected by TBI. Figure 1 shows a clean capture of hippocampal pyramidal CA1, CA2 and CA3 cell layers. We are able to laser capture individual neurons by adjusting the spot size of the infrared laser (Fig. 2). We were also able to use LCM to obtain neurons from the brain region known as the suprachiasmatic nucleus (SCN) which functions as the master circadian pacemaker in the body. We showed for the first

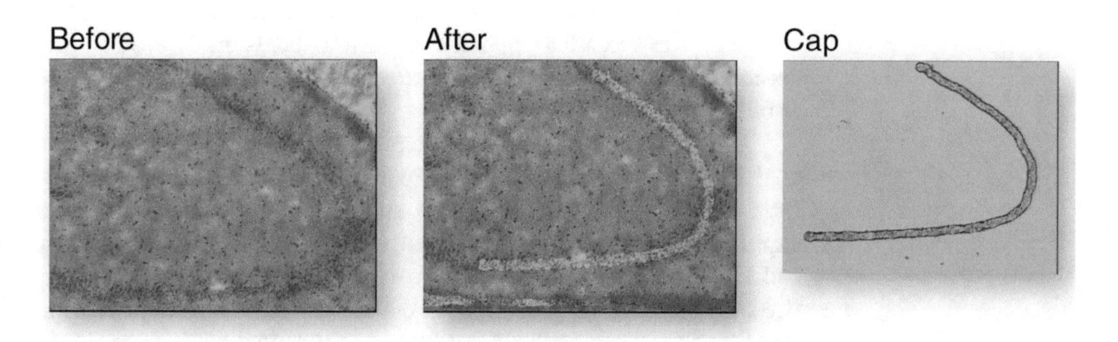

Fig. 1 Laser capture microdissection (LCM) of hippocampal CA1–CA3 pyramidal neuron layers. The before, after, and cap images of the LCM swaths were taken using the Arcturus PixCell IIe LCM system

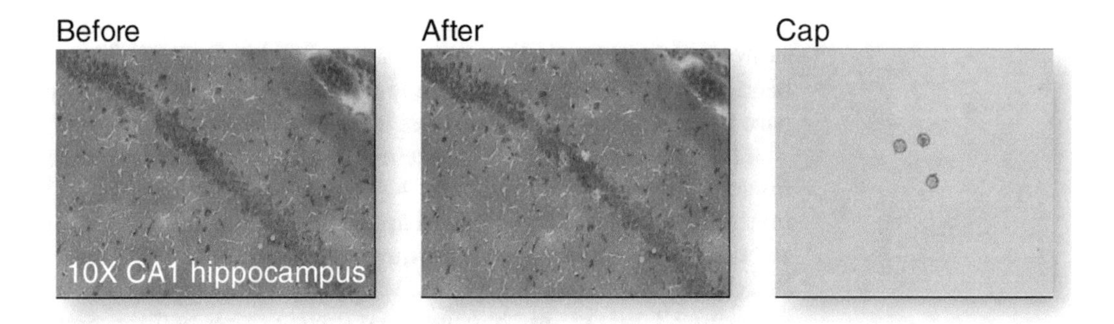

Fig. 2 Single cell laser capture microdissection (LCM) of three individual hippocampal CA1 pyramidal neurons. The before, after and cap images of the single cell captures were taken using the Arcturus PixCell IIe LCM system

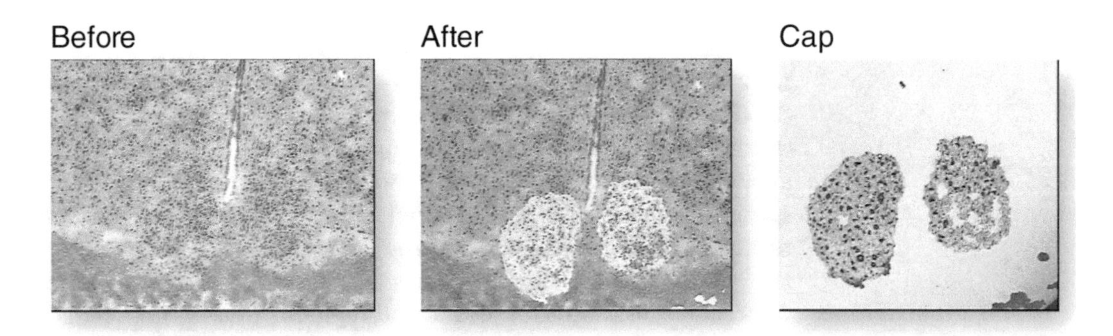

Fig. 3 Laser capture microdissection (LCM) of neurons in the suprachiasmatic nucleus, the master circadian pacemaker of the body. The before, after and cap images of the single cell captures were taken using the Arcturus PixCell IIe LCM system

time that TBI dysregulates expression of genes that generate daily circadian rhythms in the SCN [47]. Laser capture of the SCN is facilitated in part because it is an anatomically distinct bilateral nucleus located above the optic chiasma (Fig. 3). A recent study reported using LCM followed by mass spectrometry analysis to compare the circadian profiles of amino acids in the SCN and cerebral cortex with liver [48], showing that this method is also applicable to metabolomic studies.

To facilitate studies of gene expression in specific neurons such as dying, degenerating hippocampal neurons, we adapted Fluoro-Jade staining protocols [49] to suit the requirements for laser capture of single Fluoro-Jade-positive (FJ+) neurons. Using single cell LCM of FJ+ pyramidal neurons in the hippocampus, we were the first to demonstrate, that TBI plus hemorrhagic hypotension specifically suppressed expression of neuroprotective genes in dying, FJ+ neurons [50]. Figure 4 shows a clean capture of four

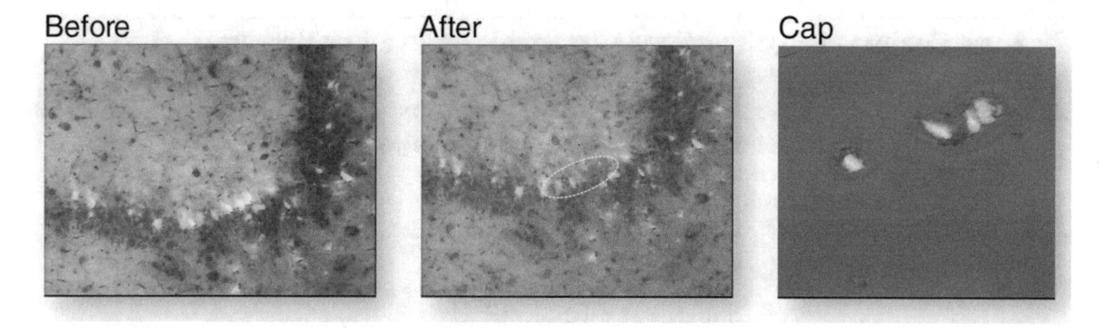

Fig. 4 Laser capture microdissection (LCM) of dying, Fluoro-Jade C-positive hippocampal CA3 pyramidal neurons. The before, after and cap images of the four single cell captures were taken using the Arcturus PixCell IIe LCM system. In the after image, dashed lines encircle the holes left by the infrared laser in the CA3 subfield

degenerating (FJ+) pyramidal neurons from the CA3 subfield from an injured rat hippocampus.

For over 14 years we have used LCM to delineate the molecular underpinnings of TBI. This technique was instrumental in helping us establish that two different drugs exerted their neuroprotective effects after TBI on common gene targets and cell signaling pathways in hippocampal neurons [51]. By using LCM to collect pools of dying and surviving hippocampal neurons for genome-wide microarray and PCR array analysis, we have shown that prosurvival gene expression is significantly decreased and prodeath gene expression is significantly increased in dying neurons after TBI [52, 53]. In our RNA interference studies, to examine the effects of knocking down (using adeno-associated virus (AAV)-siRNA) expression of neuronal nitric oxide synthase and glutathione peroxidase-1, two genes involved in neurodegeneration and neuroprotection, respectively, in the injured rat brain, we used LCM to capture AAV infected hippocampal neurons for genome-wide microarray analysis [54].

Recently, we have been using LCM to study the effects of TBI on expression of microRNAs (miRNAs) that negatively regulate the expression of gene targets involved in the pathogenesis of TBI; we showed that TBI-induced miRNAs suppress the expression of hundreds of pro-survival gene targets in dying, FJ+ hippocampal neurons obtained by laser capture [55]. In this same study, we also used the technique of in situ-hybridization-guided LCM [56] to obtain dying, FJ+ neurons that expressed miR-15b for microfluidic qPCR analysis [55].

2 Laser Capture Microdissection in Traumatic Brain Injury Research

2.1 Materials

2.1.1 Animals

Adult male Sprague-Dawley rats, 300–400 g are obtained from vendor Charles Rivers Laboratories.

2.1.2 Tissue Preparation

1. CM 1850 cryostat (Leica, IL).
2. RNase Zap.
3. Superfrost microscope glass slides, uncharged.
4. Arcturus PEN membrane glass slides (Applied Biosystems, CA).
5. Single-edge razor blades.
6. Disposable Vinyl Specimen Molds 25 mm × 20 mm × 5 mm.
7. Optimum Cutting Temperature "OCT" compound.
8. Disposable, low profile blades.

2.1.3 Staining of Brain Tissue with Nissl Stain and Fluoro-Jade C

1. Eliminase.
2. RNase Zap.
3. Cresyl Violet.
4. Fluoro-Jade C (Histo-Chem, AR).
5. ETOH/Xylene.
6. Staining dishes and racks.
7. Stericups 0.2 μm filters.

2.1.4 Laser Capture Microdissection of Single, Injured, FJ+ Cells and Hippocampal Regions

1. ArcturusXT Laser Capture Microdissection System (Applied Biosystems, CA).
2. Arcturus CapSure Macro Caps (Applied Biosystems, CA).
3. Gene Amp (0.5 ml) thin walled reaction tube (Applied Biosystems, CA).
4. Lysis buffer from RNAqueous Micro Kit (Ambion, TX).

2.1.5 RNA Isolation and qPCR

1. miRVana RNA Isolation Kit (Ambion)—For whole tissue.
2. miScript Single Cell Kit (Qiagen)—for laser-captured single cells and regions.
3. Roche Lightcycler 96.

2.2 Animal Preparation

For TBI we use adult, male Sprague Dawley rats (300–400 × *g*, Charles River Laboratories, Houston, TX). Rats are subjected to experimental fluid percussion injury (FPI), as described in Shimamura et al., and Rojo et al., [45, 52]. In brief, animals are anesthetized with 4% isoflurane in an anesthetic chamber, intubated, and mechanically ventilated with 1.5–2.0% isoflurane in O_2/room air

(30:70) using a volume ventilator (NEMI Scientific: New England Medical Instruments, Medway, MA). Rats are positioned in a stereotaxic head holder, after which a midline incision of the skin is performed and skull exposed. A craniotomy is performed lateral to the sagittal suture, midway between lambda and bregma sutures and the bone flap is removed exposing intact dura at the site. A modified 20-gauge needle hub is secured over the craniotomy site with cyanoacrylic adhesive and cemented in place with hygienic dental acrylic. FPI is administered through a trauma device consisting of a Plexiglas cylinder 60 cm long and 4.5 cm in diameter filled with isotonic saline connected to a pressure transducer (Statham PA856-100, Data Instruments, Acton, MA). The transducer is connected to the rat by a plenum tube cemented to a craniotomy trephined in the skull. TBI is induced when a 4.8-kg steel pendulum struck the piston after being dropped from a variable height that determines the intensity of injury. Pressure pulse recordings are taken on a storage oscilloscope triggered photoelectrically by the descent of the pendulum and are used to calibrate the injury severity between animals. Animals receive a moderate-to-severe injury (2.3 atm) and severity is assessed by the time it takes the animal to right itself from a supine to prone position, otherwise known as a righting reflex. At 24 h post-FPI (most of our data is obtained from animals sacrificed at 24 h but other post-injury time points are also studied in our laboratory) animals are anesthetized with 4% isoflurane in an anesthetic chamber and sacrificed. Brains are quickly removed and frozen on dry ice and stored at −80 °C until used for LCM.

2.3 Sectioning for Laser Capture Microdissection

Before LCM of tissue to be used for transcriptional analysis it is imperative that all surfaces be thoroughly cleaned with an RNase eliminating detergent (Eliminase, RNase-Zap) to reduce the risk of RNA degradation. This includes the areas used for staining, sectioning, and around the LCM device as well as all dishware. First, we retrieve excised brains from storage and place into a cryostat (CM 1850, Leica) to equilibrate to −20 °C. After equilibration (about 10 min) the brain is placed ventral side up on a gauze sheet on the cryostat stage. Using a clean, RNase-free razor blade, we separate the posterior portion just rostral to the cerebellum (can be saved or discarded) and the portion just anterior to the optic chiasm. The anterior portion contains two brain regions of interest, frontal cortex and nucleus accumbens, and the other portion contains the hippocampus and suprachiasmatic nucleus. A small amount of optimal cutting temperature (OCT) medium is placed into two separate cryomolds. The brain containing the frontal association cortex and nucleus accumbens core is then placed anterior side down into the OCT medium and more OCT is added to completely cover the tissue. We then repeat this step with the brain tissue containing the hippocampus (HP) and suprachiasmatic nucleus (SCN), placing the tissue anterior side as well and covering with OCT. The OCT medium should completely freeze after approximately 20 min inside the cryostat at −20 °C.

Once frozen, a mounting head can then be placed onto the cryomold with a small amount of OCT and secured to the head attachment of the cryostat. The angle of the mounted cryomold may need to be adjusted in relation to the cryostat blade so that the sections are even and symmetrical when cut. The thickness of the section will depend on the specific cell type and/or region of interest. For example, if you are intending to capture individual cells with only the IR laser then the sectioning thickness should not exceed 20 μm, whereas for brain region collection using UV laser cutting, up to 100 μm sections can be used. For the purposes of our studies, section thickness was set to 30 μm for collection. Regions are identified while sectioning by referring to a stereotactic brain atlas (Paxinos and Watson, The Rat Brain Atlas) [28] and sectioned onto polyethylene napthalate (PEN) membrane slides (Life Technologies). Sections taken for single cell laser capture are mounted on precleaned (with 100% ethanol) uncoated glass slides (Fisher Scientific, Pittsburgh, PA). Note that to minimize the risk of RNA degradation, we cut 25–100 sections (25 sections per rack) and keep the racks in the cryostat at −20 °C until they are stained. However, for optimum results and to maintain RNA integrity, we recommend cutting 25 sections at a time, staining, performing laser capture then repeating: section, stain, LCM until the desired numbers of cells are laser-captured.

2.4 Staining Protocol for Brain Region and Single Cell Laser Capture Microdissection

For LCM of brain regions, warm a rack of brain sections at room temperature for approximately 30 s and then immediately place in 75% ETOH for 1 min. After fixation, rinse slides in RNase-free water (1 min), stain with 1% Cresyl Violet (1 min), rinse in RNase-free water (3 × 1 min), dehydrate with 95% ETOH (2 × 30 s), 100% ETOH (2 × 30 s), and xylene (2 × 3 min). Airdry the rack for no more than 10 min at room temperature in a chemical fume hood.

A modified staining protocol is used for identification of individual dying and surviving neurons. This protocol incorporates Fluoro Jade-C, a neuronal marker of cell death that binds preferentially to dying cells [55]. As before, a rack of slides is warmed at room temperature for no more than 30 s and then immediately placed in 75% ETOH for 1 min. After fixation, slides are rinsed in RNase-free water (1 min), stained with 1% Cresyl Violet (20 s), rinsed in RNase-free water (2 × 30 s), stained with 0.001% Fluoro-Jade C (4 min), rinsed in RNase-free water (3 × 1 min), then dehydrated with 95% ETOH (30 s), 100% ETOH (30 s), and xylene (2 × 3 min).

2.5 Laser Capture Microdissection

LCM of single cells and hippocampal swaths is performed as shown in Boone et al., (https://www.ncbi.nlm.nih.gov/pubmed/23603738) [27] and described in Rojo et al. [52]. LCM of identified brain regions is performed as shown in Weisz et al. (https://www.ncbi.nlm.nih.gov/pubmed/28930995) [28]. In our studies of regional gene expression in the hippocampus, prefrontal cortex and SCN, approximately 25 sections each of hippocampal pyramidal layers (CA1, CA2, and CA3), dentate granule cell layers, prefrontal cortex cells, or SCN cells are serially collected on LCM Macro Caps and immediately placed into a 0.5 mL RNase-free tube with 100 μL of RNAqueos lysis buffer (Ambion), vortexed to ensure complete lysis and then placed at −80 °C until RNA is isolated. Total RNA is isolated from laser-captured cell pools using the RNAqueous kit (Ambion, ThermoFisher) following manufacturer's protocols.

3 Application of LCM Methods to Study of TBI

3.1 Using Laser Capture Microdissection to Address Cell Heterogeneity in Complex Nervous System Tissues after Experimental Traumatic Brain Injury

Despite numerous treatments successfully tested in experimental models of traumatic brain injury (TBI), none have translated into FDA approved treatment [57]. Contributing factors include the heterogeneity of clinical TBI [58] and the varied response of animal brain tissue to both TBI and subsequent interventions. The mammalian brain is composed of hundreds to thousands of cell types [59]. Commonly employed methods to examine patterns of gene or protein expression after experimental TBI typically utilize manually dissected regions of tissue. This form of sampling has the potential to mask transcripts that are expressed at low levels in individual injured cells or to imply a large signal is representative of all cells. To address this tissue heterogeneity, we [52] and others [22] have performed expression profiling of hippocampal subregions using LCM. This approach validates the method as a valuable technique for accurate assessment of gene expression in anatomically distinct hippocampal subfields. Here, we further extend these studies by identifying similarities and disparities between expression of small, noncoding miRNAs in the rat hippocampus- whole tissue, LCM regions, and individual injured neurons after TBI.

We recently showed that TBI-induced miRNAs, by suppressing expression of pro-survival gene targets, are strongly associated with hippocampal neuronal cell death [55]. Our long-term goals are to therapeutically target miRNAs linked to neurodegeneration and to alter miRNA target gene expression in injury-associated pathways to minimize post-TBI damage; however, we are concerned that analysis of expression levels in a heterogeneous mix of cell types, e.g., laser-captured swaths of hippocampal neurons or manually dissected brain regions, may inadvertently select miRNAs not biologically relevant. Thus, our objective in this study was to characterize heterogeneity of miRNA expression after TBI in the hippocampus, a brain region important for learning and memory, by comparing whole hippocampus with individual and swaths of hippocampal pyramidal neurons obtained by LCM.

Following LCM of single cells, RNA isolation, reverse transcription, and preamplification from approximately 50 FJ ± cells was performed using the miScript Single Cell qPCR System (Qiagen) following manufacture's protocol utilizing a universal primer for both cDNA synthesis and preamplification. Total RNA was isolated from laser-captured brain regions using the RNAqueous kit (Ambion) follow manufacturer's protocol. Whole tissue RNA isolation was completed using the miRVana RNA isolation Kit (Ambion) following manufacturer's protocol. RNA from whole dissected and LCM hippocampal pyramidal layers was assessed for integrity and concentration using an Agilent Bioanalyzer (Agilent Technologies). Approximately 1 ng of total RNA isolated from whole hippocampus or laser-captured hippocampal pyramidal layers was used in lieu of tissue as specified in the miScript Single Cell Kit protocol. Quality control PCRs using a synthetic *C. elegans* miR-39 spike-in were performed to ensure RNA isolation and reverse transcription efficiency. Quantitative real-time PCR was performed using preamplified cDNA samples on miScript miFinder PCR Arrays (Qiagen) following manufacturer's protocols on a Roche Lightcyler 96 (Roche Industries).

A Venn diagram of miRNA expression in whole hippocampus, swathes of pyramidal neurons and groups of individual laser-captured neurons 24 h after TBI shows surprisingly little overlap (Fig. 5), contrary to the expectation that miRNAs differentially expressed in the single cells or cell layers would also be differentially expressed in the whole hippocampal tissue. The miFinder PCR array analysis (Table 1) compares whole hippocampus (containing multiple cell types), laser-captured swathes of hippocampal pyramidal neurons from the CA1–CA3 subfields (a mix of dying, Fluoro-Jade positive [FJ+] and surviving, Fluoro-Jade negative [FJ−] neurons), and 50 cell pools of individual, FJ+ cells. Notably, individual dying cells have transcriptional profiles not represented in the whole injured hippocampus. In fact, expression of multiple differentially expressed (compared to controls) miRNAs in whole hippocampus or swaths of CA1–CA3 regions were not significant in pools of dying (FJ+) neurons. Based on these data, we suggest that whole tissue analysis masks or erroneously accentuates the altered expression of microRNAs in specific cell types that could be biologically relevant in disease and injury.

The implications of these results can be profound. For instance, miR-17-5p, a tumor suppressor (i.e., promotes apoptosis) or oncogene (i.e., promotes cell proliferation) depending on cellular context [60] regulates both pro-survival and pro-death target genes. It is significantly upregulated in the whole hippocampus but not in

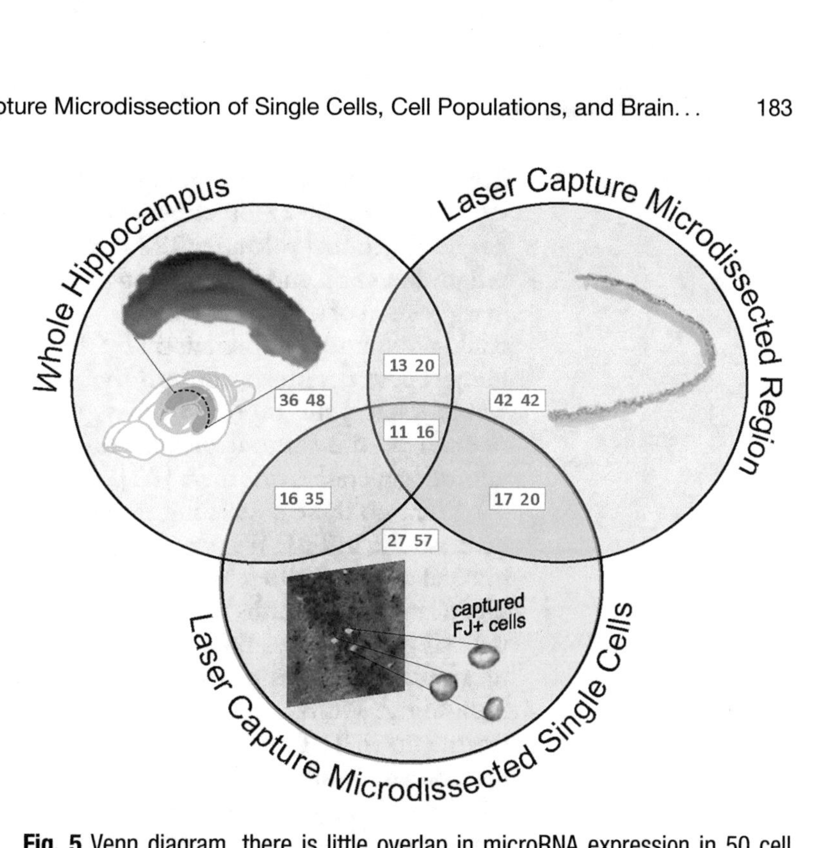

Fig. 5 Venn diagram, there is little overlap in microRNA expression in 50 cell pools of Fluoro-Jade C-positive hippocampal neurons, laser capture microdissected swaths of hippocampal CA1–CA3 subregions, and manually dissected hippocampal tissue. Red (numbers of upregulated miRNAs), Blue (numbers of downregulated miRNAs)

Table 1
Contrasting miRNA expression in FJC cells, hippocampal cell swaths, and manually dissected whole hippocampus

	miRNA fold changes		
	Whole hippocampus	Pyramidal layer	Single cell
miR-21-5p	*2.20	*2.80	1.25
miR-27a-3p	*1.32	1.29	1.40
miR-150-5p	*0.65	0.84	1.06
miR-92a-3p	*0.68	1.20	1.29
miR-191a-5p	*0.88	0.91	0.88
miR-17-5p	*1.49	1.46	1.19
miR-20a-5p	*1.45	*1.78	1.18
miR-146a-5p	*1.43	0.91	1.15
miR-15b-5p	*1.40	1.74	1.92
miR-223-3p	*7.37	10.63	2.06
miR-142-3p	2.63	*3.81	1.19

All statistical modeling was completed in *R* using a Student's *t*-test for within group comparisons, an * denotes a statistically significant difference when compared to control (naïve) tissue dying, FJ+ neurons. Depending on the source (cells, whole tissue), we can infer miR-17-5p either promotes cell death or survival or has no functional role after TBI. Likewise, miR-21 regulates both cell proliferation and apoptosis [61], therefore, significant changes in expression of miR-21 in whole hippocampus, but not single cells, could confound interpretation of TBI's influence on neuronal viability. These data raise concern that studies reporting changes in whole tissue gene expression may generate data not biologically relevant to the survival of individual neurons, a critical issue in neurodegenerative disorders [62].

What do these conflicting results mean in terms of therapeutic drug studies in TBI? We previously determined the effects of experimental drugs on a heterogeneous mix of cell types, as in the whole hippocampus or swaths of laser-captured neurons [51]. If individual cell types express different levels of genes and miRNAs, then measurements of gene expression in heterogeneous tissues may be misleading. We speculate the failure of all clinical trials of drug treatments in TBI may, in part, be attributable to erroneous interpretation of results from experimental studies in animal models that failed to identify mechanisms of injury in specific vulnerable cell populations. Brain heterogeneity may also have contributed to the disappointing failures of drug trials in other neurodegenerative disorders [63].

Our data suggest examining discrete populations of cells obtained by LCM can facilitate development of more targeted and effective therapies for TBI and other neurodegenerative disorders. Factors such as age and species-specific genetic heterogeneity can also confound the biological and translational relevance of brain injury studies in rodent models of TBI. For instance, age-related changes in gene expression differ greatly between mice and humans, requiring caution in generalizing conclusions about aging from mouse studies alone [64]. We conclude that traditional analysis of gene/microRNA expression from large swathes or chunks of tissue may provide misleading information about transcriptional dynamics. Examination of discrete populations of laser-captured cells, using recently developed methods of single cell analysis [65] is likely to be more accurate in providing mechanistic information thus eventually leading to increased accuracy of translating the results of preclinical studies into treatments for TBI and other human diseases.

3.2 Laser Capture Microdissection of Multiple Brain Regions from Individual Brains for microRNA Analysis

In our LCM studies, we are investigating molecular dynamics in several rat brain regions adversely impacted by TBI. We focus on regions in rats that have synonymous structures in human TBI patients which are associated with higher order executive functioning (i.e. frontal cortex [66]) and TBI comorbidities including; depression (nucleus accumbens [67]) and circadian rhythm disorders (suprachiasmatic nucleus [68]). LCM has allowed us to study the region-specific molecular changes induced after TBI by genome-wide microarray, PCR array and individual TaqMan and digital qPCR analyses.

The UV-cutting and IR lasers in the Arcturus XT system allow for precise microdissection of desired brain regions with guidance by stereotaxic coordinates from a rat brain atlas to identify and laser capture each brain region. Sectioning for the frontal cortex begins as soon as the cerebrum is exposed when trimming the OCT on the cryomolds. Sections are collected until the Secondary Motor Cortex (M2) is reached at Bregma 5.16 mm. For the nucleus accumbens, we continue sectioning until the anterior commissure is visible in both hemispheres at Bregma 1.80 mm and collect sections until the lateral ventricles appear to connect to the anterior commissure at Bregma 0.84 mm. Extreme care must be taken when collecting the SCN. We sectioned only until the optic chiasm appears flattened and the third ventricle becomes apparent at Bregma −0.48 mm. Sections are collected from that point until the supraoptic decussation begins at Bregma −0.72 mm. The SCN is easily passed over during sectioning, so it may be necessary to collect more sections encompassing the optic chiasm to ensure that this region of interest is obtained. Hippocampus collection begins when the horns of the granule cell layer of the dentate gyrus are visibly apparent at Bregma −3.00 mm and continue until the hippocampal CA subfields are fused at Bregma −4.78 mm.

We examined expression of miRNAs in these four brain regions in TBI rats. Expression of miR-15b, a miRNA that we have shown suppresses the expression of several pro-survival genes, including *Bdnf* and *Bcl-2* in the injured rat brain, is strikingly different in all four brain areas (Fig. 6); notably it is highly upregulated in the hippocampus and nucleus accumbens but downregulated in the SCN and prefrontal cortex which suggests that there is differential expression of pro-survival gene targets in different brain regions

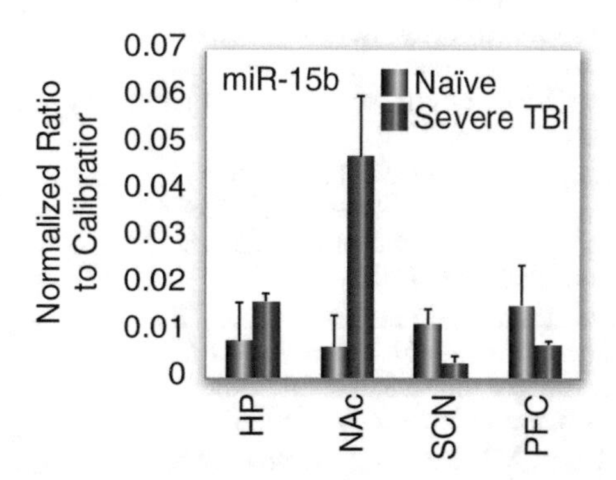

Fig. 6 Quantitative real-time PCR analysis of microRNA expression in laser capture microdissected cells from four brain regions affected by traumatic brain injury. *HP* hippocampus. *NAc* nucleus accumbens. *SCN* suprachiasmatic nucleus. *PFC* prefrontal cortex

after TBI. Currently, we are using LCM to confirm the negative regulation of pro-survival and pro-death gene targets by TBI-dysregulated miRNAs and to study the effects of therapeutic drugs and noninvasive neuroprotective treatments on region-specific gene expression (manuscript in press, Esenaliev et al., *J Neurotrauma*).

3.3 Methodological Considerations

Reporting the details of RNA isolation and RNA quality analysis (RNA integrity) is always necessary since the preservation of RNA is of prime importance in LCM experiments that segue into molecular studies. The quality of RNA is a key determining factor in whether we obtain accurate and biologically relevant genomic data from our samples. Small reported differences [69] in section thickness, staining, and processing procedures in the LCM literature are not as critical as maintaining RNA integrity during these procedures. Thus, we have empirically determined that in addition to the precautions detailed here on establishing and maintaining an RNase-free environment and preparing all reagents with RNase-free water, the key determinant of a successful LCM experiment is speed, especially in the actual laser capture process. Speed and accuracy can only be acquired by extensive practice and experience. Therefore, it is unreasonable to expect to obtain samples that yield good quality RNA in the beginning of an LCM project until all aspects of LCM are perfected. Our JoVE video articles were produced to enable visualization of the laser capture process and enhance learning of this sophisticated and incredibly useful technique in brain trauma research.

References

1. Cahoy JD, Emery B, Kaushal A, Foo LC, Zamanian JL, Christopherson KS, Xing Y, Lubischer JL, Krieg PA, Krupenko SA, Thompson WJ, Barres BA (2008) A transcriptome database for astrocytes, neurons, and oligodendrocytes: a new resource for understanding brain development and function. J Neurosci 28:264–278
2. Emmert-Buck MR, Bonner RF, Smith PD, Chuaqui RF, Zhuang Z, Goldstein SR, Weiss RA, Liotta LA (1996) Laser capture microdissection. Science 274:998–1001
3. Bonner RF, Emmert-Buck M, Cole K, Pohida T, Chuaqui R, Goldstein S, Liotta LA (1997) Laser capture microdissection: molecular analysis of tissue. Science 278:1481–1483
4. Banks RE, Dunn MJ, Forbes MA, Stanley A, Pappin D, Naven T, Gough M, Harnden P, Selby PJ (1999) The potential use of laser capture microdissection to selectively obtain distinct populations of cells for proteomic analysis-- preliminary findings. Electrophoresis 20:689–700
5. Webb T (2000) Laser capture microdissection comes into mainstream use. J Natl Cancer Inst 92:1710–1711
6. Nagle RB (2001) New molecular approaches to tissue analysis. J Histochem Cytochem 49:1063–1064
7. Rubin MA (2001) Use of laser capture microdissection, cDNA microarrays, and tissue microarrays in advancing our understanding of prostate cancer. J Pathol 195:80–86
8. Taatjes DJ, Palmer CJ, Pantano C, Hoffmann SB, Cummins A, Mossman BT (2001) Laser-based microscopic approaches: application to cell signaling in environmental lung disease. Biotechniques 31:880–888. 890, 892
9. Torres-Munoz J, Stockton P, Tacoronte N, Roberts B, Maronpot RR, Petito CK (2001) Detection of HIV-1 gene sequences in hippocampal neurons isolated from postmortem AIDS brains by laser capture microdissection. J Neuropathol Exp Neurol 60:885–892
10. Trogan E, Choudhury RP, Dansky HM, Rong JX, Breslow JL, Fisher EA (2002) Laser capture

microdissection analysis of gene expression in macrophages from atherosclerotic lesions of apolipoprotein E-deficient mice. Proc Natl Acad Sci U S A 99:2234–2239

11. Luzzi V, Holtschlag V, Watson MA (2001) Expression profiling of ductal carcinoma in situ by laser capture microdissection and high-density oligonucleotide arrays. Am J Pathol 158:2005–2010
12. Suarez-Quian CA, Goldstein SR, Pohida T, Smith PD, Peterson JI, Wellner E, Ghany M, Bonner RF (1999) Laser capture microdissection of single cells from complex tissues. BioTechniques 26:328–335
13. Datta S, Malhotra L, Dickerson R, Chaffee S, Sen CK, Roy S (2015) Laser capture microdissection: Big data from small samples. Histol Histopathol 30:1255–1269. https://doi.org/10.14670/HH-11-622. HH-11-622 [pii]
14. Burgess JK, Hazelton RH (2000) New developments in the analysis of gene expression. Redox Rep 5:63–73
15. Rekhter MD, Chen J (2001) Molecular analysis of complex tissues is facilitated by laser capture microdissection. Cell Biochem Biophys 35:103–113
16. Hamburg MA, Collins FS (2010) The path to personalized medicine. N Engl J Med 363:301–304. https://doi.org/10.1056/NEJMp1006304. NEJMp1006304 [pii]
17. Pasinetti GM, Fivecoat H, Ho L (2010) Personalized medicine in traumatic brain injury. Psychiatr Clin North Am 33:905–913. https://doi.org/10.1016/j.psc.2010.09.003. S0193-953X(10)00078-X [pii]
18. Fend F, Kremer M, Quintanilla-Martinez L (2000) Laser capture microdissection: methodical aspects and applications with emphasis on immuno-laser capture microdissection. Pathobiology 68:209–214
19. Fend F, Emmert-Buck MR, Chuaqui R, Cole K, Lee J, Liotta LA, Raffeld M (1999) Immuno-LCM: laser capture microdissection of immunostained frozen sections for mRNA analysis. Am J Pathol 154:61–66
20. Murakami H, Liotta L, Star RA (2000) IF-LCM: laser capture microdissection of immunofluorescently defined cells for mRNA analysis rapid communication. Kidney Int 58:1346–1353
21. Bernard R, Burke S, Kerman IA (2011) Region-specific in situ hybridization-guided laser-capture microdissection on postmortem human brain tissue coupled with gene expression quantification. In: Murray GL (ed) Laser capture microdissection: methods in moloecular biology. Springer Science & Business Media LLC, New York, pp 345–361
22. Datson NA, Meijer L, Steenbergen PJ, Morsink MC, van der Laan S, Meijer OC, de Kloet ER (2004) Expression profiling in laser-microdissected hippocampal subregions in rat brain reveals large subregion-specific differences in expression. Eur J Neurosci 20:2541–2554. https://doi.org/10.1111/j.1460-9568.2004.03738.x. EJN3738 [pii]
23. Segal JP, Stallings NR, Lee CE, Zhao L, Socci N, Viale A, Harris TM, Soares MB, Childs G, Elmquist JK, Parker KL, Friedman JM (2005) Use of laser-capture microdissection for the identification of marker genes for the ventromedial hypothalamic nucleus. J Neurosci 25:4181–4188. https://doi.org/10.1523/JNEUROSCI.0158-05.2005. 25/16/4181 [pii]
24. Paulsen SJ, Larsen LK, Jelsing J, Janssen U, Gerstmayer B, Vrang N (2009) Gene expression profiling of individual hypothalamic nuclei from single animals using laser capture microdissection and microarrays. J Neurosci Methods 177:87–93. https://doi.org/10.1016/j.jneumeth.2008.09.024. S0165-0270(08)00577-3 [pii]
25. Craven RA, Banks RE (2001) Laser capture microdissection and proteomics: possibilities and limitation. Proteomics 1:1200–1204
26. Liao L, Cheng D, Wang J, Duong DM, Losik TG, Gearing M, Rees HD, Lah JJ, Levey AI, Peng J (2004) Proteomic characterization of postmortem amyloid plaques isolated by laser capture microdissection. J Biol Chem 279:37061–37068. https://doi.org/10.1074/jbc.M403672200. M403672200 [pii]
27. Boone DR, Sell SL, Hellmich HL (2012) Laser capture microdissection of enriched populations of neurons or single neurons for gene expression analysis after traumatic brain injury. J Vis Exp 74:1–7
28. Weisz HA, Boone DR, Sell SL, Hellmich HL (2017) Stereotactic atlas-guided laser capture microdissection of brain regions affected by traumatic injury. J Vis Exp 11. https://doi.org/10.3791/56134
29. Wang X, Pal R, Chen XW, Kumar KN, Kim OJ, Michaelis EK (2007) Genome-wide transcriptome profiling of region-specific vulnerability to oxidative stress in the hippocampus. Genomics 90:201–212. https://doi.org/10.1016/j.ygeno.2007.03.007. S0888-7543(07)00068-7 [pii]
30. Ko Y, Ament SA, Eddy JA, Caballero J, Earls JC, Hood L, Price ND (2013) Cell type-specific genes show striking and distinct

patterns of spatial expression in the mouse brain. Proc Natl Acad Sci U S A 110:3095–3100

31. Okaty BW, Sugino K, Nelson SB (2011) Cell type-specific transcriptomics in the brain. J Neurosci 31:6939–6943
32. Drexel M, Puhakka N, Kirchmair E, Hortnagl H, Pitkanen A, Sperk G (2015) Expression of GABA receptor subunits in the hippocampus and thalamus after experimental traumatic brain injury. Neuropharmacology 88:122–133. https://doi.org/10.1016/j.neuropharm.2014.08.023. S0028-3908(14)00306-2 [pii]
33. Majak K, Dabrowski M, Pitkanen A (2009) Epileptogenesis alters gene expression pattern in rats subjected to amygdala-dependent emotional learning. Neuroscience 159:468–482. https://doi.org/10.1016/j.neuroscience.2008.12.060. S0306-4522(09)00004-9 [pii]
34. Mahmood A, Wu H, Qu C, Mahmood S, Xiong Y, Kaplan D, Chopp M (2014) Downregulation of Nogo-A by collagen scaffolds impregnated with bone marrow stromal cell treatment after traumatic brain injury promotes axonal regeneration in rats. Brain Res 1542:41–48. https://doi.org/10.1016/j.brainres.2013.10.045. S0006-8993(13)01455-8 [pii]
35. Li S, Overman JJ, Katsman D, Kozlov SV, Donnelly CJ, Twiss JL, Giger RJ, Coppola G, Geschwind DH, Carmichael ST (2010) An age-related sprouting transcriptome provides molecular control of axonal sprouting after stroke. Nat Neurosci 13:1496–1504. https://doi.org/10.1038/nn.2674. nn.2674 [pii]
36. Ferraiuolo L, Higginbottom A, Heath PR, Barber S, Greenald D, Kirby J, Shaw PJ (2011) Dysregulation of astrocyte-motoneuron cross-talk in mutant superoxide dismutase 1-related amyotrophic lateral sclerosis. Brain 134:2627–2641. https://doi.org/10.1093/brain/awr193. awr193 [pii]
37. Xu Q, Walker D, Bernardo A, Brodbeck J, Balestra ME, Huang Y (2008) Intron-3 retention/splicing controls neuronal expression of apolipoprotein E in the CNS. J Neurosci 28:1452–1459. https://doi.org/10.1523/JNEUROSCI.3253-07.2008. 28/6/1452 [pii]
38. Mehta P, Premkumar B, Morris R (2016) Production of high quality brain-derived neurotrophic factor (BDNF) and tropomyosin receptor kinase B (TrkB) RNA from isolated populations of rat spinal cord motor neurons obtained by Laser Capture Microdissection (LCM). Neurosci Lett 627:132–138. https://doi.org/10.1016/j.neulet.2016.05.063. S0304-3940(16)30393-7 [pii]
39. Narantuya D, Nagai A, Sheikh AM, Masuda J, Kobayashi S, Yamaguchi S, Kim SU (2010) Human microglia transplanted in rat focal ischemia brain induce neuroprotection and behavioral improvement. PLoS One 5:e11746. https://doi.org/10.1371/journal.pone.0011746
40. Huusko N, Pitkanen A (2014) Parvalbumin immunoreactivity and expression of GABAA receptor subunits in the thalamus after experimental TBI. Neuroscience 267:30–45. https://doi.org/10.1016/j.neuroscience.2014.02.026. S0306-4522(14)00138-9 [pii]
41. Kigerl KA, Ankeny DP, Garg SK, Wei P, Guan Z, Lai W, McTigue DM, Banerjee R, Popovich PG (2012) System x(c)(−) regulates microglia and macrophage glutamate excitotoxicity in vivo. Exp Neurol 233:333–341. https://doi.org/10.1016/j.expneurol.2011.10.025. S0014-4886(11)00397-9 [pii]
42. Saxena T, Karumbaiah L, Gaupp EA, Patkar R, Patil K, Betancur M, Stanley GB, Bellamkonda RV (2013) The impact of chronic blood-brain barrier breach on intracortical electrode function. Biomaterials 34:4703–4713. https://doi.org/10.1016/j.biomaterials.2013.03.007. S0142-9612(13)00289-5 [pii]
43. Martinez-Galvez G, Zambrano JM, Diaz Soto JC, Zhan WZ, Gransee HM, Sieck GC, Mantilla CB (2016) TrkB gene therapy by adeno-associated virus enhances recovery after cervical spinal cord injury. Exp Neurol 276:31–40. https://doi.org/10.1016/j.expneurol.2015.11.007. S0014-4886(15)30121-7 [pii]
44. Lefebvre dC, Harry GJ (2005) Molecular profiles of mRNA levels in laser capture microdissected murine hippocampal regions differentially responsive to TMT-induced cell death. J Neurochem 93:206–220. https://doi.org/10.1111/j.1471-4159.2004.03017.x. JNC3017 [pii]
45. Shimamura M, Garcia JM, Prough DS, Hellmich HL (2004) Laser capture microdissection and analysis of amplified antisense RNA from distinct cell populations of the young and aged rat brain: effect of traumatic brain injury on hippocampal gene expression. Mol Brain Res 17:47–61
46. Shimamura M, Garcia JM, Prough DS, DeWitt DS, Uchida T, Shah SA, Avila MA, Hellmich HL (2005) Analysis of long-term gene expression in neurons of the hippocampal subfields following traumatic brain injury in rats. Neuroscience 131:87–97
47. Boone DR, Sell SL, Micci MA, Crookshanks JM, Parsley MA, Uchida T, Prough DS,

DeWitt DS, Hellmich HL (2012) Traumatic brain injury-induced dysregulation of the circadian clock. PLoS One 7:e46204
48. Fustin JM, Karakawa S, Okamura H (2017) Circadian profiling of amino acids in the SCN and cerebral cortex by laser capture microdissection-mass spectrometry. J Biol Rhythms 32:609–620. https://doi.org/10.1177/0748730417735922
49. Schmued LC, Albertson C, Slikker W Jr (1997) Fluoro-Jade: a novel fluorochrome for the sensitive and reliable histochemical localization of neuronal degeneration. Brain Res 751:37–46
50. Hellmich HL, Garcia JM, Shimamura M, Shah SA, Avila MA, Uchida T, Parsley MA, Capra BA, Eidson KA, Kennedy DR, Winston JH, DeWitt DS, Prough DS (2005) Traumatic brain injury and hemorrhagic hypotension suppress neuroprotective gene expression in injured hippocampal neurons. Anesthesiology 102:806–814
51. Hellmich HL, Rojo DR, Micci MA, Sell SL, Boone DR, Crookshanks JM, DeWitt DS, Masel BE, Prough DS (2013) Pathway analysis reveals common pro-survival mechanisms of metyrapone and carbenoxolone after traumatic brain injury. PLoS One 8:e53230
52. Rojo DR, Prough DS, Boone DR, Micci MA, Kahrig KM, Crookshanks JM, Jimenez A, Uchida T, Cowart JC, Hawkins BE, Avila M, DeWitt DS, Hellmich HL (2011) Influence of stochastic gene expression on the cell survival rheostat after traumatic brain injury. PLoS One 6:e23111
53. Boone DR, Micci MA, Taglialatela IG, Hellmich JL, Weisz HA, Bi M, Prough DS, DeWitt DS, Hellmich HL (2015) Pathway-focused PCR array profiling of enriched populations of laser capture microdissected hippocampal cells after traumatic brain injury. PLoS One 10:e0127287. https://doi.org/10.1371/journal.pone.0127287. PONE-D-14-52985 [pii]
54. Boone DR, Leek JM, Falduto MT, Torres KEO, Sell SL, Parsley MA, Cowart JC, Uchida T, Micci MA, DeWitt DS, Prough DS, Hellmich HL (2017) Effects of AAV-mediated knockdown of nNOS and GPx-1 gene expression in rat hippocampus after traumatic brain injury. PLoS One 12: e0185943. https://doi.org/10.1371/journal.pone.0185943. PONE-D-17-25338 [pii]
55. Boone DK, Weisz HA, Bi M, Falduto MT, Torres KEO, Willey HE, Volsko CM, Kumar AM, Micci MA, DeWitt DS, Prough DS, Hellmich HL (2017) Evidence linking microRNA suppression of essential prosurvival genes with hippocampal cell death after traumatic brain injury. Sci Rep 7:6645. https://doi.org/10.1038/s41598-017-06341-6
56. Bernard R, Kerman IA, Meng F, Evans SJ, Amrein I, Jones EG, Bunney WE, Akil H, Watson SJ, Thompson RC (2009) Gene expression profiling of neurochemically defined regions of the human brain by in situ hybridization-guided laser capture microdissection. J Neurosci Methods 178:46–54
57. Doppenberg EM, Choi SC, Bullock R (2004) Clinical trials in traumatic brain injury: lessons for the future. J Neurosurg Anesthesiol 16:87–94
58. Schouten JW (2007) Neuroprotection in traumatic brain injury: a complex struggle against the biology of nature. Curr Opin Crit Care 13:134–142
59. Bota M, Dong HW, Swanson LW (2003) From gene networks to brain networks. Nat Neurosci 6:795–799
60. Cloonan N, Brown MK, Steptoe AL, Wani S, Chan WL, Forrest AR, Kolle G, Gabrielli B, Grimmond SM (2008) The miR-17-5p microRNA is a key regulator of the G1/S phase cell cycle transition. Genome Biol 9:R127. https://doi.org/10.1186/gb-2008-9-8-r127. gb-2008-9-8-r127 [pii]
61. Xu LF, Wu ZP, Chen Y, Zhu QS, Hamidi S, Navab R (2014) MicroRNA-21 (miR-21) regulates cellular proliferation, invasion, migration, and apoptosis by targeting PTEN, RECK and Bcl-2 in lung squamous carcinoma, Gejiu City, China. PLoS One 9:e103698. https://doi.org/10.1371/journal.pone.0103698. PONE-D-13-40415 [pii]
62. Jellinger KA (2001) Cell death mechanisms in neurodegeneration. J Cell Mol Med 5:1–17
63. Beauchamp K, Mutlak H, Smith WR, Shohami E, Stahel PF (2008) Pharmacology of traumatic brain injury—where is the "golden bullet"? Mol Med 14:731–740
64. Galatro TF, Holtman IR, Lerario AM, Vainchtein ID, Brouwer N, Sola PR, Veras MM, Pereira TF, Leite REP, Moller T, Wes PD, Sogayar MC, Laman JD, den DW, Pasqualucci CA, Oba-Shinjo SM, Boddeke EWGM, Marie SKN, Eggen BJL (2017) Transcriptomic analysis of purified human cortical microglia reveals age-associated changes. Nat Neurosci 20:1162–1171. https://doi.org/10.1038/nn.4597. nn.4597 [pii]
65. Livak KJ, Wills QF, Tipping AJ, Datta K, Mittal R, Goldson AJ, Sexton DW, Holmes CC (2012) Methods for qPCR gene expression profiling applied to 1440 lymphoblastoid single cells. Methods 59:71–79

66. Caeyenberghs K, Leemans A, Leunissen I, Gooijers J, Michiels K, Sunaert S, Swinnen SP (2014) Altered structural networks and executive deficits in traumatic brain injury patients. Brain Struct Funct 219:193–209. https://doi.org/10.1007/s00429-012-0494-2
67. Bewernick BH, Kayser S, Sturm V, Schlaepfer TE (2012) Long-term effects of nucleus accumbens deep brain stimulation in treatment-resistant depression: evidence for sustained efficacy. Neuropsychopharmacology 37:1975–1985
68. Karatsoreos IN (2012) Effects of circadian disruption on mental and physical health. Curr Neurol Neurosci Rep 12(2):218–225. https://doi.org/10.1007/s11910-012-0252-0
69. Wang WZ, Oeschger FM, Lee S, Molnar Z (2009) High quality RNA from multiple brain regions simultaneously acquired by laser capture microdissection. BMC Mol Biol 10:69. https://doi.org/10.1186/1471-2199-10-69. 1471-2199-10-69 [pii]

Chapter 12

Rat Microglia Isolation and Characterization Using Multiparametric Panel for Flow Cytometric Analysis

Naama E. Toledano Furman, Karthik S. Prabhakara, Supinder Bedi, Charles S. Cox, Jr., and Scott D. Olson

Abstract

In the field of traumatic brain injury (TBI), characterization of microglia is traditionally done using imaging-based methods as immunohistochemistry (IHC), while flow cytometry (FC) is a supporting method that reassures the results. However, often, the flow cytometry method used, especially in the commonly used rat models, is not microglial specific, not comprehensive, and certainly not standardized. Here we describe the full, optimized procedure for isolating rat microglia, proper sample preparation for FC run and provide essential analysis guidelines. This protocol allows the specific identification of microglia, by discriminating them from other myeloid cells, with high-resolution observation of microglial subpopulations and their activation markers expression.

Key words Rat microglia, M1 polarized microglia, M2 polarized microglia, Traumatic brain injury, Spinal cord injury, Neuro-immuno-phenotyping, Flow cytometry

1 Introduction

The method detailed in this chapter was developed to utilize flow cytometry to phenotypically characterize microglial cells isolated from brain or spinal cord following traumatic brain injury (TBI) or spinal cord injury (SCI) in rat model. This method is an alternative for the traditional immunohistochemistry (IHC) acceptable in the field of neurobiology, by offering fast, accurate, and quantitative sample processing and data acquiring with additional resolution in data analysis. All of these aspects of the method will be discussed along the chapter.

1.1 Background

Central nervous system (CNS) injury immediately results in a robust inflammatory cell-mediated response. Microglial cells, the resident immune cell of the CNS, play critical role in the host response to the injury. In routine, the nonprimed microglia are ramified and express characteristic protein profile, which alters

Amit K. Srivastava and Charles S. Cox, Jr. (eds.), *Pre-Clinical and Clinical Methods in Brain Trauma Research*, Neuromethods, vol. 139, https://doi.org/10.1007/978-1-4939-8564-7_12,

upon injury. The ramified cell becomes amoeboid and overexpresses proteins to support its new acquired role in the inflamed tissue on the injury site. CNS injuries also lead to infiltration of myeloid cells, as monocytes and macrophages, to the injury site. Both microglia and the infiltrating monocytes/macrophages share the expression of surface markers such as CD45, CD11b/c. The problem arises with the common way of identifying microglia using flow cytometry by applying gating on CD11b/c^{+} CD45^{low} cell population [1–3]. This flow cytometry-based method allows differentiation between monocytes/macrophages and resident microglia, and also defines phenotypic changes in microglia in rat model. Using the combination of CD45, CD11b/c, and p2y12 receptor we are able to specifically identify microglia. The relatively new p2y12 receptor is an APD-responsive G protein-coupled receptor. In the CNS, its expression is restricted to microglia. It is now known that the p2y12 receptor is selectively expressed in both nonprimed and activated (primed) microglia and mediates process motility during early injury responses [4]. Anti-p2y12 antibody was produced and studied well in both mouse and human models [1], but less so in rat models. As far as we know, only one anti-rat p2y12 antibody (Alomone Labs) is currently available for flow cytometry purposes. During the development of this method, we validated and optimized its use on the isolated microglia.

Similar to the pro- and anti-inflammatory polarization of macrophages in non-neuronal tissues, microglia present varied phenotypes when activated. The classical pro-inflammatory (M1) activation of microglia, typically in response to TNF-α and IFN-γ, involves phagocytosis, ability to kill pathogens and ROS release [5]. As antigen presenting cells, microglia communicate with T cells, and thus activated microglia will upregulate their cell surface markers such as MHC-II and co-stimulatory molecule CD86 [6]. The anti-inflammatory (M2) classification can be directed in three alternative pathways. Stimulation of microglia with IL4 or IL13 directs cells in an alternative pathway that enhances immunity against parasites, Th2 cell recruitment, and tissue repair. At these conditions, microglia are defined as M2a-polarized. Next, in response to IL10, glucocorticoids, or uptake of apoptotic cells microglia can be directed at a path of "acquired deactivation"/M2c. In this path microglia are involved in tissue remodeling processes. Our method evaluated the expression CD163, the membrane-bound scavenger receptor for haptoglobin/hemoglobin complexes, which is overexpressed in M2c [7–9]. Finally, the M2b phenotype has characteristics of both pro- (M1) and anti-inflammatory (M2) and is associated with memory immune response [10].

1.2 Development of the Method

Microglial function studies in mice and human models share common markers that differ in rat microglial. Therefore, rat microglia requires its own unique panel of markers to be evaluated. Only recently, rat microglial markers have become commercially available, allowing us to perform comprehensive analysis of these cells. This flow cytometry-based method offers standardized procedure to evaluate rat microglia activation that is quick and quantitative. The method was inspired and developed based on mouse microglia isolation and evaluation, by adjusting and optimizing procedures that were previously established in our lab [11]. This method allows the evaluation of rat microglia originating from CNS-related tissues following an injury, as well as in other neurodegenerative condition/diseases studied on rat model. The protocol details the procedure for acquiring single cell suspension from freshly harvested rat brain or spinal cord tissue. It is enriched to purify the myeloid and microglia population by using CD11 positive cells selection. The flow cytometry multiparametric panel detailed here was optimized and validated [12]. We will also detail sample preparation for flow cytometric processing and briefly discuss the analysis of those samples to best evaluate the cells.

2 Materials

2.1 List of Reagents

1. Phosphate buffered saline (PBS) (Gibco cat # 10010-049).
2. Hank's balanced salt solution (HBSS) (Gibco cat # 14170112).
3. Percoll (GE Healthcare cat # 17-0891-01).
4. Neural tissue dissociation kit (P) (Miltenyi Biotec cat # 130-092-628).
5. MACS SmartStrainer (70 μm) (Miltenyi Biotec cat # 130-098-462).
6. C tubes (Miltenyi Biotec cat # 130-096-334).
7. LS columns (Miltenyi Biotec cat # 130-042-401).
8. 50 ml tubes.
9. 15 ml tubes.
10. CD11b/c (microglia) microbeads, rat (Miltenyi Biotec cat # 130-105-634).
11. Microglia cell media (Sciencell cat # 1901).
12. Cell staining buffer (Biolegend cat # 420201).
13. Falcon® 5 ml Polystyrene Round Bottom Tube 12 × 75 mm style (Corning cat # 352008).
14. Ghost Violet 510 Live/Dead Reagent (Tonbo 13-0870-T500).
15. Cyto-Cal™ Count Control (Thermo Scientific 09-980-698).
16. Antibodies: **Note: the table below (Table 1) was adapted with permission from OMIP-041: Optimized multicolor*

Table 1
List of antibodies

Specificity	Clone	Fluorochrome	First antibody Vendor/Cat #	Dilution of first antibody	Second antibody Vendor/Cat #	Dilution of second antibody
CD45	OX-1	APC-Cy7	Biolegend (304014)	1.5:100	–	–
CD11b/c	OX-42	PE-Cy7	BD Biosciences (562222)	2.5:100	–	–
P2Y12	Polyclonal	BV421	Alomone Labs (APR-020)	5:100	Biolegend (406410)	2:100
CD32	D34-485	PE	BD Biosciences (562189)	0.06:100	–	–
CD86	24-F	APC	Biolegend (200314)	0.125:100	–	–
CD200R	OX-102	PE	Abcam (34135)	2.5:100	–	–
RT1B	OX-6	Alexa Fluor 647	BD Biosciences (562223)	10:100	–	–
CD163	ED2	PerCP Cy5.5	BioRad (MCA342B)	2.5:100	Biolegend (405214)	2:100

immunofluorescence panel rat microglial staining protocol. Toledano Furman, N. E., Prabhakara, K. S., Bedi, S., Cox, C. S., Jr. & Olson, S. D. Cytometry A, (2017).

2.2 List of Equipment

* *Note: list of equipment is what we used and can be easily adjusted (see Notes).*

1. Analytic balance.
2. Forceps.
3. Scalpel.
4. GentleMACS Dissociator ((Miltenyi Biotec cat # 130-090-235).
5. MACSmix™ tube rotator (Miltenyi Biotec cat # 130-090-753).
6. QuadroMACS™ Separator (Miltenyi Biotec cat # 130-091-051).
7. Incubator 37 °C, 5% CO_2.
8. Flow cytometer, LSRII (BD Biosciences) or Gallios™ (Beckman coulter) equipped with eight color filter settings.
9. Centrifuge (to fit, 50 ml tubes, 15 ml tubes and 5 ml flow cytometry tubes).

3 Methods

Male Sprague Dawley Rats (225–250 g, Harlan Labs) were the source of CNS tissue. The usage of the animals was approved by the Animal welfare committee at University of Texas Health Science Center at Houston, Texas, protocol: AWC16-0046 and AWC14-0023. Animals were handled in accordance with the standards of the American Association for the Accreditation of Laboratory Animal Care (AAALAC). The procedure is optimized to sample a single brain hemisphere or a 1-in. long section of spinal cord tissue. To evaluate microglial activation we developed this panel based upon an existing mouse microglia panel that was previously developed and utilized in our lab [11]. The rationale was to first identify the main microglia population using CD45, CD11b/c and p2y12 antibodies. Next, we chose markers that are available for rat that could indicate the polarization state a cell is in, see Table 1.

3.1 Brain Tissue Dissolving

1. Using a sterile scalpel remove the cerebellum.
2. Split the brain to its hemispheres (ipsilateral and contralateral).
3. Weigh the brain hemispheres * *to determine absolute cell number per mg tissue. As microglial expansion is an important parameter in the activation process* [13].

4. *Neural dissociation kit usage*: For samples ranging 400–600 mg.
   (a) Add each brain hemisphere tissue to a C tube that contains prewarmed 3800 μl of buffer *x* and 100 μl of Enzyme P.
   (b) Carefully place the C tubes in the MACS Gentle. Run program m_brain_01. **At this point the tissue is not fully dissolved; just make sure that no brain chunks are stuck between the tube and the lid.*
   (c) Rotate on the tube rotor inside the incubator for 15 min.
   (d) Run samples on program m_brain_02.
   (e) Add 40 μl of buffer y and 20 μl of Enzyme A.
   (f) Rotate on the tube rotor inside the incubator for 10 min.
   (g) Run samples on program m_brain_03.
   (h) Rotate on the tube rotor inside the incubator for 10 min. **At this point, sample is usually a white solution, tissue is fully dissolved.*
5. *Sample washing and myelin removal* *myelin debris can activate the microglia. Therefore washing and cleaning of the samples from myelin is important.
   (a) Filter the samples through a sterile MACS SmartStrainer 70 μm, into a 50 ml tube.
   (b) Rinse each tube with 10 ml of HBSS and collect it through the same filter (**to wash the C tube and the filter that way*).
   (c) Centrifuge at 400 *g* for 10 min. **After centrifuging you'll have a large pellet, which is the myelin and the cells together.*
   (d) In the meanwhile prepare 15 ml tubes with 2 ml of 30% Percoll, 70% HBSS, each.
   (e) Aspirate the supernatant.
   (f) Add 1 ml of the 30% Percoll in HBSS to each sample and suspend the pellet.
   (g) Add the suspension to the appropriate 15 ml tube (with 2 ml of the 30% Percoll).
   (h) Centrifuge for 10 min in 700 g w/no break to separate the myelin from the cells.
   (i) Carefully aspirate the myelin—tilt the tube while aspirating for quick and better results.
   (j) To wash the cells from the Percoll—complete the volume to 12 ml with HBSS. **Percoll at high concentrations is toxic to the cells* [3]. *Therefore washing step is very important.*
   (k) Centrifuge for 10 min, at 500 *g*.
   (l) Carefully aspirate the supernatant.

6. *CD11b/c positive cell enrichment.*
    - (a) Add 120 μl of Buffer to each sample and suspend it.
    - (b) Add 40 μl of beads to each sample and gently suspend by pipetting. DO NOT VORTEX!
    - (c) Incubate in 4 °C for 15 min.
    - (d) In the mean while prepare the LS columns by washing each one with 3 ml of buffer and set the MACS separator with 15 ml tube rack underneath.
    - (e) Add the sample to the LS column. Let it flow.
    - (f) Add 1 ml of buffer to the column and let it flow.
    - (g) Add another 1 ml of buffer to the column and let it flow.
    - (h) Remove the column and place it on a new 15 ml tube.
    - (i) Add 5 ml of buffer and use the piston to push the cells from the column. **Carefully hold the tube and column so it will not flip.*
    - (j) Centrifuge the enriched cells in 400 *g* for 10 min.
    - (k) In the meanwhile—prepare six well tissue culture plate for both portions.
    - (l) Suspend each sample in 2 ml of the Rat Microglia Media.
    - (m) Incubate for overnight.

3.2 Cell Staining and Sample Preparation for Flow Cytometry

1. Collect microglia into 15 ml falcon tube by gentle pipetting the cells.
2. Wash cells twice with PBS and suspend cells at 1–10 million cells per 1 ml of PBS. Add 1 μl of Ghost reagent for every 1 ml of cells and incubate in the dark for 30 min*.
3. Wash cells: dilute ×5 with staining buffer and centrifuge at 300 *g* for 5 min.
4. Suspend each cell sample in 200 μl staining buffer.
5. Divide 100 μl per sample for both M1 and M2 tubes.
6. Add antibody cocktails for M1 and M2, according to Table 1. Incubate 30 min at RT in the dark.
7. Dilute cells to 1 ml with staining buffer and centrifuge at 300 *g* for 5 min.
8. Resuspend cells in 100 μl staining buffer.
9. Add secondary antibodies, according to the Table 1, to the appropriate tubes. Incubate for 20 min.
10. Wash cells and suspend each sample in 300 μl of staining buffer.
11. Add Cyto-Cal counting beads.
12. Read the samples on BD LSRII or Beckman Coulter Gallios.

3.3 Results Analysis

1. Define the main population of cells using forward scatter (FSC) and side scatter (SSC) filters. **Microglial activation results in morphological changes along with protein expression. The data of the FSC vs. SSC should imply on the morphological changes the cells are going through. Injury should result in shift in both axes.*
2. Exclude cell debris and cell aggregates using FSC-A vs. FSC-H (or if given FSC-W, as in the Gallios).
3. Define the live cells under Ghost negative cells.
4. Under live cell gating, identify the CD45$^+$ cells.
5. Under CD45$^+$, plot CD11 vs. p2y12 dot plot. Cells that are CD11$^+$p2y12$^+$ are microglia; cells that are CD11$^+$p2y12$^-$ are other myeloid cells. **According to the time point, these non-microglial cells can be monocytes, macrophages, or other myeloid cells.*
6. Apply gating of positive population of the activation markers as CD32, CD86 or CD200R, CD163 and RT1B for the M1 or M2 panels, respectively. *At times.
7. Use the Cyto-Cal counting beads to calculate the absolute number of microglia. Volume of the beads and cell number calculation are according to manufacturer instructions.

4 Notes

1. The procedure is optimized for one hemisphere of rat brain. For spinal cord of 1 in. long use half amounts.
2. Our empiric data suggests that a rat with an average weight of approximately 300 g will typically have an average of 570 mg hemisphere. Typically injury would weigh more, due to edema of the tissue, meaning the ipsilateral would weigh slightly more than the contralateral. Shams are not expected to follow that rule.
3. Prior to using the Ghost dye, remove it from freezer and allow equilibrating to room temperature. Also, Ghost dye is an amine reactive viability dye that irreversibly binds free amines available on the cell surface as well as intracellular free amines exposed in cells with compromised cell membranes. Prior to cell staining wash cells thoroughly to remove serum proteins and allow proper binding of the dye.

References

1. Bennett ML, Bennett FC, Liddelow SA, Ajami B, Zamanian JL, Fernhoff NB, Mulinyawe SB, Bohlen CJ, Adil A, Tucker A, Weissman IL, Chang EF, Li G, Grant GA, Hayden Gephart MG, Barres BA (2016) New tools for studying microglia in the mouse and human CNS. Proc Natl Acad Sci U S A 113(12): E1738–E1746. https://doi.org/10.1073/pnas.1525528113
2. Davies LC, Jenkins SJ, Allen JE, Taylor PR (2013) Tissue-resident macrophages. Nat

Immunol 14(10):986–995. https://doi.org/10.1038/ni.2705
3. Goslin GBK (1998) Culturing nerve cells. MIT Press, Cambridge, MA
4. Sipe GO, Lowery RL, Tremblay ME, Kelly EA, Lamantia CE, Majewska AK (2016) Microglial P2Y12 is necessary for synaptic plasticity in mouse visual cortex. Nat Commun 7:10905. https://doi.org/10.1038/ncomms10905
5. Cherry JD, Olschowka JA, O'Banion MK (2014) Neuroinflammation and M2 microglia: the good, the bad, and the inflamed. J Neuroinflammation 11:98. https://doi.org/10.1186/1742-2094-11-98
6. Taylor PR, Martinez-Pomares L, Stacey M, Lin HH, Brown GD, Gordon S (2005) Macrophage receptors and immune recognition. Annu Rev Immunol 23:901–944. https://doi.org/10.1146/annurev.immunol.23.021704.115816
7. Pey P, Pearce RK, Kalaitzakis ME, Griffin WS, Gentleman SM (2014) Phenotypic profile of alternative activation marker CD163 is different in Alzheimer's and Parkinson's disease. Acta Neuropathol Commun 2:21. https://doi.org/10.1186/2051-5960-2-21
8. Wilcock DM (2014) Neuroinflammatory phenotypes and their roles in Alzheimer's disease. Neurodegener Dis 13(2–3):183–185. https://doi.org/10.1159/000354228
9. Colton CA (2009) Heterogeneity of microglial activation in the innate immune response in the brain. J Neuroimmune Pharmacol 4(4):399–418. https://doi.org/10.1007/s11481-009-9164-4
10. Mantovani A, Sica A, Sozzani S, Allavena P, Vecchi A, Locati M (2004) The chemokine system in diverse forms of macrophage activation and polarization. Trends Immunol 25(12):677–686. https://doi.org/10.1016/j.it.2004.09.015
11. Bedi SS, Smith P, Hetz RA, Xue H, Cox CS (2013) Immunomagnetic enrichment and flow cytometric characterization of mouse microglia. J Neurosci Methods 219(1):176–182. https://doi.org/10.1016/j.jneumeth.2013.07.017
12. Toledano Furman NE, Prabhakara KS, Bedi S, Cox CS Jr, Olson SD (2017) OMIP-041: optimized multicolor immunofluorescence panel rat microglial staining protocol. Cytometry A 93(2):182–185. https://doi.org/10.1002/cyto.a.23267
13. Wang G, Zhang J, Hu X, Zhang L, Mao L, Jiang X, Liou AK, Leak RK, Gao Y, Chen J (2013) Microglia/macrophage polarization dynamics in white matter after traumatic brain injury. J Cereb Blood Flow Metab 33(12):1864–1874. https://doi.org/10.1038/jcbfm.2013.146

Chapter 13

The Principles of Experimental Design and the Determination of Sample Size When Using Animal Models of Traumatic Brain Injury

Michael F. W. Festing

Abstract

There is a crisis in preclinical research involving laboratory animals. Too many experiments, in all disciplines, have been found to give results which turn out to be irreproducible. And, unfortunately, the results of clinical trials of potential treatments of traumatic brain injury (TBI), suggested by extensive animal research, have so far been disappointing. Faulty experimental design could be one of several possible causes. This chapter briefly covers the basic principles of experimental design, with emphasis on completely randomized, randomized block (cohort), and factorial designs. The determination of sample size is discussed together with some aspects of the statistical analysis and presentation of the results. The Standardized Effect Size (Cohen's *d*) is introduced as a description of the magnitude of a response.

Key words Traumatic brain injury, Preclinical, Experimental design, Sample size, Statistical analysis

1 Introduction

The study of animal models of traumatic brain injury and the search for treatments is one of the most difficult and technically demanding areas in which laboratory animals are used. It is important that this research is done to the highest possible standards and that it produces results which are reproducible and useful in suggesting clinical treatments.

The precedents from other disciplines are not good. Too many experiments are producing results which turn out to be irreproducible. For example, an attempt to reproduce the results from 53 "landmark" papers in cancer research was only successful with six (11%) of them [1]. Scientists at Beyer Pharmaceuticals [2] were only able to reproduce the results 22% of 67 papers of potential interest to them. More than 50 papers claimed to have identified agents which alleviated symptoms in the standard transgenic mouse model of ALS, but none of them were effective in humans.

Amit K. Srivastava and Charles S. Cox, Jr. (eds.), *Pre-Clinical and Clinical Methods in Brain Trauma Research*, Neuromethods, vol. 139, https://doi.org/10.1007/978-1-4939-8564-7_13, © Springer Science+Business Media, LLC, part of Springer Nature 2018

A detailed study of these papers identified a number of confounding factors which needed to be controlled when using this model. When these 50 drugs and an additional 20 drugs were retested, using an improved experimental design using a power analysis to determine sample size, none of them proved to be effective in the mouse model; all of them were false positive results [3].

It has been estimated that the waste of scientific resources due to irreproducibility of preclinical experiments amounts to about $28 billion per annum in the USA alone [4].

The causes of this lack of reproducibility are not fully understood. They may include bias due to failure to randomize or to randomie correctly and/or failure of blinding when outcomes are measured [5]. Other possible causes include technical and measurement errors, incorrect statistical analysis, animal strain differences in response which are not taken into account, and fraud.

It has also been suggested [6] that excessively small sample size, leading to false negative results, could also lead to excessive numbers of false positive results if the negative results remain unpublished. This is because the proportion of published false positives normally expected when using a 5% significance level would be increased. So it is important that negative results as well as positive results are published.

It is important that investigators using animal models of traumatic brain injury (TBI) should clearly understand the principles of experimental design. This includes choice of an appropriate design, correct randomization and blinding, rational determination of sample size, and the statistical analysis of the resulting data. This chapter briefly discusses the principles of experimental design including methods of determining sample size and introduces the "standardized effect size" as a research tool for specifying the magnitude of a treatment response. Ideally, a statistician should be included in the research team.

2 The Basic Principles of Experimental Design

Randomized, controlled experiments were largely developed by R.A. Fisher and his colleagues when he was the statistician at Rothamsted Agricultural Experimental Station in the UK in the 1920s [7]. The first randomized clinical trial, on the effect of streptomycin on tuberculosis [8], was not carried out until 1946.

The principles are deceptively simple. All experiments compare two or more groups which differ as a result of some "treatment" applied to one or more of them. There needs to be *independent* replication of the experimental subjects so that inter-individual variation can be assessed in the statistical analysis. And the treatments need to be assigned to the experimental subjects at random *in such a way that they are intermingled in time and space during the course of the experiment*. In order to avoid bias, investigators measuring the outcome of the experiment need to be "blinded" so that they do not know the treatment to which an animal was assigned.

The results need to be statistically analyzed using a method which was planned *before* the experiment was started. This analysis will determine the probability that any observed differences could be attributed to chance rather than the effect of a treatment. The results need to be presented in such a way that they can easily be interpreted and used by subsequent investigators. Basic descriptive statistics such as means, standard deviations and numbers per group should be presented, in addition to any other charts and diagrams needed to illustrate the results. A more extensive coverage of the design and statistical analysis of animal experiments is given elsewhere [9] and in many statistical textbooks.

3 Formal Experimental Designs

There are a number of formal experimental designs which can be used, where appropriate. These include "Completely randomized" (CR), "Randomized block" (RB), "Latin square," "Split plot," "Crossover," and "Within-subject" designs. All these designs have at least one fixed (i.e. controllable) effect "factor" often called "treatment." The treatments are determined by the investigator. There can be any number of levels. For example, a CR experiment may have a factor called "treatment" which could have three levels designated "Sham," "TBI," and "TBI-treated." Sometimes they may also have a second fixed effect factor, such as gender, making them into a "factorial" design. In this case half the animals would be male and half female.

Virtually all TBI experiments seem to have been done using a simple CR design. *But this is may not be an optimum design to use in TBI research because if the randomization is not done correctly and no account is taken of environmental variation, it could easily lead to false positive results.*

A Randomized Block (RB) design also has one or more fixed effect factors, such as "Treatment," "gender," or even "time to measure outcome," which can similarly have any number of levels determined by the investigator. But the RB design also has a "random effect" factor usually called "block." A "blocked" experiment is one in which the whole experiment is split up into a number of "mini-experiments" with a single individual on each of the treatments. The term "block" reflects the origin of experimental design in agricultural research where a block was an area of land on which there were a number of plots each receiving, for example, a different treatment. The block is a "random effect factor" because the difference between blocks is not under the control of the investigator.

In animal research the term "block" might be better described as a "cohort" or group of animals. Each block/cohort should be composed of subjects that are as similar as possible in age, weight, and gender, but there can be larger differences between the blocks. Each member of the block/cohort will receive a different treatment. For example, there might be four treatments designated "Sham," "TBI," "TBI-low dose," and "TBI-High dose." So the cohort/block size is four and there will be "*N*" cohorts, where "*N*" is the sample size. Differences between cohorts are removed in the statistical analysis. So, with four treatments, cohort 1 could be the set of four animals treated on day 1 (each receiving a different treatment) and cohort N could consist of the four animals receiving the treatments on day "*N*."

The term "block" will be retained here in conformity with the literature, but it should be regarded as being synonymous with "cohort."

Randomized Block experiments are statistically analyzed by a two-way analysis of variance *without* interaction. This analysis removes differences between cohorts which are of no interest and would otherwise lead to increased inter-individual variation and a reduction in statistical power (the ability of the experiment to detect an effect). An RB design with two or more factors will require a mixed analysis of variance with only the block effect being analyzed "without interaction."

4 Animal Experiments Versus Clinical Trials

For practical reasons, clinical trials nearly always use a CR design. Patients are difficult to recruit and trials are started gradually as patients become available. It is not practical to gather a cohort of similar patients to use in an RB design.

In contrast, RB designs are easily the most widely used in agricultural and industrial research. Blocking is an easy and convenient way of splitting the experiment into small, more manageable parts. So an experiment can be set up over a period of time at the investigator's convenience. The designs are usually more powerful than CR designs due to better control of inter-individual variation within each block. If blocks are separated in time, there is even some in-built repeatability.

The failure to use randomized block designs more widely in laboratory animal research is probably because much of the research is done by clinicians who are more familiar with clinical trials than with agricultural research. Randomized block designs are widely used in in-vitro experiments although investigators are often not aware that they are doing so. They often say that they "repeated the experiment three times," with the "experiment" being one culture

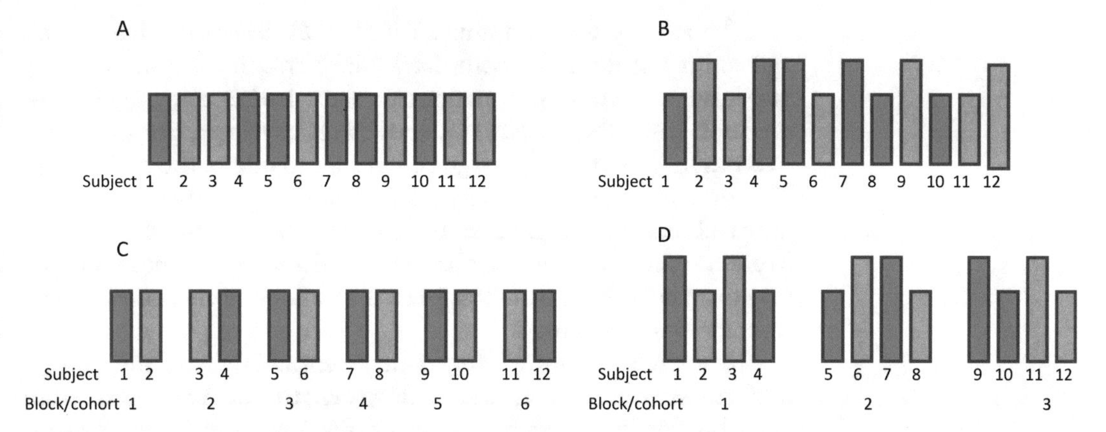

Fig. 1 (**a**) Diagram of a completely randomized (CR) design with two treatments (colors). Notice that the treatments are in random order and treatment cannot be deduced from the subject number. (**b**) A two treatments (colors) × two genders (short or tall) completely randomized factorial design. Note that no extra animals are needed when including both sexes. (**c**) A randomized block (cohort) (RB) experiment with two treatments (colors) and six blocks. Randomization is done separately for each block. (**d**) A randomized block (cohort) two (treatment) × two (gender) factorial experiment with three blocks

dish of cells assigned to each treatment. Unfortunately the correct statistical analysis is rarely done.

CR and RB designs are shown diagrammatically in Fig. 1. Assuming that individual animals are the subject of the research, a completely randomized design requires a *uniform* batch of animals to which the treatments are assigned at random in such a way that *the treatments are in random order with respect to the animal number*. This is discussed below.

In contrast, RB designs can tolerate more variation in the available animals, provided cohorts of "*T*" similar animals can be assembled, where "*T*" is the number of treatments, with "*N*" cohorts where *N* is the sample size. So even if the available animals are somewhat variable in age or weight it is still possible to use them in a powerful RB experiment. A skeleton analysis of variance table for these four experiments is given in Table 2.

5 Randomization in a CR Design

Randomization is of fundamental importance in avoiding bias and false positive results. Evidence [5] suggests that a large proportion of irreproducible studies, noted above, could be due to faulty randomization.

It is notable that few TBI papers mention randomization. Where they do so, they usually state that the animals were assigned to the treatments at random, but rarely state the order in which the animals were treated and maintained.

The reasons for randomization of a CR design in clinical trials and animal studies are completely different. In a clinical trial a physician will examine a patient to see whether they qualify for the trial. If so, the physician will open an envelope, prepared by a statistician, to randomly assign the patient to one of the treatments. This cannot be left to the physician because experience shows that they will tend to assign patients with more severe symptoms to the treated, rather than the placebo or control group, assuming that the new drug is likely to be better than the old one. This would lead to severely biased results.

In animal studies all the animals should be healthy and of uniform weight and age, so assigning them to treatments at random probably has little benefit. However, the animal research environment is heterogeneous. Animals and the investigators are subject to circadian rhythms, so animals and surgical interventions may not be the same when carried out in the morning or afternoon. Investigators can become more proficient as the experiment progresses, or maybe they over-indulged the night before and are tired. The same problems can arize when the outcomes, such as behavior, are measured. Experience shows that it is difficult to replicate a set of numerical observations made at different times. So if there are, say, three treatments (Sham, TBI, TBI-treated) with eight animals per treatment it would be completely wrong to label the cages 1–8 (Sham), 9–16 (TBI) and 17–24 (TBI treated). Such an experiment may give false positive results because the animals will nearly always be housed, treated, and the outcomes measured in numerical order. "Blinding" will also be impossible because investigators will remember that the Sham group is in cages 1–8.

Table 1 shows how such an experiment could be randomized using EXCEL. In this example, the randomization was not ideal because five of the six TBI-treated animals are in the first half of the table. It is permissible to redo the randomization if it appears to be unbalanced.

6 Randomizing an RB Design

RB designs are less likely to lead to biased results. With three treatments and a sample size of ten, each block or "cohort" would have one animal of each treatment (Sham, TBI, TBI treated), and there would be ten blocks or cohorts. In such a design the cages would be labeled with the block number. The treatments within a block would be assigned to each animal at random. This could be done, for example, with three playing cards which would be shuffled for each cohort. In the animal house the three cages in each block would be housed together on the same shelf, and treated in every way as similarly as possible.

Table 1
Example of the use of the random number function "=Rand()" in EXCEL to randomize a CR experimental design with three treatments and a sample size of six

Treatment un-randomized	Random number (=rand()) pasted back as "values"	Random number sorted with column 1	Treatment randomized (carried along with sorted random number)	Cage/ animal number
Sham	0.415	0.022	TBI-treated	1
Sham	0.960	0.084	Sham	2
Sham	0.312	0.122	TBI-treated	3
Sham	0.749	0.213	TBI-treated	4
Sham	0.084	0.256	TBI-treated	5
Sham	0.607	0.312	Sham	6
TBI	0.522	0.375	TBI-treated	7
TBI	0.517	0.415	Sham	8
TBI	0.694	0.517	TBI	9
TBI	0.631	0.522	TBI	10
TBI	0.962	0.607	Sham	11
TBI	0.905	0.631	TBI	12
TBI-treated	0.213	0.694	TBI	13
TBI-treated	0.122	0.749	Sham	14
TBI-treated	0.862	0.862	TBI-treated	15
TBI-treated	0.375	0.905	TBI	16
TBI-treated	0.022	0.960	Sham	17
TBI-treated	0.256	0.962	TBI	18

Columns 1 and 2 were sorted on column 2 (copied back as values) and the cage numbers were then added
When the outcome is being measured investigators should only know the animal number, but not the treatment

Cohorts (blocks) are treated and the outcomes measured one cohort at a time, spaced out in any way that is convenient. Any differences in the measured outcomes between the cohorts, which may be substantial, are removed in the two-way ANOVA without interaction. If both sexes were to be included (using a factorial design), then the block size would be doubled, but the sample size could be halved giving the same total number as if only a single sex were to be used. For example, if there are three treatments and a sample size of eight, then the experiment could be done using eight blocks each consisting of three males (24 animals in total). If both sexes are to be included then the block size would be increased to six (three males and three females) in four blocks (24 animals total). This is briefly explained in Fig. 1.

Table 2
"Skeleton" analysis of variance tables for the four designs shown in Fig. 1

A		B		C		D	
Source	**DF**	**Source**	**DF**	**Source**	**DF**	**Source**	**DF**
Treatment	1	Treatment (T)	1	Treatment	1	Treatment	1
Error	10	Gender (G)	1	Block	5	Gender	1
Total	11	TXG	1	Error	5	TxG	1
		Error	8	Total	11	Block	2
		Total	11			Error	6
						Total	11

TxG is the treatment × gender interaction, showing whether the two sexes respond in the same way. If only one gender were to be used, then the factor "Time" (to termination) could be used. Additional factors could be added with the advice of a statistician

7 Statistical Analysis

The assumptions of homogeneity of variances (i.e. the variation being about the same in each treatment group) with the residuals (deviation of each observation from the mean of its group) having a normal distribution should always be checked. Some computer software, such as MINITAB, and R-Commander (a package based on the free R statistical software package) allows the assumptions to be checked graphically. However, the ANOVA is quite tolerant of small departures from these assumptions.

Table 2 shows a "skeleton" analysis of variance (showing source of variation and degrees of freedom) for each of the examples of designs is shown in Fig. 1. With three or more treatment groups differences among groups are usually assessed using *post-hoc* comparisons. Dunnett's test is usually used if two or more treatment means are to be compared with a control group mean.

Note that with an RB design there is only a single pooled estimate of the standard deviation, which is obtained as *the square root of the error mean square* in the analysis of variance table. With this design error bars on bar charts will all be the same length.

8 In Conclusion, the Main Advantages of the RB Design

1. The experiment is logistically more convenient than the CR design. It can be done one block at a time spread over any amount of time. Each block has a single individual on each treatment. So if time factors are important, as is likely to be the case in TBI research, one or two blocks could be set up each

day or week for a period of several days or weeks, depending on the required sample size (the number of blocks). Measuring the outcomes would also be distributed over time at the convenience of the investigator. This repetition in time also provides some assurance that the experimental results are repeatable.

2. RB designs generally give better control of inter-individual variation because the animals can be matched for initial weight or age, time of use and location in the animal house during the experiment. Each block is independent and there are fewer time pressures on the investigators.
3. Animals are assigned to treatments separately in each block so bias due to faulty or unlucky randomization is much less likely than when using a CR design.

9 The Factorial Design

CR and RB designs with *two or more* fixed effect factors (types of treatment) are called "factorial designs." These provide extra information, often at no extra cost. For example, the NIH now requires investigators to use both sexes of animals in the experiments which they fund. An experiment might be designed with a fixed effect factor called "Treatment" with levels of "Treated" and "Control" and a second fixed effect factor called "Gender" with levels "Male" and "Female," giving a total of four treatments.

Assuming that the original experiment was a CR design with 12 treated and 12 control male animals, it could be altered to a factorial design using both sexes by using 6 control and 6 treated females and similar numbers of control and treated males. This would be a 2 × 2 CR factorial experiment as shown in Fig. 1. It should be statistically analyzed using a two-way analysis of variance *with* interaction, as shown in Table 2. The effect of the treatment would be estimated, averaging across gender (so $N/\text{group} = 12$), the effect of gender would be estimated averaging across treatments (so $N/\text{group} = 12$) and a gender × treatment interaction (also $N = 12$) would be estimated showing whether males and females respond differently. If there were three treatments, then the resulting experiment would be a 3 (treatments) × 2(gender) factorial design. Adding extra factors does not necessarily require an increase in sample size unless the experiment is already very small (as in the examples in Fig. 1 where the degrees of freedom for the error term in examples C and D are now excessively small). Methods of determining sample size are discussed below. A study of the effect of an atmosphere with alcohol vapor versus air in TBI and Sham rats [10] is a nice example of a 2 × 2 factorial experiment in TBI research. This is discussed in Table and associated text.

RB designs can also be factorial designs as shown in Fig. 1. The block size must be increased because with a 2 × 2 design there are four instead of two treatments, but the number of blocks can be reduced. In a 2 × 2 factorial RB design a three-way analysis of variance is used with the block being a random effect factor but treatment and gender being fixed effects involving interaction.

A protocol for a TBI experiment involving four treatments and two termination times in a 4 × 2 factorial design with five blocks is suggested in the appendix.

10 The Determination of Sample Size

TBI experiments are likely to be both time-consuming and expensive. They also involve important ethical considerations. Investigators should be fully aware of the need to take account of the "Three Rs": *Replacement, Refinement and Reduction* [11]. Animals should only be used if there is no alternative. "Refinement" requires the investigator to ensure that the animals suffer the minimum pain and distress. A veterinary surgeon should normally be consulted when planning the experiment and there will normally be some oversight by the local ethics committee. "Reduction" can best be achieved by doing high quality research, so that neither animals nor scientific resources are wasted, *not* by reducing sample size. Three methods of determining sample size are discussed below:

10.1 "Common Sense" or "Tradition"

Investigators often use the sample size that previous investigations have already used with apparent success. This has support from some eminent statisticians who state that "Except in rare instances..., a decision on the size of the experiment is bound to be largely a matter of judgment and some of the more formal approaches to determining the size of the experiment have spurious precision" [12]. Sir David Cox has written two books on experimental design and is the first winner of the "International Statistics prize." There are few statisticians in the world who are as highly respected. He and Dr. Reid are clearly referring to the power analysis when they mention "spurious precision."

Of course this approach assumes that there is a body of well-designed experiments which are producing good quality results on which to base future experiments.

10.2 The "Resource Equation"

This very simple formula is based on past experience [13]. It is $E =$ (total number of animals) − (number of treatments). E should be between about 10 and 20. Where there are just two groups it suggests that sample sizes should usually be between 6 and 11 animals per group. In general, it ensures that the resulting ANOVA has between 10 and 20 degrees of freedom for error thereby avoiding excessively small or large experiments. It is in general agreement

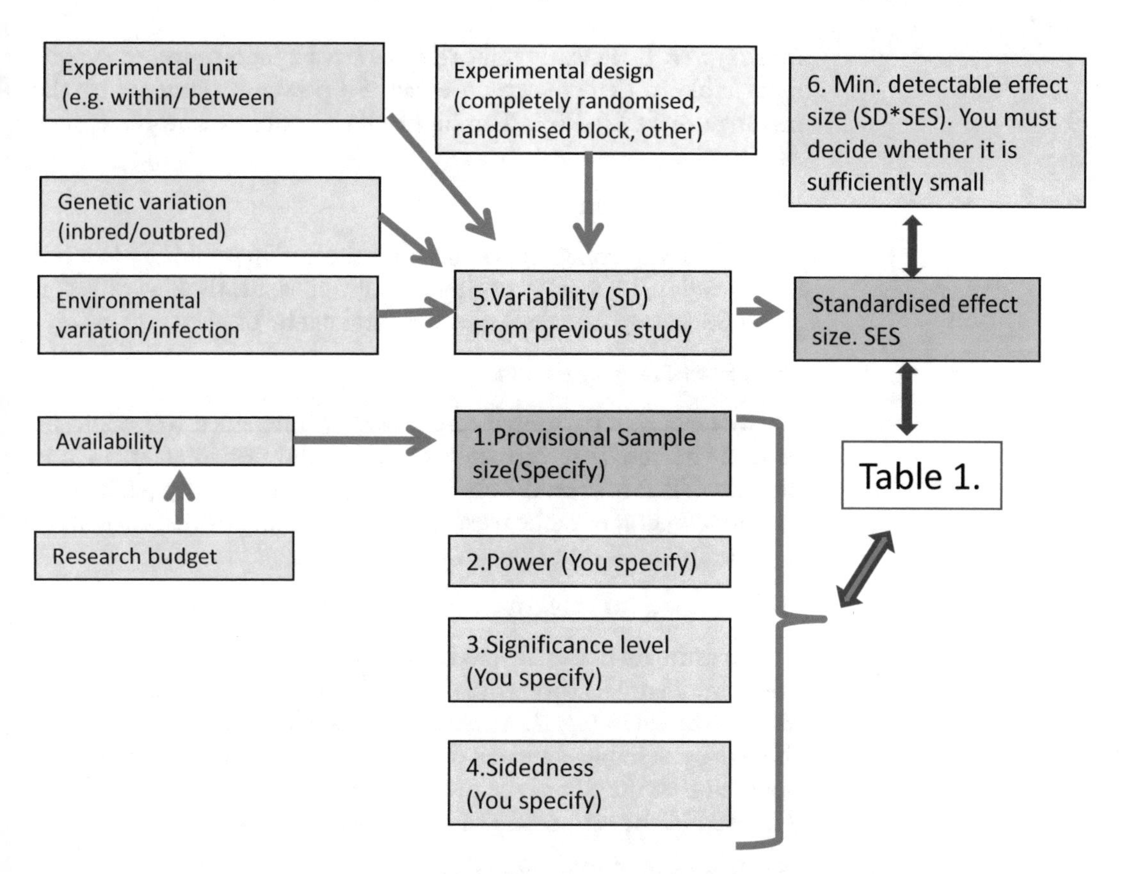

Fig. 2 The six variables used in a power analysis are labeled 1–6. The investigator should make a provisional estimate of sample size (1) and look up the corresponding SES in Table 3 for specified values of variables 2–4. Multiply the SES by an estimate of the SD (5) from a previous study to obtain the minimum detectable effect size (6). Then decide if this is sufficiently small. See text for details

with "common sense," and is probably the method used in most agricultural research.

10.3 The "Power Analysis"

This method was developed largely by Jacob Cohen in the 1960s [14]. It is the method used in clinical trials and is often the method suggested by funding organizations and IACUCs. It applies both to binary data and measurement outcomes. Only the latter is discussed here. It depends on a mathematical relationship between six variables. If five of these are specified by the investigator the sixth one (usually sample size) can be estimated. These variables are shown, labeled 1–6, in Fig. 2, together with some of the factors which influence them. There are, for example, many factors which influence the standard deviation.

Assuming two treatment groups the variables are:

1. *The required power*

This is the probability that the experiment will be able to detect a difference between the means of the two groups *specified by the investigator*. It is clearly important to have a high powered experiment, able to detect such an effect. So power is often set by the investigator at 80–90%. The higher level requires a larger sample size.

2. *The significance level*

This is the probability of rejecting the null hypothesis when it is true, leading to a false positive result. It is nearly always set at $p = 0.05$ but other values may occasionally be used.

3. *The sidedness of the test*

By default, a two-sided test is used. A one-sided test could be used if the response can only go, or is only of interest, in one direction. But it should only be used if it can be fully justified in advance. It should *not* be used as a way of making a difference more statistically significant (known as "p-hacking").

4. *The standard deviation of the outcome variable*

An estimate of the SD has to be obtained from a previous study. There are many factors which can influence inter-individual variability as shown in Fig. 2. These need to be taken into account when choosing a suitable standard deviation for the power analysis. It is assumed that the SD in the proposed experiment will be similar to that used in the estimation of sample size.

5. *The specified effect size (ES)*

This is the smallest difference between the means of the two groups *thought likely to be of scientific interest.* This requires a subjective assessment by the investigator, based on familiarity with the literature. It is this subjective assessment that makes a power analysis a procedure that is not fully objective.

Investigators often find it difficult to specify the effect size. So an alternative *iterative* approach is suggested below. This involves choosing a *provisional* sample size and using it to determine the minimum predicted detectable ES. If this predicted detectable ES seems to be reasonable then the provisional sample size is accepted. If the investigator decides that they would like to be able to detect a smaller ES, then a larger provisional sample size should be chosen with the calculations being repeated.

6. *The sample size*

This is the number of subjects in *each* group. In the method suggested here, a provisional sample size is chosen based on previous studies and common sense.

7. *Combining these variables*

If five of these variables are specified, the sixth one, classically sample size, can be estimated. Dedicated software is usually needed but it is widely available and free. The investigator inputs values for each of the variables in order to estimate the required sample size. However, the iterative approach, suggested below, does not require access to software.

11 An Iterative Approach to Determining Sample Size Using a Power Analysis

The power analysis asks the investigator to specify the minimum effect size (difference between the two means) likely to be of biological importance. An open-ended question like this is difficult to answer. The alternative, used here, is to choose a *provisional* sample size and estimate the effect size that it is likely to be able to detect. The investigator can then decide whether this is acceptable. If not a larger or smaller sample size can be chosen, with the calculations being repeated.

The method uses the "standardized effect size" (SES also known as Cohen's *d*) which is the difference between the means of the treated and control groups divided by the pooled standard deviation. It is labeled "7" in Fig. 2. No software is needed, only reference to Table 3.

The relationship between sample size, statistical power and the SES is shown in Fig. 2, for four levels of power, assuming a two-sided test and a 5% significance level. For example, this shows that a sample size of six animals per group will have a 90% power to detect an SES of 2.5 SDs, an 80% chance of detecting an SES 1.8 SDs, a 70% chance of detecting an effect of 1.6 SDs and a 60% chance of detecting an effect of 1.4 SDs, and so on.

The SESs multiplied by the standard deviation (SD) of a character obtained from a previous study, (SES × SD) provides an estimate of the *effect size* (ES) likely to be detectable for a specified power, significance level, and sidedness, *in the original units.*

11.1 The Procedure Is

1. Obtain an estimate of the standard deviation (SD) of the character of interest from a previous study. Ideally this should come from the investigator's own work and be based on a reasonably large sample size.
2. Make a *provisional* estimate of a suitable sample size based on previous studies and "common sense."
3. In Table 3 find the SES for the provisional sample size for the required power, significance level, and sidedness of the test.
4. Multiply the SES by the estimated SD. This provides an estimate of the "minimum detectable effect size" (MDES) in the

Table 3
Cohen's *d* (SESs) for sample sizes of 4–34 subjects per group assuming 80% and 90% power, a 5% significance level and a one-sided or two-sided test

Sample size	80% one-sided	80% two-sided	90% one-sided	90% two-sided
4	2.00	2.38	2.35	2.77
5	1.72	2.02	2.03	2.35
6	1.54	1.80	1.82	2.08
7	1.41	1.63	1.66	1.89
8	1.31	1.51	1.54	1.74
9	1.23	1.41	1.44	1.63
10	1.16	1.32	1.36	1.53
11	1.10	1.26	1.29	1.45
12	1.05	1.20	1.23	1.39
13	1.00	1.15	1.18	1.33
14	0.97	1.10	1.14	1.27
15	0.93	1.06	1.10	1.23
16	0.90	1.02	1.06	1.18
17	0.87	0.99	1.03	1.15
18	0.85	0.96	1.00	1.11
19	0.82	0.93	0.97	1.08
20	0.80	0.91	0.94	1.05
21	0.78	0.89	0.92	1.03
22	0.76	0.86	0.90	1.00
24	0.73	0.83	0.86	0.96
26	0.70	0.79	0.82	0.92
28	0.67	0.76	0.79	0.88
30	0.65	0.74	0.76	0.85
32	0.63	0.71	0.74	0.82
34	0.61	0.69	0.72	0.80

original units that the experiment is likely to be able to detect for that sample size, power, etc.

5. Decide whether this MDES is acceptable. If so, accept the provisional sample size. If not, the calculations should be repeated with a different provisional sample size.

6. If more than one outcome variable is to be measured, then the calculations should be repeated for each of them and an assessment made of whether the MDES is acceptable in all cases.

11.2 An Example

1. Zhang et al. [15] studied brain water content (BWC) in rats following fluid percussion TBI. Averaged across a 1-week period the mean and SD of BWC was 83 ± 5.5% and the sample size was $n = 10$/group.
2. Suppose an investigator wishes to study the effect of a new drug on BWC.
3. She decides to choose a *provisional* sample size of 15 rats per group and assumes an SD of 5.5%

   From Table 3 the SES for a sample size of 15, with a 90% power, a 5% significance level and a two-sided test is 1.23.
4. So for brain water the minimal detectable effect size (MDES) in her proposed experiment is SD $\times$ SES $= 5.5 \times 1.23 = 6.7\%$ with a 90% power.
5. She would also have an 80% power to detect an effect size of $1.1 \times 5.5 = 6.0\%$
6. Suppose that she would like to be able to detect a 5% difference between the two groups. This sample size is not large enough with these power levels. She must decide whether to go ahead with this MDES (minimal detectable effect size) or choose a larger sample size and repeat the calculations. In fact she would need a sample size of about 26 rats/group to detect a 5% change at the 90% level. This assumes that the SD in her experiment will be the same as that found by Zhang et al.

If she decides to only use 15 rats per group then in the Materials and Methods section of the report, according to item 10 in the ARRIVE Guidelines [16] or when applying for funds she could make a statement (adapted to her conclusions) such as:

"A power analysis was used to determine sample size. Assuming a 5% significance level, a two-sided test, and a standard deviation of 5.5% a sample size of 15 animals per group will have a 90% power to detect a difference between the control and treated rats of 6.7%."

11.2.1 Minimizing the SD During the Experiment

The estimated sample sizes assume that the experiment will have a standard deviation at least as low as that used in the power calculations. So minimizing variation in severity levels during the experiment should have a high priority. Detailed studies of the operating characteristics of the various types of apparatus can be helpful [17], and staff should be well trained and given adequate time to perform their tasks.

There may be rat and mouse strain differences in response to drugs and chemicals [18] and even in tolerance of the injury. Inbred strains (such as C57BL/6 mice or F344 rats) are more uniform than outbred stocks (such as CD-1 mice or Sprague-Dawley rats) although they are less robust, which could be a disadvantage. F1 hybrids are genetically uniform though heterozygous, and are robust, but special breeding may be required to produce them. Bone shape is under strong genetic control [19] and the shape and thickness of the skull will vary between individuals and strains of mice and rats. Age and gender may also be important variables. Investigators funded by the NIH will be required to use both sexes unless they can make a strong case for not doing so. If both sexes are used, a factorial design will probably be necessary in order to avoid having to use extra animals.

Variation caused by circadian rhythms [20] may be minimized by, for example, restricting the surgery to the morning or afternoon. The animal house environment is not homogeneous. Animals housed on the top shelves will have more light and a warmer temperature than those on lower shelves. It would be a serious error to house treated and control animals on different shelves as this could lead to biased results. If a completely randomized design is used the animals must be maintained in random order in the animal house. If randomized block (cohort) designs are used, each block should be kept as a group on the same shelf.

The written paper should be checked against the ARRIVE guidelines [21] to ensure that no important details have been omitted. Clinical trials should only ever be done when the results of a paper can be independently and critically repeated by investigators in other laboratories. So every effort should be made to provide sufficient details to make it easy for other investigators to repeat the experiment.

Standard descriptive statistics including means, standard deviations and "n" per group should always be presented in addition to any graphical presentation. These are required for planning future experiments (as above) as well as for systematic reviews and meta-analyses which are being increasingly widely used in animal studies.

11.2.2 The Standardized Effect Size (Cohen's d) as a Standardized Measure of Effect Size

The strategy currently used in biomedical research, dubbed "NHST" or "Null Hypothesis Significance Testing" in which the results of randomized controlled experiments are statistically analyzed to determine whether the observed differences are "statistically significant" (are unlikely to be due to chance) has been highly successful in progressing biomedical research for nearly a century. But it has come under increasing criticism recently. The editors of *Basic and Applied Social Psychology* (*BASP*) have even announced that the journal would no longer publish papers containing *P* values "because the statistics were too often used to support lower-quality research."

One problem is that *p*-values and statistical significance have no direct biological meaning. Two experiments can have exactly the same means and standard deviations, but in one of them the difference is "statistically significant" while in the other one the difference is "not statistically significant": the difference between them being only due to sample size. And p-values only provide a very poor estimate of the *magnitude* of a treatment response. Even trivial differences between groups can be statistically "significant" if the sample size is sufficiently large.

There is also an obvious problem with the word "significant." In the English language it is synonymous with "important." But in statistics it only means that the observed effect is unlikely to be due to chance. The word "detected" might have been a better choice. ("The difference was/was not statistically *detected* at $p = 0.XX$"). The word "significant" is (excessively?) widely used in TBI research. In a small informal random sample of ten TBI papers the word "significant" was used from 3 to 52 times, with an average of 20.2 times per paper. Often it is unclear whether the authors consider the results to be both important and unlikely to be due to chance. Perhaps the word "important" should be reserved for when the effect really is considered to be important. And authors might try using the word "detected" instead of "significant" if referees and journal editors will allow them to do so!

Another criticism is that NHST tends to lead to binary thinking. An effect is either present (significant) or absent (not significant). Bar charts do show the differences between the means, but the actual differences are rarely calculated, recorded, and interpreted.

There is now a growing literature on the use of effect sizes [22]. The SES (Cohen's d), used in a previous section in the estimations of sample sizes, is the difference between two means divided by the pooled standard deviation. So it is the effect size in units of standard deviations. It is used in meta-analysis [23] as a standardized measure of the magnitude of an effect. A great advantage is that two SESs can be compared even if they are based on different units of measurement. The SES is also directly related to sample size as shown in Fig. 3. Table 4 shows the SESs for 11 characters following TBI in animals maintained in a normal atmosphere or one with alcohol vapor [10]. In most cases the alcohol increased the severity, as shown by the SESs.

12 Small, Medium and Large Standardized Effect Sizes

Jacob Cohen was a psychologist working with human subjects. He suggested that values of d (the SES) of 0.2, 0.5 and 0.8 SDs in an experiment would represent small, moderate and large treatment responses requiring sample sizes of 394, 64, or 26 subjects per group, respectively, assuming an 80% power, a 5% significance level, and a two-sided t-test.

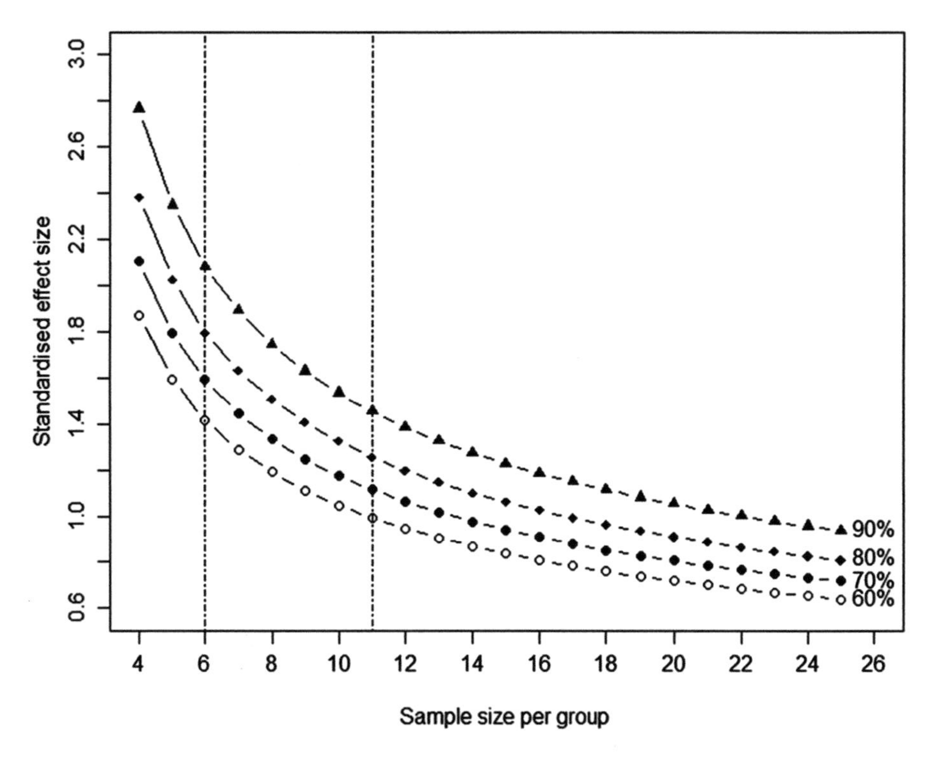

Fig. 3 Standardized effect size (SES or Cohen's *d*) as a function of sample size (per group) for four levels of power (60–90%) assuming a two-sized *t*-test with a 5% significance level, two groups and a quantitative dependent variable. The vertical dotted lines show the range of sample sizes using the "resource equation" method of determining sample size

Table 4
Standardized effect sizes (SESs) for 11 characters following fluid-impact TBI in rats exposed to air or alcohol vapor

Character	Air	Alcohol
NSS	2.58	5.23
Beam only	3.04	5.24
Body weight	0.39	0.84
Open filed % cent. Sq.	2.50	2.35
Locomotion	1.49	2.50
NOR familiar	1.33	0.46
NOR novel	0.94	2.69
NOR discrimination	0.92	1.03
Astrocytes	2.87	8.53
ED-1 stain	3.74	11.16
HMGB1	2.05	12.24

Data from Teng et al. (2015)

However, laboratory animals are more uniform than humans so SESs are usually larger. Accordingly, values of SES of 1.1, 1.5 and 2.0 SDs, designated "extra-large" "gigantic" and "awesome" could be added to take account of laboratory animal experiments, including in-vitro studies using animal cells or extracts. Detecting SESs of these magnitudes would require sample sizes of 17, 8, and 5 per group, respectively with the same assumptions.

12.1 Calculating the SES

The pooled standard deviation can be calculated in various ways. If the data has been analyzed using an ANOVA, then the *error mean square* is an estimate of the pooled variance, so the square root of that is the pooled SD. Alternatively, if means and standard deviations are available, then the pooled SD is the square root of the mean of the two variances.

Table 5 sets out the calculations for a Neural Sensitivity Score used by Teng et al. [10]. Now all the characters are in the same units (standard deviations) and their individual magnitudes can be compared. Body weight was a poor indicator of severity, particularly in the group not exposed to alcohol. In absolute terms many of the SESs are "extra-large" to "awesome." However, the ED-1 and HMGB1 values are off the scale. They represent tissue damage at the injury site, so should probably be classified as a qualitative rather than a quantitative change. Whether SESs estimated from quantitative variation in such characters could be developed as a useful scale of severity might be worth exploring in more detail. For example, the SESs for brain water content in some papers were as high as eight standard deviations. This could imply that the TBI was excessively severe and unlikely to respond to a drug treatment.

Note that if the combined sample sizes are less than 12, the SES is biased and needs to be multiplied by the factors shown in Table 6.

Although the SES is not (yet) widely used in research with laboratory animals, it has been used to summarizing the results of toxicity tests, where many characters are measured on each individual. These may include hematology, clinical biochemistry, and even some behavior variables. Classically each character is statistically analyzed separately. But this leads to multiple tests and excessive numbers of false positive outcomes. However, if the results are converted to SESs then all characters are in the same units of (standard deviations) they can be compared and combined to give an average response. The graphical presentation of the results is also improved [24].

Table 5
Calculation of SESs for neural sensitivity score (NSS)

Character	Treatment	TBI	SD	SHAM	SD	Pooled SD[a]	SES
NSS (neural sensitivity score	Air	1.65	0.42	0.83	0.16	0.317805	2.58

Data from Teng et al. (2015)
[a]Pooled SD for "Air" group = $\text{sqrt}((0.42^2 + 0.16^2)/2) = 0.318$. SES = $(1.65 - 0.83)/0.318 = 2.58$.

Table 6
Correction for bias when estimating SES with small sample sizes (see text for details)

DF	Multiplier
3	0.73
4	0.80
5	0.84
6	0.87
7	0.89
8	0.90
9	0.91
10	0.92

As an example in TBI research, Teng et al. [10] used a 2 × 2 factorial experimental design to explore the effect of mild focal TBI versus Sham treatment of rats maintained either in an atmosphere of air or alcohol vapor (Table 4). They scored the rats for 11 variables. The resulting data was statistically analyzed as a 2 × 2 factorial design and the means and standard deviations were reported for each character and group (as recommended here).

If an investigator is able to specify the SES (instead of the Effect Size) that they would like to be able to detect, based on SESs found in previous studies, then determination of a suitable sample size is very simple. All that they need to do is look up that SES in the body of Table 3 for the chosen power and sidedness of the test, and read off the required sample size. For example, it they would like their experiment to be able to detect an SES of 1.5 or more SDs with a 90% power, a two-sided test and a significance level of 5% they would find from the body of Table 3 that they need a sample size of ten animals per group.

13 Discussion

As stated in the introduction, there is a crisis in preclinical research. Too many authors are publishing papers with results that turn out to be irreproducible. The results of clinical trials of drugs found to be effect in animal models of TBI have also been disappointing [25]. Are these two facts related? This chapter should not be taken as a review of the statistical quality of animal TBI research, but it is clear that there is room for improvement. In particular very few of the papers quoted here mentioned randomization and blinding. Where randomization was mentioned, it was usually stated that the animals were assigned to the treatment groups at random. But the animals are usually uniform. It is the research environment which is heterogeneous. In no case was it suggested that the treatment groups were intermingled in space and time as explained in Table 1.

This, combined with the universal use of completely randomized experimental designs (which were not correctly randomized!) can easily lead to false positive results. *A better alternative would be to use randomized block (cohort) designs.* Such designs are less likely to lead to bias, they are generally more powerful and logistically more convenient than CR designs. Examples of the use of such designs are given elsewhere [9], but they seem to be rare in TBI research so some pioneering work would be needed to demonstrate their value. This should, preferably, be done with the assistance of a statistician.

Although most papers seem to use appropriate statistical procedures, it is unclear whether the authors had checked the assumptions (homogeneity of variances and normal distribution of the residuals) required for a parametric statistical analysis.

There also seems to be an unhealthy emphasis on the word "significant." One paper used the word 73 times. Could it be replaced by the words "detected" and "important" as appropriate? Would any journal editor be prepared to promote such a change?

There also seems to be insufficient importance attached to the *magnitude* of the response to a treatment. This might be partly because of the binary thinking arising from significance testing and *p*-values. The SES is a standardized measure of the magnitude of a response. It should be of value in TBI research when comparing the results from different experiments, particularly if some treatments are only be suitable for mild but not for serious injury.

The SES was not conceived by Cohen as a tool for the quality control of experimental data, but it could be useful as such. For example Table 7 shows SESs for brain water content in Sham versus TBI rats at six time points following the TBI in two publications. The mean BWC in the Sham rats was 72.3% and 79.2% in the two publications [15, 26], respectively. Are such large differences among Sham animals common? Is it due to the methods, the rats which are used, or is it an artifact? The SES for the increase in BWC following TBI averaged 7.0 SDs in one paper compared with a mean of only 0.8 SDs in the other paper, averaged over six time points with eight animals per group. This implies that the severity of the TBI differed very considerably between the two studies as measured by BWC. What are the implications when screening drug treatments? There was little change in BWC over the six time points in either paper. If this is the case then two or three time points would be sufficient to show the response, releasing resources and animals to do twice as many experiments. Good experimental design is not just about designing individual experiments. It should also take into account the efficient use of the available scientific resources.

Table 7
Standardized effect sizes (SESs) of differences in brain water content in Sham and TBI rats in two publications

Zhang (2015) [15][a]		Jin (2016) [26][b]	
Time (h)	SES	Time	SES
1	0.02	2	3.34
6	0.52	6	8.06
24	1.23	12	8.80
48	0.84	24	7.42
72	1.29	72	8.15
168	0.69	168	7.46
Mean	0.77		7.21

The mean brain water content in the sham rats was 72.3% (Jin et al.) and 79.2% (Zhang et al.)
[a]TBI using a modified version of Feeney's method and a PinPoint™ Precision Cortical Impactor. It is unclear whether this is the BWC for the whole brain or just the damaged side
[b]Used fluid percussion TBI. Whole brain water was measured

13.1 Recommendations

1. Investigators should be given much better training in the principles and practice of good experimental design, including the available designs such as completely randomized, randomized block, and factorial designs and the determining sample size. It should be recognized that clinical trials and animal research require different designs.
2. Investigators should be made more aware of the factors which can influence the variation in response of animals to experimental treatments. These include the strain, gender, proximal environment, microbiological status, circadian and other rhythms, and social interactions.
3. Much more attention should be paid to the magnitude of effects with less emphasis on "statistical significance." When discussing results it should be made clear whether "significant" means "important" or "unlikely to be due to chance." The Standardized Effect Size (SES or Cohen's *d*) could be used to measure the magnitude of an effect although further research is needed to "calibrate" it for TBI research.
4. There should be more input into TBI research from statisticians, particularly ones with a good background in animal research and experimental design.
5. A scientist trained in meta-analysis should be recruited to work on TBI research. He or she would help to make better use of the literature, and set standards for biomarkers of severity, using standard effect sizes and other meta-analysis techniques.

Acknowledgment

I thank Dr. Johnny Lifshitz for helpful advice in the preparation of this chapter.

Appendix

An example of a possible protocol for testing a proposed treatment using a randomized block/cohort design. This could easily be modified to investigate different treatments or times.

The protocol might have two factors "Treatment" and "Time following treatment."

Four treatments might be:

1. Sham
2. TBI + vehicle
3. TBI + Treatment-low dose
4. TBI + Treatment-high dose

The factor "Time" might have two levels, say 12 h and 48 h following the treatments.

Each block/cohort would therefore involve eight animals (4 treatments × 2 times post treatment).

Six of the rats in each block would have TBI and two would be shams. It is assumed that the treatments could be done conveniently in one day.

Various outcomes could be measured in the live animals including, say, a "neural sensitivity score," and other types of behavior. It is assumed that all eight rats in a block could be measured in a relatively short period. Histological and anatomical outcomes would be made after the animals are euthanized at the two time points.

With, say, five blocks/cohorts the protocol would involve 40 animals. In this case the ANOVA table to be used for all the outcome variables such as behavior and brain water content would be as follows:

Source	DF
Blocks	4
Treatments	3
Times	1
Trt. × Time	3
Error	28
Total	39

In the absence of a Treatment × Time interaction sample sizes for comparing the four groups will be $n = 10$. If a power analysis justification is needed for that sample size, the SES for a sample size of 10 can be looked in Table 3. It is 1.53 SDs. This is the predicted detectable effect size in units of standard deviations. So multiplied by the SD of a character, it predicts the effect size in original units that the experiment will be able to detect for a 90% power and a 5% significance level. The sample size for comparing the times will be $N = 20$. This with 28 degrees of freedom in the error term will provide plenty of power. If a Treatment × Time interaction is detected it implies that the observed treatment differences are not equally valid at the two time points. An experiment like this one is really much more powerful than suggested by these sample sizes because there is a much better estimate of the error (28 DF) which is above the suggested maximum using the Resource Equation method.

References

1. Begley CG, development ELMD (2012) Raise standards for preclinical cancer research. Nature 483(7391):531–533
2. Prinz F, Schlange T, Asadullah K (2011) Believe it or not: how much can we rely on published data on potential drug targets? Nat Rev Drug Discov 10(9):712
3. Scott S, Kranz JE, Cole J, Lincecum JM, Thompson K, Kelly N et al (2008) Design, power, and interpretation of studies in the standard murine model of ALS. Amyotroph Lateral Scler 9(1):4–15
4. Freedman LP, Cockburn IM, Simcoe TS (2015) The economics of reproducibility in preclinical research. PLoS Biol 13(6): e1002165
5. Bebarta V, Luyten D, Heard K (2003) Emergency medicine animal research: does use of randomization and blinding affect the results? Acad Emerg Med 10(6):684–687
6. Button KS, Ioannidis JP, Mokrysz C, Nosek BA, Flint J, Robinson ES et al (2013) Power failure: why small sample size undermines the reliability of neuroscience. Nat Rev Neurosci 14(5):365–376
7. Fisher RA (1960) The design of experiments. Hafner Publishing Company, Inc, New York
8. Hill AB (1967) Principles of medical statistics, 8th edn. The Lancet, London
9. Festing MFW (2016) The design of animal experiments, 2nd edn. Sage Publications, Thousand Oaks
10. Teng SX, Katz PS, Maxi JK, Mayeux JP, Gilpin NW, Molina PE (2015) Alcohol exposure after mild focal traumatic brain injury impairs neurological recovery and exacerbates localized neuroinflammation. Brain Behav Immun 45:145–156
11. Russell WMS, Burch RL (1959) The principles of humane experimental technique, special edition. Universities Federation for Animal Welfare, Potters Bar, England
12. Cox DR, Reid N (2000) The theory of the design of experiments. Chapman and Hall/CRC, Boca Raton, FL
13. Mead R (1988) The design of experiments. Cambridge University Press, Cambridge, NY
14. Cohen J (1988) Statistical power analysis for the behavioral sciences. Lawrence Erlbaum Associates, Hillsdale, NJ
15. Zhang C, Chen J, Lu H (2015) Expression of aquaporin-4 and pathological characteristics of brain injury in a rat model of traumatic brain injury. Mol Med Rep 12(5):7351–7357
16. Kilkenny C, Altman DG (2010) Improving bioscience research reporting: ARRIVE-ing at a solution. Lab Anim 44(4):377–378
17. Lin YP, Jiang RC, Zhang JN (2015) Stability of rat models of fluid percussion-induced traumatic brain injury: comparison of three different impact forces. Neural Regen Res 10 (7):1088–1094
18. Festing MFW (2016) Genetically defined strains in drug development and toxicity testing. In: Proetzel, Wiles MV (eds) Mouse models of drug discovery, 2nd edn. Humana, New York, pp 1–17

19. Festing MFW (1972) Mouse strain identification. Nature 238:351–352
20. Schwartz WJ, Zimmerman P (1990) Circadian timekeeping in BALB/c and C57BL/6 inbred mouse strains. J Neurosci 11:3685–3694
21. Kilkenny C, Browne W, Cuthill IC, Emerson M, Altman DG (2010) Animal research: reporting in vivo experiments: the ARRIVE guidelines. Br J Pharmacol 160 (7):1577–1579
22. Ellis PD (2010) The essential guide to effect sizes. Cambridge University Press, Cambridge
23. Cumming G (2012) Understanding the new statistics. Routledge, Abingdon. 12 A.D.
24. Festing MFW (2014) Extending the statistical analysis and graphical presentation of toxicity test results using standardized effect sizes. Toxicol Pathol 42(8):1238–1249. https://doi.org/10.1177/0192623313517771
25. Stein DG (2015) Embracing failure: what the phase III progesterone studies can teach about TBI clinical trials. Brain Inj 29(11):1259–1272
26. Jin H, Li W, Dong C, Ma L, Wu J, Zhao W (2016) Effects of different doses of levetiracetam on aquaporin 4 expression in rats with brain Edema following fluid percussion injury. Med Sci Monit 22:678–686

Chapter 14

Advanced Informatics Methods in Acute Brain Injury Research

Jude P. J. Savarraj, Mary F. McGuire, Ryan Kitagawa, and Huimahn Alex Choi

Abstract

The failure of several clinical trials in traumatic brain injury (TBI) targeting thresholds of physiologic variables challenges the clinical utility of existing approaches and highlights the need for new patient management strategies in acute brain trauma. Medical informatics is a multidisciplinary field that focuses on solving clinical problems using techniques from various quantitative disciplines including engineering, statistics, and computer science. They are proven to be useful in solving problems in the omics domains, but less popular in TBI research. In this chapter, some of the applications of medical informatics in TBI research are discussed. First, we discuss the need for patient-specific threshold of physiologic metrics rather than population based metrics. The role of cerebral autoregulation (CAR) and recent techniques to measure a patient's CAR status and the utility of CAR in clinical practice are discussed. Second, we focus on two important subfields of informatics – supervised and unsupervised machine learning – and their applications in ABI research. Machine learning (ML) is a powerful tool that is widely used in the discovery of patterns in large datasets and in the development of predictive models. We discuss some of the applications of ML in TBI research; ranging from the prediction of ICP hypertension to discovery of patterns. Also, recent advancements in our ability to store large quantities of data and the availability of cheap processing power has spawned the era of 'big data'. 'Big data' made huge impacts on different fields including genomics and proteomics. We discuss recent advancements in this field and the need for large databases for TBI research. As the condition of each patient is unique, there is a need for a transition to a 'personalized-medicine' approach, where the management protocol for each patient is unique and is based on the patient's status, rather than a guideline-based approach. Techniques from informatics applied to TBI research can aid this transition.

Key words Continuous physiological monitoring, Machine learning, Data mining, Bioinformatics, Signal processing, Multimodal monitoring

1 Introduction

Traumatic brain injury (TBI) is a devastating disease affecting 1.7 million people in the United States annually and is a significant burden to society. Several high profile clinical trials targeting thresholds of physiologic measures to avoid secondary brain injury

Amit K. Srivastava and Charles S. Cox, Jr. (eds.), *Pre-Clinical and Clinical Methods in Brain Trauma Research*, Neuromethods, vol. 139, https://doi.org/10.1007/978-1-4939-8564-7_14, © Springer Science+Business Media, LLC, part of Springer Nature 2018

after TBI have failed. The findings from these trials have challenged the clinical utility of threshold-based treatments and have questioned our usage of traditional statistical methods in acute brain injury (ABI) research. Traditional approaches are reductionist i.e., attempting to address a complex disease like TBI by controlling a single physiological variable or pathway. These failures have highlighted a need for the development of systematic approaches that accounts for the complexities inherent in the disease. Secondly, we need better methods of categorizing ABI patients on a patient-specific and injury-specific level. Each injury is unique in type and severity, therefore a 'personalized medicine' approach that is tailored to the specifics of managing a particular patient is required. Finally, to make advancements in ABI research, hardware and software infrastructure to collect, store and analyze large amounts of data is needed.

In this chapter we discuss methods from the field of informatics and its applications in ABI research. Informatics is a science that focuses on developing techniques for efficient storing, retrieval, and processing/analyzing data with the goal of deriving useful and actionable information. It is a cross-disciplinary field involving ideas from computational, signal processing, engineering, mathematical, and statistical methods. The application of informatics techniques to address problems in medicine is called 'medical informatics'. There are several subfields within medical informatics, and a detailed treatise on all of them is beyond the scope of this chapter. We will discuss specifically applications of informatics in the analysis of continuously obtained high-frequency continous physiological data/variables (CPV), discovery of patterns in data and development of predictive models. First, we will discuss the limitations in existing techniques that are used in the analysis of physiological variables and the recent advancements in the analysis of CPVs. Second, we will discuss machine learning (ML), a field of artificial intelligence that focuses on discovery of patterns in datasets and the development of mathematical models that can predict adverse complications or outcomes in patients. The final part of the chapter will discuss the data infrastructure, software tools and unmet needs of ABI research.

2 Threshold Based Approach vs. Personalized Medicine Approach

The primary trauma event can trigger secondary responses such as vascular engorgement, obstruction to cerebrospinal fluid, or cerebral edema, which can result in an increase in the intracranial pressure (ICP) which can further exacerbate injury spread and worsen clinical outcomes. Parenchymal ICP monitors and extra ventricular catheters are typically used to measure ICP. 5–15 mm of Hg is considered the optimal ICP; an ICP of >20 mm Hg for more than 5 min is considered intracranial hypertension and requires clinical intervention. However, recent clinical trials on managing ICP have not shown promising outcomes. The *Benchmark Evidence from South American Trials: Treatment of Intracranial Pressure* (*BEST:TRIP*) trial reported that TBI subjects who had an ICP monitor and were managed for ICP crisis did not do significantly better than a control group that was only managed with imaging and clinical examination without ICP monitoring. In another clinical trial—Decompressive craniectomy in diffuse traumatic brain injury (DECRA)—155 severe TBI subjects were randomized to either undergo a bifrontotemporoparietal decompressive craniectomy (to manage raised ICP) or standard care. The group assigned to the craniectomy procedure did have decreased ICP; however, the subjects had unfavorable outcomes although the mortality rates between the two groups were similar [1]. The failure of the BEST and DECRA trials have led to the questioning of threshold-based metrics (for e.g., 20 mm of Hg) and suggests a need for a more patient-specific approach. What is considered an 'optimal range' for a physiological parameter can be different for each patient. A shift in research focus to a patient-specific target range—rather than absolute thresholds—could lead to discoveries of better strategies in managing ICP in ABI patients.

Individualized-ICP approach: Several informatics techniques have been used to address the limitations of threshold-based therapies. The ICP waveform is a continuously fluctuating signal, and research has focused on analyzing these waveforms using signal processing techniques [2, 3]. Different techniques including individualized thresholds, morphological feature extraction, and advanced spectral analysis methods have been developed for the analysis of ICP waveform.

Individualized thresholds of ICP based on the cerebral autoregulatory status (which is quantified by the PRx—the pressure reactivity index) has been proposed as an alternative to universal thresholds. The PRx is the moving correlation between the ICP and MAP and takes values between −1 and +1. PRx values close to zero is indicative of an intact autoregulatory status. By plotting the ICP and the PRx, the values of PRx for different measures of ICP were visualized and the ICP range for which the PRx was <0.2 is considered an optimal value. Deviations from the calculated individualized ICP-thresholds were better predictors of mortality and 6-month outcomes than the universal threshold-based metrics [4].

The morphological technique involves analyzing the shape of the biologic signal of an ICP waveform which consists of: arterial pulsation (P1), intracranial (P2) compliance, and aortic valve closure (P3). An advanced automatic informatics method called the 'Morphological Clustering Analysis Algorithm' utilizes the morphological characteristics of these waveforms and can forecast ICP hypertension.

The algorithm used 24 extracted variables—including mean ICP value, the slope of the waveforms, the decay time, and the distances between the pulses—and was able to predict ICP hypertension a few minutes before the onset (with specificity of 0.97 and sensitivity of 0.8) [5]. Predicting the occurrence of ICP hypertension prior to its onset will allow for physicians to initiate prophylactic treatments with the goal to preventing it.

The spectral (or frequency) analysis of the waveform provides information on repetitive (frequency) components of the ICP waveform. Since ICP waveform is a continuous and a periodic signal, appropriate signal transformation techniques can be used to convert the time-domain signal into a frequency-domain signal and isolate the 'frequency' components in the waveform. In a study investigating the spectrum of the ICP, the authors identified three major frequency bandwidths (0.2–2.6, 2.6–4.0, and 4.0–15 Hz) in patients with hydrocephalous and were able to leverage the spectral power in the frequency bands to predict shunt failure [6]. These techniques have shown promise in TBI research as well [7]. The ICP waveform has rich dynamics and multiple techniques to analyze the waveforms offer better alternatives to threshold-based targets.

Cerebral autoregulation: Cerebral autoregulation (CAR) is the ability of the brain's vasculature to adapt its resistance to ensure consistent blood flow irrespective of the mean arterial pressure (MAP). However, CAR can be compromised in TBI resulting in hypoperfusion or hyperperfusion which can cause secondary neurological injuries and poor outcomes [8]. The target cerebral perfusion pressure (CPP)—the difference between the MAP and ICP—is typically around 50–70 mmHg in adults [9] and 45 mmHg in pediatric TBI population [10]. However, this universal target for CPP may not necessarily be the optimal value for all patients. The optimal CPP (CPP_{opt}) is both a patient-specific and a dynamic value; maintaining the CPP close to CPP_{opt} has been hypothesized to result in better patient outcomes [11] and is increasingly becoming an important component of neuromonitoring [12]. Different concepts and methods of CPP_{opt} has been proposed [12–15]. The pressure reactivity Index (PRx) is the most commonly used CAR measure [16]. PRx is defined as the continuous Pearson's correlation coefficient of the 10 s windows of ICP and ABP over a moving window with 80% overlap [17]. Unfavorable outcomes in brain trauma patients were directly associated with the magnitude of the deviation from CPP and PRx-derived CPP_{opt} [18]. Since some institutions do not have the capacity to collect continuous high-frequency data, several low-frequency minute-by-minute [19] measures of the PRx have been proposed and have been found to be as effective as the high-frequency PRx-based CPP_{opt} measures [15]. There is increasing evidence from practice that optimizing cerebral regulation can result in good outcomes [11, 20] and CAR measures provides a very concise, meaningful and precise description of the patient's cerebral vasculature.

3 Machine Learning in ABI Research

Due to the ease in which data can be stored and retrieved, large datasets often involving hundreds of patients and thousands of variables are increasingly available to clinicians and researchers. The variables within these datasets could have complex associations with each other and traditional statistical tools are often inadequate in deciphering "hidden" patterns and thus limiting our understanding of the disease. These datasets could contain valuable knowledge that go undiscovered without the application of appropriate techniques. Another challenge is the development of computational models that could learn from previous datasets to identify a future patient's risk of developing secondary injuries and complications prior to the onset of clinical symptoms. Such a computational model employed by the bedside could potentially warn the physician of a patient's impending risk of a complication, prior to its onset, thus providing the physician valuable time to intervene and administer prophylactic care that can prevent the onset of the complication. Machine Learning (ML) is a branch of artificial intelligence that deals with developing computational models that are useful in discovering "knowledge" from larger datasets and make predictions on impending complications. There are two main subtypes of machine learning models: unsupervised and supervised. Unsupervised learning (or data mining) is a type of ML model that is useful in drawing inferences from high-dimensional datasets without labels. These are useful is extracting "knowledge" from large datasets by uncovering complex associations between the variables. Supervised learning is a type of ML model that is useful in developing models that can predict an impending event. The applications of both types of models/algorithms in ABI are discussed.

Unsupervised Learning: Unsupervised learning algorithms are a class of machine learning algorithms that are useful in discovering patterns and clusters within the dataset. The input to an unsupervised model is an unlabeled dataset which may consist of several patients and many variables. Figure 1 shows the output of a 'hierarchical clustering' algorithm (an unsupervised learning method) whose input was 28 patients with 18 variables. The model "discovered" three groups of patients (A, B and C) who had similar physiological characteristics (Fig. 1: left dendrogram). It also grouped the physiological variables that were similar to each other (Fig. 1: top dendrogram).

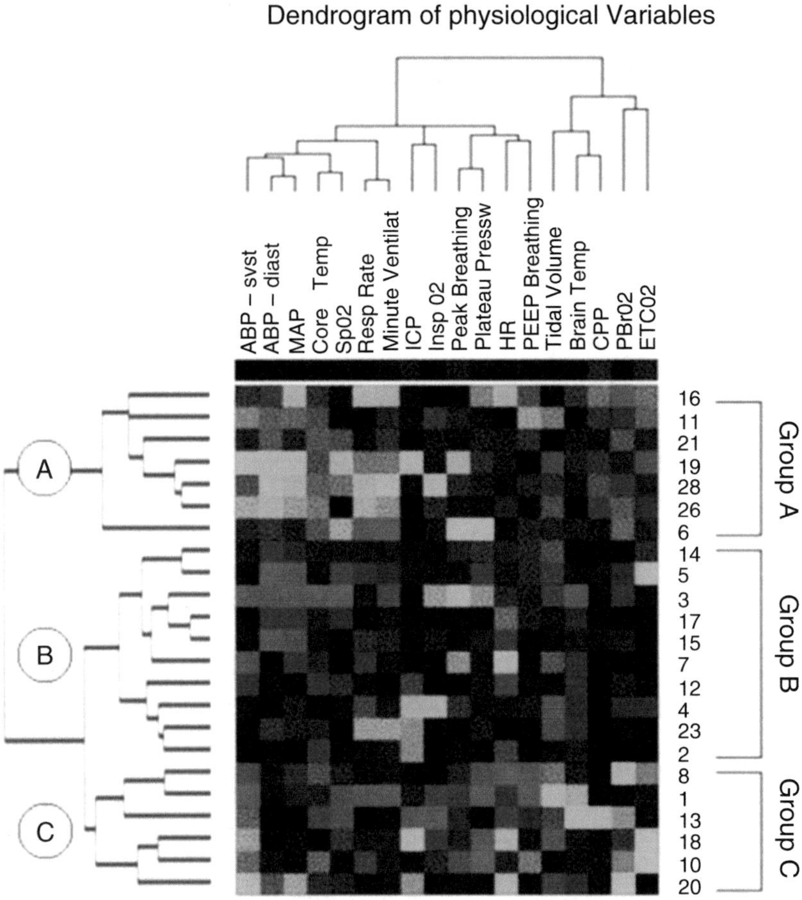

Fig. 1 A heat-map based on a self-organizing hierarchical clustering algorithm groups patients who have similar physiological characteristics'. This technique is routinely used in genetics to cluster genes with similar functionality. But in this particular heat-map, the physiological variables from 28 patients were used. The algorithm discovered three groups of patients (A, B and C). The patients within each cluster were similar in their physiological profile. Such analysis is useful in deciphering high-dimensional data sets with a lot of variables and very few patients; as in the case of neurocritical informatics. Abbreviations: *ABP* arterial blood pressure, *CPP* cerebral perfusion pressure, *ETCO*$_2$ end tidal carbon dioxide, *ICP* intracranial pressure, *MAP* mean arterial blood pressure, *O*$_2$ oxygen, *PbtO*$_2$ brain tissue oxygen tension, *PEEP* positive end-expiratory pressure, *SpO*$_2$ systemic oxygen saturation. *Sorani, M.D.* et al. *Neurocrit Care 7, 45–52 (2007)*

The user can then deconstruct the results to reason out why some patients and some physiological variables were grouped together. For instance, the model, as expected grouped ABP (systolic and diastolic) and MAP together as they are physiologically similar. However, the rationale for the grouping of the patient may not be straightforward. The user may have to reason the algorithm's decision to cluster these patients and through the reasoning process may gain insights that are not straightforward. There is no limit to the number of patients or the number of variables the clustering algorithm can handle; however, if the dataset is very large, the output of the algorithm and the heat-map will be challenging to interpret.

Correlational network analysis is another type of informatics tool that is useful in deciphering relationships in systems involving very large number of variables. This technique has been previously used in the analysis of omics data from other domains. For instance, inflammation after brain injury is an active area of research and the role of several cytokine biomarkers and their relationships to clinical outcomes are being investigated. One major limitation in several studies exploring inflammatory biomarkers is that the statistical methods typically used to compare concentrations of one cytokine across different outcome groups (e.g., good vs. poor) are reductionistic. However, the inflammatory response after injury is a complex process involving several associations between the cytokines. A nontraditional informatic technique called weighted correlation network analysis (WGCNA) [21] is useful in elucidating the associations between large number of variables. The WGCNA technique represents associations between cytokines as a network comprising of nodes and edges. The cytokines are represented as the nodes and the Pearson's correlation coefficient between any two cytokine that is above 0.6 is denoted as an edge. In our unpublished study investigating peripheral inflammatory activity after TBI, we collected serum samples from 87 TBI patients within 24 h after admission and determined the concentration levels of 40 pro- and anti-inflammatory cytokines using a multiplex assay. We compared the concentration levels of the different cytokines between patients who had poor and good outcomes. We found that at <24 h after TBI, serum concentrations of seven cytokines—EGF, IL10, IL13, IL6, MCP3, MIP1α, TNFβ—were significantly higher in patients who proceeded to have poor clinical outcomes. Additionally, to understand the complex interactions between these cytokines, we used WCNA analysis to plot the 'inflammatory network' for the good (Fig. 2, left) and poor (Fig. 2, right) outcome groups. On visual inspection, the differences in the 'inflammatory network' between the two groups were stark. As seen in Fig. 2, the number of correlations in the poor outcome group (red labels) were much higher than the good (green labels) outcome group suggesting an increased inflammatory response in patients with poor outcomes.

Interestingly, many cytokines which did not show up as significant in the statistical analysis were in fact highly correlated with several other cytokines in the networks signifying its importance. Similarly, several cytokines formed 'tightly knit' groups within each outcome group (not shown in figure) suggesting the activation of distinct inflammatory pathways across the groups. Our results suggest that in addition to examining the expression levels of individual cytokines, investigating the associations between cytokines and identifying clusters offers more insight into the TBI pathophysiology.

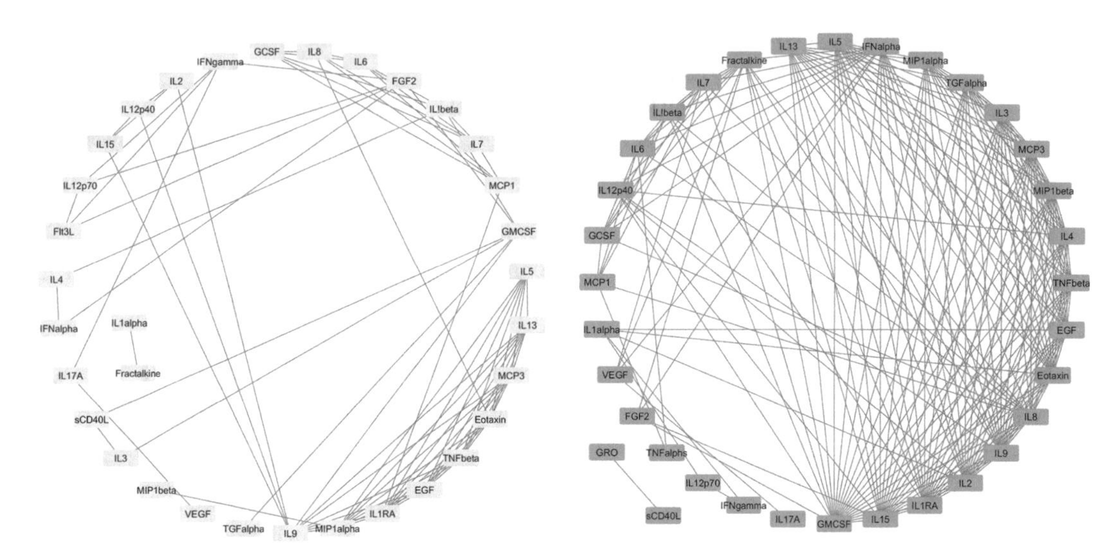

Fig. 2 The serum cytokine networks of TBI patients with good (green) and poor clinical outcomes (red) at discharge. Serum samples were drawn from 87 patients at $<$24 h of admission. The serum was analyzed using a multiplex assay to determine concentration levels of several cytokines implicated in the pathophysiology of acute brain injury. Using WCNA analysis, cytokine inflammatory networks were constructed to visualize correlations between the cytokines. Subjects with poor clinical outcomes (right) have very high network connectivity compared to subjects with good clinical outcomes (left). Network-based approaches are typically used in other domains where a large number of variables are under investigation (e.g., cancer research). The adaptation of such approaches in TBI research could accelerate our understanding of the disease and lead to discovery of inflammatory pathways associated with TBI pathophysiology (Savarraj et al., unpublished)

Supervised learning algorithms: Supervised learning algorithms are a class of ML algorithms that can automatically learn from a dataset and make future predictions on unseen data. The variables can be obtained from any data source including radiographic images, physiological monitors, labs and clinical information. The labels are usually adverse and preventable clinical events (such as onset of seizure, ICP crisis, ischemia and edema) that are adjudicated by an expert. It is preferable that the 'variables' chosen for a machine learning training procedure should be intuitively associated with the event being predicted. For instance, the variables that are most relevant in a model predicting a patient's risk of heart disease are high blood pressure, high cholesterol and smoking; the inclusion of these features in the model will enhance model performance. The algorithms themselves have several parameters that are self-adjustable. When presented with the training data, the parameters of the algorithms change in iterative steps (a process which we call iterative learning), becomes better and better and finally converges into an optimal set of values that enables the model to make best predictions. These data-driven algorithms are not constrained by either the number of variables or labels; they have been successfully used in diverse domains such as handwriting recognition, email spam filtering, and self-driving cars. Since ML algorithms can execute quickly, they can be trained to make continuous assessments in real-time as well [22].

ML algorithms are increasingly being used in critical care medicine, particularly in the field of cardiac monitoring [23–25]. Recently, ML algorithms have shown promise in identifying patients at-risk of sepsis and septic shock using a variety of data sources [26–33]. Within the field of neurocritical care, ML algorithms have been used in proof-of-concept studies to predict delayed cerebral ischemia (DCI), a very serious secondary complication in patients with subarachnoid hemorrhage [34–36]. For example, ML algorithms used variability measures from continuous ECG obtained <48 h of admission to predict DCI 24 h prior to onset with high sensitivity (87%) and good specificity (66%) [34]. In TBI research, ML studies have previously focused on identifying mild TBI from radiological images [37–39] and predicting long-term recovery and outcomes after discharge [37, 40, 41]. These models are being increasingly used in the analysis of continuous signals as well. In particular, the ML models have been used in the prediction of ICP hypertension. In a multicenter study involving 264 TBI patients across 11 countries, minute-by-minute ICP and MAP data was used to predict episodes of ICP crisis 30 min prior to crisis onset [42]. In a study involving 817 subjects with severe TBI, an algorithm using only five variables was able to predict ICP hypertension and $PbtO_2$ crisis 30 min prior to onset with an AUC of 0.86 and 0.91 [43].

Generalizability of an algorithm is an important criterion in the assessment of its performance. Algorithms may work well with the dataset they were trained upon (perhaps obtained from a single center) but fail to achieve good performance on unseen data. Generalizability can be achieved be using larger data sets from multiple centers—larger data sets have more variance. The choice of algorithm and variables is crucial. Unfortunately, there is little *a priori* knowledge as to which algorithms and variables are optimal for each situation. Computational expertise and repeated trial-and-error approach are required to determine algorithms that are most suited for a particular prediction problem. The choice of variables is typically decided by a clinical domain-expert to reduce redundant and meaningless variables. Implementation of a successful ML project is a multidisciplinary effort requiring active collaboration among physicians and informaticians. A schematic of a typical ML pipeline is show in Fig. 3.

The performance of the algorithms and consequently the quality of the results obtained is dependent on the quantity of data available for training. Larger the dataset available, better are the developed models. Due to the rapid decrease in the cost of storage and networking components, large quantities of medical data are

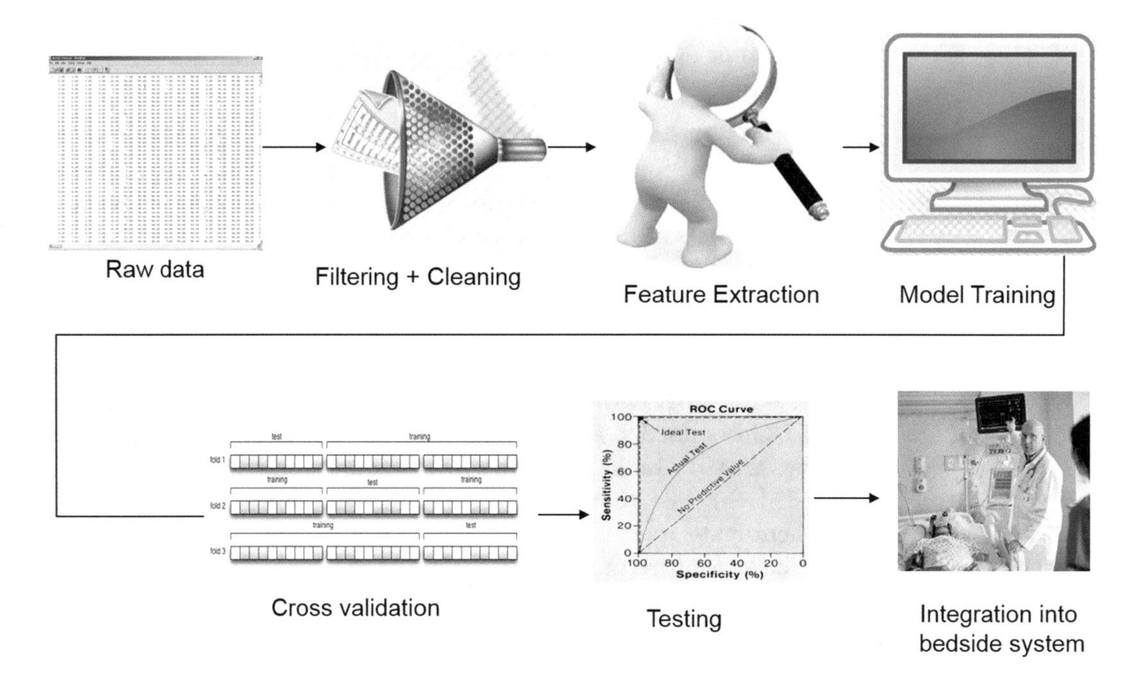

Fig. 3 A typical machine learning pipeline begins from the acquiring of raw data from several patients that is validated by a physician, followed by data cleaning and normalization procedures. The feature extraction often required clinical-domain expertise and the training of the computational model requires technical expertise. The developed model is then cross-validated using the appropriate cross-validation schemes and the models are tested to evaluate its performance prior to integration into a bedside system

readily available for use. The field of 'Big data' refers to the efforts that focus on the collection, organization, storage—and the research and development of tools—to extract knowledge and insights from large quantities of data. The discovery of new knowledge from large datasets could improve efficiency, lower costs and change practices. 'Big Data' is characterized by the five Vs: the *V*olume, *V*elocity, *V*ariety, *V*eracity, and *V*alue of the data. The neuro-ICU is a source for a large *V*olume of data that includes medical and pharmaceutical records, radiological images, physiological signals from monitors and more. The advent of EMR has tremendously aided the digitization and archival of large patient records which can be extracted easily for further research and analysis [44]. The *V*elocity of the data, especially from the invasive and noninvasive physiological monitors (including EEG) that continuously monitor the physiological status of the patient, is very high. Efforts to implement infrastructure that can handle streams of high velocity data in the neuro-ICU is underway [45, 46]. The *V*ariety of the data available in the neuro ICU is increasing; data from a plethora of sources including imaging, clinical status and physiological monitors are available. As the push towards personalized medicine grows, in the future, genomic, proteomic and metabolomic data will influence the clinical decision making process [47, 48]. The *V*eracity of clinical data is a significant problem [49]. As with any human-driven efforts, the arduous process of data collection can be characterized by errors. In addition, variability in practice and management protocols among institutions may further exacerbate these challenges. However, efforts to standardize data collection in ABI are underway [50]. The *V*alue that "big data" could bring to the neuro-ICU is still unclear. While data-driven approaches are in the rise, clear and definite goals and metrics should be used to evaluate the utility of the technology. These issues provide challenges and opportunities to change the field of neurocritical monitoring and ABI research [51].

4 Neuromonitoring Software and Data Repositories

Several commercially available software systems are used for neuromonitoring and research. The University of Cambridge based ICM+ is one of the oldest and most widely used real-time neuromonitoring software [52]. Another 'plug-and-play' approach for real time monitoring without the need for complicated setup has been developed by Moberg ICU Solutions [53]. It is a standalone product that is both easy to use and a has a highly customizable platform that can be adapted to suit individual patient monitoring needs, match study goals, and archive data for off-line analysis. It is being used clinically for individualized therapy [54], to monitor cortical spreading depression [55], ICP monitoring [56] and more. IBM healthcare, a division of IBM corporation, has developed a streaming analytics tool called the Real Time Analytic Processing (RTAP) with *InfoSphere Streams* that can use specialized applications to analyze physiological data in acute brain trauma [57]. Recently, some specialized tools have been developed that are useful for both patient monitoring and large-scale data archiving and analysis. DECISIO is an ICU monitoring platform that is both customizable and useful for off-line analysis of collected data as well (www.decisiohealth.com). It has a modifiable display platform that can be tailored to suit the monitoring needs of that patient (Fig. 4).

"Sickbay" is a product that employs clusters of distributed servers to store terabytes of continuous physiological data for easy retrieval and analysis (http://medicalinformaticscorp.com/). One major challenge in analytics is the rapid deployment and adaptation of an algorithm from one institution to another. For instance, if one research group discovers an efficient way to monitor a physiologic variable or predict complications, it is often required to test this application in other centers to prove generalizability prior to universal adaptation. Currently, an ecosystem that can facilitate this does not exist. An "open source" development platform that would

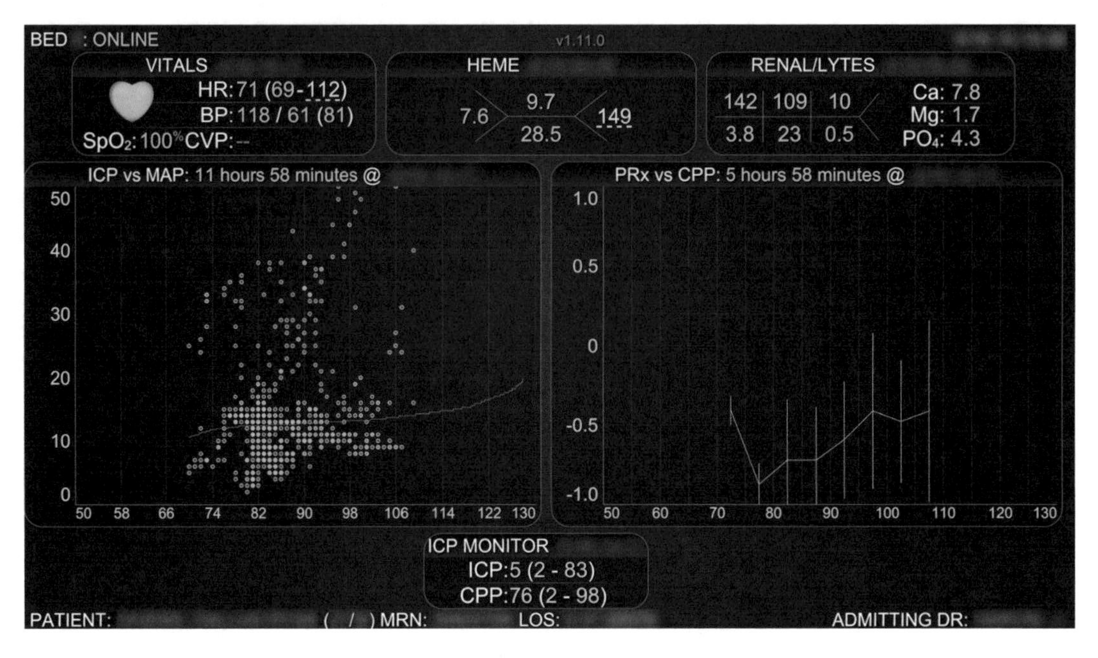

Fig. 4 Development of a bedside integrated visualization of a TBI patient's autoregulatory status developed in collaboration with the DECISIO monitoring system. (Left) the plot between ICP and MAP over a period of 12 h fitted with a Lowess curve. The continuous evaluation of CPP_{opt} using the PRx vs. CPP plot over a period of 12 h (right). For this TBI patient, during the most recent 12 h of the ICU stay (shown in the figure), the CPP_{opt} was around 78 mmHg. The management protocol was not altered to maintain the patient's CPP close to 78 mmHg (CPP_{opt}) as the purpose of the collaboration was a proof-of-concept of the rapid deployment from research to bedside deployment of advanced monitoring algorithms

allow users from different institutions to use a common platform to share data and develop their own "apps" is required. Such ecosystems are in place for smartphones where an app developed by one user can be easily used by several others provided that the app meets a certain standard. Such an ecosystem in TBI research would accelerate the prototyping-to-bedside deployment cycle.

TBI data repositories: The availability of large-scale, multisite, open source data repositories has advanced research in many diseases. The Alzheimer's Disease Neuroimaging Initiative—an open-source repository of common data elements including clinical information, neuroimages, biospecimens, and genetic data—has greatly advanced research in Alzheimer's disease and has led to better design of clinical trials as well [58, 59]. Several data repositories for TBI research exist including the TRACK-TBI [60], the European-based CENTER-TBI [61], IMPACT [62] and FITBIR [63]. These databases contain comprehensive datasets including patient demographics, outcome assessment, images and biomarkers. However, continuous physiological signals are not available, limiting the potential to discover new physiological biomarkers. Physionet is a publicly available database that contains continuous physiological signals from the ICU, but it is not limited to any specific pathology [64]. The brain trauma population within Physionet is small and can lack clinical information, images and biospecimens. A more TBI-specific repository within Physionet is under development [65]. The only exclusive database containing physiological variables from TBI patients is the BrainIT project from Institute of Neurological Sciences in Glasgow which maintains a data repository of over 250 brain trauma patients from 22 cities and 11 EU countries [66]. A large-scale data repository solely for TBI research that includes physiological signals as well would greatly advance brain trauma research.

Conclusion: The failure of clinical trials targeting universal thresholds of physiological measures has emphasized a need for "precision-based" approaches in managing ABI patients. Informatics approaches are useful in the development of physiological targets that are injury specific. Unsupervised learning methods and network visualization can uncover patterns and extract knowledge from large datasets. Supervised machine learning models can be trained to predict secondary complications prior to symptom onset. There is an unmet need for data repositories for ABI patients that include continuous physiological signals.

References

1. Cooper DJ, Rosenfeld JV, Murray L et al (2011) Decompressive craniectomy in diffuse traumatic brain injury. N Engl J Med 364:1493–1502. https://doi.org/10.1056/NEJMoa1102077
2. Czosnyka M, Smielewski P, Timofeev I et al (2007) Intracranial pressure: more than a number. Neurosurg Focus 22:E10
3. Wagshul ME, Eide PK, Madsen JR (2011) The pulsating brain: a review of experimental and clinical studies of intracranial pulsatility. Fluids Barriers CNS 8:5. https://doi.org/10.1186/2045-8118-8-5
4. Lazaridis C, Smielewski P, Menon DK et al (2016) Patient-specific thresholds and doses of intracranial hypertension in severe traumatic brain injury. Acta Neurochir Suppl 122:117–120. https://doi.org/10.1007/978-3-319-22533-3_23
5. Hu X, Xu P, Scalzo F et al (2009) Morphological clustering and analysis of continuous intracranial pressure. IEEE Trans Biomed Eng 56:696–705
6. Kim D-J, Kim H, Jeong E-J et al (2016) Spectral analysis of intracranial pressure: is it helpful in the assessment of shunt functioning in-vivo? Clin Neurol Neurosurg 142:112–119. https://doi.org/10.1016/j.clineuro.2016.01.023
7. Kvandal P, Sheppard L, Landsverk SA et al (2013) Impaired cerebrovascular reactivity after acute traumatic brain injury can be detected by wavelet phase coherence analysis of the intracranial and arterial blood pressure signals. J Clin Monit Comput 27:375–383. https://doi.org/10.1007/s10877-013-9484-z
8. Czosnyka M, Smielewski P, Piechnik S et al (2001) Cerebral autoregulation following head injury. J Neurosurg 95:756–763. https://doi.org/10.3171/jns.2001.95.5.0756
9. Aiolfi A, Benjamin E, Khor D et al (2017) Brain trauma foundation guidelines for intracranial pressure monitoring: compliance and effect on outcome. World J Surg 41:1543–1549
10. Mehta A, Kochanek PM, Tyler-Kabara E et al (2010) Relationship of intracranial pressure and cerebral perfusion pressure with outcome in young children after severe traumatic brain injury. Dev Neurosci 32:413–419. https://doi.org/10.1159/000316804
11. Lazaridis C, Smielewski P, Steiner LA et al (2013) Optimal cerebral perfusion pressure: are we ready for it? Neurol Res 35:138–148.

https://doi.org/10.1179/1743132812Y.0000000150
12. Prabhakar H, Sandhu K, Bhagat H et al (2014) Current concepts of optimal cerebral perfusion pressure in traumatic brain injury. J Anaesthesiol Clin Pharmacol 30:318–327. https://doi.org/10.4103/0970-9185.137260
13. Czosnyka M, Miller C, Participants in the International Multidisciplinary Consensus Conference on Multimodality Monitoring (2014) Monitoring of cerebral autoregulation. Neurocrit Care 21(Suppl 2):S95–S102. https://doi.org/10.1007/s12028-014-0046-0
14. Brady KM, Lee JK, Kibler KK et al (2007) Continuous time-domain analysis of cerebrovascular autoregulation using near-infrared spectroscopy. Stroke 38:2818–2825. https://doi.org/10.1161/STROKEAHA.107.485706
15. Depreitere B, Güiza F, Van den Berghe G et al (2014) Pressure autoregulation monitoring and cerebral perfusion pressure target recommendation in patients with severe traumatic brain injury based on minute-by-minute monitoring data. J Neurosurg 120:1451–1457. https://doi.org/10.3171/2014.3.JNS131500
16. Czosnyka M, Dias C (2015) Role of pressure reactivity index in neurocritical care. In: Uchino H, Ushijima K, Ikeda Y (eds) Neuroanesthesia and cerebrospinal protection. Springer, Tokyo, pp 223–236. https://doi.org/10.1007/978-4-431-54490-6_21
17. Czosnyka M, Smielewski P, Kirkpatrick P et al (1997) Continuous assessment of the cerebral vasomotor reactivity in head injury. Neurosurgery 41:11–17; discussion 17-19
18. Dias C, Silva MJ, Pereira E et al (2015) Optimal cerebral perfusion pressure management at bedside: a Single-center Pilot Study. Neurocrit Care 23:92–102. https://doi.org/10.1007/s12028-014-0103-8
19. Hu X, Xu P, Asgari S et al (2010) Forecasting ICP elevation based on prescient changes of intracranial pressure waveform morphology. IEEE Trans Biomed Eng 57:1070–1078. https://doi.org/10.1109/TBME.2009.2037607
20. Lee JK, Poretti A, Perin J et al (2017) Optimizing cerebral autoregulation may decrease neonatal regional hypoxic-ischemic brain injury. Dev Neurosci 39:248–256
21. Zhang B, Horvath S (2005) A general framework for weighted gene co-expression network analysis. Stat Appl Genet Mol Biol 4:Article17. https://doi.org/10.2202/1544-6115.1128
22. Deo RC (2015) Machine learning in medicine. Circulation 132:1920–1930. https://doi.org/10.1161/CIRCULATIONAHA.115.001593
23. Hijazi S, Page A, Kantarci B, Soyata T (2016) Machine learning in cardiac health monitoring and decision support. Computer 49:38–48. https://doi.org/10.1109/MC.2016.339
24. Boursalie O, Samavi R, Doyle TE (2015) M4CVD: mobile machine learning model for monitoring cardiovascular disease. Proc Comput Sci 63:384–391. https://doi.org/10.1016/j.procs.2015.08.357
25. Pinsky MR, Clermont G, Hravnak M (2016) Predicting cardiorespiratory instability. Crit Care 20:70. https://doi.org/10.1186/s13054-016-1223-7
26. Thottakkara P, Ozrazgat-Baslanti T, Hupf BB et al (2016) Application of machine learning techniques to high-dimensional clinical data to forecast postoperative complications. PLoS One 11:e0155705. https://doi.org/10.1371/journal.pone.0155705
27. Taylor RA, Pare JR, Venkatesh AK et al (2016) Prediction of in-hospital mortality in emergency department patients with sepsis: a local big data-driven, machine learning approach. Acad Emerg Med Off J Soc Acad Emerg Med 23:269–278. https://doi.org/10.1111/acem.12876
28. Scott H, Colborn K (2016) Machine learning for predicting sepsis in-hospital mortality: an important start. Acad Emerg Med Off J Soc Acad Emerg Med 23:1307. https://doi.org/10.1111/acem.13009
29. Giannini HM, Chivers C, Draugelis M et al (2017) Development and implementation of a machine-learning algorithm for early identification of sepsis in a multi-hospital academic healthcare system. Am J Resp Crit Care Med 195:A7015. D15 Crit. Care we have Cryst. Ball Predict. Clin. Deterioration outcome Crit. Ill patients. Am Thoracic Soc
30. Horng S, Sontag DA, Halpern Y et al (2017) Creating an automated trigger for sepsis clinical decision support at emergency department triage using machine learning. PLoS One 12:e0174708. https://doi.org/10.1371/journal.pone.0174708
31. Danner OK, Hendren S, Santiago E et al (2017) Physiologically-based, predictive analytics using the heart-rate-to-systolic-ratio significantly improves the timeliness and accuracy of sepsis prediction compared to SIRS. Am J Surg 213:617–621. https://doi.org/10.1016/j.amjsurg.2017.01.006
32. Shashikumar SP, Stanley MD, Sadiq I et al (2017) Early sepsis detection in critical care patients using multiscale blood pressure and heart rate dynamics. J Electrocardiol 50:739–749. https://doi.org/10.1016/j.jelectrocard.2017.08.013

33. Taneja I, Reddy B, Damhorst G et al (2017) Combining biomarkers with EMR data to identify patients in different phases of sepsis. Sci Rep 7:10800. https://doi.org/10.1038/s41598-017-09766-1
34. Schmidt JM (2016) Heart rate variability for the early detection of delayed cerebral ischemia. J Clin Neurophysiol Off Publ Am Electroencephalogr Soc 33:268–274. https://doi.org/10.1097/WNP.0000000000000286
35. Kumar G, Elzaafrani K, Nakhmani A (2017) Machine learning approach to automate detection of cerebral vasospasm using transcranial Doppler monitoring (S23.004). Neurology 88:S23.004
36. Roederer A, Holmes JH, Smith MJ et al (2014) Prediction of significant vasospasm in aneurysmal subarachnoid hemorrhage using automated data. Neurocrit Care 21:444–450. https://doi.org/10.1007/s12028-014-9976-9
37. Molaei S, Korley FK, Soroushmehr SMR, et al (2016) A machine learning based approach for identifying traumatic brain injury patients for whom a head CT scan can be avoided. In: 2016 38th Annu. Int. Conf. IEEE Eng. Med. Biol. Soc. EMBC. pp 2258–2261
38. Chong S-L, Liu N, Barbier S, Ong MEH (2015) Predictive modeling in pediatric traumatic brain injury using machine learning. BMC Med Res Methodol 15:22. https://doi.org/10.1186/s12874-015-0015-0
39. Mitra J, Shen K, Ghose S et al (2016) Statistical machine learning to identify traumatic brain injury (TBI) from structural disconnections of white matter networks. NeuroImage 129:247–259. https://doi.org/10.1016/j.neuroimage.2016.01.056
40. Vergara VM, Mayer AR, Damaraju E et al (2016) Detection of mild traumatic brain injury by machine learning classification using resting state functional network connectivity and fractional anisotropy. J Neurotrauma 34:1045–1053. https://doi.org/10.1089/neu.2016.4526
41. Celtikci E (2017) A systematic review on machine learning in neurosurgery: the future of decision making in patient care. Turk Neurosurg 28:167–173. https://doi.org/10.5137/1019-5149.JTN.20059-17.1
42. Güiza F, Depreitere B, Piper I et al (2013) Novel methods to predict increased intracranial pressure during intensive care and long-term neurologic outcome after traumatic brain injury: development and validation in a multicenter dataset. Crit Care Med 41:554–564. https://doi.org/10.1097/CCM.0b013e3182742d0a
43. Myers RB, Lazaridis C, Jermaine CM et al (2016) Predicting intracranial pressure and brain tissue oxygen crises in patients with severe traumatic brain injury. Crit Care Med 44:1754–1761. https://doi.org/10.1097/CCM.0000000000001838
44. Menachemi N, Collum TH (2011) Benefits and drawbacks of electronic health record systems. Risk Manag Healthc Policy 4:47–55. https://doi.org/10.2147/RMHP.S12985
45. Fartoumi S, Emeriaud G, Roumeliotis N et al (2016) Computerized decision support system for traumatic brain injury management. J Pediatr Intensive Care 5:101–107
46. Singh MP, Hoque MA, Tarkoma S (2016) A survey of systems for massive stream analytics. ArXiv Prepr. ArXiv160509021
47. Herr TM, Bielinski SJ, Bottinger E et al (2015) A conceptual model for translating omic data into clinical action. J Pathol Inform 6:46. https://doi.org/10.4103/2153-3539.163985
48. Shukla SK, Murali NS, Brilliant MH (2015) Personalized medicine going precise: from genomics to microbiomics. Trends Mol Med 21:461–462. https://doi.org/10.1016/j.molmed.2015.06.002
49. Bowman S (2013) Impact of electronic health record systems on information integrity: quality and safety implications. Perspect Health Inf Manag 10:1c
50. Maas AIR, Harrison-Felix CL, Menon D et al (2011) Standardizing data collection in traumatic brain injury. J Neurotrauma 28:177–187. https://doi.org/10.1089/neu.2010.1617
51. Sivaganesan A, Manley GT, Huang MC (2014) Informatics for Neurocritical care: challenges and opportunities. Neurocrit Care 20:132–141. https://doi.org/10.1007/s12028-013-9872-8
52. Smielewski P, Czosnyka Z, Kasprowicz M et al (2012) ICM+: a versatile software for assessment of CSF dynamics. Acta Neurochir Suppl 114:75–79. Intracranial Press. Brain Monit. XIV. Springer
53. Mertz L (2014) Saving lives and money with smarter hospitals: streaming analytics, other new tech help to balance costs and benefits. IEEE Pulse 5:33–36
54. Makarenko S, Griesdale DE, Gooderham P, Sekhon MS (2016) Multimodal neuromonitoring for traumatic brain injury: a shift towards individualized therapy. J Clin Neurosci 26:8–13
55. Carlson AP, William Shuttleworth C, Mead B et al (2017) Cortical spreading depression

occurs during elective neurosurgical procedures. J Neurosurg 126:266–273. https://doi.org/10.3171/2015.11.JNS151871
56. Le Roux P (2016) Intracranial pressure monitoring and management. In: Laskowitz D, Grant G (eds) Translational research in traumatic brain injury. CRC, Boca Raton
57. (2016) IBM research streaming analytics solution saves time and lives—IBM. http://researcher.watson.ibm.com/researcher/view_group.php?id=1775. Accessed 25 Sep 2017
58. Weiner MW, Veitch DP, Aisen PS et al (2017) The Alzheimer's Disease Neuroimaging Initiative 3: continued innovation for clinical trial improvement. Alzheimers Dement 13:561–571. https://doi.org/10.1016/j.jalz.2016.10.006
59. Weiner MW, Veitch DP, Aisen PS et al (2017) Recent publications from the Alzheimer's Disease Neuroimaging Initiative: reviewing progress toward improved AD clinical trials. Alzheimers Dement 13:e1–e85. https://doi.org/10.1016/j.jalz.2016.11.007
60. Yue JK, Vassar MJ, Lingsma HF et al (2013) Transforming research and clinical knowledge in traumatic brain injury pilot: multicenter implementation of the common data elements for traumatic brain injury. J Neurotrauma 30:1831–1844. https://doi.org/10.1089/neu.2013.2970
61. Maas AIR, Menon DK, Steyerberg EW et al (2015) Collaborative European NeuroTrauma Effectiveness Research in Traumatic Brain Injury (CENTER-TBI): a prospective longitudinal observational study. Neurosurgery 76:67–80. https://doi.org/10.1227/NEU.0000000000000575
62. Marmarou A, Lu J, Butcher I et al (2007) IMPACT database of traumatic brain injury: design and description. J Neurotrauma 24:239–250
63. Ivory M (2015) Federal interagency traumatic brain injury research (FITBIR) bioinformatics platform for the advancement of collaborative traumatic brain injury research and analysis
64. Goldberger AL, Amaral LAN, Glass L et al (2000) PhysioBank, PhysioToolkit, and PhysioNet: components of a new research resource for complex physiologic signals. Circulation 101:e215–e220. https://doi.org/10.1161/01.CIR.101.23.e215
65. Kim N, Krasner A, Kosinski C et al (2016) Trending autoregulatory indices during treatment for traumatic brain injury. J Clin Monit Comput 30:821–831. https://doi.org/10.1007/s10877-015-9779-3
66. Piper I, Citerio G, Chambers I et al (2003) The BrainIT group: concept and core dataset definition. Acta Neurochir 145:615–629

Chapter 15

Rapid Detection and Monitoring of Brain Injury Using Sensory-Evoked Responses

Jonathan A. N. Fisher and Cristin G. Welle

Abstract

There is currently a dearth of quantitative biomarkers for traumatic brain injury (TBI) that can be rapidly acquired and interpreted in active field environments. Clinical imaging, via computed tomography (CT) scan or magnetic resonance imaging (MRI), in combination with a clinical examination, is currently the "gold standard" for diagnosing TBI. These technologies, however, require extended imaging sessions and are rarely available during the peak therapeutic window following injury. Moreover, mild TBI (mTBI) often does not present with structural damage that can be detected by CT or MRI imaging. Techniques that probe neurophysiological function, however, present an opportunity to directly and rapidly assess brain health following head impact. One of the most basic roles of the CNS is to register and parse sensory stimuli from the environment. This process relies on an intricate feedback network that involves a multitude of widely distributed brain structures, and subtle perturbation in brain health can have a dramatic effect on afferent relay and processing of sensory information. In this chapter, we describe recent preclinical approaches for rapidly detecting and monitoring TBI using sensory-evoked physiological biomarkers, particularly somatosensory-evoked electrophysiological and hemodynamic responses. With an eye toward clinical implementation, we focus our discussion on measurements that can be achieved noninvasively.

Key words Traumatic brain injury, Biomarkers, Somatosensory-evoked potentials, Epidermal electronics, Diffuse correlation spectroscopy, Cerebral blood flow, Animal models

1 Introduction: Sensory Systems Following Traumatic Brain Injury: "Canary in the coalmine"

In emergency medicine, there is a well-known principle of the "golden hour," the 60 min following an out-of-hospital traumatic injury within which rapid medical intervention can have a positive impact on medical outcome [1]. Although the precise duration is debated [2, 3], timely transport to a hospital after a head injury is essential for preventing sequelae such as edema, increased intracranial pressure, and cerebral dysautoregulation [4, 5]. Clinical exam and CT imaging are currently the gold standard for traumatic brain injury (TBI) diagnosis, but these typically take place within a clinical setting, hours to days following injury. There remains a need for

Amit K. Srivastava and Charles S. Cox, Jr. (eds.), *Pre-Clinical and Clinical Methods in Brain Trauma Research*, Neuromethods, vol. 139, https://doi.org/10.1007/978-1-4939-8564-7_15, © Springer Science+Business Media, LLC, part of Springer Nature 2018

rapid, point of care determination of injury to take place outside of the clinical environment to allow for immediate diagnostic and triage decisions. In active field settings such as professional sports or military operations, helmet-mounted sensors that measure acceleration, pressure, or impact force [6–8] offer possibilities for detecting and quantifying events that could lead to head injury. Although monitoring mechanical force can inform decisions on whether an exposed individual should be removed from action, mechanical signals alone cannot unambiguously report on injury.

Techniques that probe neurophysiological function present an opportunity to directly and rapidly assess brain health following head impact. Afferent sensory transmission and processing relies on an intricate feedback network that involves a variety of brain structures as well as interhemispheric interaction. Subtle perturbation in brain health due to injury can have a dramatic effect on these sensory networks. Alterations in long-range functional connectivity have been traced with magnetoencephalography [9–11] and functional magnetic resonance imaging (fMRI) [12, 13]. The integrity of these systems can be probed through sensory-evoked responses, which may therefore serve as a "canary in a coalmine," following the metaphor of the use of a physiologically sensitive sentinel species for detecting hazardous conditions. In this chapter, we review recent findings on the effects of TBI on sensory-evoked responses, the underlying cellular and biochemical cascades that lead to these dysfunctions, and discuss methodology that is capable of noninvasively detecting these functional biomarkers for TBI in vivo.

2 Pathophysiology of Sequelae within the First 24 h of a TBI

A major challenge associated with quantitative detection of TBI, particularly in the case of mTBI, is that the resulting pathological processes evolve rapidly and the window for early detection is narrow. The event of a TBI initiates a complex set of cellular and molecular reactions that can be broadly divided into an acute phase, which occurs at the moment of injury, and a secondary chronic phase that involves cascades of processes initiated by the original neurological insult [14, 15]. In the acute phase, the rapid energy depletion associated with disrupted cerebral blood flow (CBF) and mechanical shearing of neurons dramatically increases extracellular K^+ and glutamate concentrations [16, 17]. Concomitant depolarization and neurotoxic effects due to elevated glutamate levels rapidly perturb network activity in the traumatized region [18, 19] and trigger a panoply of rapidly progressing adverse events. For example, initial cytotoxic edema is followed by the onset of vasogenic edema within the first few hours [20, 21]. On a longer timescale, TBI also evokes a neuroinflammatory response, which involves resident and peripherally derived inflammatory cells that respond to injury and participate in repair [22, 23]. Ultimately, the inflammatory response, which upregulates pro-inflammatory cytokines and chemokines [24, 25], can lead to axonal and neuronal degeneration, and the accumulation of tau protein [14, 15, 22]. Blast-exposed mice, for example, demonstrate phosphorylated tauopathy, myelinated axonopathy, microvasculopathy, chronic neuroinflammation, and neurodegeneration in the absence of macroscopic tissue damage or hemorrhage [26, 27]. Overall, TBI therefore poses challenges for clinicians at all phases of injury management, from neurosurgical intervention within the first few hours of head injury up through later phase, outpatient assessments, and therapy months or years after the primary injury.

3 Sensory-Evoked Electrophysiological Potentials

The electroencephalogram (EEG) is a powerful potential tool for rapidly diagnosing brain injury [28]. While electrophysiological signals are far from being considered a gold standard for the diagnosis of traumatic brain injury, their value as quantitative indicators of brain injury has been explored in a research setting for decades [29, 30]. Quantitative EEG (qEEG) exploits both spectral and spatiotemporal dynamics of ongoing EEG activity recorded at multiple locations over the scalp following a head injury. Although clinical implementation remains controversial [31], commercial devices that utilize qEEG for diagnosing TBI are used in clinical investigation [32] and are gaining increased acceptance. qEEG algorithms produce a discriminate score with a sensitivity of >90% and specificity of ~60%, significantly better than diagnosis based on hematoma detection on a CT scan [33, 34]. Recent studies suggest increased diagnostic accuracy for detection of mild injury, even with no hematoma visible on CT images [35–37].

A limitation of using ongoing EEG to monitor brain health is that it requires a relatively controlled environment in order to minimize external noise sources that can distort electrical signals. An alternative, time-domain metric for probing the integrity of sensory systems is by means of sensory-evoked electrophysiological potentials, or EPs. EPs consist of a series of electrical "waves" of alternating polarity that are triggered by sensory stimuli and reflect sequential activation of sensory processing structures in the nervous system. EPs such as somatosensory-evoked potentials (SSEPs), auditory brainstem responses (ABRs), and visual-evoked responses (VERs), are standard diagnostic clinical assays. Because these potentials are triggered, this affords the ability to increase the signal-to-noise ratio by stimulus-locked analysis, thereby alleviating the impact of artifact due to factors such as motion. This permits responses to be resolved in severe conditions and environments that

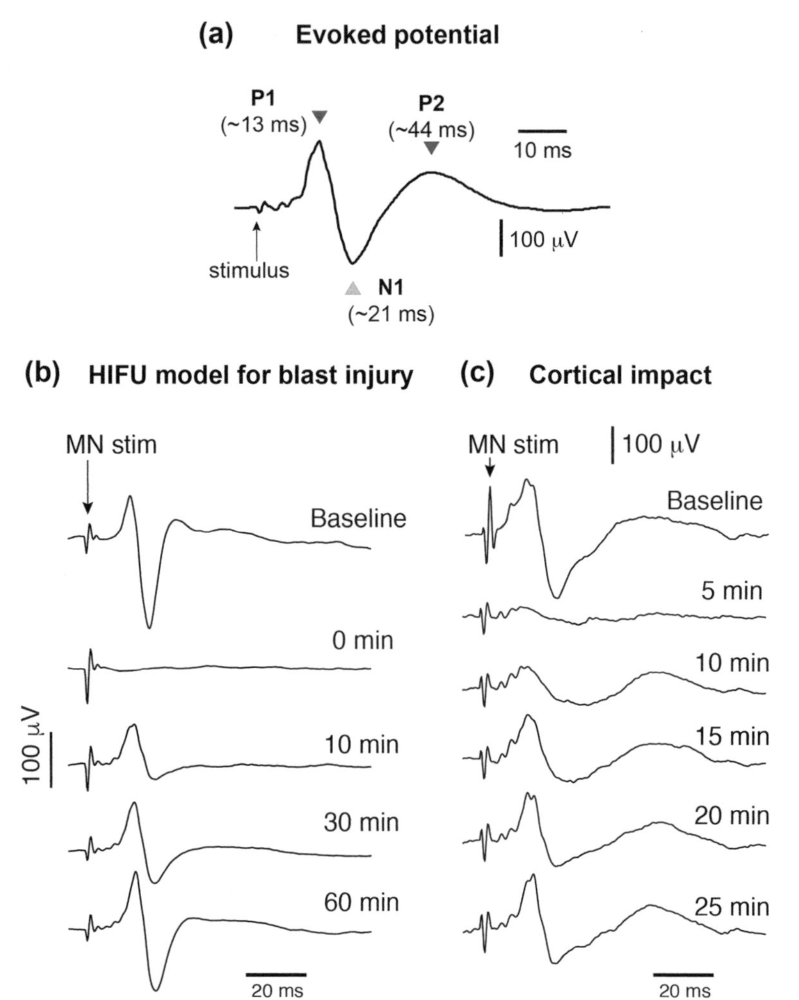

Fig. 1 The effect of TBI on somatosensory-evoked potentials (SSEPs). (**a**) SSEP recorded from a mouse under light anesthesia (0.5% isoflurane) upon electrical stimulation of the median nerve. The prominent peaks P1, N1, and P2 occur at temporal latencies that are well-established in the literature. The trace represents the average of 120 individual-evoked responses, and electrical stimulation was delivered at 3 Hz. (**b**) SSEPs recorded before, immediately after, 10, 30, and 60 min following delivery of a series of high-intensity focused ultrasound (HIFU) pulses administered to the frontal association cortex. The recording location was directly above the region of somatosensory cortex where the forelimbs are represented. (**c**) SSEPs recorded before and after controlled cortical impact (CCI) administered in frontal association cortex

feature significant electrical, mechanical, and physiological noise. A prototypical example of an SSEP, recorded from a mouse, is shown in Fig. 1a. EPs are used for aiding neurological assessments to confirm and localize neural abnormalities, to identify clinically "silent" lesions, and to monitor functional recovery over time

[31, 38]. Following severe TBI, abnormalities in EPs correlate with a poor prognosis for survival [39, 40] and are in fact more sensitive predictors of an unfavorable outcome than pupillary reflex, EEG, and even CT [41].

Recently, the use of cortical SSEPs to detect and subsequently monitor head injury has been explored in preclinical models of blast injury [42, 43] and controlled cortical impact [44]. Following either TBI model, the average amplitude of evoked peaks of latencies less than 100 ms fell more than an order of magnitude immediately following injury. The peaks gradually reemerged above the noise level over the course of ~60 min (Fig. 1b, c), yet with some differences between the two injury models. In both models, P2 displayed a significant delay post-injury post injury which did not recover within 1 h. However, CCI additionally induced significant temporal delays in both P1 and N1; these latencies recovered to baseline values within 30 min.

All sensory systems in the brain feature extensive networking and myriad layers of feedback. Consequently, injury can alter sensory responses recorded at locations distal to the primary mechanical insult. This is advantageous for real-time detection of injury given that sensor placement will not necessarily coincide with a site of primary impact. As an example, Fisher et al. (2016) observed hemispheric asymmetries in the SSEPs that became apparent rapidly following injury. Specifically, SSEPs evoked on the hemisphere contralateral to injury were larger in amplitude (albeit of altered waveform) (Fig. 2). This asymmetry became more pronounced over the course of an hour after injury, and ultimately the uninjured hemisphere exhibited SSEP amplitudes that surpassed baseline responses. Such interhemispheric asymmetries in SSEP recovery may be due to suppression of transcallosal inhibition (TCI), which normally tempers cortical excitability [45]. In the cortex, GABA-ergic projection neurons provide inhibitory input to corresponding areas of contralateral cortex via the corpus callosum [46]. Damage to one hemisphere can compromise TCI, as a loss of action potential generation on the injured hemisphere results in contralateral depolarization [47]. This effect has been observed in human subjects following TBI and stroke [48].

4 Hemodynamic Biomarkers for TBI

Functional cerebral blood flow (CBF) offers another window for probing the integrity of sensory systems following injury. Sensory-evoked CBF and local tissue oxygenation dynamics reflect an interplay between CBF and changes in the cerebral metabolic rate of oxygen ($CMRO_2$), both of which are susceptible to alteration following acute injury. A major repercussion of traumatic injury in the brain is an ensuing mismatch between oxygen supply and the

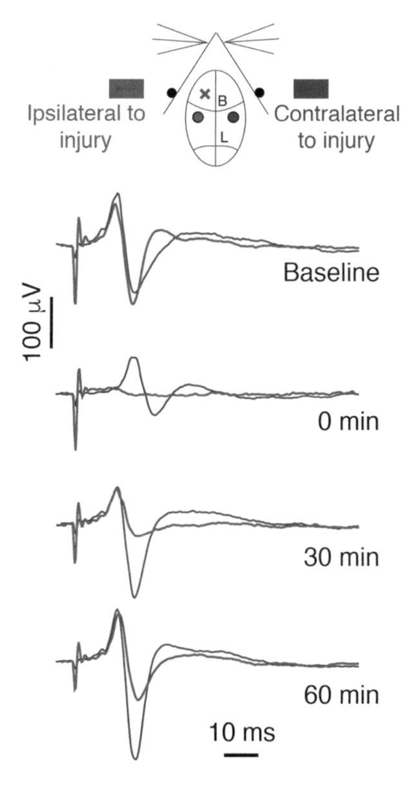

Fig. 2 SSEPs recorded on the ipsilateral (red) and contralateral (blue) cortical hemispheres, relative to the location of injury (using HIFU model of blast blast injury). The traces represent responses to stimulation of the contralateral limb, relative to recording location (stimulus ipsilateral to the recording locations yielded no significant response). Both channels are referenced to an anterior electrode. The recordings represent the average of 20 trials and are from one representative experiment

considerable neural metabolic demand [49, 50]. The primary mechanical injury, for example, can alter the vascular network that delivers oxygen to the parenchyma and, simultaneously, can alter neural activity and metabolic rate through a wide spectrum of mechanisms, including mechanical shearing [51] or excitotoxicity due to damaged glia [52]. Given that the brain consumes a large fraction of the body's total energy budget [53], even slight changes in the efficacy of energy delivery can have grave consequences for sensory-evoked responses. Additionally, measurements of CBF can aid in early detection post-injury [54], relate to the severity of post-concussive symptoms [55] and have prognostic value for TBI outcomes [56, 57].

An example of the CBF response to brief somatosensory stimuli is shown in Fig. 3a. The hemodynamic response features an initial increase in blood flow that peaks ~9 s after the initial stimulus (electrical stimulation of the median nerve, in this case), followed

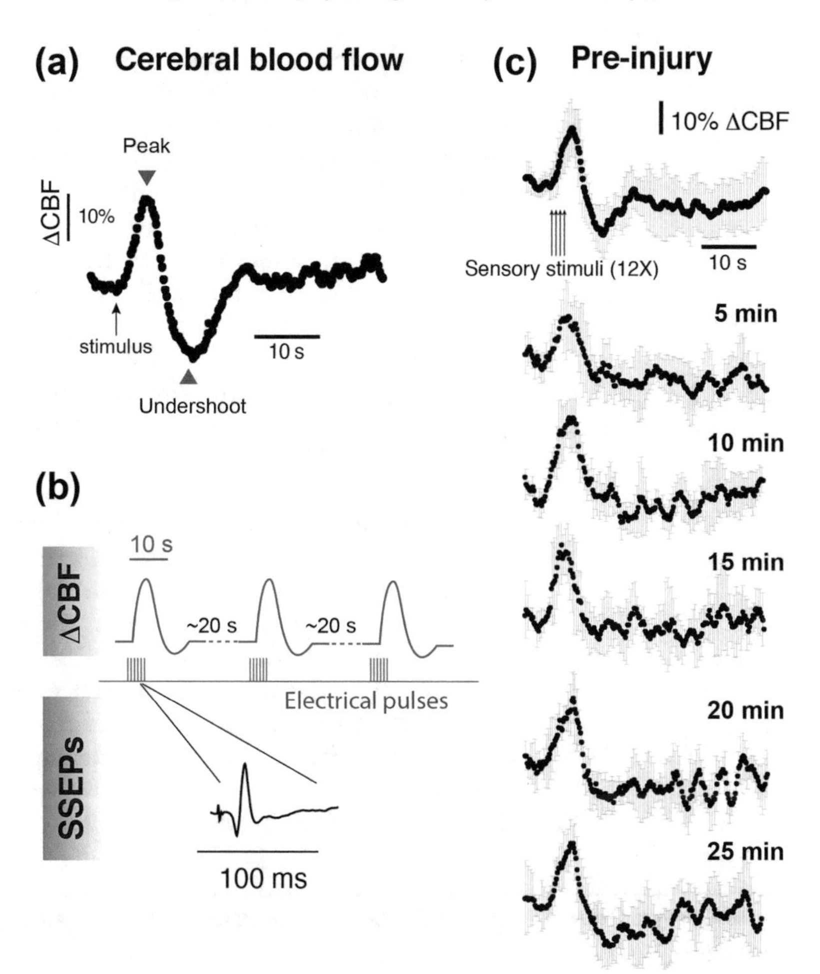

Fig. 3 Somatosensory-evoked hemodynamic responses. (**a**) Optically measured changes in relative cerebral blood flow (CBF) following electrical stimulation of the median nerve. The trace represents the average of ten trials in one experiment; blood flow is normalized to the average of CBF values recorded in the 4 s prior to the stimulus onset, which is indicated by the arrow. (**b**) Snapshot of the measurement timelines during event-related measurements. The red traces depict simulated hemodynamic signals, which are broad compared with SSEPs (note the timescale difference in the insets). (**c**) Sensory-evoked CBF responses from forepaw stimulation for a representative experiment before and after CCI. The CBF traces shown here represent the average of multiple stimulations (average of ten trials pre-injury and five trials for each time point post-injury). The black line is the average and the red error bars represents standard deviation

by a subsequent undershoot, which reaches a negative peak ~17 s after the stimulus. The initial CBF peak reflects the influx of blood flow recruited by local changes in neural activity [58], and, on the timescale of days, its amplitude has been found to be reduced following traumatic injury [59] and stroke [60].

Using noninvasive optical techniques, Jang et al. (2017) recently explored how TBI affects sensory-evoked CBF following CCI. The authors monitored CBF during and after TBI using diffuse correlation spectroscopy (DCS), which takes advantage of the dynamic scattering properties of red blood cells to directly measure flow [61]. Unlike laser Doppler measurements of flow, which is the most common measurement modality for CBF, DCS is particularly sensitive to flow in the cortical microvasculature due to the high absorption (and thus low probability of photon escape) in larger blood vessels. DCS has been used to measure functional hemodynamics associated with sensory stimuli and motor tasks [62, 63], and has also been used to track baseline CBF following brain trauma [64].

Compared side-by-side, Jang et al. found that CCI acutely reduces the amplitude and alters the waveforms of sensory-evoked hemodynamic (Fig. 3c) and electrophysiological responses (Fig. 1c). Injury additionally alters the latencies of other waveform features, such as the CBF undershoot. Despite being more sensitive to injury than other features of the ΔCBF waveform, the initial peak recovered at a relatively fast rate, and within 5–10 min following injury, the peak amplitude had essentially recovered completely. In contrast, the SSEP peaks are generally significantly slower to recover, reflecting the extreme sensitivity of neural function to slight perturbations in CBF.

5 Wearable Devices for Monitoring Sensory-Evoked Responses

While real-time monitoring of physiological signals, in particular EPs, provides critical information regarding neurological health, current measurement devices are typically too cumbersome to be worn continuously. Flexible epidermal electronics [65], however, are an alternative measurement strategy that offers the potential for monitoring with a negligible level of user interface burden (Fig. 4). The ease with which they can be applied, for instance via adhesives in which electronic components are embedded, also makes them attractive for potential use in emergency medical situations because they can be disseminated potentially on large scale at low cost [66]. Furthermore, their mechanical pliability makes them ideal wearables for constant health monitoring for active, at-risk individuals such as athletes or Military Service members. In human experiments, flexible electronics have indeed demonstrated stability for detecting both rapid, sensory related potentials such as P300 [67], mismatch negativity [68], as well as for long-term recording of ongoing EEG signals [69].

In general, SSEP recordings that are obtained with invasive metal electrodes yield higher signal amplitudes than can be obtained by flexible epidermal electrodes, because skin and skull

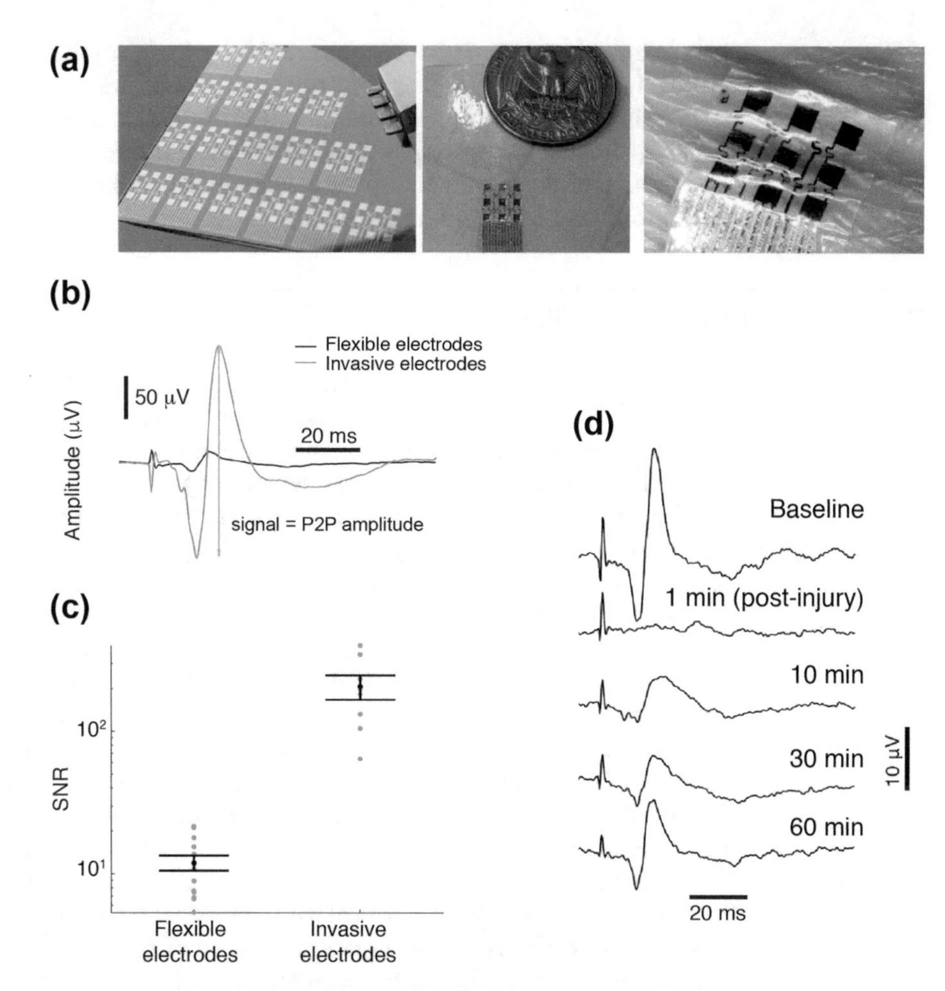

Fig. 4 Flexible epidermal electronics for detecting acute injury. (**a**) (left) Multiple flexible epidermal electrode arrays on a silicon wafer during fabrication; (middle) A completed array on 3M Tegaderm™ medical transparent film dressing; (right) A flexible epidermal array adhered to the skin, demonstrating its conformal properties. (**b**) A representative average SSEP recorded with flexible epidermal electrodes (black, an average of 25 evoked trials) and conventional invasive electrodes (orange, an average of 20 evoked trials). We define the noise as the maximum peak-to-peak amplitude within the pre-stimulation period. (**c**) Signal-to-noise ratio (SNR) comparison of flexible epidermal electrodes and conventional invasive electrodes. Signal and noise levels of SSEPs, one for each animal. (Flexible: $N = 14$; Invasive: $N = 8$) The error bars represent the standard error of the mean. SNR = (peak-to-peak amplitude of the SSEP)/(peak-to-peak amplitude of pre-stimulation baseline). (**d**) A representative time course of SSEPs recorded with a flexible electrode array before and after HIFU, which served as a model of blast injury. Each trace is the average of 25 evoked trials. Electrical stimulation pulses were delivered at 0.5 Hz

both degrade and attenuate the signals detected at the scalp surface. Figure 4b, c show a signal-to-noise ratio (SNR) comparison of flexible epidermal electrodes and conventional invasive electrodes. The noise levels recorded with the invasive electrodes were within the same order of magnitude of those with flexible epidermal electrodes, despite the invasive electrodes' signal levels being approximately one order of magnitude higher; as a result, invasive electrodes' overall SNR was one order of magnitude higher. As shown in Fig. 4d, using the same HIFU model as Fisher et al. (2016), dynamic alterations in SSEPs were observed upon and after injury, and the trends were largely identical to those obtained with more invasive epidural electrodes.

6 Outlook for Clinical Implementation

While preclinical work is advantageous because it permits high-throughput, controlled measurements before and after injury, validation must ultimately be performed through human studies. There is currently strong evidence supporting such utility, given that event related potentials [70] and mismatch negativity [71, 72] are chronically altered in subjects with mTBI. Additionally, somatosensory- and motor-evoked potentials have been found to be suppressed in the acute phase following severe injury [73–76].

Beyond demonstrating promise for wearable TBI detection strategies, an attractive feature that emerges from recent preclinical work is the possibility of quantitative mTBI diagnostics that do not rely on pre-injury "baseline" measurements. During the period of roughly 60 min following impact, even mild injury induces profound, dynamic alterations in electrophysiological and hemodynamic responses that are highly distinct from normal conditions. These functional biomarkers are therefore well-suited for implementation in the context of critical early transport/triage decisions when responding to a head injury of uncertain severity. Though technically difficult, future experiments with human subjects in active field settings are necessary for assessing the sensitivity of SSEPs and sensory-evoked CBF to acute mTBI.

Acknowledgements

Supported by National Science Foundation awards 1541612 and 1641133 (J.A.N.F), internal recruitment funds at New York Medical College (J.A.N.F), Boettcher Webb-Waring Research Award (C.G.W.), internal recruitment funds at University of Colorado (C.G.W.).

References

1. Ruff R (2005) Two decades of advances in understanding of mild traumatic brain injury. J Head Trauma Rehabil 20:5–18. https://doi.org/10.1097/00001199-200501000-00003
2. Lerner EB, Moscati RM (2001) The golden hour: scientific fact or medical "urban legend"? Acad Emerg Med 8:758–760. https://doi.org/10.1111/j.1553-2712.2001.tb00201.x

3. Dinh MM, Bein K, Roncal S, Byrne CM, Petchell J, Brennan J (2013) Redefining the golden hour for severe head injury in an urban setting: the effect of prehospital arrival times on patient outcomes. Injury 44:606–610. https://doi.org/10.1016/j.injury.2012.01.011
4. Becker DP, Miller JD, Ward JD, Greenberg RP, Young HF, Sakalas R (1977) The outcome from severe head injury with early diagnosis and intensive management. J Neurosurg 47:491–502. https://doi.org/10.3171/jns.1977.47.4.0491
5. Stiver SI, Manley GT (2008) Prehospital management of traumatic brain injury. Neurosurg Focus 25:E5. https://doi.org/10.3171/FOC.2008.25.10.E5
6. Steward W, Jones N, Schneider W (1999) Helmet system including at least three accelerometers and mass memory and method for recording in real-time orthogonal acceleration data of a head
7. Duma SM, Manoogian SJ, Bussone WR, Brolinson PG, Goforth MW, Donnenwerth JJ, Greenwald RM, Chu JJ, Crisco JJ (2005) Analysis of real-time head accelerations in collegiate football players. Clin J Sport Med 15:3
8. Ouckama R, Pearsall DJ (2011) Evaluation of a flexible force sensor for measurement of helmet foam impact performance. J Biomech 44:904–909. https://doi.org/10.1016/j.jbiomech.2010.11.035
9. Dimitriadis SI, Zouridakis G, Rezaie R, Babajani-Feremi A, Papanicolaou AC (2015) Functional connectivity changes detected with magnetoencephalography after mild traumatic brain injury. Neuroimage Clin 9:519–531. https://doi.org/10.1016/j.nicl.2015.09.011
10. Dunkley BT, Da Costa L, Bethune A, Jetly R, Pang EW, Taylor MJ, Doesburg SM (2015) Low-frequency connectivity is associated with mild traumatic brain injury. Neuroimage Clin 7:611–621. https://doi.org/10.1016/j.nicl.2015.02.020
11. Alhourani A, Wozny TA, Krishnaswamy D, Pathak S, Walls SA, Ghuman AS, Krieger DN, Okonkwo DO, Richardson RM, Niranjan A (2016) Magnetoencephalography-based identification of functional connectivity network disruption following mild traumatic brain injury. J Neurophysiol 116:1840–1847. https://doi.org/10.1152/jn.00513.2016
12. Mayer AR, Mannell MV, Ling J, Gasparovic C, Yeo RA (2011) Functional connectivity in mild traumatic brain injury. Hum Brain Mapp 32:1825–1835. https://doi.org/10.1002/hbm.21151
13. Sharp DJ, Beckmann CF, Greenwood R, Kinnunen KM, Bonnelle V, De Boissezon X, Powell JH, Counsell SJ, Patel MC, Leech R (2011) Default mode network functional and structural connectivity after traumatic brain injury. Brain 134:2233–2247. https://doi.org/10.1093/brain/awr175
14. Chiu C-C, Liao Y-E, Yang L-Y, Wang J-Y, Tweedie D, Karnati HK, Greig NH, Wang J-Y (2016) Neuroinflammation in animal models of traumatic brain injury. J Neurosci Methods 272:38–49. https://doi.org/10.1016/j.jneumeth.2016.06.018
15. Karve IP, Taylor JM, Crack PJ (2016) The contribution of astrocytes and microglia to traumatic brain injury. Br J Pharmacol 173:692–702. https://doi.org/10.1111/bph.13125
16. Katayama Y, Becker DP, Tamura T, Hovda DA (1990) Massive increases in extracellular potassium and the indiscriminate release of glutamate following concussive brain injury. J Neurosurg 73:889–900. https://doi.org/10.3171/jns.1990.73.6.0889
17. Nilsson P, Hillered L, Pontén U, Ungerstedt U (1990) Changes in cortical extracellular levels of energy-related metabolites and amino acids following concussive brain injury in rats. J Cereb Blood Flow Metab 10:631–637. https://doi.org/10.1038/jcbfm.1990.115
18. Prado GR, Ross JD, DeWeerth SP, LaPlaca MC (2005) Mechanical trauma induces immediate changes in neuronal network activity. J Neural Eng 2:148. https://doi.org/10.1088/1741-2560/2/4/011
19. Goforth PB, Ren J, Schwartz BS, Satin LS (2011) Excitatory synaptic transmission and network activity are depressed following mechanical injury in cortical neurons. J Neurophysiol 105:2350–2363. https://doi.org/10.1152/jn.00467.2010
20. Nag S, Manias JL, Stewart DJ (2009) Pathology and new players in the pathogenesis of brain edema. Acta Neuropathol (Berl) 118:197–217. https://doi.org/10.1007/s00401-009-0541-0
21. Donkin JJ, Vink R (2010) Mechanisms of cerebral edema in traumatic brain injury: therapeutic developments. Curr Opin Neurol 23:293–299. https://doi.org/10.1097/WCO.0b013e328337f451
22. Corps KN, Roth TL, McGavern DB (2015) Inflammation and neuroprotection in traumatic brain injury. JAMA Neurol 72:355–362. https://doi.org/10.1001/jamaneurol.2014.3558

23. Lozano D, Gonzales-Portillo GS, Acosta S, de la Pena I, Tajiri N, Kaneko Y, Borlongan CV (2015) Neuroinflammatory responses to traumatic brain injury: etiology, clinical consequences, and therapeutic opportunities. Neuropsychiatr Dis Treat 11:97–106. https://doi.org/10.2147/NDT.S65815
24. Papa L, Brophy GM, Welch RD, Lewis LM, Braga CF, Tan CN, Ameli NJ, Lopez MA, Haeussler CA, Mendez Giordano DI, Silvestri S, Giordano P, Weber KD, Hill-Pryor-C, Hack DC (2016) Time course and diagnostic accuracy of glial and neuronal blood biomarkers GFAP and UCH-L1 in a large cohort of trauma patients with and without mild traumatic brain injury. JAMA Neurol 73:551–560. https://doi.org/10.1001/jamaneurol.2016.0039
25. Bogoslovsky T, Wilson D, Chen Y, Hanlon D, Gill J, Jeromin A, Song L, Moore C, Gong Y, Kenney K, Diaz-Arrastia R (2017) Increases of plasma levels of glial fibrillary acidic protein, Tau, and amyloid β up to 90 days after traumatic brain injury. J Neurotrauma 34:66–73. https://doi.org/10.1089/neu.2015.4333
26. Goldstein LE, Fisher AM, Tagge CA, Zhang X-L, Velisek L, Sullivan JA, Upreti C, Kracht JM, Ericsson M, Wojnarowicz MW, Goletiani CJ, Maglakelidze GM, Casey N, Moncaster JA, Minaeva O, Moir RD, Nowinski CJ, Stern RA, Cantu RC, Geiling J, Blusztajn JK, Wolozin BL, Ikezu T, Stein TD, Budson AE, Kowall NW, Chargin D, Sharon A, Saman S, Hall GF, Moss WC, Cleveland RO, Tanzi RE, Stanton PK, McKee AC (2012) Chronic traumatic encephalopathy in blast-exposed military veterans and a blast neurotrauma mouse model. Sci Transl Med 4:134ra60. https://doi.org/10.1126/scitranslmed.3003716
27. Daneshvar DH, Goldstein LE, Kiernan PT, Stein TD, McKee AC (2015) Post-traumatic neurodegeneration and chronic traumatic encephalopathy. Mol Cell Neurosci 66:81–90. https://doi.org/10.1016/j.mcn.2015.03.007
28. Chen J, Xu X-M, Xu ZC, Zhang JH (2012) Electrophysiological approaches in traumatic brain injury—Springer. Humana, New York
29. Trudeau DL, Anderson J, Hansen LM, Shagalov DN, Schmoller J, Nugent S, Barton S (1998) Findings of mild traumatic brain injury in combat veterans with PTSD and a history of blast concussion. J Neuropsychiatry Clin Neurosci 10:308–313. https://doi.org/10.1176/jnp.10.3.308
30. Thatcher RW, North DM, Curtin RT, Walker RA, Biver CJ, Gomez JF, Salazar AM (2014) An EEG severity index of traumatic brain injury. http://neuro.psychiatryonline.org/doi/10.1176/jnp.13.1.77. Accessed 24 Feb 2015
31. Nuwer MR, Hovda DA, Schrader LM, Vespa PM (2005) Routine and quantitative EEG in mild traumatic brain injury. Clin Neurophysiol 116:2001–2025. https://doi.org/10.1016/j.clinph.2005.05.008
32. Naunheim RS, Treaster M, English J, Casner T, Chabot R (2010) Use of brain electrical activity to quantify traumatic brain injury in the emergency department. Brain Inj 24:1324–1329. https://doi.org/10.3109/02699052.2010.506862
33. Ayaz SI, Thomas C, Kulek A, Tolomello R, Mika V, Robinson D, Medado P, Pearson C, Prichep LS, O'Neil BJ (2015) Comparison of quantitative EEG to current clinical decision rules for head CT use in acute mild traumatic brain injury in the ED. Am J Emerg Med 33:493–496. https://doi.org/10.1016/j.ajem.2014.11.015
34. Prichep LS, Naunheim R, Bazarian J, Mould WA, Hanley D (2015) Identification of hematomas in mild traumatic brain injury using an index of quantitative brain electrical activity. J Neurotrauma 32:17–22. https://doi.org/10.1089/neu.2014.3365
35. O'Neil B, Prichep LS, Naunheim R, Chabot R (2012) Quantitative brain electrical activity in the initial screening of mild traumatic brain injuries. West J Emerg Med 13:394–400. https://doi.org/10.5811/westjem.2011.12.6815
36. Rapp PE, Keyser DO, Albano A, Hernandez R, Gibson DB, Zambon RA, Hairston WD, Hughes JD, Krystal A, Nichols AS (2015) Traumatic brain injury detection using electrophysiological methods. Front Hum Neurosci 9:11. https://doi.org/10.3389/fnhum.2015.00011
37. Hanley D, Prichep LS, Badjatia N, Bazarian J, Chiacchierini R, Curley KC, Garrett J, Jones E, Naunheim R, O'Neil B, O'Neill J, Wright DW, Huff JS (2017) A brain electrical activity (EEG)-based biomarker of functional impairment in traumatic brain injury: a multi-site validation trial. J Neurotrauma 35:41–47. https://doi.org/10.1089/neu.2017.5004
38. Gaetz M (2004) The neurophysiology of brain injury. Clin Neurophysiol 115:4–18. https://doi.org/10.1016/S1388-2457(03)00258-X
39. Walser H, Emre M, Janzer R (1986) Somatosensory evoked potentials in comatose patients: correlation with outcome and neuropathological findings. J Neurol 233:34–40. https://doi.org/10.1007/BF00313989

40. Carter BG, Butt W (2005) Are somatosensory evoked potentials the best predictor of outcome after severe brain injury? A systematic review. Intensive Care Med 31:765–775. https://doi.org/10.1007/s00134-005-2633-1
41. Paxinos G, Franklin KBJ (2004) The mouse brain in stereotaxic coordinates. Gulf Professional Publishing, Amsterdam
42. Fisher J, Huang S, Ye M, Nabili M, Wilent W, Krauthamer V, Myers M, Welle C (2016) Real-time detection and monitoring of acute brain injury utilizing evoked electroencephalographic potentials. IEEE Trans Neural Syst Rehabil Eng 24:1003–1012. https://doi.org/10.1109/TNSRE.2016.2529663
43. Huang S, Fisher JAN, Ye M, Kim YS, Ma R, Nabili M, Krauthamer V, Myers MR, Coleman TP, Welle CG (2017) Epidermal electrode technology for detecting ultrasonic perturbation of sensory brain activity. IEEE Trans Biomed Eng. https://doi.org/10.1109/TBME.2017.2713647
44. Jang H, Huang S, Hammer DX, Wang L, Rafi H, Ye M, Welle CG, Fisher JAN (2017) Alterations in neurovascular coupling following acute traumatic brain injury. Neurophotonics 4:045007. https://doi.org/10.1117/1.NPh.4.4.045007
45. Takechi U, Matsunaga K, Nakanishi R, Yamanaga H, Murayama N, Mafune K, Tsuji S (2014) Longitudinal changes of motor cortical excitability and transcallosal inhibition after subcortical stroke. Clin Neurophysiol 125:2055–2069. https://doi.org/10.1016/j.clinph.2014.01.034
46. Schwarzbach E, Bonislawski DP, Xiong G, Cohen AS (2006) Mechanisms underlying the inability to induce area CA1 LTP in the mouse after traumatic brain injury. Hippocampus 16:541–550. https://doi.org/10.1002/hipo.20183
47. DeWitt DS, Prough DS, Taylor CL, Whitley JM (1992) Reduced cerebral blood flow, oxygen delivery, and electroencephalographic activity after traumatic brain injury and mild hemorrhage in cats. J Neurosurg 76:812–821. https://doi.org/10.3171/jns.1992.76.5.0812
48. McIntosh TK, Vink R, Noble L, Yamakami I, Fernyak S, Soares H, Faden AL (1989) Traumatic brain injury in the rat: characterization of a lateral fluid-percussion model. Neuroscience 28:233–244. https://doi.org/10.1016/0306-4522(89)90247-9
49. Werner C, Engelhard K (2007) Pathophysiology of traumatic brain injury. Br J Anaesth 99:4–9. https://doi.org/10.1093/bja/aem131
50. Toth P, Szarka N, Farkas E, Ezer E, Czeiter E, Amrein K, Ungvari Z, Hartings JA, Buki A, Koller A (2016) Traumatic brain injury-induced autoregulatory dysfunction and spreading depression-related neurovascular uncoupling: pathomechanisms, perspectives, and therapeutic implications. Am J Physiol Heart Circ Physiol 311:H1118–H1131. https://doi.org/10.1152/ajpheart.00267.2016
51. Dietrich WD, Alonso O, Halley M (1994) Early microvascular and neuronal consequences of traumatic brain injury: a light and electron microscopic study in rats. J Neurotrauma 11:289–301. https://doi.org/10.1089/neu.1994.11.289
52. Yi J-H, Hazell AS (2006) Excitotoxic mechanisms and the role of astrocytic glutamate transporters in traumatic brain injury. Neurochem Int 48:394–403. https://doi.org/10.1016/j.neuint.2005.12.001
53. Herculano-Houzel S (2011) Scaling of brain metabolism with a fixed energy budget per neuron: implications for neuronal activity, plasticity and evolution. PLoS One 6:e17514. https://doi.org/10.1371/journal.pone.0017514
54. Vavilala MS, Farr CK, Watanitanon A, Clark-Bell BC, Chandee T, Moore A, Armstead W (2017) Early changes in cerebral autoregulation among youth hospitalized after sports-related traumatic brain injury. Brain Inj 32:269–275. https://doi.org/10.1080/02699052.2017.1408145
55. Albalawi T, Hamner JW, Lapointe M, Meehan WP, Tan CO (2017) The relationship between cerebral vasoreactivity and post-concussive symptom severity. J Neurotrauma 34:2700–2705. https://doi.org/10.1089/neu.2017.5060
56. Zeiler FA, Cardim D, Donnelly J, Menon D, Czosnyka M, Smielewski P (2017) Transcranial Doppler systolic flow index and ICP derived cerebrovascular reactivity indices in TBI. J Neurotrauma 35:314–322. https://doi.org/10.1089/neu.2017.5364
57. Ziegler D, Cravens G, Poche G, Gandhi R, Tellez M (2017) Use of transcranial Doppler in patients with severe traumatic brain injuries. J Neurotrauma 34:121–127. https://doi.org/10.1089/neu.2015.3967
58. Buxton RB, Wong EC, Frank LR (1998) Dynamics of blood flow and oxygenation changes during brain activation: the balloon model. Magn Reson Med 39:855–864. https://doi.org/10.1002/mrm.1910390602
59. Niskanen J-P, Airaksinen AM, Sierra A, Huttunen JK, Nissinen J, Karjalainen PA, Pitkänen A,

Gröhn OH (2013) Monitoring functional impairment and recovery after traumatic brain injury in rats by fMRI. J Neurotrauma 30:546–556. https://doi.org/10.1089/neu.2012.2416

60. D'Esposito M, Deouell LY, Gazzaley A (2003) Alterations in the BOLD fMRI signal with ageing and disease: a challenge for neuroimaging. Nat Rev Neurosci 4:863–872. https://doi.org/10.1038/nrn1246
61. Durduran T, Yodh AG (2014) Diffuse correlation spectroscopy for non-invasive, micro-vascular cerebral blood flow measurement. Neuroimage 85(Part 1):51–63. https://doi.org/10.1016/j.neuroimage.2013.06.017
62. Durduran T, Burnett MG, Yu G, Zhou C, Furuya D, Yodh AG, Detre JA, Greenberg JH (2004) Spatiotemporal quantification of cerebral blood flow during functional activation in rat somatosensory cortex using laser-speckle Flowmetry. J Cereb Blood Flow Metab 24:518–525. https://doi.org/10.1097/00004647-200405000-00005
63. Zhou C, Yu G, Furuya D, Greenberg J, Yodh A, Durduran T (2006) Diffuse optical correlation tomography of cerebral blood flow during cortical spreading depression in rat brain. Opt Express 14:1125. https://doi.org/10.1364/OE.14.001125
64. Zhou C, Eucker SA, Durduran T, Yu G, Ralston J, Friess SH, Ichord RN, Margulies SS, Yodh AG (2009) Diffuse optical monitoring of hemodynamic changes in piglet brain with closed head injury. J Biomed Opt 14:034015. https://doi.org/10.1117/1.3146814
65. Kim D-H, Lu N, Ma R, Kim Y-S, Kim R-H, Wang S, Wu J, Won SM, Tao H, Islam A, Yu KJ, Kim T-I, Chowdhury R, Ying M, Xu L, Li M, Chung H-J, Keum H, McCormick M, Liu P, Zhang Y-W, Omenetto FG, Huang Y, Coleman T, Rogers JA (2011) Epidermal electronics. Science 333:838–843. https://doi.org/10.1126/science.1206157
66. Kang DY, Kim Y-S, Ornelas G, Sinha M, Naidu K, Coleman TP (2015) Scalable microfabrication procedures for adhesive-integrated flexible and stretchable electronic sensors. Sensors 15:23459–23476. https://doi.org/10.3390/s150923459
67. Gil-da-Costa R, Fung R, Kim S, Mesa D, Ma R, Kang D, Bajema M, Albright TD, Coleman TP (2013) A novel method to assess event-related brain potentials in clinical domains using frontal epidermal electronics sensors. In: Society of Neuroscience Proceedings, San Diego, CA
68. Coleman TP, Ma R, Fung R, Bajema M, Albright TD, Rogers J, Gil-da-Costa R (2012) Epidermal electronics capture of event-related brain potentials (ERP) signal in a "real-world" target detection task. In: Society of Neuroscience Proceedings, New Orleans, LA
69. Harbert MJ, Rosenberg SS, Mesa D, Sinha M, Karanjia NP, Nespeca M, Coleman TP (2013) Demonstration of the use of epidermal electronics in neurological monitoring. Wiley-Blackwell, Hoboken, pp S76–S77
70. Duncan CC, Summers AC, Perla EJ, Coburn KL, Mirsky AF (2011) Evaluation of traumatic brain injury: brain potentials in diagnosis, function, and prognosis. Int J Psychophysiol 82:24–40. https://doi.org/10.1016/j.ijpsycho.2011.02.013
71. Larson MJ, Kaufman DAS, Schmalfuss IM, Perlstein WM (2007) Performance monitoring, error processing, and evaluative control following severe TBI. J Int Neuropsychol Soc 13:961–971. https://doi.org/10.1017/S1355617707071305
72. Larson MJ, Fair JE, Farrer TJ, Perlstein WM (2011) Predictors of performance monitoring abilities following traumatic brain injury: the influence of negative affect and cognitive sequelae. Int J Psychophysiol 82:61–68. https://doi.org/10.1016/j.ijpsycho.2011.02.001
73. Lew HL, Dikmen S, Slimp J, Temkin N, Lee EH, Newell D, Robinson LR (2003) Use of somatosensory-evoked potentials and cognitive event-related potentials in predicting outcomes of patients with severe traumatic brain injury. Am J Phys Med Rehabil Assoc Acad Physiatr 82:53–61.; quiz 62–64, 80. https://doi.org/10.1097/01.PHM.0000043771.90606.81
74. Houlden DA, Taylor AB, Feinstein A, Midha R, Bethune AJ, Stewart CP, Schwartz ML (2010) Early somatosensory evoked potential grades in comatose traumatic brain injury patients predict cognitive and functional outcome. Crit Care Med 38:167–174. https://doi.org/10.1097/CCM.0b013e3181c031b3
75. Xu W, Jiang G, Chen Y, Wang X, Jiang X (2012) Prediction of minimally conscious state with somatosensory evoked potentials in long-term unconscious patients after traumatic brain injury. J Trauma Acute Care Surg 72:1024–1029. https://doi.org/10.1097/TA.0b013e31824475cc
76. Schorl M, Valerius-Kukula S-J, Kemmer TP (2014) Median-evoked somatosensory potentials in severe brain injury: does initial loss of cortical potentials exclude recovery? Clin Neurol Neurosurg 123:25–33. https://doi.org/10.1016/j.clineuro.2014.05.004

Chapter 16

Advanced Neuroimaging Methods in Traumatic Brain Injury

Jenifer Juranek

Abstract

Advanced neuroimaging tools have become increasingly available to clinical MRI scanners in hospital settings. Applying these noninvasive MRI tools to patients with traumatic brain injury, particularly in the acute phase following injury, is not only clinically feasible, but is also the best way to capture information about the pathophysiological processes underway in different brain regions. A high resolution multi-weighted and multimodal MRI protocol for TBI can be completed within 25 min of scanner time. Such a protocol would provide quantitative information about the extent and location of edema (cytotoxic vs vasogenic), diffuse axonal injury, bleeds, and cerebral perfusion deficits. Future development of new targeted therapies to prevent secondary injuries from molecular cascades which can have long-term effects on patient outcome measures will likely be driven by characterizing each patient's own neuroimaging phenotype. After all, the current standard of care is to *medically manage* the overt signs of robust secondary brain injury such as an elevated neuroinflammatory response with subsequent cerebral edema leading to elevated intracranial pressure and cerebral perfusion deficits. Since high quality multi-weighted and multi-modal neuroimaging data have rarely been acquired in the acute phase following the primary brain injury, the prognostic value of these sequences yielding relevant imaging biomarkers has yet to be explored. However, precision medicine approaches continue to emerge across other disease states, so why not TBI? The development of viable therapeutic targets (and elimination of therapies unlikely to be effective) in treating secondary brain injuries (not just managing them) should benefit from multivariate analyses of each individual patient's imaging phenotype, particularly during the acute stage when interruption/disruption of biomolecular cascades underlying long-term impairments can potentially have the greatest impact on preserving brain structure and function.

Key words TBI, MRI, DTI, SWI, pCASL, Multimodal image integration

1 Introduction

Neuroimaging methods of the brain have rapidly advanced over the last 5 years. Even clinical MRI scanners in hospital settings are benefiting from these technological developments. Whereas most clinical scanners were unable to run high quality "research" protocols due to rigid time constraints inherent with a hospital setting, recent advances are making research quality MRI protocols clinically feasible to aid in diagnosis, progress monitoring, and evaluation of treatment efficacy over time. Multi-weighted and

Amit K. Srivastava and Charles S. Cox, Jr. (eds.), *Pre-Clinical and Clinical Methods in Brain Trauma Research*, Neuromethods, vol. 139, https://doi.org/10.1007/978-1-4939-8564-7_16, © Springer Science+Business Media, LLC, part of Springer Nature 2018

multimodal MRI protocol sessions which used to take an hour or more can currently acquire high resolution images in ~30 min of clinical scanner time. This milestone has significantly changed how physicians can think about (1) assessing their clinical TBI patients, (2) developing a treatment plan, and (3) evaluating treatment efficacy via longitudinal progress monitoring which includes neuroimaging as a noninvasive tool. In this era of precision medicine, determining a TBI patient's prognosis and optimal treatment plan has the potential to benefit the most from multivariate evaluation of clinically relevant variables. Previously, neuroimaging variables have been underexplored in a multivariate context primarily due to technical challenges of acquiring and analyzing such highly dimensional data in a reasonable timeframe. TBI patients are particularly challenging due to the uniqueness of their primary brain injury (extent and locations of the brain impacted) and the complex pathophysiological processes underlying a cascade of secondary brain injuries, the sequelae of which can be quite devastating in terms of long-term outcomes.

The advanced neuroimaging methods discussed below are not just state-of-the-art, but importantly they are clinically feasible in hospital settings. With less than 30 min of scanner time, clinicians can acutely and noninvasively obtain quantitative information regarding the location and extent of the following features of brain injury which are likely to occur in TBI patients: edema type (cytotoxic vs vasogenic), diffuse axonal injury, bleeds, and cerebral perfusion deficits. Furthermore, this information can assist physicians with their selection of the optimal treatment plan (and lead to the development of new targeted therapies) for the best prognosis given the neuroimaging phenotype of each individual TBI patient.

2 Materials

1. 3T MRI scanner with at least 30 min time slot for image acquisition and a multi-channel phased array head coil for implementing parallel imaging and sequence acceleration techniques for faster acquisition times.
2. High capacity storage device for acquired and processed images (~2 GB per MRI session).
3. Linux or Mac workstation for image processing and analyses.
4. DICOM to NIFTI converter software (dcm2niix; https://github.com/rordenlab/dcm2niix/releases).
5. FMRIB's software library (FSL; https://fsl.fmrib.ox.ac.uk/fsl/fslwiki/).

3 Methods

3.1 MRI Acquisition Protocol: Multi-weighted and Multimodal Images

Depending on the MRI scanner platform (e.g., Siemens, Philips, GE), model (e.g., Prisma, Ingenia, Discovery), and software version release installed on the scanner console, certain aspects of specified parameters for each sequence will need to be "tuned" to each individual MRI scanner to obtain optimal signal-to-noise and contrast-to-noise images for quantitative analyses, with "research keys" providing access to unlock more features and values than systems without research keys. However, the most important characteristics of a solid research MRI acquisition protocol include the following: (1) stable acquisition parameters over time, (2) vigilant monitoring for indications of subject motion throughout acquisition session, (3) repetition of sequences when motion is observed, (4) acquisition of multiple imaging modalities in the same MRI session, (5) minimizing the duration of imaging session (e.g., <25 min or ~35 min). Details of each sequence are available in Table 1.

For TBI studies, an exemplar <35 min acquisition protocol would include the following sequences:

1. Localizer
2. Isotropic 3D T1-weighted (anatomy)
3. Isotropic 3D T2-weighted (CSF)

Table 1
Clinical 3T Philips Ingenia (no research keys)

	32ch head coil			
Sequence	**NSA**	**Duration**	**VoxDims**	**FoV (mm)**
[a]3Dsag T1-w	1	4′50″	1 mm^3	256 × 256
[a]3Dsag T2-w	1	2′37″	1 mm^3	256 × 256
[a]3Dsag T2-FLAIR	2	6′00″	1 mm^3	256 × 256
3Dax SWI	1	3′42″	0.8 × 0.8 × 1.0	240 × 195
2Dax DWI	1	1′37″	2.2 × 2.2 × 2.2	240 × 240
Ax pCASL	1	4′42″	2.75 × 2.75 × 5	240 × 240
		23′28″		
Add-ons				
Ax DTI_32dir	1	8′06″	2.2 × 2.2 × 2.2	240 × 240
Ax Distortion maps	1	0′40″	2.2 × 2.2 × 2.2	240 × 240
		8′46″		

[a]Identical geometry; common prescription (including shim volume)

4. Isotropic 3D T2-weighted FLAIR (white matter injury)
5. 3D SWI (bleeds, microhemorrhages)
6. DWI (edema)
7. pCASL (perfusion, cerebral blood flow)
8. DTI (tissue integrity, tractography of major white matter fiber bundles)
9. Reversed polarity spin echo distortion maps (A-P and P-A)

NOTE: Following the localizer, the prescription for the first 3D isotropic sequence (e.g., T1-weighted) can be set up to obtain whole brain coverage. Subsequently, one can copy and paste parameters (including shim volume) to the other 3D isotropic sequences since they all share the same matrix and field of view (FoV). On Philips scanners, all three sequences can be geo-linked to achieve the same outcome.

3.2 MRI Processing and Analyses

3.2.1 DICOM to NIFTI Data Conversion

Once the MRI acquisition has been completed, the images need to be exported from the scanner console to an adequate storage device for data retrieval. Utilizing DICOM export protocols for this process ensures that specific standards have been met regarding organization and labeling of metadata in the image headers. Even though each MRI scanner vendor may have some proprietary information hidden in the headers with special tags, information about acquisition parameters and image reconstruction are accessible to third party software packages such as DICOM readers and converters. Essentially, DICOMs are the rawest form of imaging data one can view away from the scanner console. However, to work with the images and analyze them, the DICOMs really need to be converted to a different file format. Based on experience, the most common file format handled by research-based image processing software is the NIFTI file format. While several different dicom to nifti converters are readily available, I advocate the use of Chris Rorden's well-supported and well-documented converter recently released as dcm2niix [1]. Although the predecessor to dcm2niix was widely popular in the neuroimaging community (e.g., dcm2nii), the newer converter is just as easy to use and handles all of the latest sequences across imaging modalities, provided the exported DICOMs adhere to the DICOM standard.

NOTE: Chris Rorden's website (http://www.mccauslandcenter.sc.edu/crnl/tools) is an excellent resource and a highly recommended primer for all who work with MRI data.

3.2.2 Multimodal Integration in Clinical Patient Populations

Due to the tremendous variability between TBI patients with respect to location, extent, and tissue types impacted by each individual's unique brain injury, quantitative analyses of neuroimaging data in this patient population can be quite challenging. I strongly recommend an image processing pipeline that works with all images (across modalities) in each individual's native space rather than transforming each subject's data to standard template space. This approach is absolutely necessary when working with brain imaging data acquired from patient populations which do not align well with standard templates created from healthy normal people. Furthermore, in this era of precision medicine, improved prognostic value of long-term outcomes is likely to emerge by characterizing each individual's "imaging phenotype" from conducting *multivariate* analyses of co-registered imaging modalities in each individual's native space.

3.2.3 High Resolution Anatomical Sequences

In severe TBI cases, gross structural changes acutely occur following the primary brain injury (Fig. 1) such as midline shift, edema, diffuse white matter damage, bleeds, and cerebral perfusion deficits. The key to maximizing the multimodal information available from each TBI patient's MRI session is to co-register all sequences to a common base sequence with identical spatial geometry (voxel dimensions and FoV). I have found the isotropic 3D T2-weighted images to be an excellent base to which the other sequences are co-registered. The T2-weighted images share similar features with the T2-FLAIR, and the DWI/DTI sequences (e.g., white matter appears dark and gray matter appears light) facilitating the co-registration process. Since all 3D anatomical sequences (e.g., T1, T2, T2-FLAIR) share the same prescribed geometry at acquisition (e.g., 1 mm^3 isotropic, 256 × 256 FoV), co-registration of these images to each subject's T2-weighted base is straightforward, provided that image distortions (e.g., artifacts) from motion or implanted devices (e.g., arch bar wires, extraventricular drain, etc.) do not interfere with the linear registration algorithm. My preference is to use FMRIB's software library (FSL) for co-registering multiple sequences to each subject's native T2 image series. From my experience, FSL is easy to use, readily scriptable for maintaining standardization of processing parameters, and reviewing the outputs of the co-registration process in FSLeyes (imaging viewer) is simply achieved with elegant tools to adjust layer transparencies which can be viewed in all three cardinal planes concurrently.

NOTE: A key step to successfully establishing and running an image processing pipeline is to *standardize* the directory structure containing each patient's imaging data as well as individual file names within that directory structure.

Once all of the 3D anatomical images (T1-weighted and T2-FLAIR) have been co-registered to the subject's native T2 image series, then one can load them into FSL's viewer (FSLeyes) *concurrently* and begin to appreciate the tremendous value added by acquiring high resolution 3D images with multiple weightings at a cost of ~14 min of scanner time (Fig. 2). White matter

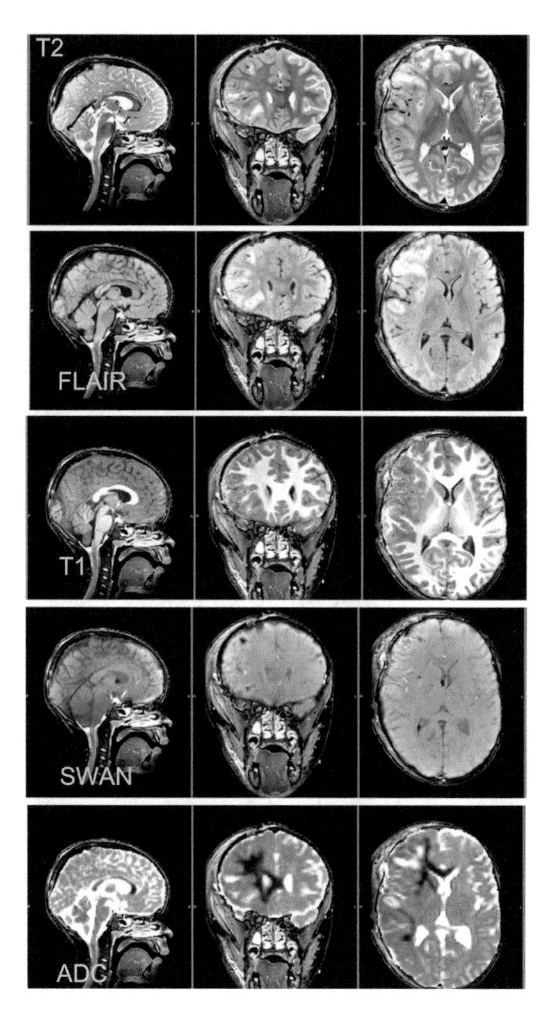

Fig. 1 Severe TBI patient imaged as soon as medically stable (7 days post injury). Midline shift and edema are clearly evident. Multimodal imaging sequences after co-registration to the base T2-images (top row). The hyperintensity visible in the right frontal lobe in the T2-images is also hyperintense in the FLAIR images (second row), indicative of edema. The corresponding area is hypointense on the T1-images (third row) without any indication of blood products which would be evident in the SWAN-images (fourth row). The last row (ADC image) demonstrates both types of edema: cytotoxic (hypointensities restricted to white matter) and vasogenic (hyperintensities colocalized with hyperintensities in FLAIR images)

hyperintensities on the T2-FLAIR series can be readily observed in periventricular areas like the corpus callosum due to the nulling of the CSF signal. Frequently the same voxels identified as white matter hyperintensities on the T2-FLAIR are also hyperintense on the T2-weighted series. Yet whether the same voxels are hyper- or hypointense on the T1-weighted series can be particularly useful for distinguishing between different types of injury. For instance,

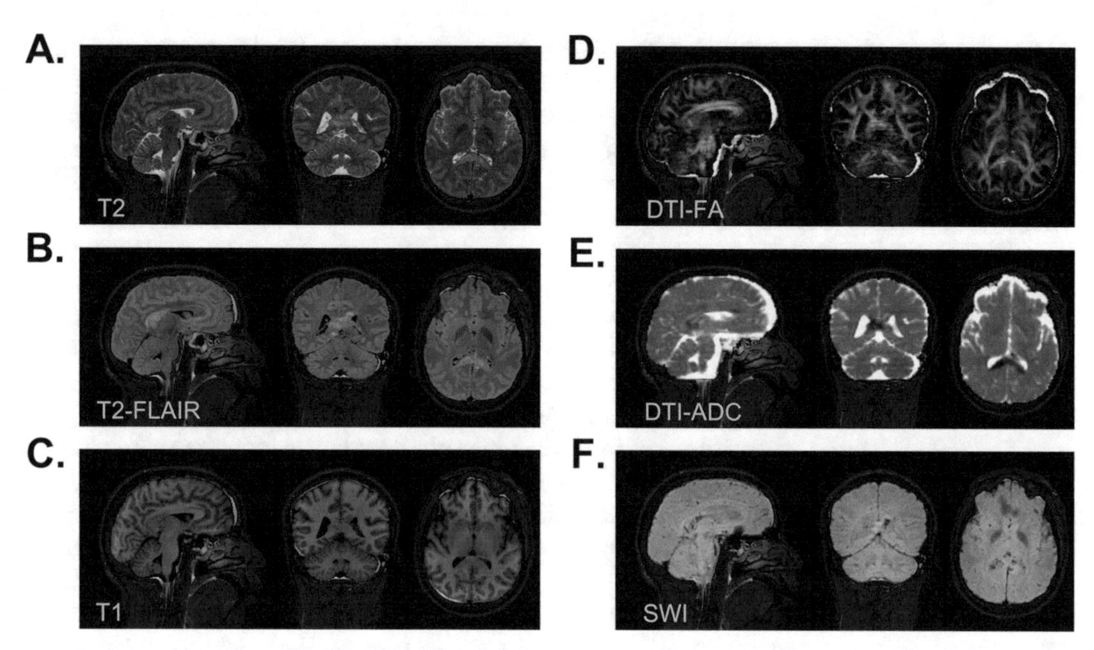

Fig. 2 Severe TBI patient imaged as soon as medically stable (8 days post injury). (**a**–**f**) Multi-weighted and multimodal imaging sequences have been co-registered to the base T2-images (**a**). Focal hyperintensity evident in the posterior region of the corpus callosum in T2 (**a**) and FLAIR (**b**) images reflect edema, a common occurrence in TBI resulting from motor vehicle accidents. The same region of the corpus callosum as in panels (**a**) and (**b**) is hypointense in the T1-images and co-localizes with some evidence of blood products evident in the SWI-images (**f**). Diffusion tensor imaging demonstrates reduced FA values (hypointensities in FA map) in the posterior corpus callosum (**d**), indicative of reduced white matter integrity in this region. Reduced ADC values (**e**; hypointensities) in the corresponding region of the corpus callosum are indicative of cytotoxic edema in this area

T2 = bright and T1 = dark can be an indication of lower protein levels than voxels which are T2 = bright and T1 = bright. Furthermore, even higher protein content can be inferred from T2 = dark and T1 = dark. Such information can be valuable for characterizing the underlying cellular and subcellular processes occurring as a function of time since injury. For example, higher protein content might possibly differentiate between two different forms of cellular death: advanced cellular lysis and necrosis vs apoptosis. As cleverly described by Bonfoco et al. [2], apoptosis represents a cellular version of “death with dignity” since the cellular membrane is not disrupted and the interstitial milieu is not polluted with the intracellular contents of the dying cell until phagocytosis “intervenes.” Furthermore, such information obtained noninvasively can be an invaluable tool for progress monitoring of treatment efficacy or elucidating mechanism of action exhibited by potential therapeutic agents. After all, the “secondary” brain injuries following severe TBI are the *viable* targets of putative therapeutic strategies. These secondary brain injuries result from complex biochemical cascades which are clinically manifested as an elevated neuroinflammatory response with subsequent cerebral edema leading to elevated intracranial pressure and impaired cerebral perfusion. Current standard of care focuses on medical management of these overt signs of robust secondary brain injury. However, disrupting the complex biochemical cascades of secondary brain injury has the potential to minimize the amount and extent of secondary brain injury.

3.2.4 Bleeds

Since bleeds associated with TBI can range from dense and focal to diffuse and punctate, I highly recommend including a susceptibility weighted imaging (SWI) sequence in the MRI acquisition protocol. Although cerebral hemorrhage is acutely assessed routinely using computed tomography (CT), CT is less sensitive than MRI to detecting and localizing microhemorrhages in white matter which often occur in TBI cases with diffuse axonal injury (DAI [3]). Previously, most MRI protocols relied on T2* Gradient Recalled Echo (GRE) images to identify and localize bleeds in TBI cases. However, numerous research studies have reported significantly increased sensitivity, and few false-positives (e.g., calcification), using an SWI sequence for this purpose [4]. Adding a minimum intensity projection (MIP) reformat step to the acquired SWI sequence further improves the detection and localization of micro-bleeds. Such information can be exceptionally helpful, as it is in stroke, to distinguish between different types of tissue abnormalities (and their responsiveness to different types of treatment interventions) throughout the recovery process.

3.2.5 Cerebral Perfusion

To evaluate perfusion deficits, we have recently added a pseudo-continuous Arterial Spin Label (pCASL) sequence to our TBI MRI protocol since it became available on our 3T Philips Ingenia clinical scanner at Memorial Hermann Hospital about a year ago. Cerebral blood flow measures have traditionally required the use of injectable tracers/contrast agents which are not advisable in children or TBI cases which have been recently medically stabilized enough to undergo an MRI. Furthermore, acquisition of images before the contrast agent, the timing of the injection, and the acquisition of images after delivery of the tracer is a complicated process which is difficult to repeat when poor quality images are obtained. Recent advances in MRI have made it possible to acquire quantitative measures of cerebral blood flow (ml 100 g^{-1} min^{-1}) without contrast agents, but instead rely on the "labeling" of arterial blood supply before it ascends into the brain with a magnetization pulse built into the MRI sequence. This method of "tagging" the blood supply to the brain is referred to as Arterial Spin Labeling (for additional notes from a neuroradiologist's point of view, see Grade et al. 2015) [5]. Total acquisition time for this quantitative sequence to obtain unlabeled and labeled sets of images for subsequent subtraction is <4 min for whole brain coverage.

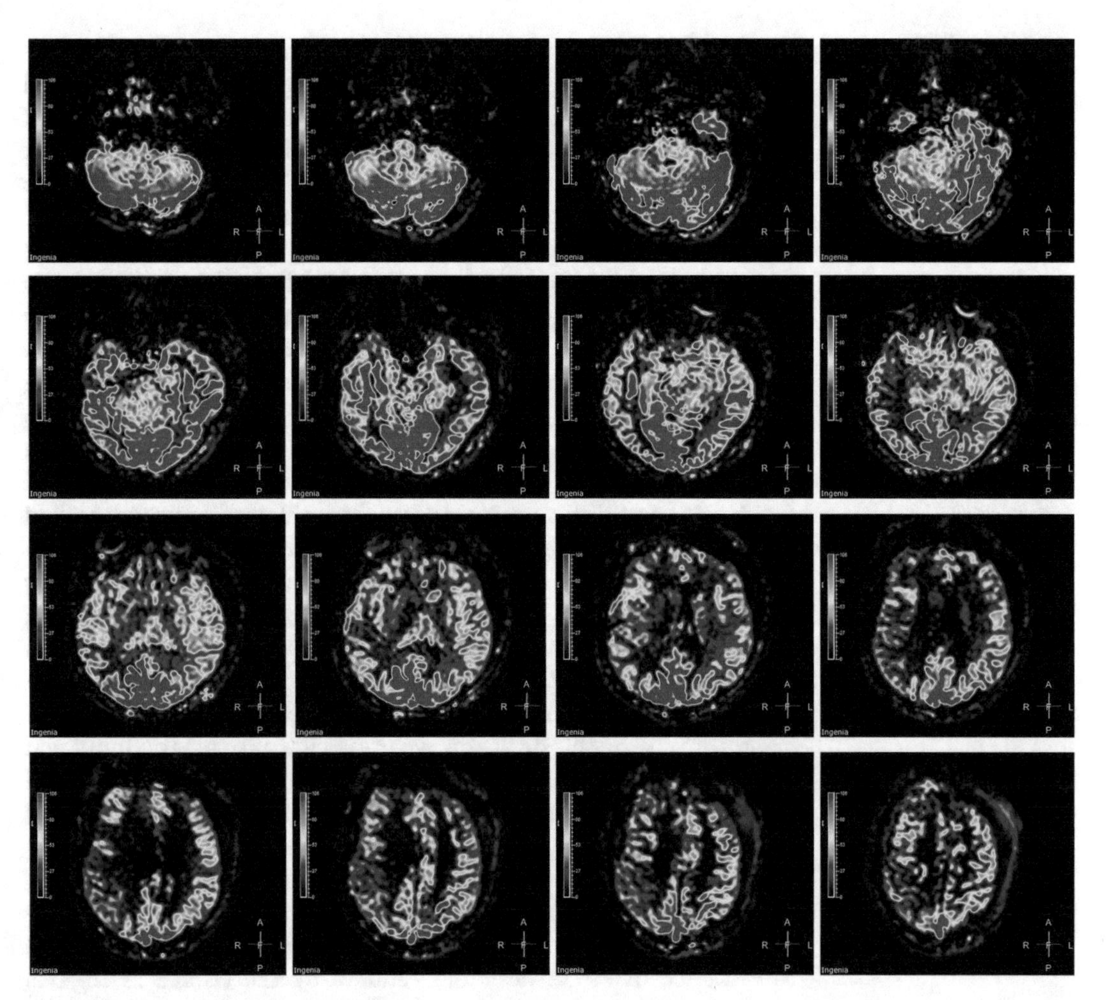

Fig. 3 Severe TBI patient imaged as soon as medically stable (16 days post injury). Arterial spin labeling indicates low cerebral perfusion in right superior cortex (cold colors)

The most important acquisition parameters to select for pCASL are the labeling duration and post-label delay. Since the selection of post-label delay time is associated with presumed blood velocity, this parameter should be tuned to the clinical population of interest. Generally, neonates and older adults require a longer post-label delay time (~2000 ms), children require a shorter post-label delay time (~1500 ms), and healthy adults require ~1800 ms post-label delay time. Furthermore, the pCASL sequence can readily be repeated if needed to obtain good quality images. The subtraction maps between the unlabeled and labeled series provide spatial information with quantitative measurements of cerebral blood flow. As shown in Fig. 3, the basal portion of the brain is well perfused while the subcortical and superior aspects (particularly in the right hemisphere) demonstrate hypoperfusion.

NOTE: There are several different variants of ASL sequences: continuous (cASL), pseudocontinuous (pCASL), and pulsed (PASL). Each ASL sequence variant has advantages and limitations. We selected pCASL for its high labeling efficiency, ease of implementation, <5 min scanner time, and availability on our non-research clinical MRI scanner at Memorial Hermann Hospital. The key to successful ASL-based labeling of the cerebral blood supply is based on mindful positioning of the labeling plane: it needs to be placed perpendicular to the ascending arterial blood vessels and it should avoid areas with susceptibility artifact such as the orbitofrontal sinuses and dental fillings or metallic surgical implants.

3.2.6 Diffusion Weighted Imaging (DWI)

If it is necessary to minimize scanner time for the MRI session (<25 min) and diffusion tensor imaging will not be acquired, then DWI is an excellent sequence to acquire in TBI patients [6]. The DWI sequence is basically a 2D echoplanar imaging based sequence (EPI), so it is quick to obtain (<2 min) and has the benefit of adding unique information about microstructural tissue properties throughout the brain, particularly when DAI has occurred [7]. The two primary quantitative metrics that emerge from analyses of DWI are the mean of the diffusion weighted images (mDWI) and the apparent diffusion coefficient (ADC). Indications of cerebral edema, vasogenic or cytotoxic, can be identified and classified with these two metrics. Although little information is currently available in the TBI literature regarding cytotoxic edema [7–11], the stroke literature has utilized DWI in MRI protocols of the brain to identify regions exhibiting restricted diffusion (low ADC values) as evidence of cytotoxic edema and vasogenic edema as regions marked by increased diffusion (high ADC values). According to the stroke literature, cytotoxic edema presents acutely, is potentially reversible, and represents the "rescuable" tissue to therapeutic intervention (recent review by Heiss 2016) [12]. The vasogenic tissue results from disrupted permeability of the blood–brain barrier and is less likely to recover over time. By adding edema imaging analyses to TBI MRI protocols, one can longitudinally quantify the amount and extent of each type of cerebral edema (cytotoxic vs vasogenic) at each imaging timepoint as part of progress monitoring and evaluation of treatment efficacy.

3.2.7 Diffusion Tensor Imaging (DTI)

If one has the capability to acquire and analyze multi-directional diffusion weighted imaging data such as DTI (~8 min to acquire on most clinical scanners), then one can probe the integrity of different tissue types (e.g., white matter and gray matter) as well as reconstruct white matter pathways using tractography methods to evaluate structural connectivity between brain regions. Using a gradient table of at least 30 non-collinear directions which are equally spaced over a sphere [13], one can obtain reliable quantitative estimates of water diffusion in the brain. Specifically, directionality is captured by the metric of fractional anisotropy (FA) with values ranging from 0 to 1 to indicate directional preference of diffusion (FA = 0 represents isotropic diffusion; FA = 1 represents preferred diffusion in a single direction, a.k.a. anisotropic diffusion). Magnitude of diffusion is captured by the metric of Mean Diffusivity (MD). Both of these metrics, FA and MD, are scalars and can be represented as spatial maps where each voxel is resampled and co-registered with the subject's own T2-weighted image series. This step results in the fusion of macrostructural information (high resolution anatomical series of T1-weighted, T2-weighted, and T2-FLAIR images) with microstructural information (tissue integrity characterized by FA and MD values). When these images are co-registered, can one begin to utilize a *multivariate* approach to elucidate the mechanisms underlying aberrant MD values (restricted diffusion due to cytotoxic edema vs increased extravascular protein bleeds; expanded diffusion due to vasogenic edema from blood–brain barrier disruption vs gross tissue disruptions due to primary injury or surgical intervention) and their impact on the structural integrity of white matter pathways connecting different brain regions.

While DTI assumes Gaussianity in its calculation of DTI metrics (FA and MD), most diffusion within the brain is non-Gaussian due to complexities in cellular microstructure. Microstructural complexities of cellular and subcellular tissues in the brain hinder water diffusion in a non-Gaussian fashion. Thus, a rapidly evolving imaging modality known as diffusion kurtosis imaging can be used to estimate microstructural tissue parameters without assuming Gaussian diffusion.

NOTE: In TBI patients, orbitofrontal and inferior temporal regions are commonly impacted by the primary injury. These same regions are particularly vulnerable to susceptibility distortions inherent to EPI-based spin echo sequences. Therefore, to estimate and correct susceptibility distortions in these regions of interest in TBI patients, a pair of spin echo sequences (matched geometry to the DTI sequence) with reversed phase encoding directions should be acquired (0′20″ each PE direction).

4 Rapidly Emerging Techniques

4.1 Diffusion Kurtosis Imaging (DKI)

Diffusion Kurtosis Imaging (DKI) is an extension of diffusion tensor imaging (DTI) which uses a special pulse sequence with a minimum of three *b*-values (compared to two *b*-values required for DTI). These multiple shells of *b*-values include higher *b*-values than typical DTI sequences as well as modified post-processing procedures of the complex imaging data obtained to assess deviation from Gaussian diffusion (extent of kurtosis), thereby quantifying structural complexity of different tissue types within the brain.

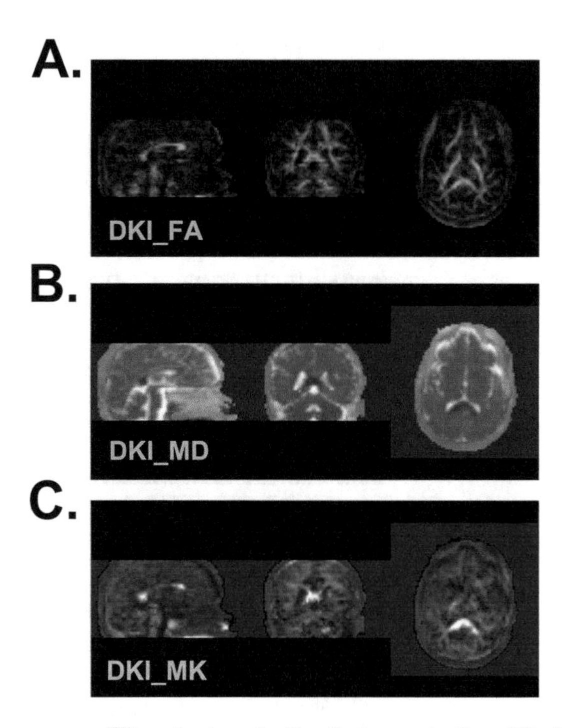

Fig. 4 Same severe TBI patient as in Fig. 2 demonstrating diffusion kurtosis images acquired during the same imaging session (8 days post injury). The mean diffusivity map (**b**) indicates low ADC values in the same posterior portion of the corpus callosum as diffusion tensor imaging in Fig. 2. Additional information is available in the mean kurtosis map (**c**) which indicates increased cellularity underlying the cytotoxic edema evident from the reduced ADC values in (**b**)

Elevated mean kurtosis values (MK) are thought to reflect increased cellularity, as one might expect intracellularly during cytotoxic edema, or extracellularly during reactive gliosis. Recent evidence in an animal model of TBI using controlled cortical impact suggests that DKI is sensitive to microstructural changes associated with reactive astrogliosis [14]. Specifically, in the sub-acute stage following injury (DTI metrics had already re-normalized), increased MK values corresponded to immunohistochemical evidence of increased reactive astrogliosis. Thus, a DKI sequence in addition to a DTI sequence has the potential to provide a more complete characterization of diffusion characteristics (and tissue microstructure) impacted by pathophysiological processes associated with secondary brain injury (Fig. 4).

4.1.1 Simultaneous Multi-slice (SMS) Acquisition

Over the past several years, parallel imaging and in-plane acceleration techniques have substantially reduced the time it takes to acquire sophisticated pulse sequences and reconstruct high resolution images on clinical MRI scanners. However, the recent development of SMS (aka multiband) acquisition methods is revolutionizing MRI protocols by reducing the time it takes to acquire high quality data by additional factors of 2–4 beyond the gains already achievable without SMS. Traditionally, MRI scanners acquire a single "slice" of the brain at a time before moving on to acquire the next "slice." SMS methods acquire several "slices" at the same time as specified by the multiband factor. Thus, if 60 slices are needed for whole brain coverage, one can use a multiband factor of 4 to acquire 4 slices at a time. In this scenario, one can obtain 60 slices in the same amount of scanner time as it would take to acquire 15 slices without SMS. These types of technological advances are narrowing the gap between "research" protocols and "clinical" protocols as clinical protocols have traditionally been "stripped down" due to pressing time constraints in order to be feasible in hospital settings.

5 Summary

Unraveling the secondary molecular cascades underlying secondary brain injury in TBI cases each is challenging, particularly when a broad range of edema, bleeds, perfusion deficits, and ischemia occurs across patients depending on the unique circumstances of each individual's injury. Precision medicine approaches to patient care should consider that targeted therapeutic action can have vastly different consequences depending on timing relative to injury and the underlying imaging phenotype of individual patients. Utilizing currently available noninvasive MRI sequences on clinical scanners in hospital settings to acquire multi-weighted and multimodal neuroimaging data in a reasonable timeframe has become not only feasible but also practical and should be part of routine clinical care of TBI patients. Not only does the integration of information from different image series assist with each physician's clinical assessment of TBI patients but also the development of the optimal treatment plan for each patient's imaging phenotype. Furthermore, these MRI protocols are feasible for progress monitoring of treatment efficacy. Thus, incorporation of these MRI protocols into routine clinical care of TBI patients will add tremendous prognostic value as we begin to utilize multivariate approaches, including quantitative neuroimaging data, to predict treatment response in each individual patient.

Acknowledgements

This work was supported by the following federal research grants awarded by the Department of Defense W81XWH-11-1-0460 and W81XWH-16-C-0040 as well as one award from NIH 5R01NS077963-05. Special thanks to Sef Romo and Raouf at Memorial Hermann Hospital in the Texas Medical Center for their outstanding MRI services and support of our research efforts.

References

1. Li X, Morgan PS, Ashburner J, Smith J, Rorden C (2016) The first step for neuroimaging data analysis: DICOM to NIfTI conversion. J Neurosci Methods 264:47–56
2. Bonfoco E, Krainc D, Ankarcrona M, Nicotera P, Lipton SA (1995) Apoptosis and necrosis: two distinct events induced, respectively, by mild and intense insults with N-methyl-D-aspartate or nitric oxide/superoxide in cortical cell cultures. Proc Natl Acad Sci U S A 92:7162–7166
3. Beauchamp MH, Ditchfield M, Babl FE, Kean M, Catroppa C, Yeates KO, Anderson V (2011) Detecting traumatic brain lesions in children: CT versus MRI versus Susceptibility Weighted Imaging (SWI). J Neurotrauma 28:915–927. https://doi.org/10.1089/neu.2010.1712
4. Tong KA, Ashwal S, Obenaus A, Nickerson JP, Kido D, Haacke EM (2008) Susceptibility-weighted MR imaging: a review of clinical applications in children. Am J Neuroradiol 29:9–17. https://doi.org/10.3174/ajnr.A0786
5. Grade M, Hernandez Tamames JA, Pizzini FB, Achten E, Golay X, Smits M (2015) A neuroradiologist's guide to arterial spin labeling MRI in clinical practice. Neuroradiology 57:1181–1202. https://doi.org/10.1007/s00234-015-1571-z
6. Ashwal S, Tong KA, Ghosh N, Bartnik-Olson-B, Holshouser BA (2014) Application of advanced neuroimaging modalities in pediatric traumatic brain injury. J Child Neurol 29:1704–1717. https://doi.org/10.1177/0883073814538504
7. Hergan K, Schaefer P, Sorensen A, Gonzalez R, Huisman T (2002) Diffusion-weighted MRI in diffuse axonal injury of the brain. Eur Radiol 12:2536–2541. https://doi.org/10.1007/s00330-002-1333-2
8. Brashdi YHA, Albayram MS (2015) Reversible restricted-diffusion lesion representing transient intramyelinic cytotoxic edema in a patient with traumatic brain injury. Neuroradiol J 28:409–412. https://doi.org/10.1177/1971400915598071
9. Hudak AM et al (2014) Cytotoxic and vasogenic cerebral oedema in traumatic brain injury: assessment with FLAIR and DWI imaging. Brain Inj 28:1602–1609. https://doi.org/10.3109/02699052.2014.936039
10. Muccio CF, Simone MD, Esposito G, De Blasio E, Vittori C, Cerase A (2009) Reversible post-traumatic bilateral extensive restricted diffusion of the brain. A case study and review of the literature. Brain Inj 23:466–472. https://doi.org/10.1080/02699050902841912
11. Newcombe VFJ et al (2013) Microstructural basis of contusion expansion in traumatic brain injury: insights from diffusion tensor imaging. J Cereb Blood Flow Metab 33:855–862. https://doi.org/10.1038/jcbfm.2013.11
12. Heiss WD (2016) Malignant MCA infarction: pathophysiology and imaging for early diagnosis and management decisions. Cerebrovasc Dis 41:1–7
13. Jones DK, Charles S, Williams R, Gasston D, Horsfield MA, Simmons A, Howard R (2002) Isotropic resolution diffusion tensor imaging with whole brain acquisition in a clinically acceptable time. Hum Brain Mapp 15:216–230. https://doi.org/10.1016/j.jneumeth.2016.03.001
14. Zhuo J, Xu S, Proctor JL, Mullins RJ, Simon JZ, Fiskum G, Gullapalli RP (2012) Diffusion kurtosis as an in vivo imaging marker for reactive astrogliosis in traumatic brain injury. NeuroImage 59:467–477. https://doi.org/10.1016/j.neuroimage.2011.07.050

Chapter 17

Dynamic Contrast-Enhanced MRI for the Analysis of Blood-Brain Barrier Leakage in Traumatic Brain Injury

Qiang Shen and Timothy Q. Duong

Abstract

Blood-brain barrier (BBB) could be impaired following traumatic brain injury. Here we describe a modified technique for noninvasive longitudinal assessment of the of BBB integrity based on dynamic contrast-enhanced magnetic resonance imaging technique to evaluate the longitudinal progression of BBB leakage following traumatic brain injury in rats.

Key words Blood-brain barrier, Magnetic resonance imaging, Traumatic brain injury, Dynamic contrast enhanced, Rodents

1 Introduction

Traumatic brain injuries (TBI) often involve vascular dysfunction that leads to long-term alterations in physiological and cognitive functions of the brain. The cerebrovascular dysfunctions caused by TBI could include cerebral blood flow alterations, autoregulation impairments, subarachnoid hemorrhage, vasospasms, blood-brain barrier disruption, and edema formation. The blood-brain barrier (BBB) plays a vital role in regulating the entry of blood-borne factors and circulating immune cells into the brain, hereby providing a highly stable biochemical environment for the normal functioning of neuronal cells. Perturbations in BBB function have been extensively documented after TBI in both experimental models and human patients [1]. Mechanical impact may induce immediate BBB damage; microbleeding, vascular inflammation, and secondary ischemia/hypoxia may induce prolonged BBB dysfunction [1–3]. The resulting BBB disruption can evolve dynamically in both time and space with studies showing multi-phasic characteristics [4]. BBB leakage can lead to an imbalance of electrolytes, edema formation, inflammation, etc., and ultimately delayed neuronal dysfunction and degeneration [1]. Studies have shown that

Amit K. Srivastava and Charles S. Cox, Jr. (eds.), *Pre-Clinical and Clinical Methods in Brain Trauma Research*, Neuromethods, vol. 139, https://doi.org/10.1007/978-1-4939-8564-7_17, © Springer Science+Business Media, LLC, part of Springer Nature 2018

posttraumatic BBB disruption is one of the major factors that contribute to increased severity of TBI [5]. BBB has also been suggested as a target for therapeutic invention [1, 6]. For example, free radical scavengers reduced Evans blue leakage in rat-controlled cortical impact [7], blocking the effects of inflammatory cytokines with neuregulin appears to decrease endothelial tight junction permeability [8], and deleting the gene for MMP9 improved outcomes in transgenic mice [9].

Quantitative assessment of BBB permeability is of particular importance for studying the disease pathophysiology and for optimizing therapeutic interventions in TBI. Traditional histological Evans Blue extravasation has been widely used to measure BBB leakage following TBI [4, 10, 11]. However, this method requires the sacrifice of the animals and does not permit for longitudinal assessments. Dynamic contrast-enhanced (DCE) MRI provides a promising alternative for noninvasive longitudinal assessment of the opening of BBB caused by many brain pathologies, such as tumors [12], multiple sclerosis [13], and acute ischemic strokes [14]. Disruption of the BBB can enable the extravasation of low-molecular weight MRI contrast agents. This accumulation of contrast material in the extravascular extracellular space (EES) of affected tissues leads to increased longitudinal relaxation rate and, therefore, increased signal intensity in T1-weighted images. Repeated acquisition of T1-weighted images following an intravenous injection of contrast agent, e.g. gadolinium-diethylenetriamine pentaacetic acid (Gd-DTPA), provides measurements of signal enhancement as a function of time. By fitting the mechanistic tracer kinetic models, DCE-MRI yields the volume transfer coefficient K^{trans}, the extracellular extravascular space v_e, etc. A consensus was reached that K^{trans} best reflects tissue permeability alterations and should be the primary end point of data fitting of DCE-MRI [15, 16].

In this paper, we describe a modified DCE-based K^{trans} measurement MRI protocol [17] and its application to study TBI in rats [18].

2 Materials

2.1 TBI Modeling

1. Rats (200–250 g) (Taconic Farms, Hudson, NY; Charles River, Wilmington, MA).
2. Anesthetics (isoflurane or pentobarbital, etc.) (VetOne, MWI, Boise, ID).
3. Common surgical tools and supplies (Fine Science Tools, Foster City, CA; Integra Miltex, Plainsboro, NJ; World Precision Instruments, Sarasota, FL; Ethicon, Somerville, NJ).

4. Pneumatic-controlled cortical impactor (Precision Systems and Instrumentation, Fairfax Station, VA).
5. Stereotaxic frame equipped with tooth and ear bars (myNeurolab.com, St. Louis, MO).
6. Bone wax (Ethicon, Somerville, NJ).
7. Antibiotic ointment (McKesson, Richmond, VA).
8. Analgesic (Buprenex) (Henry Schein, Melville, NY).
9. Warm pad, temperature feedback monitoring, and other monitoring equipment to ensure normal animal physiology (Fisher Scientific, Pittsburgh, PA; Cole-Palmer, Vernon Hills, IL).
10. Cresyl violet acetate for Nissl staining (Sigma, St. Louis, MO).

2.2 MRI

1. Bruker 7 T scanner (Billerica, MA).
2. 40-G/cm BGA12 gradient insert (ID = 12 cm, 120-μs rise time).
3. Animal holder (custom-made).
4. Custom-made RF transmitter and receiver coils for brain imaging.
5. Custom-made RF transmitter coil for arterial spin labeling.
6. Actively decoupled switch box to detune RF coils.
7. Other magnet, gradient, RF coil configurations should also work.

2.3 Peripheral MRI Compatible Monitor Equipment and Animal Supports

1. Oximetry (heart rate, arterial oxygen saturation)—(Mouse Ox, STARR Life Sciences, Oakmont, PA).
2. Blood pressure (invasive with artery catheterization)—(Biopac/Acknowledge, Goleta, CA).
3. Respiration rate via force transducer—(Biopac/Acknowledge, Goleta, CA).
4. Forepaw stimulation device—Home-made device or Grass stimulators.
5. Circulating warm water bath (Haake water bath, Rheology Solutions, Bacchus Marsh, Victoria, Australia; Cole Palmer, Vernon Hills, IL).
6. Temperature feedback regulator (Digi-Sense, Cole-Palmer, Vernon Hills, IL).
7. Anesthetic delivery, such as vaporizer—(Universal Vaporizer Support, Foster City, CA).

3 Methods

3.1 TBI Surgery [19–24]

1. Male rats (250–350 g) are anesthetized with isoflurane (~2%). Other anesthetics can be also be used. Male rats are often used to avoid the effects of female hormones on outcome. Female rats are also widely studied and some female hormones have been found to have neuroprotective effects.
2. Aseptic preparations (betadine and ethanol washes) should be performed to prevent infection and immunological responses that could affect outcome.
3. The animal is secured in a stereotaxic frame with ear and tooth bars and an incision is made at the level of the cerebellum as posterior from the impact site as possible to prevent artifacts during MRI acquisition. The periosteum is removed over the impact site. A Ø 5 mm craniotomy is created over the left S1FL (+0.25 mm anterior and 3.5 mm lateral to bregma), exposing the dura matter. The intact dura matter is impacted using a pneumatic-controlled cortical impactor (Precision Systems and Instrumentation, LLC, Fairfax Station, VA) fitted with a Ø 3 mm tip (5.0 m/s, 250 μs dwell time, 1 mm depth) to produce a mild focal TBI. Following the impact, the cranial opening is sealed with bone wax, the scalp sutured closed, and antibiotic ointment applied. Saline is injected under the skin to facilitate the removal of air pockets between the scalp and the skull to minimize artifacts during MRI acquisition (*see* **Note 1**). Buprenex (0.05 mg/kg) is given subcutaneously every 12 hours for 3 days as needed for pain.
4. The right femoral artery is catheterized for blood-gas sampling, continuous blood pressure and heart rate monitoring. These physiological parameters are important, because deviations could affect TBI outcome, increasing statistical scatters.
5. Rats are secured in a supine position on an MR-compatible rat stereotaxic headset, anesthesia is reduced to ~1.1% isoflurane. Rats breathe spontaneously. Mechanical ventilation can also be used. Rectal temperature should be maintained at 37.0 ± 0.5 °C. It is strongly suggested that heart rate, respiration rate, mean arterial blood pressure, oxygen saturation (from oximetry) are monitored. Blood gas should be sampled once during a break between imaging scans. All recorded physiological parameters are within normal physiological ranges.
6. Rectal temperature is maintained at 36.5–37.5 °C and respirations are recorded throughout the study. Body core temperature is critical, because it could affect outcome.

3.2 MRI

MRI was acquired on the day of the TBI procedure, at 1 and 3 h post TBI and on days 1, 2, 3, and 7 following TBI onset.

K^{trans} MRI data are acquired using a 2D multi-slice fast low angle shot (FLASH) sequence.

1. Prescan module (7.4 mins): to determine the flip angle and M0 distribution, includes three FLASH scans with different TRs: 64 ms (scan 1), 200 ms (scan 2) and 3000 ms (scan 3). The rest of imaging parameters are: five 1.0-mm coronal slices, TE = 2 ms, FOV = 2.2 × 2.2 cm^2, 128 × 128 data matrix, and 30° nominal flip angle.
2. Dynamic scans: TR = 64 ms, and otherwise identical sequence parameters as prescans. After baseline data are acquired for 2 mins, a bolus (0.2 mL/kg) of gadodiamide (GE Healthcare, USA) is injected intravenously through the tail vein, during which the dynamic scan is continued. A total of 90 dynamic images are acquired with a temporal resolution of 8 s, lasting 12 mins total.
3. Before K^{trans} MRI, cerebral blood flow (CBF) map can be measured using continuous arterial spin labeling (cASL) technique. Apparent diffusion coefficient (ADC) and fractional anisotropy (FA) can be measured using diffusion tension imaging (DTI) sequence. T_2 can be measured using multi-echo fast spin-echo sequence. Please see detail protocols in [18].

3.3 K^{trans} Mapping

The steady-state spoiled gradient echo (GRE, acquired using the 2D FLASH sequence) signal amplitude for a given TR (M_{TR}) can be related to M_0, R_1 and α as follows:

$$M_{TR} = M_0(r, s_{coil}) \cdot \sin\alpha \frac{1 - \exp(-TR \cdot R_1)}{1 - \exp(-TR \cdot R_1) \cdot \cos\alpha} \cdot \exp\left(-TE/T_2^*\right) \quad (1)$$

where M_0 is a function of both spatial location r and coil sensitivity s_{coil}. To minimize the R_2^* signal decay, a single-echo GRE scan was used with minimum TE and a low dose of gadolinium. As an approximation, $\exp\left(-TE/T_2^*\right) \approx 1$ was used in the subsequent calculation. A pre-scan module composed of three GRE scans was used to determine flip angle (α) and tissue magnetization (M_0).

For flip angle mapping, simulations using Eq. 1 show that the ratio of GRE magnitudes at TRs of 200 ms and 64 ms depends strongly on α, but very weakly on T_1. Given the narrow T_1 distribution of brain tissue, a raw α (or B_1^+) map can be obtained from the ratio map with a fixed T_1 value. Assuming α is a smooth function of location, the final α map can be obtained after smoothing.

For M_0 determination, according to Eq. 1, the GRE signal becomes less dependent on T_1 with increasing TR. With a long TR of 3000 ms, GRE magnitude depends primarily on sin(α) for small to medium flip angles (α <45°), but very weakly on T_1. Given the normal flip angle of 30°, the actual flip angle of the brain tissue is within the range of 15–45°; therefore, M_0 can be determined from the long TR scan and the flip angle distribution as follows:

$$M_0 = M_{3000\mathrm{ms}} / \sin \alpha \tag{2}$$

Given the smooth α distribution, high-SNR M_0 map, and dynamic scans using sensitive surface coil at high field, dynamic R_1 map can be obtained using Eq. 1 with sufficient SNR. The baseline R_1 was subtracted from the time series to calculate the ΔR_1 maps, which are linearly related to the changes of contrast agent concentration.

The AIF was determined using the data and the scaling approach by Ewing and colleagues [25, 26]. Briefly, the mean AIF ($\mathrm{AIF}_{\mathrm{mean}}$) was measured in a group of male rats (approximately 300 g) using the custom-synthesized radiolabeled Gd-DTPA [25]. Assuming the plasma volume was 1% and there is no BBB leakage in the contralesional side of caudate-putamen (CPU), the AIF was determined using the following scaling:

$$\mathrm{AIF}(t) = \mathrm{AIF}_{\mathrm{mean}} \cdot \left[100 \cdot \int_{t=3\mathrm{min}}^{9\mathrm{min}} R_{1,\mathrm{CPU}}(t)\mathrm{d}t\right] \Big/ \left[\int_{t=3\mathrm{min}}^{9\mathrm{min}} \mathrm{AIF}_{\mathrm{mean}}\mathrm{d}t\right] \tag{3}$$

The R_1 and the AIF were then used to fit the extended Kety model:

$$C_{\mathrm{t}}(t) = K^{\mathrm{trans}} \int_{\tau=0}^{t} C_{\mathrm{p}}(\tau)\mathrm{e}^{-k_{\mathrm{ep}}(t-\tau)} d\tau + v_{\mathrm{p}} C_{\mathrm{p}}(t) \tag{4}$$

where $C_{\mathrm{t}}(t)$ and $C_{\mathrm{p}}(t)$ are tissue and plasma gadolinium concentrations, and k_{ep} is the reversible mass transfer coefficient. In this study, the R_1 and scaled AIF values were directly used for data fitting without further conversion to concentrations using the relativity of contrast agent. Since it is difficult to obtain acceptable K^{trans} maps by simultaneously fitting all three parameters, we used the model selection approach by Ewing and colleagues [26] to select a simpler model that could sufficiently describe the dynamic contrast change. Data was fit to the extended Kety model with the two following assumptions as described by Ewing and colleagues [26]:

Irreversible leakage (two-parameter model, or Model 2):

$$K^{\mathrm{trans}} > 0, \qquad k_{\mathrm{ep}} = 0 \tag{5}$$

Reversible leakage (three-parameter model, or Model 3):

$$K^{\mathrm{trans}} > 0, \qquad k_{\mathrm{ep}} > 0$$

For model selection, the irreversible model is selected, unless the reversible model yields a statistically significant better fit. The cut-off criteria use the F-statistic, which can be calculated using the summed squared residues (SSE) and the number of samples (N) is as follows:

$$F = [(\mathrm{SSE}_3 - \mathrm{SSE}_2)/1]/[\mathrm{SSE}_3/(N - 3)] \tag{6}$$

where the subscript represents the corresponding model. The final selection mask is determined as:

$$M_{\mathrm{select}} = F > F_0 \tag{7}$$

The threshold for F-statistic (F_0) was set as 10 ($P < 0.05$). The final K^{trans} map was then determined as:

$$K^{\mathrm{trans}}_{\mathrm{final}} = K^{\mathrm{trans}}_{\mathrm{Model2}} \cdot (1 - M_{\mathrm{select}}) + K^{\mathrm{trans}}_{\mathrm{Model3}} \cdot M_{\mathrm{select}} \tag{8}$$

All the calculations were performed using Matlab R2011b (Mathworks, Natick, MA).

Figure 1 shows representative single slice, K^{trans}, CBF, T_2, ADC, and FA images from a single animal at 1 and 3 h, 1, 2,

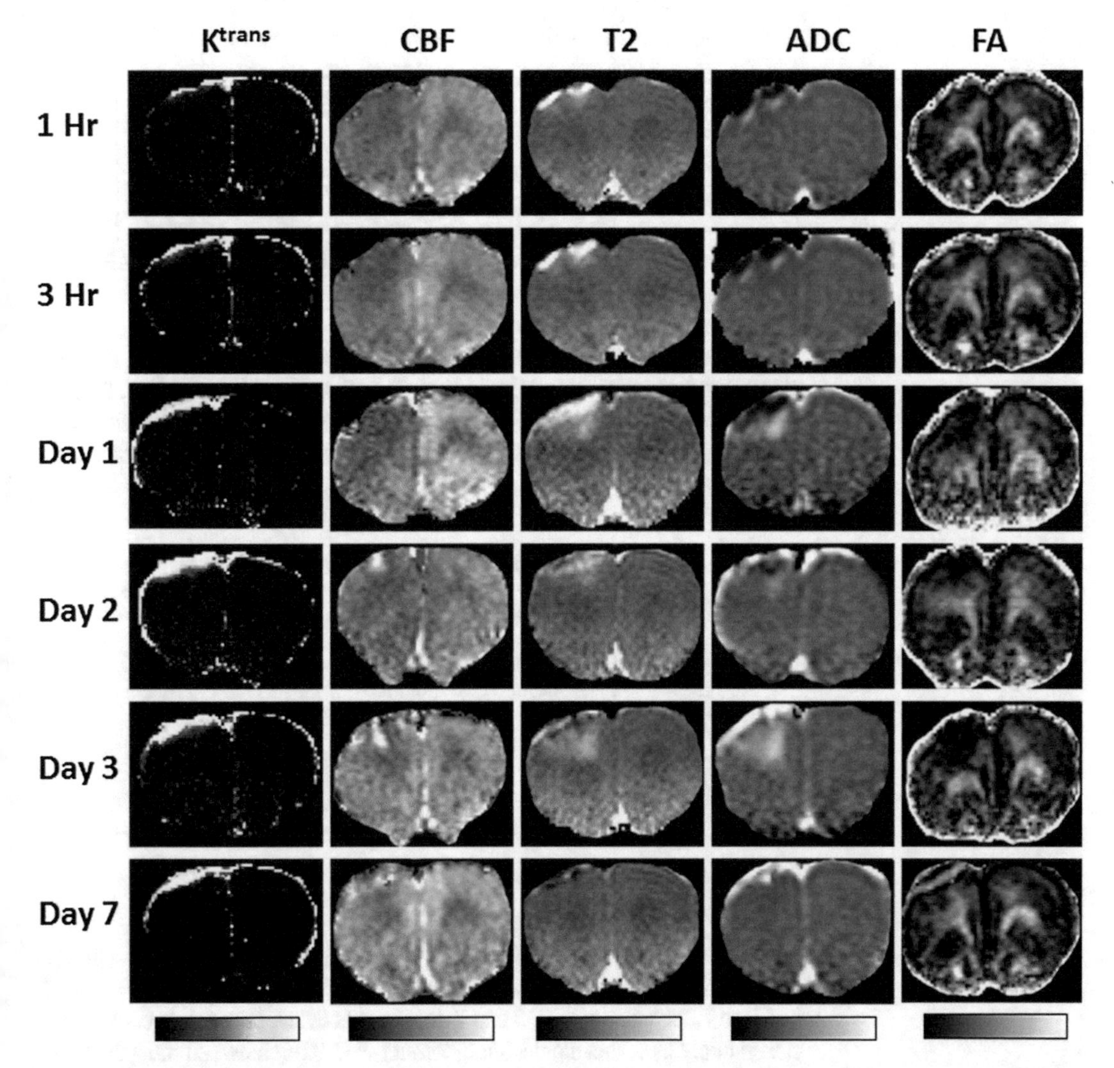

Fig. 1 The representative maps of K^{trans}, CBF, T_2, ADC, and FA at different time points. Color Scale Bar for K^{trans}: 0–0.05 min^{-1}; Gray-scale Bar for CBF: 0–2.5 mL/g/min, T_2: 30–100 ms, ADC: 0.0004–0.0013 mm^2/s, FA: 0.1–0.6

3 and 7 days post-TBI. In the ipsilateral hemisphere, K^{trans} map showed hyperintensity, most prominent on the cortical surface of the impacted area (bright yellow and white pixels), indicative of disrupted BBB integrity. Abnormal K^{trans} was apparent at 1 h, and peaked on day 2–3 post-TBI and returning toward (but did not reach) normal on day 7.

Spatial and temporal characteristics of K^{trans} in TBI rats can also be analyzed. Group analysis results were show in Figs. 2 and 3. The ipsilesional K^{trans} changes were localized to the superficial layers from the surface to approximately 1 mm in depth. There were

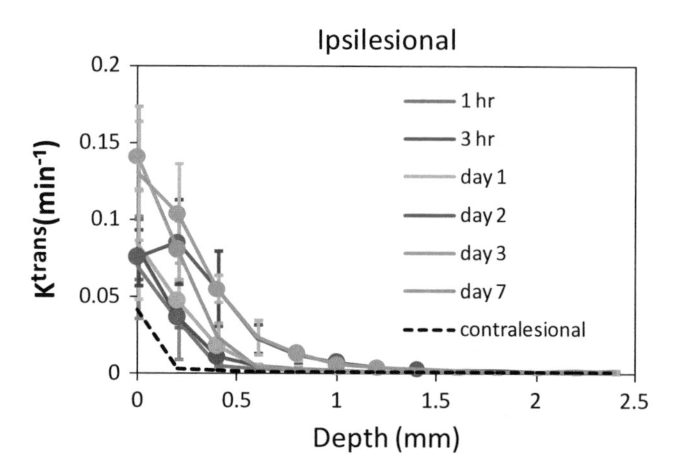

Fig. 2 Spatial profiles from cortical surface to corpus callosum of K^{trans}. Data was represented as mean ± SEM. $n = 5$, 4, 7, 6, 5, and 7, for 1 h, 3 h, 1, 2, 3 and 7 days, respectively. Data points displayed as closed circle indicates significant difference ($P < 0.05$) between the ipsilesional and contralesional cortex at the corresponding time points

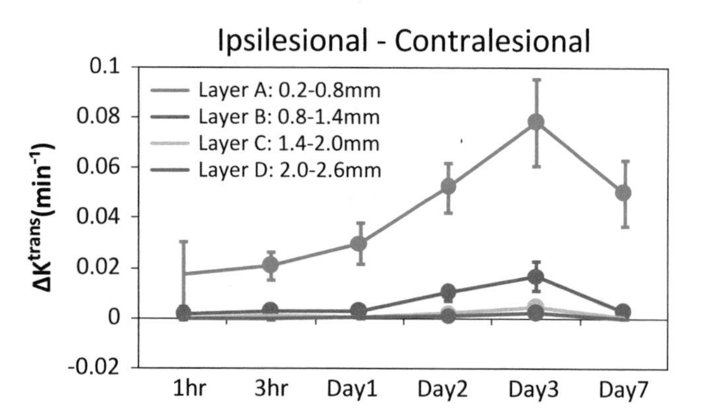

Fig. 3 Temporal profiles of ΔK^{trans} (difference between ipsilesional and contralesional regions). Data was represented as mean ± SEM. $n = 5$, 4, 7, 6, 5, and 7, for 1 h, 3 h, 1, 2, 3 and 7 days, respectively. Data points displayed as closed circle indicates significant different ($P < 0.05$) between the ipsilesional and contralesional cortex at the corresponding time points

significant differences between ipsilesional and contralesional data points as displayed as closed circle ($P < 0.05$) at the corresponding time points. The K^{trans} for the ROI at the superficial layer (0–0.6 mm) was significantly different from zero at 1 h ($P < 0.05$), and continued to increase peaking on day 3 ($P < 0.05$) and then subsequently decreased toward normal at day 7. The K^{trans} for the ROI immediately underneath it (0.6–1.2 mm) started from no difference in K^{trans}, peaked on day 3, and end with no difference in K^{trans} on day 7. The K^{trans} of two subsequent layers (1.2–2.4 mm) showed no difference in K^{trans} at all the time points.

4 Notes

1. Following the TBI impact, the cranial opening needs to be sealed with bone wax and then the scalp sutured closed. The blood, if any, needs to be cleared. Saline should be injected under the skin to facilitate removal of air pockets between the scalp and the skull to minimize artifacts during MRI acquisition. These steps (using bone wax and saline) are very important to avoid MRI artifacts.
2. Gd-DTPA or gadodiamide will change the relaxation time (T_1 and T_2) of blood and tissue. Therefore CBF will not be accurately obtained using cASL after using Gd-DTPA or gadodiamide.
3. For the flip angle mapping and the following dynamic scans, the same sequence should be used to avoid different distortions associated with different sequences.
4. Accurate determination of AIF for preclinical MRI is nontrivial, given the fast arterial blood flow and the small dimensions of rodent brains. As such, different strategies have been proposed to mitigate this difficulty. For example, AIF determined from other imaging modalities has been used for fitting K^{trans} [27], alternative reference region approaches have been developed to eliminate the use of AIF [28], and venous blood signals have also been employed to determine the vascular input function (VIF) instead of AIF [29]. In this study, we adopted the established approach by Ewing and colleagues [25, 26], and used a scaled group average AIF determined previously from radio-labeled Gd-DTPA. Although we could obtain similar VIF profiles from the veins to cross-validate the scaling approach, it should be noted that the accuracy of the VIF was poor due to the wash-in effects and its reproducibility could not be guaranteed, so we chose to utilize the scaled group AIF in the model fitting.

5. To achieve the desirable image quality for K^{trans}, we used model selection as described by Ewing and colleagues [26]. For TBI, the reversible leakage model was necessary for the focal lesion. Significant underestimation was observed if the reversible leakage was neglected. This result indicated that the K^{trans} was sufficiently large, so that the contrast agent can accumulate to a significant amount in the extracellular extravascular space to allow a significant reversible flux of contrast agent back to the plasma. In other region, the irreversible leakage model was sufficient, which is consistent with the existing knowledge that intact brain tissue has negligible contrast agent leakage. Overall, the model selection approach was necessary to ensure that K^{trans} was not underestimated in regions with significant reversible leakage and provided sufficient SNR for the whole brain.
6. Look-Locker type of sequences have also been previously used in DCE-MRI studies [30]. While such methods allow independent R_1 mapping at each time point, it has much lower temporal resolution and lower SNR compared with the gradient-echo methods used in this and other studies [29, 31]. While some recent studies achieved higher temporal resolution (4–6 s per image), the 8-s temporal resolution for the dynamic scan in this study was sufficient for TBI studies, since the BBB leakage is generally much lower than in tumors [29, 31]. The slightly lower temporal resolution allows higher spatial resolution and more spatial coverage, which is desirable for TBI studies.
7. Our motivation to further develop the K^{trans} MRI method using a surface coil to quantify BBB permeability was to obtain a method that would also be compatible with arterial spin labeling measurements. Arterial spin labeling uses a separate neck coil to measure cerebral blood flow [32–34]. There is substantial evidence that blood flow is markedly perturbed following TBI [18, 21] but CBF is seldom measured in TBI. Acute and chronic perfusion abnormality could have a negative impact on tissue viability following TBI.

5 Conclusion

We developed a K^{trans} MRI method for measuring BBB permeability using a surface coil and a gradient-echo-based pulse sequence. The obtained K^{trans} map provided excellent SNR, high spatial resolution and spatial coverage. This technique was applied to evaluate the longitudinal progression of BBB leakage following TBI in rats. Spatiotemporal evolution of K^{trans} was evaluated. The K^{trans} increase was primarily localized to the superficial cortical layers that were adjacent to the impact site, which were different from the much deeper and widespread changes of CBF disturbance, edema formation, and diffusion abnormalities. Temporally, the K^{trans} increase was present at 1 h, increased progressively over time, peaked at day 3 and then recovered towards normal at 7 days.

References

1. Chodobski A, Zink BJ, Szmydynger-Chodobska J (2011) Blood–brain barrier pathophysiology in traumatic brain injury. Transl Stroke Res 2(4):492–516
2. Donkin JJ, Vink R (2010) Mechanisms of cerebral edema in traumatic brain injury: therapeutic developments. Curr Opin Neurol 23 (3):293–299. https://doi.org/10.1097/WCO.0b013e328337f451
3. Alves JL (2014) Blood-brain barrier and traumatic brain injury. J Neurosci Res 92 (2):141–147. https://doi.org/10.1002/jnr.23300
4. Başkaya MK, Muralikrishna Rao A, Doğan A, Donaldson D, Dempsey RJ (1997) The biphasic opening of the blood–brain barrier in the cortex and hippocampus after traumatic brain injury in rats. Neurosci Lett 226 (1):33–36
5. Neuwelt E, Abbott NJ, Abrey L, Banks WA, Blakley B, Davis T, Engelhardt B, Grammas P, Nedergaard M, Nutt J (2008) Strategies to advance translational research into brain barriers. Lancet Neurol 7(1):84–96
6. Shlosberg D, Benifla M, Kaufer D, Friedman A (2010) Blood–brain barrier breakdown as a therapeutic target in traumatic brain injury. Nat Rev Neurol 6(7):393–403
7. Smith SL, Andrus PK, Zhang JR, Hall ED (1994) Direct measurement of hydroxyl radicals, lipid peroxidation, and blood-brain barrier disruption following unilateral cortical impact head injury in the rat. J Neurotrauma 11 (4):393–404. https://doi.org/10.1089/neu.1994.11.393
8. Lok J, Zhao S, Leung W, Seo JH, Navaratna D, Wang X, Whalen MJ, Lo EH (2012) Neuregulin-1 effects on endothelial and blood-brain-barrier permeability after experimental injury. Transl Stroke Res 3(Suppl 1): S119–S124. https://doi.org/10.1007/s12975-012-0157-x
9. Wang X, Jung J, Asahi M, Chwang W, Russo L, Moskowitz MA, Dixon CE, Fini ME, Lo EH (2000) Effects of matrix metalloproteinase-9 gene knock-out on morphological and motor outcomes after traumatic brain injury. J Neurosci 20(18):7037–7042
10. Adelson PD, Whalen M, Kochanek P, Robichaud P, Carlos T (1998) Blood brain barrier permeability and acute inflammation in two models of traumatic brain injury in the immature rat: a preliminary report. Acta Neurochir Suppl 71:104–106
11. Dempsey RJ, Baskaya MK, Dogan A (2000) Attenuation of brain edema, blood-brain barrier breakdown, and injury volume by ifenprodil, a polyamine-site N-methyl-D-aspartate receptor antagonist, after experimental traumatic brain injury in rats. Neurosurgery 47 (2):399–406
12. Singh A, Haris M, Rathore D, Purwar A, Sarma M, Bayu G, Husain N, Rathore RK, Gupta RK (2007) Quantification of physiological and hemodynamic indices using T (1) dynamic contrast-enhanced MRI in intracranial mass lesions. J Magn Reson Imaging 26 (4):871–880. https://doi.org/10.1002/jmri.21080
13. Jelescu IO, Leppert IR, Narayanan S, Araujo D, Arnold DL, Pike GB (2011) Dual-temporal resolution dynamic contrast-enhanced MRI protocol for blood-brain barrier permeability measurement in enhancing multiple sclerosis lesions. J Magn Reson Imaging 33(6):1291–1300. https://doi.org/10.1002/jmri.22565
14. Kassner A, Roberts T, Taylor K, Silver F, Mikulis D (2005) Prediction of hemorrhage in acute ischemic stroke using permeability MR imaging. AJNR Am J Neuroradiol 26 (9):2213–2217
15. Tofts PS (2010) T1-weighted DCE imaging concepts: modelling, acquisition and analysis. Signal 500(450):400
16. Leach MO, Brindle K, Evelhoch J, Griffiths JR, Horsman MR, Jackson A, Jayson GC, Judson IR, Knopp M, Maxwell RJ (2005) The assessment of antiangiogenic and antivascular therapies in early-stage clinical trials using magnetic resonance imaging: issues and recommendations. Br J Cancer 92(9):1599–1610
17. Li W, Long JA, Watts LT, Jiang Z, Shen Q, Li Y, Duong TQ (2014) A quantitative MRI method for imaging blood-brain barrier leakage in experimental traumatic brain injury. PLoS One 9(12):e114173. https://doi.org/10.1371/journal.pone.0114173
18. Li W, Watts L, Long J, Zhou W, Shen Q, Jiang Z, Li Y, Duong TQ (2016) Spatiotemporal changes in blood-brain barrier permeability, cerebral blood flow, T2 and diffusion following

mild traumatic brain injury. Brain Res 1646:53–61. https://doi.org/10.1016/j.brainres.2016.05.036
19. Watts LT, Long JA, Chemello J, Van Koughnet S, Fernandez A, Huang SL, Shen Q, Duong TQ (2014) Methylene blue is neuroprotective against mild traumatic brain injury. J Neurotrauma 31(11):1063–1071. https://doi.org/10.1089/neu.2013.3193
20. Long JA, Watts LT, Chemello J, Huang SL, Shen Q, Duong TQ (2015) Multiparametric and longitudinal MRI characterization of mild traumatic brain injury in rats. J Neurotrauma 32(8):598–607. https://doi.org/10.1089/neu.2014.3563
21. Long JA, Watts LT, Li W, Shen Q, Muir ER, Huang S, Boggs RC, Suri A, Duong TQ (2015) The effects of perturbed cerebral blood flow and cerebrovascular reactivity on structural MRI and behavioral readouts in mild traumatic brain injury. J Cereb Blood Flow Metab 35(11):1852–1861. https://doi.org/10.1038/jcbfm.2015.143
22. Talley Watts L, Long JA, Manga VH, Huang S, Shen Q, Duong TQ (2015) Normobaric oxygen worsens outcome after a moderate traumatic brain injury. J Cereb Blood Flow Metab 35(7):1137–1144. https://doi.org/10.1038/jcbfm.2015.18
23. Talley Watts L, Shen Q, Deng S, Chemello J, Duong TQ (2015) Manganese-enhanced magnetic resonance imaging of traumatic brain injury. J Neurotrauma 32(13):1001–1010. https://doi.org/10.1089/neu.2014.3737
24. Watts LT, Long JA, Boggs RC, Manga H, Huang S, Shen Q, Duong TQ (2015) Methylene blue improves lesion volume, multiparametric quantitative MRI measurements, and behavioral outcome following TBI. J Neurotrauma 33(2):194–202. https://doi.org/10.1089/neu.2015.3904
25. Nagaraja TN, Karki K, Ewing JR, Divine GW, Fenstermacher JD, Patlak CS, Knight RA (2010) The MRI-measured arterial input function resulting from a bolus injection of Gd-DTPA in a rat model of stroke slightly underestimates that of Gd-[14C]DTPA and marginally overestimates the blood-to-brain influx rate constant determined by Patlak plots. Magn Reson Med 63(6):1502–1509. https://doi.org/10.1002/mrm.22339
26. Ewing JR, Bagher-Ebadian H (2013) Model selection in measures of vascular parameters using dynamic contrast-enhanced MRI: experimental and clinical applications. NMR Biomed 26(8):1028–1041. https://doi.org/10.1002/nbm.2996
27. Durukan A, Marinkovic I, Strbian D, Pitkonen M, Pedrono E, Soinne L, Abo-Ramadan U, Tatlisumak T (2009) Post-ischemic blood–brain barrier leakage in rats: one-week follow-up by MRI. Brain Res 1280 (0):158–165. https://doi.org/10.1016/j.brainres.2009.05.025
28. Yankeelov TE, Cron GO, Addison CL, Wallace JC, Wilkins RC, Pappas BA, Santyr GE, Gore JC (2007) Comparison of a reference region model with direct measurement of an AIF in the analysis of DCE-MRI data. Magn Reson Med 57(2):353–361. https://doi.org/10.1002/mrm.21131
29. Pike MM, Stoops CN, Langford CP, Akella NS, Nabors LB, Gillespie GY (2009) High-resolution longitudinal assessment of flow and permeability in mouse glioma vasculature: sequential small molecule and SPIO dynamic contrast agent MRI. Magn Reson Med 61 (3):615–625. https://doi.org/10.1002/mrm.21931
30. Ewing JR, Brown SL, Lu M, Panda S, Ding G, Knight RA, Cao Y, Jiang Q, Nagaraja TN, Churchman JL (2005) Model selection in magnetic resonance imaging measurements of vascular permeability: Gadomer in a 9L model of rat cerebral tumor. J Cereb Blood Flow Metab 26(3):310–320
31. Aryal MP, Nagaraja TN, Keenan KA, Bagher-Ebadian H, Panda S, Brown SL, Cabral G, Fenstermacher JD, Ewing JR (2014) Dynamic contrast enhanced MRI parameters and tumor cellularity in a rat model of cerebral glioma at 7 T. Magn Reson Med 71(6):2206–2214. https://doi.org/10.1002/mrm.24873
32. Duong TQ (2007) Cerebral blood flow and BOLD fMRI responses to hypoxia in awake and anesthetized rats. Brain Res 1135 (1):186–194. https://doi.org/10.1016/j.brainres.2006.11.097
33. Shen Q, Ren H, Fisher M, Bouley J, Duong TQ (2004) Dynamic tracking of acute ischemic tissue fates using improved unsupervised ISODATA analysis of high-resolution quantitative perfusion and diffusion data. J Cereb Blood Flow Metab 24(8):887–897
34. Shen Q, Fisher M, Sotak CH, Duong TQ (2004) Effects of reperfusion on ADC and CBF pixel-by-pixel dynamics in stroke: characterizing tissue fates using quantitative diffusion and perfusion imaging. J Cereb Blood Flow Metab 24(3):280–290

Chapter 18

Assessments for Quantifying Neuromotor Functioning After Repetitive Blast Exposure

Christopher K. Rhea, Nikita A. Kuznetsov, W. Geoffrey Wright, F. Jay Haran, Scott E. Ross, and Josh L. Duckworth

Abstract

Blast exposure may result in associated head trauma throughout the Traumatic Brain Injury (TBI) spectrum—ranging from weapon fire resulting in sub-concussive exposure and a mechanotransductive physiologic response to improvised explosive device (IED) detonation resulting in moderate and severe TBI associated with tissue-level disruption. Head trauma—regardless of severity—can result in changes to neurological functioning, which may alter neuromotor performance. Thus, measurement of neuromotor performance has been commonly used as a way to assess and track changes in functioning after head trauma. A number of subjective assessments have been developed over the years to help clinicians and researchers measure changes in neuromotor performance. In recent years, technological advances have led to more portable and cost-effective tools to objectively measure neuromotor performance, reducing the human error associated with subjective assessment. This chapter reviews relevant subjective and objective neuromotor assessments commonly used with populations who have head trauma.

Key words Neuromotor, Balance, Postural control, Gait, Concussion, Sub-concussion, mTBI, Blast exposure

1 Introduction

A traumatic brain injury (TBI) is defined by the Centers for Disease Control and Prevention (CDC) as trauma to the head that results in the disruption of normal brain functioning. There are several classifications that describe the head trauma based on level of injury. If the head trauma event primarily only caused a mechanotransductive physiological disruption, then it is classified as a sub-concussive event. If the head trauma event caused ultrastructural dysfunction, it is classified as a mild traumatic brain injury (mTBI) or concussion. Lastly, if tissue-level injury occurred, then it is classified as a moderate or severe TBI.

The mechanisms of injury to brain tissue and/or physiology depends on the etiology of the head trauma. When the head is

Amit K. Srivastava and Charles S. Cox, Jr. (eds.), *Pre-Clinical and Clinical Methods in Brain Trauma Research*, Neuromethods, vol. 139, https://doi.org/10.1007/978-1-4939-8564-7_18,

exposed to a blast, the primary blast wave propagates through the brain. While the exact effects of the primary blast wave on brain structure and physiology are not fully detailed, it is postulated that the transmission of the blast wave results in protein conformational changes and alterations in the viscoelastic properties of the cell and cell membrane, such as the cell membrane along the neuronal axon, resulting in loss of responsive compensation by the cell, which effectively increases the potential for injury at any given level of secondary acceleration or tertiary impact forces. Alternatively, blunt-force head trauma is defined as the head suddenly accelerating or decelerating, causing the brain to slosh within the skull [1]. This sudden movement of the brain inside the skull is associated with primary and secondary injury mechanisms [2–5]. Primary injury occurs due to the immediate mechanical effects of relative movement of brain tissues leading to diffuse axonal injury within subcortical tracts as well as blood vessel damage. Secondary injury takes place due to a physiological cascade of processes in response to the primary trauma, such as bruising and inflammation, alteration of neuronal cell physiology and subsequent neuronal degeneration, alteration of blood flow and cerebrospinal fluid flow, and increased glial network activity, among many other processes (for a review, see [2]).

While blunt-force trauma is the leading cause of TBI (including mTBI) in the general population, there are a number of blast-exposed professions associated with military occupations that can lead to head trauma, such as Special Forces, Infantry, Military Police, Transportation, and Explosive Ordinance Disposal (EOD) personnel, leading to mTBI being characterized as the "signature wound" of recent conflicts [6]. Exposure to a blast can be intentional, such as someone who works with explosive weapons as part of their job, or unintentional, such as the blast wave from an improvised explosive device (IEDs). Over 70% of military personnel who experienced head trauma while being deployed for a year between 2001 and 2008 reported that the mechanism of injury was from a blast [7]. In addition, service members also experience repetitive low-level blast exposure during training exercises [8–12]. The prevalence of blast exposure and role of blasts on brain tissue and human behavior has been studied [5, 13–19], but it is also uniquely challenging [20]. This has led to high-fidelity simulations to better understand the relation between the blast load and brain injury [21], and area of study that continues to grow. Given the large number of military personnel exposed to head trauma, significant resources have been devoted to study TBI from both blunt-force and blast etiologies, including the NCAA•DoD Grand Alliance CARE Consortium [22], the TBI Endpoints Development (TED) Initiative [23], and the Chronic Effects of Neurotrauma Consortium (CENC) [24].

Due to the mechanics of how blast waves interact with the brain, what has previously been learned from blunt-force TBI (primarily in civilian athletes) may not universally apply to blast-related head trauma [25, 26]. Blast exposure presents the potential risk for periventricular injury, rather than the diffuse axonal injury from direct impact that results from blunt-force trauma, which can cause a different neural cascade and may potentially result in a different presentation of sequelae [26–28]. Further, recent research has shown that blast exposure can lead to scarring across multiple interfaces in the brain, including the subpial glial plate, penetrating cortical blood vessels, grey-white matter junctions, and in the structures lining the ventricles [29]. Taken together, this evidence suggests that TBI from blast exposure may have a different neural cascade, time course, and/or outcomes relative to blunt-force head trauma.

However, there are some similarities between blast-related and blunt-force head trauma. One study showed strong similarities in chronic traumatic encephalopathy (CTE)—a neuropathology describing an ensemble of brain lesions from previous head trauma [30]—between post-mortem veterans with a history of blast exposure and elite athletes with a history of mTBI (amateur football and pro wrestling) [31]. Additionally, there is no strong evidence that injury due to blast-exposure mechanisms results in categorically different types or levels of symptoms than other head injury mechanisms [32–34]. Cognitive, affective, sensory, and somatic measures have been examined, largely using self-report questionnaires and this evidence suggests that the similarities are much greater than the differences when comparing patients with blast-related versus blunt-force trauma. There is some evidence of marginally increased incidence of post-traumatic stress disorder for blast-related mTBI [19], which highlights the importance of using diagnostic measures that are not confounded by neuropsychological comorbidities. In light of current evidence that suggests blast versus non-blast related mTBI symptoms overlap considerably, and the lack of evidence to suggest that sensorimotor deficits differ between head trauma etiologies, neuromotor assessments that were developed to help quantify and track TBI symptoms stemming from blunt-force head trauma are likely also appropriate for blast-related TBI populations.

1.1 Concussive vs. Sub-concussive Head Trauma

There is much literature describing and tracking the negative changes in behavioral and mechanistic variables after a concussion [22, 35–38]. This is certainly an important area of research, as we are beginning to understand the short-term and long-term consequences of concussions. Recent research has shown that just one concussion can lead to a suicide rate three times higher than those with no history of concussion [39]. This justifies large-scale efforts to better characterize the role of concussions on health-related behavior, such as the NCAA•DoD CARE study [22].

While concussions have received the majority of media and research attention, the role of sub-concussive head trauma is a growing area of interest [8, 40–46]. This is due to the fact that sub-concussive head trauma in sporting activities and occupational environments are much more prevalent than concussive head trauma. For example, male collegiate football players receive approximately 1000 head impacts throughout a season and only a very small number of them lead to a concussion, classifying the majority of their head trauma as sub-concussive [40]. Both short-term [8] and long-term [47] negative consequences on behavior have been identified after sub-concussive head trauma, justifying the need to include this type of head trauma in TBI research. It should be noted that while sub-concussive and repetitive low-level blast exposure is conceptually similar to repetitive blunt-force trauma, it has distinct physiological and pathological features. However, clinical presentation between the two etiologies may be similar, such as neuromotor dysfunction that can be tested with balance and gait tasks.

1.2 The Role of Sex and Age on Behavior After Head Trauma

Head trauma research has been primarily focused on males. However, sex-specific studies are warranted due to factors that typically separate men and women that can influence the response to head trauma. For example, female hormones can affect recovery from head trauma [48] and females have been shown to have a greater cortical thickness relative to men [49]. These observations may partially account for the lack of homogenous outcomes between the sexes after head trauma [50–52].

It is also well established that the brain and nervous system change with age, and increased age in adults can lead to increased sensitivity to head trauma [53]. It has been suggested that repeated sub-concussive head trauma may contribute to a decline in cognitive function similar to that which is seen with aging [54]. Thus, it is plausible that the interaction between age and sex may lead to a synergistic effect that negatively influences health-related outcomes after head trauma [52]. However, there is very little data documenting the role of sex and age on behavior after head trauma, despite an increase in the number of TBIs received by both males and females across ages, and a more acute increase in the 25–44 year old age range for both sexes over the past decade [55]. This is especially true for people exposed to blast-related trauma, as virtually no sex or age-related head trauma studies exist for this population. With the military recently announcing that females are eligible for combat-related jobs and an increase in the number of females entering occupations where blunt-force or blast-related head trauma may occur, such as federal and local law enforcement [56], examining the role of sex and age on behavior after head trauma is ripe for future research.

1.3 Behavioral and Physiological Changes After Head Trauma

Changes in behavior after head trauma are precipitated by a post-trauma pathophysiological neural cascade that has an acute (immediately after injury), subacute (up to 3 month post-injury), and chronic (greater than 3 month post-injury) time course [57, 58]. Immediately after head trauma, changes occur at the level of neuron neurophysiology (neurotransmitter release, across-membrane ion distribution, and free-radical interference), as well as glucose metabolism depression, which leads to a depletion of ATP required for neuronal function [2]. In the acute phase, this neural cascade can lead to behavioral changes, depending on the severity or repetition of the head trauma and the resulting neural response, and these behavioral changes can persist into the subacute and chronic phases. These physiological changes characterized by the neural cascade can led to behavioral changes that may present in a variety of ways. A behavioral change that is a cardinal symptom of head trauma is altered neuromotor functioning, commonly assessed through a static or dynamic balance test. For this reason, the latest Consensus Statement on Concussion in Sport recommends assessments of balance to help with concussion diagnosis and return-to-activity decisions [59]. The next section highlights common ways that neuromotor performance has been measured in the context of assessment and tracking of behavior after head trauma.

1.4 Neuromotor Assessment

Neuromotor performance reflects the integration and processing of sensory information to complete a motor task. Balance tasks have a long history of being used to measure neuromotor performance due to their strong sensitivity to neurological injury and disease [60–62]. Depending on the context, there are a number of characteristics that should be considered when selecting the appropriate neuromotor assessment.

First, the administrator must decide if they desire an objective or subjective assessment. Many clinical tests of balance are subjective, relying on the clinician to judge a patient's balance ability. Alternatively, neuromotor performance can be assessed objectively by quantifying balance control through the use of a device—typically a force plate, accelerometer, or inertial sensor. Objective assessment removes the element of human bias and increases accuracy of the measurement, so long as the device is adequately calibrated. Thus, objective neuromotor assessment may be appropriate if measurement accuracy is desired. However, objective assessment typically requires time, financial, and/or human resources that may not be available in all settings. Therefore, subjective assessment, which relies on little-to-no equipment and minimal training, may be more appropriate for some settings, though it should be noted that most subjective assessments become more reliable when more training and experience is provided to the administrator. A challenge for in-theater environments in military settings is that many of the medics/corpsmen will have limited training and even more limited experience assessing service members in a battle zone, creating a need for objective and portable assessment tools.

Second, regardless of whether the administrator selects an objective or subjective assessment, it should be reliable (consistently measure performance the same way) and valid (accurately measure performance) in order to have clinical utility. The reliability and validity of some subjective tests of neuromotor performance have been questioned [63–65], primarily due to within- and between-administrator rating issues. Objective neuromotor assessments are typically more reliable, but do require rigorous testing and comparison with other existing methods to ensure their validity. For example, the reliability and validity of a custom smartphone application (app) designed to objectively measure dynamic balance performance was recently examined [66], and other smartphone apps have been similarly examined to ensure they are a reliable and valid way to measurement neuromotor performance [67–71].

Third, portability should be considered. Traditionally, objective balance assessments were not portable due to their reliance on equipment that was too large to move or that needed stable (i.e., flat and non-vibrating) environments. Recent technological advances have led to portable devices, such as the force plate offered by BTrackS [65] and the smartphone app offered by Sway Medical [72, 73]. However, both of these portable solutions rely on a static balance test, which may mask functional changes in behavior due to low task difficulty [74]. Thus, a dynamic balance test requiring participants to maintain balance during movement may be desired for some settings.

Lastly, cost and accessibility should be considered. Most subjective neuromotor assessments have low cost due to their reliance on little-to-no equipment and minimal training. The cost of objective neuromotor tests using research grade sensors has traditionally been in the thousands of dollars. However, in the rapidly changing technology market, the cost of equipment to objectively test neuromotor performance has substantially dropped in recent years. Portable force plates can be purchased for a fraction of the previous costs [65, 75] and there are a number of smartphone apps that are cost-effective or even free. However, before adopting one of these cost-effective objective neuromotor assessments, it is highly recommended that the administrator investigates whether scientific evidence supports clinical utility of the assessment to ensure it is reliable and valid [76].

2 Materials

2.1 Subjective Assessment Materials

By their very nature, subjective tests require little-to-no materials, making them attractive to clinicians. For example, the Balance Error Scoring System (BESS) [77] only requires a foam mat. The BESS was further simplifed when the mBESS was developed and removed the need for a foam mat [78]. Previous work has shown that more errors (i.e., poorer performance) occur on the BESS after mTBI [52, 79, 80–82]. However, the subjectivity of the BESS score can lead to less than optimal reliability [63–65], as well as validity [83]. The Berg Balance Scale (BBS) [84] has been used with a variety of clinical populations with neuromotor dysfunction [85, 86], including TBI [87, 88]. The BBS requires a stopwatch, chair with arm rests, measuring tape, an object to pick up off the floor, and a step stool. The Functional Gait Assessment (FGA) [89] is a subjective dynamic balance assessment that is a revised test based on the Dynamic Gait Index (DGI) [90] and it has been used with a TBI population [91]. The FGA requires a stopwatch, marked area 20 ft long by 12 in. wide, a 9 in. tall obstacle, and a set of steps with railings. Lastly, the Community Balance and Mobility (CB&M) Scale is a subjective assessment of dynamic balance specifically designed for TBI populations [92]. The CB&M requires an 8 m track, stopwatch, laundry basket, 2 and 7 lb weights, a visual target consisting of paper circle 20 cm in diameter with a 5 cm in diameter black circle in the middle, and a bean bag. While some of these assessments seem to require a long litany of equipment, most of these items can be easily acquired at a low cost and the assessments themselves do not require extensive training to conduct.

2.2 Objective Assessment Materials

A force plate has long been used as an objective assessment of balance in research laboratories by measuring how a person controls their balance through quantifying changes in the center of pressure (CoP) at the foot–ground interface. As a person leans forward, their CoP moves forward in order to keep their center of mass (CoM) within their base of support (BoS), which allows for the maintenance of upright stance. The force plate measures the movement over time of the CoP in a horizontal plane defined by two directions [anterior-posterior (AP) and medial-lateral (ML)], allowing researchers and clinicians to see how balance control evolves over time. Trials are typically 30–60 s and recorded at a high frequency with a high spatial resolution. This type of assessment method has been used extensively to characterize balance control after TBI [22, 40, 80, 82, 83, 93–103].

Similarly, the NeuroCom Sensory Organization Test (SOT) (Natus Medical, Inc., San Carlos, CA) uses a force plate to measure balance control, and the SOT has been used with TBI populations [82, 98, 100, 104, 105]. The NeuroCom SOT tests balance under six conditions, allowing for a measurement of the contributions of the different sensory systems (visual, vestibular, and somatosensory) to neuromotor performance. While the NeuroCom SOT allows for a deeper investigation into neuromotor performance, it does require equipment that is not portable and may not be cost-effective for all field-based settings.

Traditionally, objective balance assessments were not portable due to their reliance on force plates that were not easy to move. However, recent technological advances have led to portable force plates designed for clinical application. For example, the Balance Tracking Systems (BTrackS, San Diego, CA) is a portable force plate and software combination that is designed to objectively measure balance practically in any setting [65]. The software stores the participant's data for future comparison should a head trauma event occur at a later date.

Other portable sensors have been used to measure balance, such as accelerometers [106–110] and inertial measurement units (IMUs) [111–114]. While these types of sensors can be used independently, they are also embedded in smartphones, leading to the growth of smartphone apps developed for medical and health monitoring [115]. There are a number of apps on the market purported to measure balance [116]. However, researchers and clinicians should only adopt an app that has sufficient scientific evidence supporting its clinical utility, of which there are several options [8, 67, 72, 73, 117–121].

Lastly, high speed motion capture cameras have been used to objectively measure neuromotor performance after TBI, typically during gait tasks [122]. These cameras measure the movement of active or passive sensors placed on the body at sub-millimeter accuracy, providing a precise measurement of neuromotor performance. In addition to steady-state walking [123–127], activities of daily living have been used to determine how "real-world" or functional task performance is altered after head trauma, such as walking over/around obstacles [127–135] or walking with divided attention [132, 134, 136–140]. Moreover, virtual reality has been used to create stimuli and/or environments for the patient to interact with to measure and/or enhance their neuromotor performance [141–143]. It has been argued that these more complex types of tasks may sometimes be necessary to identify neuromotor dysfunction, especially if the deficits are more subtle, such as after sub-concussive head trauma [66].

3 Methods

3.1 Subjective Assessment Methods

A large number of subjective neuromotor assessment methods have been developed over the years. Some have been tailored for a specific population, whereas others are designed for more general use. There are several online resources available that outline the different assessment methods and their uses. One of the most thorough is the Rehab Measures database developed by the Rehabilitation Institute of Chicago through the support of a grant from the National Institute on Disability and Rehabilitation Research (NIDRR), part of the U.S. Department of Education. Readers are referred to this database for more information on subjective assessments not covered in this chapter. Below are some of the most commonly used methods in the context of neuromotor performance after head trauma.

3.1.1 Balance Error Scoring System (BESS)

The BESS is a six item test, with lower scores indicating better performance. Each test uses three stances (single leg, double leg, and tandem) on two surfaces (hard and foam surfaces), leading to six conditions (Fig. 1). The modified BESS (mBESS) was shortened by removing the foam surface testing, allowing the mBESS to be administered without equipment [78]. Participants are asked to stand in one of the three stances with their eyes closed and hands on their hips for 20 s. An error is counted each time the participant makes any of the following movements: (1) moves hands off the hips, (2) opens the eyes, (3) takes a step, stumble, or fall, (4) abducts or flexes the hip beyond 30°, (5) lifts the forefoot or heel off the testing surface, or (6) remains out of the initial testing position for more than 5 s. Errors are tallied from all trials and a total BESS score is used for comparison to either normative data [144, 145] or to the baseline assessment of that individual in order to measure within-person changes after head trauma.

3.1.2 Clinical Test of Sensory Interaction and Balance (CTSIB)

The CTSIB is similar to the BESS in that the goal is to stand still while the administrator evaluates the person's performance [146]. The CTSIB has six conditions: (1) stand on firm surface with eyes open, (2) stand on firm surface with eyes closed, (3), stand on firm surface while wearing a visual conflict dome, (4) stand on foam surface with eyes open, (5) stand on foam surface with eyes closed, and (6) stand on foam surface while wearing a visual conflict dome. The visual conflict dome induces sensory conflict by removing peripheral vision and moving the visual field in a manner that is consistent with the individual's sway. Thus, they cannot use visual information to correct their postural sway. Each condition is performed for 30 s and the test is stopped if the individual's feet or arms change position. The total score is derived by adding up the total number of seconds the individual was able to compete each of the tasks. A modified version (mCTSIB) was developed that omitted the visual dome conditions.

3.1.3 Berg Balance Score (BBS)

The BBS is a 14 item test, with higher scores indicating better performance. Each item is scored from a 0 (lowest) to 4 (highest),

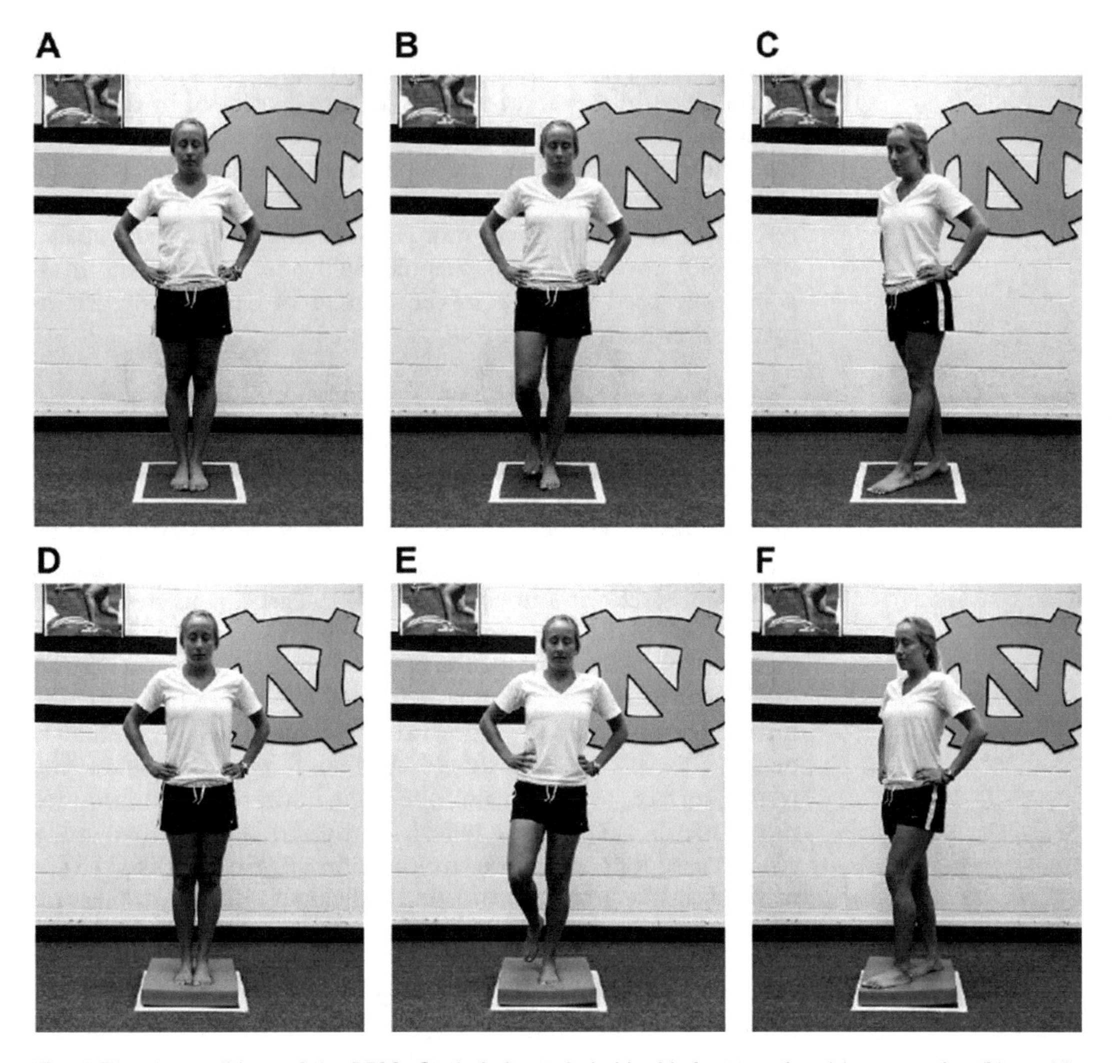

Fig. 1 The six conditions of the BESS. Static balance is held with feet together (**a**), on one leg (**b**), and in tandem stance (**c**) while on a firm surface (**a–c**) and on a foam surface (**d**, **e**). Image used with permission from [169]

with 56 as the maximum score. The items include: (1) sitting-to-standing, (2) standing unsupported, (3) sitting unsupported, (4) standing to sitting, (5) transfers, (6) standing with eyes closed, (7) standing with feet together, (8) reaching forward with an outstretched arm, (9) retrieving an object from the floor, (10), turning to look behind them, (11) turning 360°, (12) placing their foot on a stool, (13) standing with one foot in front of them, and (14) standing on one foot. Scoring guidance is provided that reflects the ability to complete the task within a set duration [147]. For example, in the sitting-to-standing task, the highest score on this item is given if the participant is able to stand using no hands and can stabilize themselves without assistance. The lowest score is given if moderate or maximal assistance is needed to stand.

3.1.4 Functional Gait Assessment (FGA)

The FGA is a ten item test, with higher scores indicating better performance. Each item is scored between 0 (lowest) to 3 (highest), leading to a maximum score of 30. The items include: (1) walking on a level surface, (2) changing gait speed, (3) walking while turning the head horizontally, (4) walking while vertically turning the head, (5) walking with a pivot turn, (6) stepping over an obstacle, (7) walking with a narrow base of support, (8) walking with eyes closed, (9) walking backwards, and (10) walking up/down steps. Scoring guidance is provided that reflects the ability to complete the task within a set duration [89].

3.1.5 Community Balance and Mobility Scale (CB&M)

The CB&M is a 13 item test, with higher scores indicating better performance. Each item is scored between 0 (lowest) to 5 (highest) and 6 of the items are scored for both sides of the body. This effectively leads to 19 scoring assessments and a maximum score of 96 (one bonus point can be awarded for item #12). The items include: (1) unilateral stance, (2) tandem walking, (3) 180° tandem pivot, (4) lateral foot scooting, (5) hopping forward, (6) crouch and walk, (7) lateral dodging, (8) walking and looking, (9) running with a controlled stop, (10) forward to backward walking, (11) walk, look, and carry, (12) descending stairs, and (13) step-ups. Scoring guidance is provided that reflects the ability to complete the task within a set duration [92].

3.2 Objective Assessment

3.2.1 Force Plate

BTrackS is a class 1 medical device that measures CoP movement while the participant stands on a force plate for four 20-s trials with eyes closed and feet shoulder width apart [65 , 148, 149]. The portable force plate is connected to a laptop or tablet to automatically record the distance traveled by the CoP (termed path length) measured in centimeters, which is an objective measurement of a person's balance ability (Fig. 2). A higher path length indicates worse performance. The clinical utility of using BTrackS to measure neurmotor deficits after a concussion has been documented [149], along with its utility when used with a fatigued population [150].

The Neurocom SOT was designed to allow researchers to measure the contribution of the visual, vestibular, and proprioceptive systems to balance control. The SOT consists of six conditions (Fig. 3), with the goal in each condition to preserve static upright posture. Each condition is repeated three times for 20 s each: (1) eyes open on a firm surface, (2) eyes closed on a firm surface, (3) eyes open with a sway reference visual surround, (4) eyes open on a sway reference support surface, (5) eyes closed on a sway referenced support surface, and (6) eyes open on a sway referenced support surface and surround. The sway referenced description refers to the walls around the force plate (visual surround) and/or the force plate itself (support surface) moving in conjunction with the person's swaying, making it difficult to use that information to correct their posture. The clinical utility of the Neurocom SOT

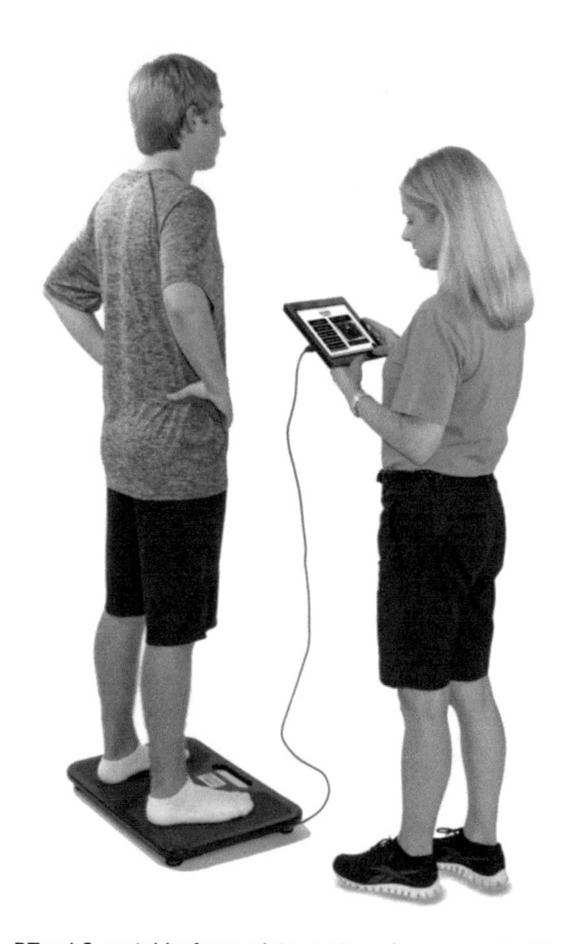

Fig. 2 The BTrackS portable force plate system. Image used with permission from [150]

with a TBI population has been demonstrated [98, 100]. However, the cost of the Neurocom SOT is a barrier for some researchers and clinicians. To address this challenge, the Virtual Environment TBI Screen (VETS) was recently developed and validated relative to the Neurocom SOT with a TBI population [105]. The VETS provides a cost-efficient and sensitive way to assess neuromotor performance by probing neurosensory contributions to balance control. It utilizes a commercially available gaming force plate, the Nintendo Wii balance board, which has been found to be reliable and valid for postural assessment [75, 151].

3.2.2 Smartphone Apps

There are a number of smartphone apps on the market to measure balance control [116]. However, most of them have little-to-no scientific documentation supporting their clinical utility. One app that is commercially available and does have documentation is the Sway Balance app offered by Sway Medical, LLC (Tulsa, OK). The app is FDA cleared and scientific studies have been publishing showing the app's validity, reliability, and clinical utility

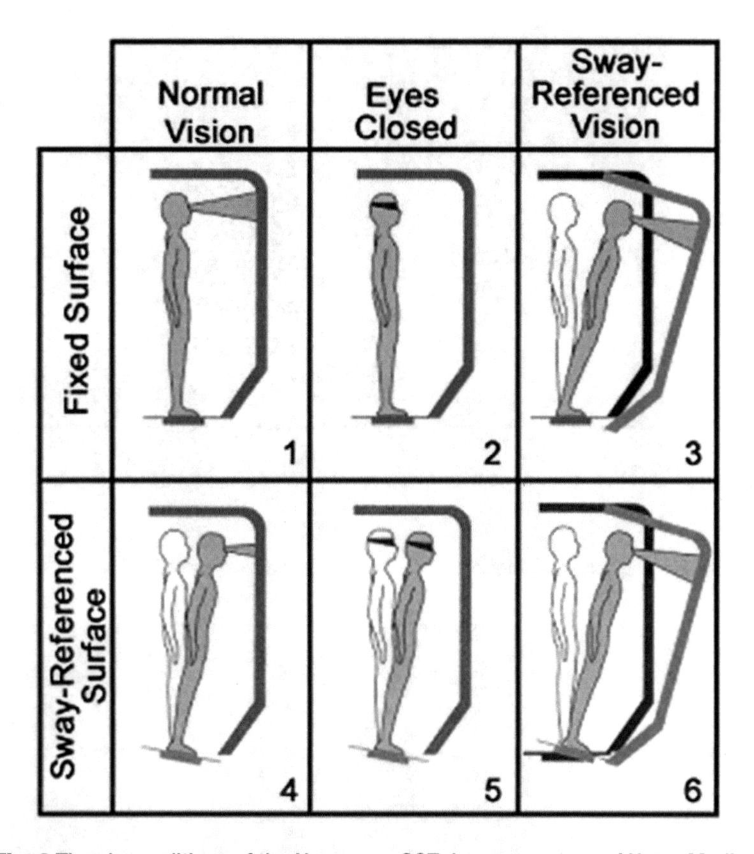

Fig. 3 The six conditions of the Neurocom SOT. Image courtesy of Natus Medical Incorporated

[72, 73, 118]. Specifically, the Sway Balance app has been compared to BESS, Athlete's Single Leg Test protocol, and the Biodex Balance System—all accepted assessments of balance control—and the data showed that the Sway app was a valid measure of balance. The Sway Balance app protocol has participants stand in five stances on a firm surface with their eyes closed for 10 s each while holding the phone against their chest (Fig. 4). The stances are: (1) feet together, (2) tandem stance (right foot behind left foot), (3) tandem stance (left foot behind right foot), (4) stance on right foot only, and (5) stance on left foot only [118]. While good clinical utility has been shown with the Sway Balance app, the protocol relies on static balance, which may mask neuromotor deficits in some cases [74]. To address this challenge, an app (termed AccWalker) was developed that uses a dynamic balance protocol. Rather than having people stand still, the AccWalker app uses a sensor fusion algorithm to collect data from the accelerometer, magnetometer, and gyroscope in order to track a person's performance during a stepping-in-place task that is completed under three conditions: eyes open, eyes closed, and while laterally rotating the head from side to side. Data from this app and dynamic balance

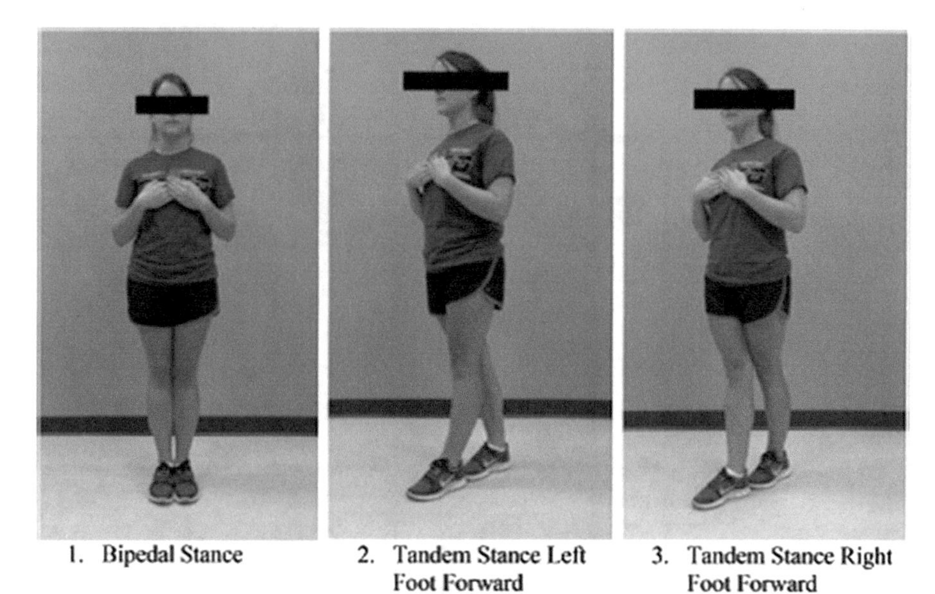

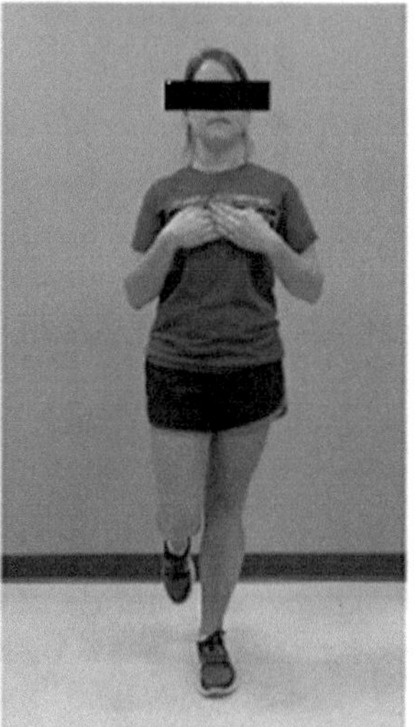

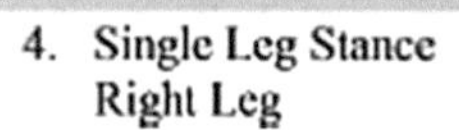

Fig. 4 The five conditions of the Sway Balance app protocol. Image used with permission from [118]

protocol have been shown to be reliable and valid [66], as well as having clinical utility in a sub-concussed [8] and concussed population [152].

3.2.3 Accelerometers/ IMUs/Motion Capture

The previous sections outlined protocols using objective neuromotor assessment tools. Depending on the context and research question, researchers sometimes adopt sensors such as accelerometers, IMUs, or motion capture to measure neuromotor performance during unique tasks. These tools are useful because they are portable and allow the researcher to measure human movement in a variety of contexts. However, there are no standardized protocols to report, only the knowledge that these tools exist to address unique movement patterns outside of standardized neuromotor assessment protocols.

3.2.4 Linear vs. Nonlinear Methods

Regardless of how the objective assessment data are collected, how the data are processed is an important consideration. A common method—commonly termed a linear or summary metric approach—is to look at the average performance (e.g., mean) or the average variation (e.g., standard deviation or coefficient of variation). While this approach has certainly been useful in clinical and research settings, it fails to characterize time-evolving patterns in the data that could be informative about the control process. To address these patterns in human movement data, a nonlinear metric framework has been developed over the past three decades [153–160], with a significant increase in popularity in recent years [161–165]. This nonlinear framework has been adopted in studies examining balance and gait control in populations with TBI [97, 99, 100, 102, 166–168] and shows promise to more fully characterize neuromotor performance after head trauma [103].

4 Conclusions

The technology available today provides clinicians and researchers with many options to objectively measure neuromotor performance. The challenge that clinicians face with regard to head injury is that the field is largely dependent on symptomology, since very few objective biomarkers exist. This is especially problematic when dealing with sub-concussion and low-level blast injury, because blast-exposed individuals are often subjectively unaware of any deficits that may have been caused. In other words, they may have no symptoms to report. The problem is further compounded by growing evidence that repetitive exposure to low-level forces or blast waves may have a cumulative effect across time, despite no single event causing a victim to have any symptoms. Therefore, one of the major dilemmas that researcher and clinician must address, is how to increase the sensitivity of objective assessments. Using new technology, new analysis techniques, and gaining a greater understanding of the pathology will all serve to improve the clinician's ability to identify underlying deficits related to repetitive blast exposure.

Some of the technology that is now available to address this challenge is relatively new. However, there are options available that have appropriate scientific validity, reliability, sensitivity, and clinical utility. Many objective assessments have been validated against the more scientifically supported subjective assessments. These objective assessments may also be tested together with subjective assessments of neuromotor performance to measure different dimensions of behavior after head trauma in order to help guide differential diagnosis following blast injury. As the Information Age continues to mature, we will likely be able to monitor more elements of human physiology and behavior, such as remotely monitoring biomarkers, heart arrhythmias, and fatigue. Combining these elements with objective measurements of neuromotor performance will help clinicians make more informed decisions about a patient's ability to return-to-duty/activity/work/learn.

Acknowledgments

The opinions or assertions contained herein are the private ones of the authors and are not to be construed as official or reflecting the views of the Department of Defense, the Uniformed Services University of the Health Sciences, or any other agency of the U.S. Government. This work was supported by funding from the United States Department of Defense to each of the following primary investigators: Christopher K. Rhea (W81XWH-15-1-0094; W91CRB-11-D-0001, subcontract P010202825; HU0001-08-1-0001, subcontract 2890; HU0001-15-2-0024, subcontract 3137), W. Geoff Wright (W81XWH-13-C-0189), F. Jay Haran (604110HP.4270.001.A1411; FY13-PH-TBI-WII-255), and to Josh L. Duckworth (HU0001-14-1-0022).

References

1. Turner RC, Naser ZJ, Bailes JE, Smith DW, Fisher JA, Rosen CL (2012) Effect of slosh mitigation on histologic markers of traumatic brain injury. J Neurosurg 117(6):1110–1118
2. Prins M, Greco T, Alexander D, Giza CC (2013) The pathophysiology of traumatic brain injury at a glance. Dis Models Mech 6 (6):1307–1315
3. Werner C, Engelhard K (2007) Pathophysiology of traumatic brain injury. Br J Anaesth 99 (1):4–9
4. Greve MW, Zink BJ (2009) Pathophysiology of traumatic brain injury. Mt Sinai J Med 76 (2):97–104
5. Kovacs SK, Leonessa F, Ling GSF (2014) Blast TBI models, neuropathology, and implications for seizure risk. Front Neurol 5:47
6. Snell FI, Halter MJ (2010) A signature wound of war: mild traumatic brain injury. J Psychosoc Nurs Ment Health Serv 48 (2):22–28
7. Hoge C, McGurk D, Thomas J, Cox A, Engel C, Castro C (2008) Mild traumatic brain injury in U.S. soldiers returning from Iraq. N Engl J Med 358(5):453–463
8. Rhea CK, Kuznetsov NA, Ross SE, Long B, Jakiela JT, Bailie JM, Yanagi MA, Haran FJ, Wright WG, Robins RK, Sargent PD, Duckworth JL (2017) Development of a portable tool for screening neuromotor sequelae from repetitive low-level blast exposure. Mil Med 182(3/4):147–154
9. Cernak I, Noble-Haeusslein LJ (2010) Traumatic brain injury: an overview of pathobiology with emphasis on military populations. J Cereb Blood Flow Metab 30(2):255–266
10. Stone JR, Tustison NJ, Wassermann EM, Polejaeva L, Tierney M, McCarron RM, LoPresti M, Carr WS (2013) Neuroimaging correlates of repetitive blast exposure in experienced military breachers. J Neurotrauma 30 (15):A120–A121

11. Carr W, Polejaeva E, Grome A, Crandall B, LaValle C, Eonta SE, Young LA (2015) Relation of repeated low-level blast exposure with symptomology similar to concussion. J Head Trauma Rehabil 30(1):47–55
12. Tate CM, Wang KK, Eonta S, Zhang Y, Carr W, Tortella FC, Hayes RL, Kamimori GH (2013) Serum brain biomarker level, neurocognitive performance, and self-reported symptom changes in soldiers repeatedly exposed to low-level blast: a breacher pilot study. J Neurotrauma 30(19):1620–1630
13. Taber K, Warden D, Hurley R (2006) Blast-related traumatic brain injury: what is known? J Neuropsychiatry Clin Neurosci 18 (2):141–145
14. Shively SB, Perl DP (2012) Traumatic brain injury, shell shock, and posttraumatic stress disorder in the military—past, present, and future. J Head Trauma Rehabil 27 (3):234–239
15. Ling G, Bandak F, Armonda R, Grant G, Ecklund J (2009) Explosive blast neurotrauma. J Neurotrauma 26(6):815–825
16. Adam O, Mac Donald CL, Rivet D, Ritter J, May T, Barefield M, Duckworth J, LaBarge D, Asher D, Drinkwine B (2015) Clinical and imaging assessment of acute combat mild traumatic brain injury in Afghanistan. Neurology 85(3):219–227
17. Mac Donald CL, Johnson AM, Cooper D, Nelson EC, Werner NJ, Shimony JS, Snyder AZ, Raichle ME, Witherow JR, Fang R (2011) Detection of blast-related traumatic brain injury in US military personnel. N Engl J Med 364(22):2091–2100
18. DePalma RG, Burris DG, Champion HR, Hodgson MJ (2005) Blast injuries. N Engl J Med 352(13):1335–1342
19. Belanger HG, Kretzmer T, Yoash-Gantz R, Pickett T, Tupler LA (2009) Cognitive sequelae of blast-related versus other mechanisms of brain trauma. J Int Neuropsychol Soc 15(1):1
20. Needham CE, Ritzel D, Rule GT, Wiri S, Young L (2015) Blast testing issues and TBI: experimental models that lead to wrong conclusions. Front Neurol 6:72
21. Wiri S, Wofford T, Dent T, Needham C (2017) Reconstruction of recoilless weapon blast environments using high-fidelity simulations. In: 30th international symposium on shock waves 2. Springer, pp 1367–1371
22. Broglio SP, McCrea M, McAllister T, Harezlak J, Katz B, Hack D, Hainline B, Investigators CC (2017) A national study on the effects of concussion in collegiate athletes and US military service academy members: the NCAA–DoD Concussion Assessment, Research and Education (CARE) consortium structure and methods. Sports Med 47 (7):1437–1451
23. Manley GT, Macdonald CL, Markowitz A, Stephenson D, Robbins A, Gardner RC, Winkler EA, Bodien Y, Taylor S, Yue JK (2017) The Traumatic Brain Injury Endpoints Development (TED) initiative: progress on a public-private regulatory collaboration to accelerate diagnosis and treatment of traumatic brain injury. J Neurotrauma. https://doi.org/10.1089/neu.2016.4729
24. Walker WC, Carne W, Franke L, Nolen T, Dikmen S, Cifu D, Wilson K, Belanger H, Williams R (2016) The Chronic Effects of Neurotrauma Consortium (CENC) multicentre observational study: description of study and characteristics of early participants. Brain Inj 30(12):1469–1480
25. Mendez MF, Owens EM, Reza Berenji G, Peppers DC, Liang L-J, Licht EA (2013) Mild traumatic brain injury from primary blast vs. blunt forces: post-concussion consequences and functional neuroimaging. NeuroRehabilitation 32(2):397–407
26. Young L, Rule GT, Bocchieri RT, Walilko TJ, Burns JM, Ling G (2015) When physics meets biology: low and high-velocity penetration, blunt impact, and blast injuries to the brain. Front Neurol 6:89
27. Dennis A, Kochanek P (2007) Pathobiology of blast injury. In: Intensive care medicine. Springer, Berlin, Heidelberg, pp 1011–1022
28. Bandak F, Ling G, Bandak A, De Lanerolle N (2014) Injury biomechanics, neuropathology, and simplified physics of explosive blast and impact mild traumatic brain injury. Handb Clin Neurol 127:89–104
29. Shively SB, Horkayne-Szakaly I, Jones RV, Kelly JP, Armstrong RC, Perl DP (2016) Characterisation of interface astroglial scarring in the human brain after blast exposure: a post-mortem case series. Lancet Neurol 15 (9):944–953
30. Iacono D, Shively SB, Edlow BL, Perl DP (2017) Chronic traumatic encephalopathy. Phys Med Rehabil Clin 28(2):301–321
31. Goldstein LE, Fisher AM, Tagge CA, Zhang X-L, Velisek L, Sullivan JA, Upreti C, Kracht JM, Ericsson M, Wojnarowicz MW (2012) Chronic traumatic encephalopathy in blast-exposed military veterans and a blast neurotrauma mouse model. Sci Transl Med 4 (134):134ra160

32. Belanger HG, Proctor-Weber Z, Kretzmer T, Kim M, French LM, Vanderploeg RD (2011) Symptom complaints following reports of blast versus non-blast mild TBI: does mechanism of injury matter? Clin Neuropsychol 25 (5):702–715
33. Lippa SM, Pastorek NJ, Benge JF, Thornton GM (2010) Postconcussive symptoms after blast and nonblast-related mild traumatic brain injuries in Afghanistan and Iraq war veterans. J Int Neuropsychol Soc 16(5):856
34. Lange RT, Pancholi S, Brickell TA, Sakura S, Bhagwat A, Merritt V, French LM (2012) Neuropsychological outcome from blast versus non-blast: mild traumatic brain injury in US military service members. J Int Neuropsychol Soc 18(3):595
35. Bazarian JJ (2016) Can serum brain proteins aid in concussion identification? Bridge Linking Eng Soc 46(1):26–30
36. Smith DH (2016) Neuromechanics and pathophysiology of diffuse axonal injury in concussion. Bridge Linking Eng Soc 46(1):79–84
37. Alberts JL (2016) A multidisciplinary approach to concussion management. Bridge Linking Eng Soc 46(1):23–25
38. Talavage TM (2016) Medical imaging to recharacterize concussion for improved diagnosis in asymptomatic athletes. Bridge Linking Eng Soc 46(1):91–97
39. Fralick M, Thiruchelvam D, Tien HC, Redelmeier DA (2016) Risk of suicide after a concussion. Can Med Assoc J 188(7):497–504
40. Gysland SM, Mihalik JP, Register-Mihalik JK, Trulock SC, Shields EW, Guskiewicz KM (2012) The relationship between subconcussive impacts and concussion history on clinical measures of neurologic function in collegiate football players. Ann Biomed Eng 40 (1):14–22
41. Abbas K, Shenk TE, Poole VN, Breedlove EL, Leverenz LJ, Nauman EA, Talavage TM, Robinson ME (2015) Alteration of default mode network in high school football athletes due to repetitive subconcussive mild traumatic brain injury: a resting-state functional magnetic resonance imaging study. Brain Connect 5(2):91–101
42. Poole VN, Breedlove EL, Shenk TE, Abbas K, Robinson ME, Leverenz LJ, Nauman EA, Dydak U, Talavage TM (2015) Sub-concussive hit characteristics predict deviant brain metabolism in football athletes. Dev Neuropsychol 40(1):12–17
43. McKee AC, Cantu RC, Nowinski CJ, Hedley-Whyte ET, Gavett BE, Budson AE, Santini VE, Lee HS, Kubilus CA, Stern RA (2009) Chronic traumatic encephalopathy in athletes: progressive tauopathy after repetitive head injury. J Neuropathol Exp Neurol 68(7):709
44. Talavage TM, Nauman EA, Breedlove EL, Yoruk U, Dye AE, Morigaki KE, Feuer H, Leverenz LJ (2014) Functionally-detected cognitive impairment in high school football players without clinically-diagnosed concussion. J Neurotrauma 31(4):327–338
45. Myer GD, Yuan W, Foss KDB, Smith D, Altaye M, Reches A, Leach J, Kiefer AW, Khoury JC, Weiss M (2016) The effects of external jugular compression applied during head impact exposure on longitudinal changes in brain neuroanatomical and neurophysiological biomarkers: a preliminary investigation. Front Neurol 7:74
46. Haran FJ, Tierney R, Wright GW, Keshner E, Silter M (2013) Acute changes in postural control after soccer heading. Int J Sports Med 34(4):350–354
47. Gavett BE, Stern RA, McKee AC (2011) Chronic traumatic encephalopathy: a potential late effect of sport-related concussive and subconcussive head trauma. Clin Sports Med 30(1):179–188
48. Wunderle K, Hoeger KM, Wasserman E, Bazarian JJ (2014) Menstrual phase as predictor of outcome after mild traumatic brain injury in women. J Head Trauma Rehabil 29 (5):E1–E8
49. Sowell ER, Peterson BS, Kan E, Woods RP, Yoshii J, Bansal R, Xu D, Zhu H, Thompson PM, Toga AW (2007) Sex differences in cortical thickness mapped in 176 healthy individuals between 7 and 87 years of age. Cereb Cortex 17(7):1550–1560
50. Covassin T, Swanik CB, Sachs ML (2003) Sex differences and the incidence of concussions among collegiate athletes. J Athl Train 38 (3):238
51. Dick R (2009) Is there a gender difference in concussion incidence and outcomes? Br J Sports Med 43(Suppl 1):i46–i50
52. Covassin T, Elbin R, Harris W, Parker T, Kontos A (2012) The role of age and sex in symptoms, neurocognitive performance, and postural stability in athletes after concussion. Am J Sports Med 40(6):1303–1312
53. Rothweiler B, Temkin NR, Dikmen SS (1998) Aging effect on psychosocial outcome in traumatic brain injury. Arch Phys Med Rehabil 79(8):881–887
54. Broglio SP, Eckner JT, Paulson HL, Kutcher JS (2012) Cognitive decline and aging: the role of concussive and subconcussive impacts. Exerc Sport Sci Rev 40(3):138–144

55. Coronado VG, Haileyesus T, Cheng TA, Bell JM, Haarbauer-Krupa J, Lionbarger MR, Flores-Herrera J, McGuire LC, Gilchrist J (2015) Trends in sports-and recreation-related traumatic brain injuries treated in US emergency departments: the National Electronic Injury Surveillance System-All Injury Program (NEISS-AIP) 2001-2012. J Head Trauma Rehabil 30(3):185–197
56. Langton L (2010) Women in law enforcement. Bureau of Justice Statistics, Washington, DC. https://www.bjs.gov/content/pub/pdf/wle8708.pdf
57. Duckworth JL, Grimes J, Ling GS (2013) Pathophysiology of battlefield associated traumatic brain injury. Pathophysiology 20 (1):23–30
58. Mayer AR, Quinn DK, Master CL (2017) The spectrum of mild traumatic brain injury. Neurology 89(6):623–632
59. McCrory P, Meeuwisse W, Dvorak J, Aubry M, Bailes J, Broglio S, Cantu RC, Cassidy D, Echemendia RJ, Castellani RJ (2017) Consensus statement on concussion in sport—the 5th international conference on concussion in sport held in Berlin, October 2016. Br J Sports Med 51(11):838–847. bjsports-2017-097699
60. Tyson S, Connell L (2009) How to measure balance in clinical practice. A systematic review of the psychometrics and clinical utility of measures of balance activity for neurological conditions. Clin Rehabil 23(9):824–840
61. Walker WC, Pickett TC (2007) Motor impairment after severe traumatic brain injury: a longitudinal multicenter study. J Rehabil Res Dev 44(7):975
62. Ragnarsdóttir M (1996) The concept of balance. Physiotherapy 82(6):368–375
63. Finnoff JT, Peterson VJ, Hollman JH, Smith J (2009) Intrarater and interrater reliability of the Balance Error Scoring System (BESS). PM&R 1(1):50–54
64. Reed-Jones RJ, Murray NG, Powell DW (2014) Clinical assessment of balance in adults with concussion. In: Seminars in speech and language, Vol 03. Thieme Medical Publishers. pp 186–195
65. Chang JO, Levy SS, Seay SW, Goble DJ (2014) An alternative to the balance error scoring system: using a low-cost balance board to improve the validity/reliability of sports-related concussion balance testing. Clin J Sport Med 24(3):256–262
66. Kuznetsov NA, Robins RK, Long B, Jakiela JT, Haran FJ, Ross SE, Wright GW, Rhea CK (2018) Validity and reliability of smartphone orientation measurement to quantify dynamic balance function. Physiological Measurement 39(2), 02NT01
67. Vohralik SL, Bowen AR, Burns J, Hiller CE, Nightingale EJ (2015) Reliability and validity of a smartphone app to measure joint range. Am J Phys Med Rehabil 94(4):325–330
68. Boissy P, Diop-Fallou S, Lebel K, Bernier M, Balg F, Tousignant-Laflamme Y (2017) Trueness and minimal detectable change of smartphone inclinometer measurements of shoulder range of motion. Telemed e-Health 23(6):503–506
69. Norris ES, Wright E, Sims S, Fuller M, Neelly K (2017) The reliability of smartphone and goniometric measurements of hip range of motion. J Rehabil Sci Res 3(4):77–84
70. Morales CR, Lobo CC, Sanz DR, Corbalán IS, Ruiz BR, López DL (2017) The concurrent validity and reliability of the Leg Motion system for measuring ankle dorsiflexion range of motion in older adults. PeerJ 5:e2820
71. Mourcou Q, Fleury A, Franco C, Klopcic F, Vuillerme N (2015) Performance evaluation of smartphone inertial sensors measurement for range of motion. Sensors 15 (9):23168–23187
72. Patterson JA, Amick RZ, Thummar T, Rogers ME (2014) Validation of measures from the smartphone sway balance application: a pilot study. Int J Sports Phys Ther 9(2):135–139
73. Patterson JA, Amick RZ, Pandya PD, Hakansson N, Jorgensen MJ (2014) Comparison of a mobile technology application with the Balance Error Scoring System. Int J Athl Ther Train 19(3):4–7
74. Baloh RW, Fife TD, Zwerling L, Socotch T, Jacobson K, Bell T, Beykirch K (1994) Comparison of static and dynamic posturography in young and older normal people. J Am Geriatr Soc 42(4):405–412
75. Clark RA, Bryant AL, Pua Y, McCrory P, Bennell K, Hunt M (2010) Validity and reliability of the Nintendo Wii Balance Board for assessment of standing balance. Gait Posture 31(3):307–310
76. Rehan Youssef A, Gumaa M (2017) Validity and reliability of smartphone applications for clinical assessment of the neuromusculoskeletal system. Expert Rev Med Devices 14 (6):481–493
77. Riemann BL, Guskiewicz KM, Shields EW (1999) Relationship between clinical and forceplate measures of postural stability. J Sport Rehabil 8(2):71–82
78. Hunt TN, Ferrara MS, Bornstein RA, Baumgartner TA (2009) The reliability of the

modified balance error scoring system. Clin J Sport Med 19(6):471–475

79. McCrea M, Guskiewicz K, Randolph C, Barr WB, Hammeke TA, Marshall SW, Powell MR, Ahn KW, Wang Y, Kelly JP (2013) Incidence, clinical course, and predictors of prolonged recovery time following sport-related concussion in high school and college athletes. J Int Neuropsychol Soc 19(1):22–33
80. Riemann BL, Guskiewicz KM (2000) Effects of mild head injury on postural stability as measured through clinical balance testing. J Athl Train 35(1):19–25
81. McCrea M, Guskiewicz KM, Marshall SW, Barr W, Randolph C, Cantu RC, Onate JA, Yang J, Kelly JP (2003) Acute effects and recovery time following concussion in collegiate football players: the NCAA Concussion Study. JAMA 290(19):2556–2563
82. Guskiewicz KM, Ross SE, Marshall SW (2001) Postural stability and neuropsychological deficits after concussion in collegiate athletes. J Athl Train 36(3):263–273
83. Rochefort C, Walters-Stewart C, Aglipay M, Barrowman N, Zemek R, Sveistrup H (2017) Balance markers in adolescents at 1 month postconcussion. Orthop J Sports Med 5 (3):2325967117695507
84. Berg KO, Wood-Dauphinee S, Williams JI, Gayton D (1989) Measuring balance in the elderly: preliminary development of an instrument. Physiother Can 41:304–311
85. Blum L, Korner-Bitensky N (2008) Usefulness of the Berg Balance Scale in stroke rehabilitation: a systematic review. Phys Ther 88 (5):559–566
86. Saether R, Helbostad JL, Riphagen II, Vik T (2013) Clinical tools to assess balance in children and adults with cerebral palsy: a systematic review. Dev Med Child Neurol 55 (11):988–999
87. Feld JA, Rabadi MH, Blau AD, Jordan BD (2001) Berg balance scale and outcome measures in acquired brain injury. Neurorehabil Neural Repair 15(3):239–244
88. Hays K, O'Dell DR, Cuthbert JP, Tefertiller C, Natale A (2015) Virtual-reality based therapy for balance deficits during traumatic brain injury inpatient rehabilitation. Arch Phys Med Rehabil 96(10):e44
89. Wrisley DM, Marchetti GF, Kurharsky DK, Whitney SL (2004) Reliability, internal consistency, and validity of data obtained with the. Funct Gait Assess Phys Ther 84 (10):906–918
90. Shumway-Cook A, Woollacott MH (1995) Motor control: theory and practical applications. Lippincott Williams & Williams, Baltimore
91. Alsalaheen BA, Whitney SL, Marchetti GF, Furman JM, Kontos AP, Collins MW, Sparto PJ (2016) Relationship between cognitive assessment and balance measures in adolescents referred for vestibular physical therapy after concussion. Clin J Sport Med 26 (1):46–52
92. Howe J, Inness E, Venturini A, Williams J, Verrier M (2006) The Community Balance and Mobility Scale—a balance measure for individuals with traumatic brain injury. Clin Rehabil 20(10):885–895
93. Guskiewicz KM, Riemann BL, Perrin DH, Nashner LM (1997) Alternative approaches to the assessment of mild head injury in athletes. Med Sci Sports Exerc 29:S213–S221
94. Lehmann J, Boswell S, Price R, Burleigh A, DeLateur B, Jaffe K, Hertling D (1990) Quantitative evaluation of sway as an indicator of functional balance in post-traumatic brain injury. Arch Phys Med Rehabil 71 (12):955–962
95. Wöber C, Oder W, Kollegger H, Prayer L, Baumgartner C, Wöber-Bingöl C, Wimberger D, Binder H, Deecke L (1993) Posturographic measurement of body sway in survivors of severe closed head injury. Arch Phys Med Rehabil 74(11):1151–1156
96. Haaland KY, Temkin N, Randahl G, Dikmen S (1994) Recovery of simple motor skills after head injury. J Clin Exp Neuropsychol 16 (3):448–456
97. Cavanaugh JT, Guskiewicz KM, Giuliani C, Marshall SW, Mercer VS, Stergiou N (2005) Detecting altered postural control after cerebral concussion in athletes with normal postural stability. Br J Sports Med 39 (11):805–811
98. Pickett TC, Radfar-Baublitz LS, McDonald SD, Walker WC, Cifu DX (2007) Objectively assessing balance deficits after TBI: role of computerized posturography. J Rehabil Res Dev 44(7):983–990
99. Gao J, Hu J, Buckley T, White K, Hass C (2011) Shannon and Renyi entropies to classify effects of mild traumatic brain injury on postural sway. PLoS One 6(9):e24446. https://doi.org/10.1371/journal.pone.0024446
100. Sosnoff JJ, Broglio SP, Shin S, Ferrara MS (2011) Previous mild traumatic brain injury and postural-control dynamics. J Athl Train 46(1):85–91
101. Powers KC, Kalmar JM, Cinelli ME (2014) Recovery of static stability following a concussion. Gait Posture 39(1):611–614

102. Fino PC, Nussbaum MA, Brolinson PG (2016) Decreased high-frequency center-of-pressure complexity in recently concussed asymptomatic athletes. Gait Posture 50:69–74
103. Buckley TA, Oldham JR, Caccese JB (2016) Postural control deficits identify lingering post-concussion neurological deficits. J Sport Health Sci 5(1):61–69
104. Kaufman KR, Brey RH, Chou L-S, Rabatin A, Brown AW, Basford JR (2006) Comparison of subjective and objective measurements of balance disorders following traumatic brain injury. Med Eng Phys 28(3):234–239
105. Wright WG, McDevitt J, Tierney R, Haran FJ, Appiah-Kubi KO, Dumont A (2017) Assessing subacute mild traumatic brain injury with a portable virtual reality balance device. Disabil Rehabil 39(15):1564–1572
106. Mayagoitia RE, Lötters JC, Veltink PH, Hermens H (2002) Standing balance evaluation using a triaxial accelerometer. Gait Posture 16 (1):55–59
107. Howell D, Osternig L, Chou L-S (2015) Monitoring recovery of gait balance control following concussion using an accelerometer. J Biomech 48(12):3364–3368
108. Brown CN, Mynark R (2007) Balance deficits in recreational athletes with chronic ankle instability. J Athl Train 42(3):367–373
109. O'Sullivan M, Blake C, Cunningham C, Boyle G, Finucane C (2009) Correlation of accelerometry with clinical balance tests in older fallers and non-fallers. Age Ageing 38 (3):308–313
110. Alberts JL, Hirsch JR, Koop MM, Schindler DD, Kana DE, Linder SM, Campbell S, Thota AK (2015) Using accelerometer and gyroscopic measures to quantify postural stability. J Athl Train 50(6):578–588
111. Soangra R, Lockhart TE (2013) Comparison of intra-individual physiological sway complexity from force plate and inertial measurement unit. Biomed Sci Instrum 49:180–186
112. Omkar S, Ganesh D, Kiran PK (2009) Standing balance: Quantification and the impact of visual sensory input. Indian J Physiother Occup Ther 3(3):44–48
113. Brown HJ, Siegmund GP, Doel KV, Cretu E, Guskiewicz K, Blouin J-S (2013) Development and validation of an objective balance assessment system. J Exerc Mov Sport 45 (1):4
114. Cohen HS, Mulavara AP, Peters BT, Sangi-Haghpeykar H, Bloomberg JJ (2014) Standing balance tests for screening people with vestibular impairments. Laryngoscope 124 (2):545–550
115. Boulos MNK, Brewer AC, Karimkhani C, Buller DB, Dellavalle RP (2014) Mobile medical and health apps: state of the art, concerns, regulatory control and certification. Online J Public Health Inform 5(3):229
116. Roeing KL, Hsieh KL, Sosnoff JJ (2017) A systematic review of balance and fall risk assessments with mobile phone technology. Arch Gerontol Geriatr 73:222–226
117. Lee B-C, Kim J, Chen S, Sienko KH (2012) Cell phone based balance trainer. J Neuroeng Rehabil 9(1):10
118. Amick RZ, Chaparro A, Patterson JA, Jorgensen MJ (2015) Test-retest reliability of the sway balance mobile application. J Mobile Technol Med 4(2):40–47
119. Mourcou Q, Fleury A, Franco C, Diot B, Vuillerme N (2015) Smartphone-based system for sensorimotor control assessment, monitoring, improving and training at home. In: International conference on smart homes and health telematics. Springer, Cham. pp 141–151
120. Shah N, Aleong R, So I (2016) Novel use of a smartphone to measure standing balance. JMIR Rehabil Assist Technol 3(1):e4
121. Burghart M, Craig J, Radel J, Huisinga J (2017) Reliability and validity of a mobile device application for use in sports-related concussion balance assessment. Curr Res Concuss 4(1):e1–e6
122. Williams G, Galna B, Morris ME, Olver J (2010) Spatiotemporal deficits and kinematic classification of gait following a traumatic brain injury: a systematic review. J Head Trauma Rehabil 25(5):366–374
123. Basford JR, Chou L-S, Kaufman KR, Brey RH, Walker A, Malec JF, Moessner AM, Brown AW (2003) An assessment of gait and balance deficits after traumatic brain injury. Arch Phys Med Rehabil 84(3):343–349
124. Ochi F, Esquenazi A, Hirai B, Talaty M (1999) Temporal–spatial feature of Fait after traumatic brain injury. J Head Trauma Rehabil 14(2):105–115
125. Perry J (1999) The use of gait analysis for surgical recommendations in traumatic brain injury. J Head Trauma Rehabil 14 (2):116–135
126. Kerrigan DC, Bang M-S, Burke DT (1999) An algorithm to assess stiff-legged gait in traumatic brain injury. J Head Trauma Rehabil 14(2):136–145
127. McFadyen BJ, Swaine B, Dumas D, Durand A (2003) Residual effects of a traumatic brain

injury on locomotor capacity: a first study of spatiotemporal patterns during unobstructed and obstructed walking. J Head Trauma Rehabil 18(6):512–525
128. Chou L-S, Kaufman KR, Walker-Rabatin AE, Brey RH, Basford JR (2004) Dynamic instability during obstacle crossing following traumatic brain injury. Gait Posture 20 (3):245–254
129. Catena RD, Donkelaar P, Halterman C, Chou LS (2009) Spatial orientation of attention and obstacle avoidance following concussion. Exp Brain Res 194(1):67–77. https://doi.org/10.1007/s00221-008-1669-1
130. Catena RD, van Donkelaar P, Chou LS (2007) Altered balance control following concussion is better detected with an attention test during gait. Gait Posture 25 (3):406–411. https://doi.org/10.1016/j.gaitpost.2006.05.006
131. Catena RD, van Donkelaar P, Chou L-S (2009) Different gait tasks distinguish immediate vs. long-term effects of concussion on balance control. J Neuroeng Rehabil 6:25
132. Chiu S-L, Osternig L, Chou L-S (2013) Concussion induces gait inter-joint coordination variability under conditions of divided attention and obstacle crossing. Gait Posture 38 (4):717–722
133. Cantin J-F, McFadyen BJ, Doyon J, Swaine B, Dumas D, Vallée M (2007) Can measures of cognitive function predict locomotor behaviour in complex environments following a traumatic brain injury? Brain Inj 21 (3):327–334
134. McFadyen BJ, Cantin J-F, Swaine B, Duchesneau G, Doyon J, Dumas D, Fait P (2009) Modality-specific, multitask locomotor deficits persist despite good recovery after a traumatic brain injury. Arch Phys Med Rehabil 90(9):1596–1606
135. Baker CS, Cinelli ME (2014) Visuomotor deficits during locomotion in previously concussed athletes 30 or more days following return to play. Physiol Rep 2(12):e12252
136. Parker TM, Osternig LR, Van Donkelaar P, Chou L (2006) Gait stability following concussion. Med Sci Sports Exerc 38 (6):1032–1040
137. Catena RD, van Donkelaar P, Chou LS (2011) The effects of attention capacity on dynamic balance control following concussion. J Neuroeng Rehabil 8:8
138. Catena RD, van Donkelaar P, Chou LS (2007) Cognitive task effects on gait stability following concussion. Exp Brain Res 176 (1):23–31. https://doi.org/10.1007/s00221-006-0596-2
139. Parker TM, Osternig LR, Lee H-J, Pv D, Chou L-S (2005) The effect of divided attention on gait stability following concussion. Clin Biomech 20(4):389–395. https://doi.org/10.1016/j.clinbiomech.2004.12.004
140. Fino PC, Nussbaum MA, Brolinson PG (2016) Locomotor deficits in recently concussed athletes and matched controls during single and dual-task turning gait: preliminary results. J Neuroeng Rehabil 13(1):65
141. Pietrzak E, Pullman S, McGuire A (2014) Using virtual reality and videogames for traumatic brain injury rehabilitation: a structured literature review. Games Health Res Dev Clin Appl 3(4):202–214
142. Ustinova K, Perkins J, Leonard W, Hausbeck C (2014) Virtual reality game-based therapy for treatment of postural and co-ordination abnormalities secondary to TBI: a pilot study. Brain Inj 28(4):486–495
143. Rábago CA, Wilken JM (2011) Application of a mild traumatic brain injury rehabilitation program in a virtual realty environment: a case study. J Neurol Phys Ther 35 (4):185–193
144. Muir B, Lynn A, Maguire M, Ryan B, Calow D, Duffy M, Souckey Z (2014) A pilot study of postural stability testing using controls: the modified BESS protocol integrated with an H-pattern visual screen and fixed gaze coupled with cervical range of motion. J Can Chiropr Assoc 58(4):361
145. Hänninen T, Tuominen M, Parkkari J, Vartiainen M, Öhman J, Iverson GL, Luoto TM (2016) Sport concussion assessment tool–3rd edition–normative reference values for professional ice hockey players. J Sci Med Sport 19(8):636–641
146. Cohen HS, Blatchly CA, Gombash LL (1993) A study of the clinical test of sensory interaction and balance. Phys Ther 73 (6):346–351
147. Berg KO, Wood-Dauphinee SL, Williams JI, Maki B (1992) Measuring balance in the elderly: validation of an instrument. Can J Public Health 83:S7–S11
148. O'Connor SM, Baweja HS, Goble DJ (2016) Validating the BTrackS Balance Plate as a low cost alternative for the measurement of sway-induced center of pressure. J Biomech 49 (16):4142–4145
149. Goble DJ, Manyak KA, Abdenour TE, Rauh MJ, Baweja HS (2016) An initial evaluation of the BTrackS balance plate and sports balance

software for concussion diagnosis. Int J Sports Phys Ther 11(2):149–155

150. Benedict SE, Hinshaw JW, Byron-Fields R, Baweja HS, Goble DJ (2017) Effects of fatigue on the BTrackS balance test for concussion management. Int J Athl Ther Train 22(4):23–28
151. Bartlett HL, Ting LH, Bingham JT (2014) Accuracy of force and center of pressure measures of the Wii Balance Board. Gait Posture 39(1):224–228
152. Rhea CK, Kuznetsov NA, Robins RK, Jakiela JT, LoJacono CT, Ross SE, MacPherson RP, Long B, Haran FJ (2017) Wright WG Dynamic balance decrements last longer than 10 days following a concussion. In: International Society of Posture & Gait Research, Ft. Lauderdale, FL
153. Glass L, Mackey MC (1979) Pathological conditions resulting from instabilities in physiological control systems. Ann N Y Acad Sci 316:214–235
154. Mackey MC, Glass L (1977) Oscillation and chaos in physiological control systems. Science 197:287–289
155. Mackey MC, Milton JG (1987) Dynamical diseases. Ann N Y Acad Sci 504:16–32
156. West BJ, Goldberger AL (1987) Physiology in fractal dimensions. Am Sci 75:354–365
157. Goldberger AL, Rigney DR, West BJ (1990) Chaos and fractals in human physiology. Sci Am 262:42–49
158. Bélair J, Glass L, an der Heiden U, Milton J (1995) Dynamical disease: identification, temporal aspects and treatment strategies of human illness. Chaos Interdiscipl J Nonlinear Sc 5(1):1–7
159. Hausdorff JM, Peng CK, Ladin Z, Wei JY, Goldberger AL (1995) Is walking a random walk? Evidence for long-range correlations in stride interval of human gait. J Appl Physiol 78(1):349–358
160. Vaillancourt DE, Newell K (2002) Changing complexity in human behavior and physiology through aging and disease. Neurobiol Aging 23(1):1–11
161. Hausdorff JM (2007) Gait dynamics, fractals and falls: finding meaning in the stride-to-stride fluctuations of human walking. Hum Mov Sci 26(4):555–589
162. Stergiou N, Decker LM (2011) Human movement variability, nonlinear dynamics, and pathology: is there a connection? Hum Mov Sci 30(5):869–888
163. Rhea CK, Kiefer AW (2014) Patterned variability in gait behavior: how can it be measured and what does it mean? In: Li L, Holmes M (eds) Gait biometrics: basic patterns, role of neurological disorders and effects of physical activity. Nova Science Publishers, Hauppauge, NY
164. van Emmerik RE, Ducharme SW, Amado A, Hamill J (2016) Comparing dynamical systems concepts and techniques for biomechanical analysis. J Sport Health Sci 5(1):3–13
165. Moon Y, Sung J, An R, Hernandez ME, Sosnoff JJ (2016) Gait variability in people with neurological disorders: a systematic review and meta-analysis. Hum Mov Sci 47:197–208
166. Cavanaugh JT, Guskiewicz KM, Giuliani C, Marshall SW, Mercer VS, Stergiou N (2006) Recovery of postural control after cerebral concussion: new insights using approximate entropy. J Athl Train 41(3):305–313
167. Cavanaugh JT, Guskiewicz KM, Stergiou N (2005) A nonlinear dynamic approach for evaluating postural control: new directions for the management of sport-related cerebral concussion. Sports Med 35(11):935–950
168. Fino PC (2016) A preliminary study of longitudinal differences in local dynamic stability between recently concussed and healthy athletes during single and dual-task gait. J Biomech 49(9):1983–1988
169. Fox ZG, Mihalik JP, Blackburn JT, Battaglini CL, Guskiewicz KM (2008) Return of postural control to baseline after anaerobic and aerobic exercise protocols. J Athl Train 43 (5):456–463

Chapter 19

Translingual Neurostimulation (TLNS): Perspective on a Novel Approach to Neurorehabilitation after Brain Injury

Yuri Danilov and Dafna Paltin

Abstract

CN-NINM technology represents a synthesis of a new noninvasive brain stimulation technique with applications in physical medicine, cognitive, and affective neurosciences. Our new stimulation method appears promising for the treatment of a full spectrum of movement disorders and for both attention and memory dysfunction associated with traumatic brain injury. The integrated CN-NINM therapy proposed here aims to restore function beyond traditionally expected limits by employing both newly developed therapeutic mechanisms for progressive physical and cognitive training while simultaneously applying brain stimulation through a portable neurostimulation device called the PoNS™. Based on our previous research and recent pilot data, we believe a rigorous in-clinic CN-NINM training program, followed by regular at-home exercises that will also be performed with CN-NINM, will simultaneously enhance, accelerate, and extend recovery from multiple impairments (e.g. movement, vision, speech, memory, attention, and mood), based on divergent but deeply interconnected neurophysiological mechanisms of neuroplasticity.

Key words Neurorehabilitation, Neuromodulation, Translingual neurostimulation, PoNS device, Targeted therapy, Cranial nerve, Neuroplasticity

Abbreviations

CN-NINM	Cranial-nerve noninvasive neurostimulation
MS	Multiple sclerosis
TBI	Traumatic brain injury
TLNS	Translingual neurostimulation
TNS	Trigeminal nerve stimulation
VNS	Vagal nerve stimulation

1 Introduction

The goal of the current chapter is to introduce our approach to neurorehabilitation called Cranial Nerve Noninvasive

Amit K. Srivastava and Charles S. Cox, Jr. (eds.), *Pre-Clinical and Clinical Methods in Brain Trauma Research*, Neuromethods, vol. 139, https://doi.org/10.1007/978-1-4939-8564-7_19,

Neuromodulation (CN-NINM) technology. CN-NINM is a method of intervention that combines Translingual Neurostimulation (TLNS), the Portable Neurostimulation Stimulator (PoNS™) device, and targeted training designed for movement control rehabilitation.

The basic principle of CN-NINM, as a platform technology, uses neurostimulation to access brain networks through cranial nerves to build a foundation for development of future directions of neurorehabilitation such as headache, tinnitus, sleep, depression, etc. It is noteworthy that the principles and corresponding treatment regimens, based on CN-NINM technology, were already successfully implemented for neurorehabilitation of other neurological conditions such as balance, gait, eye movement control, speech, and cognitive functions [1, 2]. Therefore, CN-NINM technology should be considered to be a practical realization of several theoretical concepts, based on recent scientific discoveries in the field of neuroscience.

First, we would like to consider abnormal neurological conditions, in the view of modern network science, that result from disruption in similar brain networks. The current understanding of neural-network organization can describe the variety of structural and functional network changes in many neurological and psychiatric diseases, especially in dementia, epilepsy, and schizophrenia, but also in traumatic brain injury (TBI), Parkinson's disease, multiple sclerosis (MS), cerebrovascular disease, coma, and many other conditions categorized as neuronal network disorders [3–5].

The complexly distributed neuronal network, with multiple cortical and subcortical components, is the physical substrate for any sensory, motor, and sensory-motor integrative system, providing, in turn, normal physiological or behavioral function such as vision, hearing, postural and eye movement control, as well as multiple others. Damage to or malfunction of any part of said functional network leads to dysfunction of the whole sensory-motor system (spatial and/or temporal abnormalities) that frequently manifests as clinical symptoms.

Second, the situation with the rehabilitation of many neurological symptoms is very similar. Neurological disorders like TBI, stroke, neurodegenerative disorders, or drug overdose (chemical trauma) can affect many distributed networks on several different levels and locations. So far, it is almost impossible to identify the exact place and extent of such damages or the extent of malfunctioning tissues as a result of abnormal connectivity with damaged areas. The abnormalities in the functional relationship between areas and structures, and the abnormalities in the spatiotemporal organization of separate neurons and clusters of neurons, are still beyond our reach for assessment and evaluation. As a result of such uncertainty, therapeutic and rehabilitation resources are significantly limited. For example, there are no effective rehabilitation programs for chronic stage patients after stroke and TBI, the majority of MS symptoms are considered nonrecoverable, and there is no effective treatment for tinnitus. Physical therapy can help these conditions to some extent, but not dramatically.

TLNS technology was originally designed to modulate complex networks for the purpose of neurorehabilitation. We started from the balance sensory-motor integration network, specifically from postural control rehabilitation after peripheral vestibular damages [6–8]. Later we extended our approach to the proprioceptive component of balance (multiple sclerosis, amputee), to gate control rehabilitation (Parkinson's disease, MS, TBI, stroke, cerebral palsy), and eye movement control. The combination of neurostimulation (using the PoNS™ device) and targeted therapy (a set of challenging exercises explicitly targeting the affected network) became the mainframe of TLNS therapy that is applicable to the rehabilitation of many neurological disorders which are so far mainly considered untreatable [1, 2, 9].

1.1 Neurostimulation

Although brain stimulation is well known from ancient Greek and Roman times' Galen and Scribonius Largus, who used electric eels to treat headaches and various other disorders, the current "explosion" of new neurostimulation methods, devices, and applications are hard to even count. Currently, more than a dozen forms of brain stimulation are undergoing development and evaluation as interventions for neurological and psychiatric disorders [10].

Neurostimulation and neuromodulation techniques are unique forms of treatment distinctly different from pharmacology, psychotherapy, or physical therapy. While these terms are often used interchangeably, for the purpose of this chapter and the benefit of this ever-expanding and dynamic field, we propose an important differentiation: *neurostimulation* refers to the physical action of stimulating the nervous system, whereas *neuromodulation* is the product or result of said stimulation.

1.2 Types of Neurostimulation

The specificity and applicability of different neurostimulation methods depend on several key factors: the anatomical location of the stimulation target, physical properties, and the spatiotemporal parameters of stimulation.

The human nervous system is a complex set of interrelated and interacting subsystems with hierarchical modularity. The modules correspond to major functional systems such as motor, sensory, and association networks. The subsystems are characterized by both their anotomic positions and their functional specificities.

At the highest level, the nervous system is divided into central and peripheral nervous systems. The central nervous system (CNS) is comprised of the brain and spinal cord and the peripheral nervous system (PNS) incorporates all the remaining neural structures found outside the CNS. The PNS is further divided functionally into the somatic (voluntary) and autonomic (involuntary) nervous systems. The PNS can also be described structurally as being comprised of afferent (sensory) nerves, which carry information toward the CNS, and efferent (motor) nerves, which carry commands away from the CNS [11].

The PNS also consists of spinal nerves and cranial nerves. Although 12 pairs of cranial nerves emerge directly from the brain (anatomically they are part of CNS), and 10 pairs of them arise from the brainstem, they are formally considered a part of the PNS.

Correspondingly, all neurostimulation systems can be distinct at the site of application: cranial, spinal cord, spinal ganglion, or sciatic nerve neurostimulation systems. It is vital to note that the stimulation of specific brain regions produces equally specific rehabilitation functions.

Neurostimulation systems can either be invasive or non-invasive. According to the National Institutes of Health, non-invasive devices can be defined as those that do not require surgery and do not penetrate the brain parenchyma. Furthermore, the devices for cranial stimulation can be segregated by type of energy source and include, but are not limited to, those used for focused ultrasound stimulation, magnetic seizure therapy, electroconvulsive therapy, static magnets, transcranial alternating current stimulation, transcranial direct current stimulation, transcranial magnetic stimulation, and electromagnetic stimulation in radio frequency range, in addition to several new emerging systems based on optical stimulation of the brain tissue, including infrared light [1, 12, 13].

It is important to note that neurostimulation can be external, exogenous, or generated outside of the neural system (e.g. transcranial magnetic stimulation (TMS), and transcranial direct current stimulation (tDCS)), and still attempt to affect excitable neuronal membranes directly by inducing or suppressing neural activity in the brain network [1]. This kind of stimulation is an *artificial* (rather than natural) activation of brain structures by electrical or magnetic fields, electrical current, light, or ultrasound (usually applied from outside the body or skull) and is fundamentally different from natural (internal, indigenous) activation.

The natural source of brain activation is neural impulses or spikes that are generated by billions of specialized natural receptors located within the skin or internal body tissues: This is internal stimulation from impulses streaming to the spinal cord and brain via nerves and distributed across multiple brain structures [1]. Engagement with natural pathways results in the activation of complex neuronal networks using naturally designed spatial and temporal patterns, which are unique for different brain structures and based on both the anatomical and physiological type of neurons, as well as the patterns of interneuron connections. Similar to these processes are neurostimulation systems that activate the specific receptors, free nerve endings or nerve trunks create the spike flow. In such a case, the primary stimulation on the periphery of the neural system is also artificial, but the real factor affecting the CNS is the flow of natural spikes, generated and distributed internally.

1.3 Peripheral Nerve Stimulation

In 1967, after the introduction of the dorsal column of spinal cord stimulation, the peripheral nerve stimulation (PNS) became a well-established and accepted method for the treatment of various painful conditions. As a pillar for the entirety of neuromodulation technology, PNS become mainly oriented on spinal ganglia (SGS), spinal nerve roots (SNRS), and nerve trunk stimulation (e.g. sciatic or phrenic nerve) and was focused on management of truncal and abdominal pain, migraine and cluster headache, fibromyalgia and lower back pain, urinary and gastrointestinal disorders, etc. [14]. In general, PNS applies mainly implantable, subcutaneous or epidural stimulation techniques. Several new non-invasive PNS methods are currently under development, such as non-invasive spinal cord stimulation (pcEmc) and transcutaneous vagal neurostimulation systems (tVNS) for activation of the auricular branch of the vagus nerve.

1.4 Cranial Nerve Stimulation and Neurorehabilitation

One of the major problems of neurorehabilitation is the complexity and diversity of the brain's damage. Acquired brain injury (ABI) and neurodegenerative disorders create multiple sites of malfunctioning or physically damaged neural tissue. As a result, various functional systems become inefficient or desynchronized; multiple symptoms develop almost simultaneously. The diversified nature of neural network malfunctions and the lack of methods for localizing such damages becomes an overwhelming complication for efficient neurorehabilitation, making full spectrum symptoms and disorders "untreatable."

The majority of existing methods of neurostimulation are limited in several ways. The functional specificity of stimulation creates an extended family of systems for the management of selected body parts or muscular groups. The anatomical specificity and localization of electrodes also restrains the efficiency of neurostimulation for functional recovery. The surgical precision of DBS stimulation and the small volume of affected tissue (only several cubic millimeters) allow changes in the activity of only a single node within a widely distributed functional network. In contrast, the amount of brain tissue affected by TMS might be extended to dozens of cubic centimeters but only activated in an unnatural manner and without functional specificity.

Cranial nerve stimulation might help to solve some of these problems. Cranial nerves are the most powerful nerves that are directly connected to the brain and spinal cord [1]. It is vital to note that all primary sensory systems stream information into the CNS. Vision, hearing, smell, taste, vestibular signal, and proprioception of the face and tongue continuously activate the whole brain via cranial nerves (either directly or indirectly). If we assume that "multidimensional" damage needs "multidimensional" rehabilitation, then cranial nerve stimulation may be the solution.

Thus, Translingual Neutostimulation (TLNS) is a unique way to directly and simultaneously activate multiple brain networks by generating natural spike flow from the periphery. We suggest that the non-invasive and safe "injection" of natural neural activity into damaged neural networks initiates the recovery process based on mechanisms of activity-dependent plasticity [15].

2 Existing Methods

The family of cranial nerve stimulation systems is small in comparison with the variety of other neurostimulation systems, and relatively young. The first U.S. FDA approval for vagal nerve stimulation (VNS, Cyberonics, Inc.) was received in 1997. It is a small wonder that the reception of all new methods of neurostimulation, in general, remains controversial and not widely accepted. Many cranial nerve neurostimulation systems are currently under development and some have yet to be explored. For example, the olfactory nerve has not been investigated for neurostimulation purposes yet. The optic and auditory nerves are primarily targeted for various sensory prosthetic devices, such as artificial retinas and cochlear implants, but rarely explored for rehabilitative or recovery treatments.

However, there is one system for retina and optic nerve stimulation that should be mentioned here: trans-corneal electrical stimulation (TcES), which involves the use of a low-intensity electrical current in the treatment of ophthalmic diseases, including injuries of optic nerve, light-induced photoreceptor degeneration, ocular ischemia, macular dystrophy, and retinitis pigmentosa.

Among the others, two pairs of cranial nerves are intensively under investigation for neurorehabilitation purposes: vagal nerve and trigeminal nerve. Both are large, mixed (sensory and motor) cranial nerves.

2.1 Vagal Nerve Stimulation (VNS)

Primary applications for VNS are epilepsy, depression, anxiety, and obesity. The target of vagal nerve stimulation (VNS) is the tenth cranial nerve that emerges from the brain at the medulla (brainstem) [13]. It is the longest cranial nerve, extending into the chest and abdominal cavity. Typically, a battery-operated generator is implanted subcutaneously in the left chest wall. An attached electrode is then tunnelled under the skin and wrapped around the left vagal nerve in the neck.

Adverse effects of VNS can be separated into those associated with the complications of the surgery and those resulting from the side effects from the stimulation. While risks associated with surgery are minimal, they remain an important consideration for both clinicians and patients [13].

There is one non-invasive method, which transcutaneously stimulates the auricular branch of the vagal nerve. It was developed for the treatment of a chronic migraine (NEMOS™, Cerbomed, Erlangen, Germany). A recent study provides evidence that stimulation using NEMOS™ at 1 Hz for 4 h daily is effective for chronic migraine prevention over 3 months [16, 17]. These results are promising and warrant further scientific and clinical investigation.

2.2 SYMPATO-CORRECTION

SYMPATOCORRECTION™ Technology is a method of dynamic correction of the activity in the sympathetic nervous system. It was developed to regulate the balance between the sympathetic and parasympathetic parts of the autonomic (i.e. vegetative) neural systems in range of the natural "homeostatic corridor" [18]. The SYMPATOCOR™ device provides transcutaneous electrical nerve stimulation (TENS) of the autonomic nervous system by delivering a space-distributed field of low-frequency current impulses (FCI) to the neck nerve-knot projections from the sympathetic part of the autonomic nervous system. The FCI signals differ from those typically used in TENS because of the spatiotemporal pattern of stimulation, which is formed between the electrodes. In the cathodes area, the current structure consists of spatially-distributed partial impulses. In the anodes area, the pulse structure consists of the spatially concentrated structure of the partial impulses. The number of impulses in the structure is equal to the number of cathodes and the impulse duration ranges from 400 to 600 μs, which corresponds to the velocity of creating the excitation in the nonmyelinated nerve fibers.

Indeed, during stimulation of the neck and tongue regions, it is possible to impact both the parasympathetic and sympathetic structures of the autonomic nervous system (ANS) and affect the ascending afferent conductive pathways, their corresponding centers in the brainstem, and subcortical and cortical areas of the brain. The neck section of the sympathetic stem consists of three nodes as well as inter-node connection branches that are located in the deep neck muscles behind the pre-spinal plate of the neck fascia.

Considering that the treatment of pathological processes that affect the ANS is challenging in itself, it is not surprising that using traditional methods to treat the subsequent autonomic disorders, and accompanied symptoms, frequently are not the targets of the major treatment at all.

To this end, SYMPATOCOR™ can be used as a supplemental tool in the rehabilitation process. It has become especially important in cases of brain polytrauma or multiple functional damages after stroke, moderate traumatic brain injury, or in the advanced stage of multiple sclerosis. Moreover, it may be the only way to noninvasively recover various autonomic functions, such as bladder and bowel control, GI motility, hypertension, etc. [19, 20].

2.3 Trigeminal Nerve Stimulation (TNS)

TNS targets the upper ophthalmic branches of the trigeminal nerve. There are two devices, NeuroSigma™ and Cephaly™, which were originally developed to treat drug resistant epilepsy and sleep disorder, respectively. Reported side effects of NeuroSigma™ were mild and included skin irritation, tingling, forehead pressure, and headache [21]. Miller et al. [22] found no side or adverse effects from using Cephaly™, which is consistent with our own experience using the PoNS™ device. This consistency is both validating and encouraging for the continued exploration of trigeminal nerve stimulation therapies.

3 PoNS™ Device

The PoNS™ device, shown in Figures 1 and 2, achieves localized electrical stimulation of afferent nerve fibers on the dorsal surface of the tongue via small surface electrodes. Because of the resulting

Fig. 1 PoNS™ device, displaying the palm-sized stimulator and tongue array

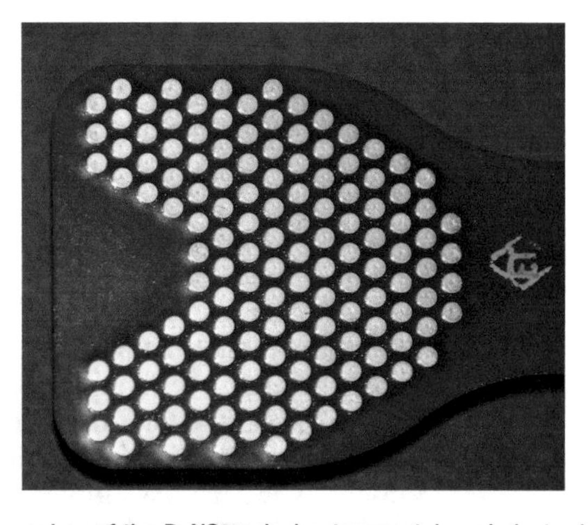

Fig. 2 A closer view of the PoNS™ device tongue-tab and electrode array

Fig. 3 A participant demonstrates how the PoNS™ device is worn around the neck and held in the mouth

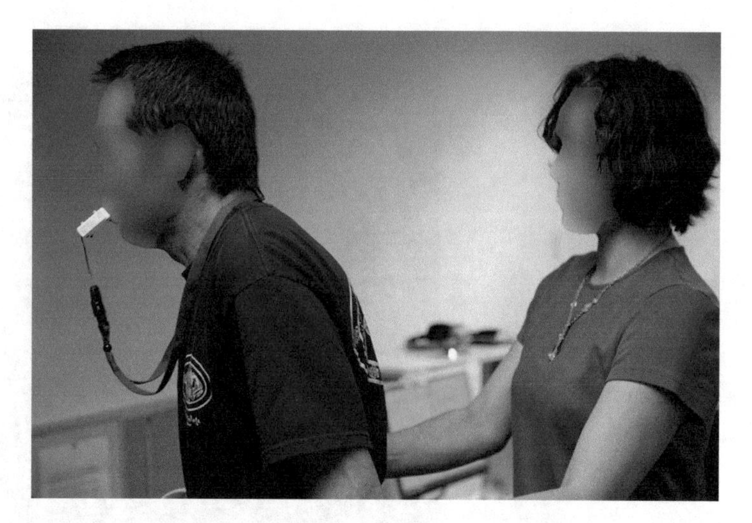

Fig. 4 The same individual as seen in Figure 3 demonstrates using the PoNS™ device during a treatment session with a physical therapist

tactile sensation, which depending on stimulation waveform typically feels like vibration, mild tingling, or pressure, it is certain that tactile nerve fibers are activated. Taste sensations are infrequently reported, although it is not known whether gustatory afferents are in fact stimulated, given the nonphysiological patterns of activation likely to result from PoNS-induced stimulation of these fibers [1]. Figures 3, 4, 5, and 6 depict two different study participants demonstrating how the device is used under the guidance of a trained physical therapist. Figure 7 shows the device in use while an individual performs physical exercise on a treadmill as part of the TLNS intervention.

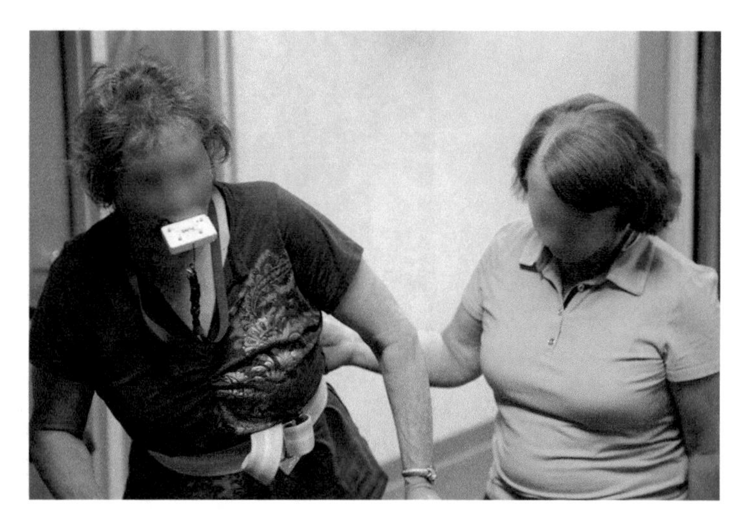

Fig. 5 A second participant demonstrates how the PoNS™ device is worn around the neck and held in the mouth

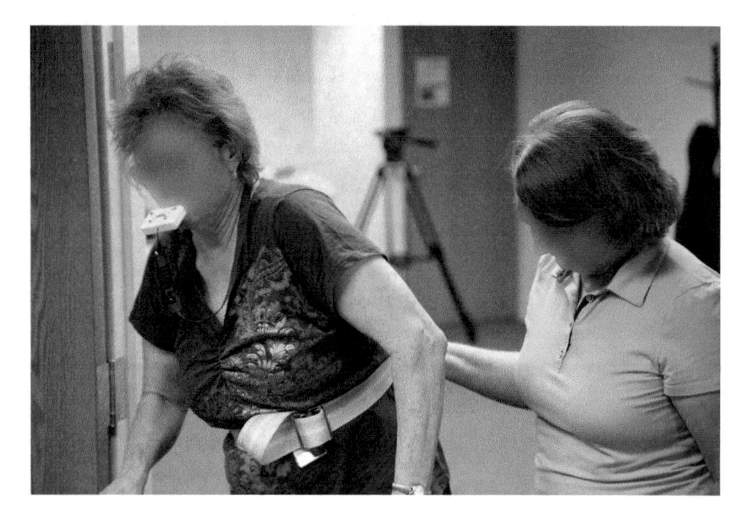

Fig. 6 The same individual as seen in Figure 5 demonstrates how the PoNS™ device is used while training with a physical therapist

All electrotactile systems, including the PoNS™ device, must adhere to a set of core principles to ensure comfortable and controllable tactile precepts, as well as safe operation. As these have already been extensively reviewed [23, 24], we will focus on the application of these principles specifically to the PoNS™ device. An expanded discussion of the waveform, electrode, and safety features appropriate for tongue stimulation has been previously published [1, 25].

Fig. 7 A participant uses the PoNS™ device under the supervision of a physical therapist while performing the exercise component of TLNS therapeutic intervention on a treadmill

3.1 Why the Tongue? Part I

Electrotactile stimulation supplanted the vibro-tactile stimulation because it is simpler, lighter, consumes less energy, and is easier to control the stimulus. Various improvements have led to the current system. It is an example of a new generation of sensory substitution devices based on computer-controlled electrical stimulation of the human skin in the most densely innervated tactile areas: the tongue and the fingers. The tongue was preferable because it affords a better environment (constant acidity level (pH), constant temperature and humidity, and low excitability thresholds) in comparison with the fingertip (variable hydration, thickness of the skin, surface contaminants, relatively limited and highly curved surface area available for stimulation, and high excitability thresholds).

3.2 Physical Construction

From around the neck, The PoNS™ device is held lightly in place using the tongue and lips by a tab that goes into the mouth and rests on the anterior, superior part of the tongue. The paddle-shaped tab of the system has a hexagonally patterned array of 143 gold-plated circular electrodes (1.50 mm diam., on 2.34 mm centers) that is created by a photolithographic process used to make printed circuit boards.

The PoNS™ device is powered by an internal battery so that it may be recharged via an external power supply that plugs into a 120-V or 240-V AC electric outlet, similar to a mobile phone.

3.3 Electrical Stimulation

The tongue electrodes deliver 19-V positive voltage-controlled pulses that are capacitively-coupled to: (A) limit maximal charge delivered in the rare event of circuit failure and to (B) ensure zero dc to the electrodes, thereby minimizing potential tissue irritation due to electrochemical reactions. However, it is important to note that such irritation has never been observed or reported using the PoNS™ device. Tongue sensitivity to positive pulses is greater than that of negative pulses. The pulse width is adjustable in 64 unequal steps from 0.4 to 60 μs by the intensity buttons. This intensity control scheme takes advantage of the steep section of the strength–duration relationship for electrical stimulation of neural tissue.

These pulses repeat at a rate of 200/s, which is within the typical physiological firing rate for tactile afferents. Because of the neural refractory period, and extrapolated from earlier single-fiber median nerve response to similar electrotactile stimuli on the rhesus monkey finger-pad [26], it is presumed that, at most, one action potentially occurs in any given afferent fiber for each stimulation pulse. To minimize sensory adaptation [26] and ensure good quality of sensation [27], every fourth pulse is removed from the pulse train, so that each electrode delivers a burst of three pulses every 20 ms.

This combination of pulse amplitude and width results in an electrotactile stimulus that may be varied by the user from well below a sensory threshold to a perceived sensation at the upper limit of comfortability.

3.4 Electrode Arrays and Pulse Sequencing

The PoNS™ electrode array is irregularly shaped to take advantage of the most sensitive regions of the tongue, and is comprised of 143 electrodes nominally organized into nine 16-electrode sectors. Within each sector, one electrode is active at any moment (pulse beginnings staggered by 312.5 μs), with unstimulated electrodes serving as the return current path. The nine sectors present simultaneous stimulation, with the intensity of each sector adjusted to compensate for the variability of tongue sensitivity to electrotactile stimuli. The sensation produced by the array has been described as similar to the feeling of drinking a carbonated beverage.

The impedance of the electrode, skin interface, presents as a resistive component of approx. 3 kΩ, in series with a resistive–capacitive network of 4–4.5 kΩ in parallel with 0.5 nF. The pulse current, therefore, contains a brief leading spike followed by an exponential decay to a plateau current of approximately 3 mA. Voltage control is used rather than the current control of typical electrotactile systems because the tongue electrode impedance is relatively stable compared to cutaneous loci otherwise used (such as the abdomen or fingertip). Use of voltage control affords circuit simplicity and therefore economy of component, space, and battery.

The electrode size and geometry were chosen to achieve a reasonable balance between the number of electrodes that may be packed into the array area and the comfortability and controllability of the electrotactile percept. The overall result of this stimulation is the comfortable and convenient presentation of almost 26 million stimulation pulses to the tongue during a typical 20-minute therapy session. How many action potentials propagate to the brain as a result of this surface stimulation at this point is unknown.

4 Translingual Neurostimulation (TLNS)

4.1 How the PoNS™ Works

In brief, TLNS uses sequenced patterns of electrical stimulation on the anterior dorsal surface of the tongue to stimulate the trigeminal and facial nerves. From a technical point of view, the electrical stimulation of the tongue skin by PoNS™ device is likely one of the safest. In each particular moment, only 1 set of 9 out of 143 PoNS™ electrodes are active, surrounded by ground electrodes (16 possible sets in total). Each electrode creates the area of activation—1.77 mm^2. The set of nine electrodes activates total area of 15.9 mm^2 or 0.16 cm^2. That is a maximum area of the tongue surface activated by PoNS™ in the single moment.

The signal pattern on each electrode is a sequence of very short rectangular impulses (fixed 19 V value) with a duration of each from 0.4 to 60 μs. For comparison, each natural neural impulse (spike duration) is 1 ms = 1000 μs (1.5–3 ms is a length of full spike waveform).

Considering the pattern of stimulation (sequence of triplet bursts), total stimulation time for single electrodes is in a range from 0.1 to 11 s, during typical PoNS™ application for 20 minutes. Due to the multiplexed nature of the stimulation, each set of 16 electrodes, therefore, delivers a total stimulation time of 1–173 s during one full session.

The depth of stimuli penetration in the PoNS™ device is fixed because perceptual intensity is regulated by signal duration, not by current or voltage. The normal thickness of the human tongue epithelia varies from 400 to 800 μm (0.4–0.8 mm). In the deeper layers, there are muscular fibers, which are moving the human tongue. Electrical stimulation of such fibers creates a very distinct sensation of jerking movements of the tongue surface. Our subjects never reach such sensation because they are instructed to limit the intensity of stimulation to their "maximal comfortable level." So, we can suggest that the depth of real activation is about 400–600 μm (0.4–0.6 mm) range. Then total volume of electrically activated tissue in each particular moment is—for one electrode—0.53–1.06 mm^3, for the set of nine electrodes—4.78–9.56 mm^3. Therefore, from a technical point of view, the set of nine electrodes PoNS™ device activates 0.16 cm^2 in area and 5–10 mm^3 of tissue volume in each particular moment, but not more than 154 s during 20 minutes' session.

4.2 Why the Tongue? Part II

The anterior dorsal surface of the tongue is a patch of the human skin with a unique innervation pattern. The relatively thin (in comparison to other skin areas) oral epithelium is saturated by a different kind of mechanic there, and taste receptors in addition to free nerve endings stratified in its depth. It is the area with the maximal density of mechanoreceptors, and, like the fovea in the retina, have the minimum two-point discrimination threshold—0.5–1 mm for mechanical stimulation [28] and 0.25–0.5 mm for electrotactile stimulation (unpublished data). The physical density, spatial distribution, size of the receptive fields and their overlapping coefficient, spatial and temporal summation properties are largely unknown, especially for electrotactile stimulation [29].

The two major nerves from the tip of the tongue deliver information streams directly to the brainstem—the lingual nerve (the texture of food) and chorda tympani (taste of food). According to our approximation, 20–25 thousand neural fibers deliver neural impulses from this area (about 7.5 cm^2) covered by PoNS™ electrode array.

4.3 CN-NINM Technology Platform

It is important to clarify that CN-NINM is a platform that consists of many technologies, all of which target cranial nerves—primarily vagal and trigeminal—with the intention of influencing the central nervous system. NeuroSigma™, Cefaly™, NEMOS™, and SYMPATOCOR™ are examples of other technologies within the CN-NINM canon. TLNS, using the PoNS™ device, is a novel class of stimulation to join this existing platform.

However, these other stimulation devices and techniques target their stimulation to nerve trunks. TLNS alternatively targets the receptors and nerve endings. In this way, TLNS is closer to natural stimulation than the other techniques because synchronous stimulation of nerve receptors is a more natural input than stimulation of nerve trunks.

5 Induced Neuroplasticity

Our hypothesis is that TLNS induces neuroplasticity by noninvasive stimulation of two major cranial nerves: trigeminal, CN-V, and facial, CN-VII. This stimulation excites a natural flow of neural impulses to the brainstem (e.g. pons varolli and medulla) and cerebellum, via the lingual branch of the cranial nerve (CN-Vc) and chorda tympani branch of CN-VII, to effect changes in the function of these targeted brain structures [30].

The spatiotemporal trains of neural activation induced in these nerves eventually produce changes of neural activity in corresponding nuclei of the brainstem—at least in the sensory and spinal nuclei of trigeminal nuclei complex (the largest nuclei in the brainstem, extending from the midbrain to the nuclei of the descending spinal tracts), and the caudal part of the nucleus tractus solitarius, cochlear, cuneate and hypoglossal nuclei and upper segment of the spine (C1–C3), where both stimulated nerves have direct projections.

Changes in neural activity were evident in the results of our pilot study, wherein we also developed a new fMRI signal processing method to yield high-resolution images of the pons, brainstem, and cerebellum beyond that previously reported, allowing observation of changes in functional activity in all of the regions of interest [31–33]. We are particularly interested in these specific changes in the pons, brainstem, and cerebellum because these neural structures are the major sensory integration and movement control centers of the brain and therefore primary targets for neuromodulation.

We postulate that the intensive activation of these structures initiates a sequential cascade of changes in neighboring and/or connected nuclei by direct projections and collateral connections, by activation of brainstem interneuron circuitries (reticular formation of the brainstem), and/or by passive transmission of biochemical compounds in the intercellular space (release of neurotransmitters in the synaptic gaps). The stream of neural impulses leads to activation of corresponding neural networks and massive release of neurotransmitters that eventually activate the glial networks of the brainstem (responsible for maintenance of the neuronal environment and synaptic gaps).

This, in turn, causes radiating therapeutic neurochemical and neurophysiological changes effecting both (a) synaptic and extrasynaptic circuitries and (b) information processing of afferent and efferent neural signals involved in movement control, including the cerebellum and nuclei of spinal motor pathways.

The temporal pattern of our observed retention effects is strikingly similar to the process well known in neuroscience literature for several decades as long-term potentiation (LTP) and depression (LTD). Both processes were tested and verified in multiple animal models by analyzing changes in brain tissue samples, and both are in intensive use in different models of human processes of learning and memory as a basic mechanism of the synaptic plasticity of the brain [34–38].

In brief, synaptic plasticity is a natural manifestation of activity-dependent processes affecting structure and function of multiple neuronal networks. As a result of such processes, multiple consequential adaptive changes are happening on different levels of brain organization (molecular, cellular, regional, and systemic), with different temporal patterns and dynamics (short and long) that reflects on multiple sensory and motor functions, cognitive performance, and behavior [39–42].

Intensive repetitive stimulation of neurons leads to the corresponding activation of synaptic contacts on the axonal tree, including the whole complex of pre- and post-synaptic neurochemical mechanisms. Multiphasic fluctuations of postsynaptic potentials, frequently described as short-term activity-dependent synaptic plasticity (in the range of milliseconds, seconds and minutes) has been shown capable of enhancing synaptic transmissions [43, 44].

In contrast, long-term potentiation (LTP) is the phenomena of synaptic structural remodeling and formation of new synaptic contacts that is activated by high-frequency stimulation [44–50]. After 10–40 min of high-frequency stimulation (50–400 Hz, range of frequencies is used in animal research) the number of synapses and proportion of multiple spine boutons can increase the efficiency of neural connections. Effects of LTP can continue for several hours and even days [51].

In our experiments using the PoNS™ device, prolonged and repetitive activation (20 min or more) of functional neuronal circuits (e.g. balance and gait) can initiate long-lasting processes of neuronal reorganization, (similar to LTP), that we can see and measure in subjects' behavior. The functional improvement after initial training sessions continues for several hours. Multiple regular sequential training sessions lead to the consistent increase of improved symptom duration and cumulative enhancement of affected functions.

This regular excitation may also increase the receptivity of numerous other neural circuitries and affect internal mechanisms of homeostatic self-regulation, according to the contemporary concept of synaptic plasticity. We cannot exclude that this induces simultaneous activation of serotonergic and noradrenergic regulation systems of the brain as well.

The result of this intervention is essentially brain plasticity on demand—a priming or up-regulating of targeted neural structures to develop new functional pathways, which is the goal of neurorehabilitation and a primary means of functional recovery from permanent physical damage caused by stroke or trauma.

The effectiveness of TLNS was demonstrated in multiple case studies (more than 300 subjects) during the last 10 years. In brief, statistically significant improvements in balance and gait were recorded in: An MS pilot study (13 subjects); MS control study (10/10 subjects); a pilot study on balance disorders (23 subjects); a pilot stroke study (5 subjects); and traumatic brain injury study (45 subjects). The results of these cases were presented at various conferences and publications are currently in preperation.

An independent control study of the effect of TLNS on balance and gait in MS subjects [52] was conducted in the Montreal Neurological Institute and Hospital (MNIH). The comparison of fMRI images before and after TNLS revealed significant changes in the activity of the cortical areas responsible for gait in the active vs. control group. Surprisingly, significant changes in blood oxygen-level dependent (BOLD) signals were also present in areas responsible for working memory (dorsolateral prefrontal cortex (DLPFC), and right anterior cingulate cortex (rACC)). Results are in press.

Significant improvements in balance and gait using TLNS stimulation was found and reported in subjects with chronic spinal cord injury [53], and in children with posterior fossa [54] and cerebral palsy [55].

6 Summary

TLNS is an effective combination of several existing neurostimulation techniques. It stimulates the trigeminal nerve, similar to eTNS and Cefaly™, but targets a different branch (V3, instead of V1, the largest branch of the trigeminal nerve). Also, it simultaneously stimulates the facial nerve (chorda tympani) branch and corresponding solitary nuclei, as VNS does, but non-invasively. Furthermore, we are observing activation of the ventral cerebellum as a result of tongue stimulation. There is solid evidence from animal research that stimulation of the anterior third of the tongue can activate the hypoglossal nuclei directly and antidromically. There is human anatomical data supporting the hypothesis that TNLS might be considered a soft, non-invasive version of DBS.

Granted, this cranial nerve neurostimulation technology is coming through its first arduous steps of development. Many more studies, controlled and blinded, should be conducted. New problems and solutions must be discovered before we have a better understanding of the mechanisms of action for TLNS.

7 Conclusions

Cranial nerve neurostimulation is a new, small, but distinct set of technologies among the wide family of peripheral nerve stimulation methods that represent a unique approach to neurorehabilitation of multiple disorders and the wide spectrum of malfunctions in the human central nervous system.

As an example of safe, non-invasive, easy-to-manage, patient-oriented technology, TLNS can be considered an alternative way to approach previously untreatable symptoms and conditions, like the chronic symptoms of traumatic brain injury (TBI). TLNS may also be applied to the spectrum of acquired brain injury manifestations, such as sport concussions, TBI, whiplash, stroke, and others with varying degrees of severity. Our position is reinforced by a recent publication, which demonstrates that trigeminal nerve stimulation improves blood perfusion throughout the brain, whereby allowing prevention of secondary brain damage [56]. This finding corroborates what we see in our clinical experiments where we observe recovery from the chronic stage of disorder.

The PoNS™ device is the ideal clinical tool. The scope of clinical applications will continue to grow due to several unique characteristics: it is multi-directional, effective, noninvasive, safe, and the stimulation is repeatable and easy to control. None of the described harms, typical of other invasive and non-invasive neurostimulation methods, are applicable to TLNS. Moreover, the majority of the side-effects produced by existing clinical devices are the targets for rehabilitation and improvement for TLNS technology. We have never observed an effect of overstimulation or "overdose" with the PoNS™ or any negative effects. However, the possible occurrence of minor, episodic discomfort or mild headache episodes during developmental or adaptive stages should be brought into consideration.

The physiological nature and network-based principles of TNLS make it a good vehicle for driving the emerging perspective on the neural network origin of many neurological disorders and the recovery of functional systems as an appropriate means of neurorehabilitation.

Acknowledgments

This research is made possible by the additional efforts of Mitchell Tyler, Kim Skinner, Kurt Kaczmareck, Jannet Ruhland, and Georgia Corner.

Disclosure

The lead author has a financial interest in Advanced NeuroRehabilitation LLC and in Helius Medical Technologies, which both have intellectual property rights in the field of use reported in this article.

References

1. Danilov Y, Kaczmarek K, Skinner K, Tyler M (2015) Cranial nerve noninvasive neuromodulation: new approach to neurorehabilitation. In: Kobeissy FH (ed) Brain neurotrauma: molecular, neuropsychological, and rehabilitation aspects. CRC, Boca Raton, pp 605–628
2. Tyler ME, Kaczmarek KA, Rust KL, Subbotin AM, Skinner KL, Danilov YP (2014) Non-invasive neuromodulation to improve gait in chronic multiple sclerosis: a randomized double blind controlled pilot trial. J Neuroeng Rehabil 11(1):79
3. Filippi M, van den Heuvel MP, Fornito A, He Y, Pol HEH, Agosta F et al (2013) Assessment of system dysfunction in the brain through MRI-based connectomics. Lancet Neurol 12(12):1189–1199
4. Pol HH, Bullmore E (2013) Neural networks in psychiatry. Eur Neuropsychopharmacol 23 (1):1–6

5. Stam CJ (2014) Modern network science of neurological disorders. Nat Rev Neurosci 15 (10):683–695
6. Ghulyan-Bedikian V, Paolino M, Paolino F (2013) Short-term retention effect of rehabilitation using head position-based electrotactile feedback to the tongue: influence of vestibular loss and old-age. Gait Posture 38(4):777–783
7. Robinson BS, Cook JL, Richburg CM, Price SE (2009) Use of an electrotactile vestibular substitution system to facilitate balance and gait of an individual with gentamicin-induced bilateral vestibular hypofunction and bilateral transtibial amputation. J Neurol Phys Ther 33 (3):150–159
8. Tyler M, Danilov Y, Bach-Y-Rita P (2003 Dec) Closing an open-loop control system: vestibular substitution through the tongue. J Integr Neurosci 2(2):159–164
9. Harbourne R, Becker K, Arpin DJ, Wilson TW, Kurz MJ (2014) Improving the motor skill of children with posterior fossa syndrome: a case series. Pediatr Phys Ther 26(4):462–468
10. Kublanov VS (2008) A hardware-software system for diagnosis and correction of autonomic dysfunctions. Biomed Eng 42(4):206–212
11. Harry JD, Niemi JB, Kellogg S, D'Andrea S. System and method for neuro-stimulation [Internet]. US9616234 B2, 2017. Available from: http://www.google.com/patents/US9616234
12. Neren D, Johnson MD, Legon W, Bachour SP, Ling G, Divani AA (2016) Vagus nerve stimulation and other neuromodulation methods for treatment of traumatic brain injury. Neurocrit Care 24(2):308–319
13. George MS, Aston-Jones G (2010) Noninvasive techniques for probing neurocircuitry and treating illness: vagus nerve stimulation (VNS), transcranial magnetic stimulation (TMS) and transcranial direct current stimulation (tDCS). Neuropsychopharmacology 35 (1):301–316
14. Slavin KV. Peripheral Nerve Stimulation. Volume 24 of Progress in neurological surgery. Karger Medical and Scientific Publishers, 2011.
15. Wolpaw JR, Tennissen AM (2001) Activity-dependent spinal cord plasticity in health and disease. Annu Rev Neurosci 24:807–843
16. Straube A, Ellrich J, Eren O, Blum B, Ruscheweyh R (2015) Treatment of chronic migraine with transcutaneous stimulation of the auricular branch of the vagal nerve (auricular t-VNS): a randomized, monocentric clinical trial. J Headache Pain 16:543
17. Straube A, Eren O, Gaul C (2016) Role of the vagal nerve in the pathophysiology and therapy of headache. MMW Fortschr Med 158 (6):74–76
18. Danilov, Y. P., Kublanov, V. S., Retjunskij, K. J., Petrenko, T. S., & Babich, M. V. (2015). Non-invasive Multi-channel Neurostimulators in Treatment of the Nervous System Disorders. In BIODEVICES (pp. 88–94).
19. Petrenko TS, Kublanov VS, Retiunskiy KY (2015) Dynamic correction of the activity sympathetic nervous system (Dcasns) to restore cognitive functions. Eur Psychiatry 30:843
20. Retyunskii KY, Kublanov VS, Petrenko TS, Fedotovskikh AV (2016) A new method for the treatment of Korsakoff's (amnestic) psychosis: neurostimulation correction of the sympathetic nervous system. Neurosci Behav Physiol 46(7):748–753
21. Soss J, Heck C, Murray D, Markovic D, Oviedo S, Corrale-Leyva G et al (2015) A prospective long-term study of external trigeminal nerve stimulation for drug-resistant epilepsy. Epilepsy Behav EB 42:44–47
22. Miller S, Sinclair AJ, Davies B, Matharu M (2016 Oct) Neurostimulation in the treatment of primary headaches. Pract Neurol 16 (5):362–375
23. Kacznaarek KA, Bach-Y-Rita P (1995) Tactile displays. Virtual Environ Adv Interface Des 55:349
24. Szeto AYJ, Saunders FA (1982) Electrocutaneous stimulation for sensory communication in rehabilitation engineering. IEEE Trans Biomed Eng BME-29(4):300–308
25. Kaczmarek KA (2011) The tongue display unit (TDU) for electrotactile spatiotemporal pattern presentation. Sci Iran 18(6):1476–1485
26. Kaczmarek KA, Tyler ME, Brisben AJ, Johnson KO (2000) The afferent neural response to electrotactile stimuli: preliminary results. IEEE Trans Rehabil Eng 8(2):268–270
27. Kaczmarek KA, Webster JG, Bach-y-Rita P, Tompkins WJ (1991) Electrotactile and vibrotactile displays for sensory substitution systems. IEEE Trans Biomed Eng 38(1):1–16
28. Vallbo AB, Johansson RS (1984) Properties of cutaneous mechanoreceptors in the human hand related to touch sensation. Hum Neurobiol 3(1):3–14
29. Johansson RS, Vallbo AB (1979) Tactile sensibility in the human hand: relative and absolute densities of four types of mechanoreceptive units in glabrous skin. J Physiol 286:283–300
30. Wildenberg JC, Tyler ME, Danilov YP, Kaczmarek KA, Meyerand ME (2010) Sustained

cortical and subcortical neuromodulation induced by electrical tongue stimulation. Brain Imaging Behav 4(3–4):199–211

31. Wildenberg JC, Tyler ME, Danilov YP, Kaczmarek KA, Meyerand ME (2011) Electrical tongue stimulation normalizes activity within the motion-sensitive brain network in balance-impaired subjects as revealed by group independent component analysis. Brain Connect 1 (3):255–265
32. Wildenberg JC, Tyler ME, Danilov YP, Kaczmarek KA, Meyerand ME (2011) High-resolution fMRI detects neuromodulation of individual brainstem nuclei by electrical tongue stimulation in balance-impaired individuals. NeuroImage 56(4):2129–2137
33. Wildenberg JC, Tyler ME, Danilov YP, Kaczmarek KA, Meyerand ME (2013) Altered connectivity of the balance processing network after tongue stimulation in balance-impaired individuals. Brain Connect 3(1):87–97
34. Bear MF, Malenka RC (1994) Synaptic plasticity: LTP and LTD. Curr Opin Neurobiol 4 (3):389–399
35. Dudek SM, Bear MF (1993) Bidirectional long-term modification of synaptic effectiveness in the adult and immature hippocampus. J Neurosci 13(7):2910–2918
36. Kirkwood A, Lee HK, Bear MF (1995) Co-regulation of long-term potentiation and experience-dependent synaptic plasticity in visual cortex by age and experience. Nature 375(6529):328–331
37. Larkman AU, Jack JJB (1995) Synaptic plasticity: hippocampal LTP. Curr Opin Neurobiol 5 (3):324–334
38. Levenes C, Daniel H, Crépel F (1998) Long-term depression of synaptic transmission in the cerebellum: cellular and molecular mechanisms revisited. Prog Neurobiol 55(1):79–91
39. Shin R-M, Tsvetkov E, Bolshakov VY (2006) Spatiotemporal asymmetry of associative synaptic plasticity in fear conditioning pathways. Neuron 52(5):883–896
40. Shin R-M, Tully K, Li Y, Cho J-H, Higuchi M, Suhara T et al (2010) Hierarchical order of coexisting pre- and postsynaptic forms of long-term potentiation at synapses in amygdala. Proc Natl Acad Sci U S A 107 (44):19073–19078
41. Tully K, Li Y, Bolshakov VY (2007) Keeping in check painful synapses in central amygdala. Neuron 56(5):757–759
42. Kublanov VS, Petrenko TS, Babich MV (2015) Multi-electrode neurostimulation system for treatment of cognitive impairments. Proc Annu Int Conf IEEE Eng Med Biol Soc 2015:2091–2094
43. Zucker RS (1999 Jun) Calcium- and activity-dependent synaptic plasticity. Curr Opin Neurobiol 9(3):305–313
44. Zucker RS, Regehr WG (2002) Short-term synaptic plasticity. Annu Rev Physiol 64:355–405
45. Buchs PA, Muller D (1996) Induction of long-term potentiation is associated with major ultrastructural changes of activated synapses. Proc Natl Acad Sci 93(15):8040–8045
46. Calverley RK, Jones DG (1990) Contributions of dendritic spines and perforated synapses to synaptic plasticity. Brain Res Brain Res Rev 15 (3):215–249
47. Engert F, Bonhoeffer T (1999) Dendritic spine changes associated with hippocampal long-term synaptic plasticity. Nature 399 (6731):66–70
48. Geinisman Y, Berry RW, Disterhoft JF, Power JM, Van der Zee EA (2001) Associative learning elicits the formation of multiple-synapse boutons. J Neurosci 21 (15):5568–5573
49. Jones DG, Itarat W, Calverley RKS (1991) Perforated synapses and plasticity. Mol Neurobiol 5(2–4):217–228
50. Toni N, Buchs PA, Nikonenko I, Bron CR, Muller D (1999) LTP promotes formation of multiple spine synapses between a single axon terminal and a dendrite. Nature 402 (6760):421–425
51. Bliss TV, Lømo T (1973) Long-lasting potentiation of synaptic transmission in the dentate area of the anaesthetized rabbit following stimulation of the perforant path. J Physiol 232 (2):331–356
52. Leonard G, Lapierre Y, Chen JK, Wardini R, Crane J, Ptito A (2017). Noninvasive tongue stimulation combined with intensive cognitive and physical rehabilitation induces neuroplastic changes in patients with multiple sclerosis: A multimodal neuroimaging study. Multiple Sclerosis Journal–Experimental, Translational and Clinical, 3(1), 2055217317690561.
53. Chisholm AE, Malik RN, Blouin JS, Borisoff J, Forwell S, Lam, T (2014). Feasibility of sensory tongue stimulation combined with task-specific therapy in people with spinal cord injury: a case study. Journal of neuroengineering and rehabilitation, 11(1), 96.
54. Harbourne R, Becker K, Arpin DJ, Wilson, TW, Kurz MJ (2014). Improving the motor

skill of children with posterior fossa syndrome: A case series. Pediatric Physical Therapy, 26(4), 462–468.

55. Ignatova T, Kolbin V, Scherbak S, Sarana A, Sokolov A, Trufanov G, Semibratov N, Ryzhkov A, Efimtsev A, Danilov Y Translingual Neurostimulation in Treatment of Children with Cerebral Palsy in the Late Residual Stage. Case Study. DOI: 10.5220/0006732403320337. In Proceedings of the 11th International Joint Conference on Biomedical Engineering Systems and Technologies (BIOSTEC 2018) - Volume 4: BIOSIGNALS, 332–337 ISBN: 978-989-758-279-0
56. Chiluwal A, Narayan RK, Chaung W, Mehan N, Wang P, Bouton CE et al (2017) Neuroprotective effects of trigeminal nerve stimulation in severe traumatic brain injury. Sci Rep 7(1):6792

Chapter 20

The Rapid Templating Process for Large Cranial Defects

Jeremy Kwarcinski, Philip Boughton, Andrew Ruys, and James van Gelder

Abstract

Cranioplasty, the reconstructive method utilized in the treatment of cranial defects, has evolved based on emerging technology, both in relation to process and material capabilities. Advances in manufacturing technology have led to the emergence of the rapid templating process, defined as *"the creation of patient specific templates, through rapid prototyping methods, which allow the creation of implants intraoperatively and minimise the lead time associated with custom implant manufacture."* This process, a fusion of traditional casting methods, emerging rapid prototyping technology, and patient-specific design, presents an effective and clinically viable system of cranial defect treatment. Here, we describe the history, methods, and capabilities of the rapid templating system, to provide the reader with an understanding of the form and function of the system and to highlight advantages and limitations.

Key words Cranioplasty, Cranial defect, Patient-specific implant, Rapid prototyping, Rapid templating, 3D Printing

1 Anatomy and Pathology: Causes and Effects of Skull Defects

1.1 Introduction

Cranial, craniofacial and craniomaxillofacial trauma refers to damage of the skull and facial bones. Such trauma, and associated injury to surrounding tissues, has high latent morbidity and mortality [1]. According to current literature, the primary causes of damage of the skull are motor vehicle accidents, workplace industrial accidents, falls, sporting injuries, and violence [2–4]. Removal of portions of the skull (bone flaps) is required in the treatment of trauma of the skull, as well as to treat infection, tumor removal, and brain swelling [5]. The anatomy of such bone flaps can be widely varied in size and shape.

1.2 Pathology Associated with Skull Defects

In many surgical procedures, significant portions of neurocranial bone may be removed from the skull to effectively treat patient symptoms [4, 5]. Without a complete skull, the patient is at an increased risk [5, 6] of traumatic brain injury, which can lead to

Amit K. Srivastava and Charles S. Cox, Jr. (eds.), *Pre-Clinical and Clinical Methods in Brain Trauma Research*, Neuromethods, vol. 139, https://doi.org/10.1007/978-1-4939-8564-7_20,

further patient condition decay and potentially death. In many cases, the aftermath of this removal is a closure of the cranial tissues with a gap in cranial anatomy, which can lead to numerous issues. Syndrome of the trephined, colloquially known as sinking skin flap syndrome, is a neurological complication which can arise after a large craniectomy. Symptoms of the condition, as highlighted in literature, include motor weakness, cognitive and language deficits, altered consciousness levels, headaches, seizures and/or electroencephalographic variation [7–10]. In addition, aesthetic results also present a significant consideration when considering skull defects, as an absence of bone and/or anatomically incorrect implant can lead to further psychosocial and psychological issues [6, 11].

1.3 Treatment Methods for Skull Defects

Treatment methods for cranial injuries vary based on the type and severity of the injury [2–5, 12]. The focus of surgical treatment is to repair skeletal anatomy, with the aim of restoring the nature architecture to the closest possible extent. There are numerous methods which are utilized to treat such trauma. These include moving broken bones—if present—back into place and securing them utilizing pins, screws, and plates, using graft materials to fill missing smaller bone cavities to promote regenerative growth, or, if the trauma is particularly severe, using an implant to act as a structural and functional replacement of lost skull structure (cranioplasty) [13–15].

Cranioplasty is defined as the repair of a skull defect through surgical means. The injury profiles of cranioplasty implants vary significantly from case to case, given the varied causes and injury profiles of head trauma cases. However, the surgical process remains relatively simple; lifting the scalp of the patient and replacing the missing hard tissue [13–15], as reflected in Fig. 1. The core benefits which drive cranioplasty procedures are protection (ensuring the brain remains shielded from harm), neurological function (preventing neurological decay through medical issues including

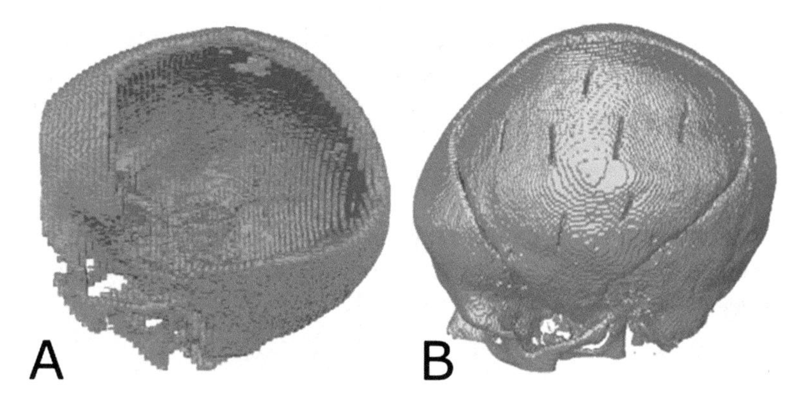

Fig. 1 Cranial reconstruction, (**a**) Damaged skull, (**b**) Reconstructed Skull with Rapid Templated Implant

syndrome of the trephined), and cosmetic outcomes (reconstruction of the natural anatomy in a way which, after healing, does not alter the appearance of the patient to a significant extent). Ultimate patient recovery depends on numerous factors, including type, severity, and cause of injury; previous cases of infection; additional surgeries; as well as many other factors which vary on an individual patient basis.

Time remains a key factor when considering cranial reconstruction. Literature highlights a consensus that better treatment outcomes—functional and cosmetic—are achieved with faster repair of such injuries [16–19]. Considering this from a patient-specific implant context, we note that time is a factor which can readily increase for complex patient cases, given the increased lead times for associated with individualized manufacture.

2 Conventional Cranioplasty Implant Systems: Current Methods of Cranial Reconstruction

2.1 Cranioplasty Implant Materials and Manufacturing Methods

There are several techniques which can be implemented to generate a cranioplasty specifically tailored to a patient. These can be differentiated based on implant material and manufacturing method.

Autologous implants can be considered the gold standard for cranioplasty applications. This is because it has the greatest regenerative capacity, which in turn can lead to the best patient outcome, complete regeneration of a skull defect; with form and function matching that of the original anatomical structure [6, 20].

However, in many cases use of an autologous bone is not possible, due to either resorption, infection, or rejection [21–23] and thus alternative materials must be used. The most common artificial materials for cranioplasty applications are polymethylmethacrylate (PMMA), Titanium, and polyetheretherketone (PEEK).

2.1.1 Polymethylmethacrylate (PMMA)

PMMA is a clinically accepted biocompatible thermosetting polymer with extensive history of medical usage, particularly in lost bone reconstruction [24–28]. PMMA has the capacity for enhancement using additive components, antibiotics representing one of the most common accompaniments [24, 29–31]. While PMMA is biocompatible, there are numerous issues which potentially lead to patient complications. Exothermic curing, with temperatures between 70 and 120 °C, can lead to thermal necrosis of tissues surrounding the PMMA [32–34]. This curing process is also thought to potentially lead to hypotension issues [35]. These issues are only present should PMMA be cured directly within the body, and thus can be removed by forming the cranioplasty external to the patient.

PMMA cranioplasty implants are formed through two distinct processes, prefabrication (fabrication of the implant prior to surgery) [22, 26, 27, 36–40] and intraoperative formation, which in turn appear in two distinct forms—hand-formed (shaped by the surgeon by hand) [28, 36, 37, 41–44] and intraoperative molding (manufactured intraoperatively using custom tools) [6, 45–48].

2.1.2 Titanium

Titanium is a biostable metal with clinical acceptance as an orthopedic implant material [20, 49, 50]. Titanium has a natural capacity for osseointegration and strength-to-weight ratio favorable for hard-tissue replacement applications [51]. Titanium requires prefabrication for cranioplasty applications and is commonly found in both a solid plate and a mesh structure. These implants are traditionally formed into the required anatomical shapes through the use of bending and pressing processes [20, 50, 52].

The emergence of additive manufacturing technologies (3D printing) allows the direct fabrication of complex metal structures, titanium included. Titanium 3D printing has been trailed clinically in the production of sternum and heel implants [35, 53, 54]. The medical 3D printing of titanium presents an emerging technology that has potential for future use in cranioplasty; however, there do not appear to be any applications in this field at this point.

2.1.3 Polyetheretherketone (PEEK)

PEEK is a biocompatible thermoplastic which is regarded as a valid alternative to metallic orthopedic implants, providing a balance between strength and rigidity in order to assist in reducing the risks associated with stress shielding [49, 55]. PEEK implants are traditionally prefabricated and are finding increased acceptance in the field of cranioplasty [56, 57].

2.2 Material Choice and Cranioplasty

When choosing an appropriate implant material for generation of a patient-specific cranioplasty, there are numerous considerations which must be examined. Defect shape and cause, surgical and clinical history, lifestyles characteristics, and other requirements present alternative outcome requirements [41, 46, 58–63], which in turn may influence the material and manufacturing method of potential reconstructive materials, as well as the associated fabrication and implantation cost.

3 Patient-Specific Implants, Manufacturing 2.0 and Rapid Templating: Shifting Trends in Implant Manufacturing

3.1 Shifting Trends in Medical Implants: Best-Fit vs Patient Specific

As medical and manufacturing technology has advanced, the ability to produce individual tailored implants has become both simplified and cost-effective [64, 65]. A major driving factor of this trend is the ability for surgical professionals to preoperatively view and examine patient-specific anatomy and geometry through accurate digital anatomical modeling. This has assisted in creating better pathways for surgical planning, practice, training, and implant design [64, 66–70], leading to an increased ease in development of tailored treatment methods and components.

This trend represents a paradigm shift regarding the current implant logistics system, as a shift occurs from traditional, generic, high-volume system, to an adaptable, fit-for-purpose, small-volume structure. This does not indicate a complete removal of traditional implant systems, rather it represents a focus shift from best-fit to tailored-fit approach, a change based on advancing technology and the ability to apply it to produce custom treatment methods, in shorter timeframes, with better functional and aesthetic end results.

3.2 Rapid Prototyping and Manufacturing 2.0: Rise of 3D Printing

3.2.1 Additive Manufacturing (3D Printing)

Initially utilized in the late 1970s and early 1980s, additive manufacturing, also known as 3D Printing, is a process which creates a 3-dimensional object through material fusion processes [71–74]. The process is referred to as "additive" due to the nature of gradual deposition of material, either in a solid, liquid, or powdered state, building a product from the ground up. This is in comparison to traditional manufacturing methods, such as CNC milling, which are subtractive, creating an object by stripping material from a large core sample [75–77].

The rapid distinction of this form of manufacturing is related directly to the decreased turnaround time, a driving characteristic of the process [72]. This is due to the absence of tooling and the creation of products directly from Computer-Aided-Design (CAD) data. As time has passed, the technology integrated within these machines has allowed them to function with increased speed, accuracy, and material capacity, as well as reducing the cost and size of machines, creating an increasingly accessible and capable technology.

3.2.2 Manufacturing 2.0 and Direct Digital Manufacturing (DDM)

Additive manufacturing is fundamental to both Manufacturing 2.0 and Direct Digital Manufacturing (DDM). Both Manufacturing 2.0 and DDM relate to the digitization of product development and production [78–80].

The nature of manufacturing is shifting from a fixed system, to one of increased flexibility and adaptability, production capable at any time and location given accessibility to additive manufacturing methods. The growing capabilities of these manufacturing methods present a transcendence of tradition supply chain processes; end users, either individuals or corporations (such as hospital) can adopt the role of manufacturer and produce required tools, guides, and implants in-house.

3.2.3 3D Printing Processes: Methods and Materials

The process of rapid additive manufacturing is relatively consistent regardless of method or material used. It involves (a) generation of a digital model which reflects the geometrical properties of the final

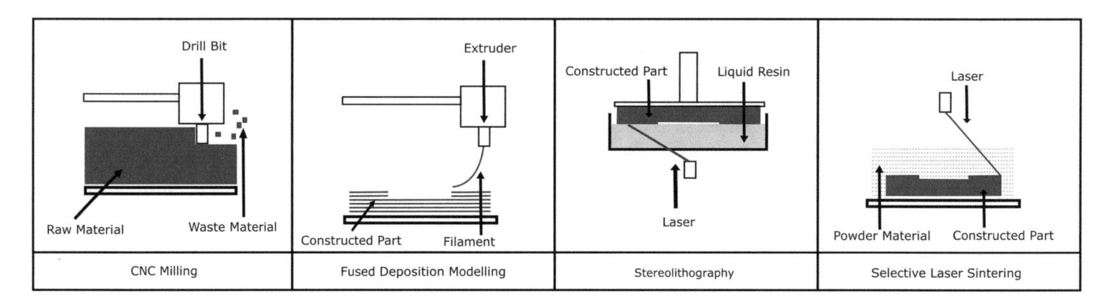

Fig. 2 Common subtractive and additive manufacture methods

device, (b) preparation of the 3D printer through calibration and material loading, (c) production of the final model, and (d) post-processing of the model as necessary [71–74, 80].

There are multitude of additive manufacturing systems which primarily differ in the methods in which they layer material to create parts, as shown in Fig. 2. These methods include:

Fused Deposition Modeling (FDM)

Fused Deposition Modeling (FDM) is one of the most widely used forms of additive manufacturing [81]. It utilizes the principle of extrusion, the production of material of a set cross-sectional area, to create a product through gradual layering of a stream of material. A material of set thickness is fed at a set rate, controlled by integrated motors, through a heat extruder nozzle, which dictates the thickness and orientation of the extruded material. Once the material leaves the nozzle, it cools and bonds, either to the set printer base or the printed component, to construct a 3D geometry layer-by-layer [77, 82, 83].

FDM technology is primarily designed to function with thermoplastic materials. Common materials include Polylactic Acid (PLA)—which is biocompatible [83]—and Acrylonitrile Butadiene Styrene (ABS) [84]. FDM machines have been used in the medical industry to create regenerative and bioresorbable scaffolds [85–87].

The limitation of FDM technologies lies in its geometrical construction capabilities. While fine resolutions can be achieved with more advanced machines, no FDM machine can construct extensive overhangs [77]. As a result, a thin support structure is required to provide support to overhanging features and allow production. These structures may be of either similar or different materials and require post-processing to remove, the difficulty of which is dictated by the potential dissolvability of the support material and or the strength and style of material bond. FDM systems are not the most accurate additive manufacturing systems and do not provide an ideal surface finish (without post-processing) for all applications. FDM technologies tend to present a cheaper option than counterpart 3D printing methods, both regarding machine operation and material sourcing [71, 77, 80].

Stereolithography (SLA)

Stereolithography (SLA) is a form of additive manufacturing which operates through the curing of photo-reactive resins using a guided UV laser. An object is created gradually by layer, with the required pattern cured by guiding the UV laser beam using a computer-guided mirror system; with the resin hardening and bonding to the print base or constructed object as soon as it is contacted by the UV light. The print base is gradually lifted upward as the model is built, with additional layers constructed until the entire model is complete and above the resin pool [71, 77, 80]. SLA technology utilized photo-polymeric resins to function and, as a result, is limited to this pool of materials. Many of these materials require post-curing, increasing the production lead time.

The process of production through SLA printing is much the same as FDM printing and, as a result, faces the same disadvantage of support material requirements [77, 80]. In addition, both the machine and material cost of SLA systems is greater than that of FDM technology, presenting the system as a costlier method of production. SLA systems are capable of production of fine detail and produce a good surface finish [77, 80].

Selective Laser Sintering (SLS) and Direct Metal Laser Sintering (DMLS)

Selective Laser Sintering (SLS) and Direct Metal Laser Sintering (DMLS) are additive manufacturing methods which utilize powerful CO_2 lasers to sinter powdered materials—plastics for SLS and metals for DMLS. Like the SLA process, an object is created gradually by layer, with directional information guided by a computer-guided mirror. However, rather than gradually rising the build platform from a liquid vat of material, the SLS process involves a gradual lowering of the build platform and the deposition of a layer of fresh powdered material for construction of each layer [77, 80].

Sintering technology allows for a wide range of materials, including both thermoplastics and metal alloys. As such, the materials choice associated with the technology is much higher than that of SLS and FDM systems. The cost of powder metal materials is greater than that of thermoplastic powder or filament, leading to a greater material cost for produced parts [77, 80].

Sintering processes create porous structures, which presents numerous opportunities for use [88]; although this may compromise the structural characteristics depending on desired product usage. It must be noted that a process called Selective Laser Melting (SLM) operates following a closely related process; however, it provides a complete melt (rather than the sintering of material) [89], leading to different functional properties and loss of porosity.

The operation of sintering machines is complex and the machines are costly to purchase and operate. Created parts will have a rough surface finish and, as a result, may require a degree of post-processing to function. However, the sintering process is capable of highly complex structures and does not require the use of support structures. In addition, it is capable of the production of high accuracy components and finer print resolutions than many FDM and SLS machines [77, 80].

3.3 Rapid Templating for Medical Implants

When we discuss the rapid templating process, it is fundamental that we examine three key concepts: templating, lead time, and rapid tooling.

Templating

Templating is a common manufacturing technique prevalent in a multitude of industries, including medical manufacturing. At its most basic it refers to the creation of objects through the generation of a "master," a core shape from which the desired form and function are obtained. This template can be used to make individual products, or be scaled for the mass manufacturing of hundreds of components; the total amount of products needed impacting potential template design and manufacture [77, 90].

Lead Time

At the core of lead time is the notion of latency, a period of delay between initiation and occurrence. In the manufacturing/production industry, lead time is a major factor in development, with implications regarding timeframe and cost. It becomes a major focus of project management in controlling and monitoring lead time to optimize the quality and quantity of output. In many cases, lead time is correlated to reliability (the value-of-reliability), indicating a trend of inventory-system cost savings associated with decreased lead-time variability. When we examine lead time from a supply chain viewpoint, whether that be for large- or small-scale operations, we note a trend highlighting that with increased volatility of demand, the marginal value of time increases. This volatility relates directly to the potential benefits associated with lead-time alterations; with a greater fluctuation in demand, it becomes beneficial to operate under shorter lead times [91]. When considering this in relation to patient-specific implants, we note high demand requirements. Each patient requires specific implant qualities and providing estimates for shape, size, volume and required material properties for patient injuries of varying location and severity is infeasible. Therefore, it becomes a fundamental requirement for any firm engaging in the production of such devices to minimize lead time to maximize the associated cost-benefits of production [92, 93].

Rapid Tooling

Rapid prototyping exists together with additive manufacturing (3D Printing). Unlike traditional tool manufacturing methods, which are subtractive in nature (removing unnecessary material to generate a desired shape), expensive and have significant lead time, the utilization of 3D printing allows generation of rapid production tools. These assist in reducing cost and lead time, through removal of tooling time and processes, leading to a decrease in development expenses and time to market. Rapid tooling is strongly suited to low-volume manufacturing, which makes it an ideal candidate for use in patent-specific implant manufacturing [94, 95].

Therefore, with these factors in mind, we define the rapid templating process as follows:

> *the creation of patient specific templates, through rapid prototyping methods, which allow the creation of implants intraoperatively and minimise the lead time associated with custom implant manufacture and sterilisation.*

4 Rapid Templating for Large Cranial Defects: Methods and Considerations

4.1 Creation of Digital Cranioplasty Implants and Implant Templates

When generating a patient-specific implant through the rapid templating process, the first step is generating a digital implant profile. Regardless of the purpose of use, the process of generating a digital image of patient-specific anatomy is as follows: (a) Medical Imaging Scans (such as CT scans or MRI scans) are taken of the patient to profile the anatomical structure of interest—in the case of cranioplasty generation CT scans of the skull defect are obtained. (b) These scans are processed using specialized software (Amira, Scan IP, 3D slicer, etc.) to convert the data into a recognizable shape and file format. (c) The implant geometry is obtained and altered using CAD software to generate an appropriate rapid template for manufacture [6, 45–47].

4.1.1 Patient-Specific Scans for Digital Implant Modeling

When we considered medical imaging, we note there is a wide variety of methods—each of which has benefits and disadvantages depending on the purpose of use and anatomy to be examined. Given the focus of this chapter on rapid templating for cranioplasty applications, CT scans are utilized given they provide significant detail for hard-tissue structures [46, 47, 96].

There is an observable relationship between ease of digital implant generation and scan resolution—greater resolution is tied to easier modeling circumstances. This issue is colloquially known as "stepping," where pronounced ridges are observable throughout the model [46]. This ridging increases the difficulty in generating the model—given the increased work time and effort required to create a smooth implant model—and presents potential accuracy issues due to the extrapolation required to bridge and smooth the surface. While it can be argued that for patient-specific implant manufacturing cases ridged guidelines are required to ensure ideal scan standards, there are feasibility issues which prevent this. Consider that scans for medical diagnostic purposes have pre-determined settings based on best practice guidelines [97] and that CT scans are associated with harmful radiation exposure and it remains imperative to limit patient exposure [98, 99]. With these factors in mind, we note that it may not be viable to obtain additional CT scans for the sole purpose of implant production—templates must be generated with available scans. That is not to say that models with greater degree of stepping are unworkable (within reason), but rather, these files required greater processing, skill, and effort to produce a model appropriate for manufacturing through the rapid templating process.

4.1.2 Digital Implant Generation

While the individual methods between institutions for digital implant/anatomical model generation vary [6, 45–48], there is commonality in the use of multiple software packages. At the core, the following is needed:

Conversion of Patient-Specific Data (CT/MRI) into a digital 3D model file capable of processing

This involves uploading DICOM files into the software package and isolating the required anatomical structures for modeling. It must be noted that depending on the requirements of the final product, it may be possible to restrict the modeled tissues to minimize processing time and effort.

To achieve this there is a requirement to apply image segmentation principles to the DICOM file. Image segmentation is the process of splitting a digital image into different segments with the purpose of highlighting features and increasing ease of analysis. From a medical imaging perspective, this allows the user to distinguish between different anatomical structures [100, 101]. To achieve this, a technique known a thresholding is utilized. Thresholding involves the differentiation of structures by converting an image into binary format; a range of pixel color values are chosen and all others are amalgamated and discounted [102, 103]. Current literature highlights that the various tissues of the body can be distinguished through different threshold values [104] although these may change depending on individual scan characteristics.

Generation of a digital model of the area of interest, injury profile and/or implant volume

This involves exporting the created geometry as a 3D file capable of being processed by 3D modeling software. Different software accepts different file formats—one of the most common being STL.

Generally, this process generates the object through a series of polygonal faces. The greater number of faces contained within the subject, the greater the object resolution and smoother the object surface [105]. However, there are numerous factors which need to be balanced to ensure effective and efficient model generation.

The first is resolution. While having a greater number of faces provides a smoother and more natural surface, it also increases the processing time, effort, and power required to generate a model. Depending on the computing power available, it may not be

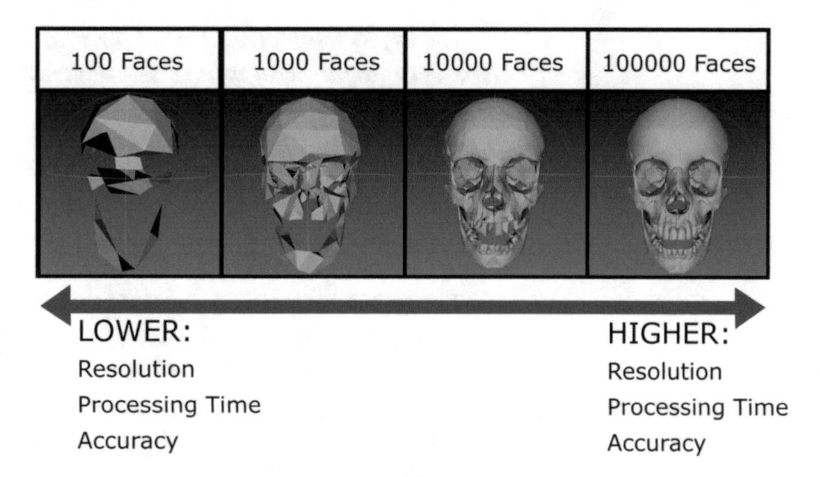

Fig. 3 Effect of remeshing and simplification on digital anatomical models

possible to alter or open the file until a degree of simplification and remeshing (restructuring the model by reducing number of faces while conserving shape, volume, and boundaries as much as possible) has occurred. The question therefor arises as to how far simplification and remeshing can be taken before compromising the integrity of the model generated. The relationship between these factors is highlighted in Fig. 3.

Based on this, it can be concluded that an ideal geometry for a digital implant model must be as high resolution as possible while ensuring it is still workable in CAD software. This is not a constant, varying based on implant shape and size and available processing power. Therefore, there exists a degree of judgment which must be taken by the engineer(s) involved to determine an appropriate workable threshold.

Generation of additional component(s)/feature(s) required to produce desired outcomes (molds, mounts, etc.)

Once a digital model has been generated, the worker (s) involved can alter the model to add/subtract components and features required to produce the desired end structure—such as placement guides, pin/screw locators, etc. This is completed using standard CAD software.

4.2 Rapid Template Design: Form and Function

4.2.1 Core Characteristics of the Rapid Templating Process

When we consider design in relation to the rapid templating process, we must highlight the differentiating aspects which characterize this method of patient-specific implant manufacture.

The output of the rapid templating process is a formation tool

While the rapid templating process is used to manufacture patient-specific implants/devices, it is not an implant. It is not designed to contact or enter the patient's body, but rather exists as a surgical tool which allows the implant to be formed and/or tailored to the patient.

The rapid templating process for medical implants is tailored to allow both prefabrication and intraoperative formation of patient-specific implants

Templating processes are common manufacturing techniques and many implants are manufactured through such methods. The distinguishing feature of the rapid templating process is the locational variance of the process—allowing external/prior formation as well as intraoperative manufacture and adjustment if necessary.

The rapid templating process provides a fast and cost-effective alternative to implant prefabrication

The absence of tooling which characterizes 3D printing as a manufacturing method provides a significant time saving—manufacturing is tied directly to print time (there are no additional requirements) [94, 95].

The costs associated with the templating method are dictated by the following:

Technician Labor Hours

The time taken to produce the digital implant template from provided CT scans, as well as time required to set up and monitor the print (taken as one-third of the total print time).

Material and Operating Cost

The total amount of material used, as well as the running costs associated with the use of 3D printers.

Production Operating Margin

A profit margin ensuring the rapid templating method remains sustainable from a health economic viewpoint. This also will include an offset to account for print failures and additional time and material use that may result.

In addition, there is little cost associated with part customization or alteration when using 3D printing methods, as there are no re-tooling requirements. Decreasing cost and technological advancements in additive manufacturing technology have increased the availability and affordability of commercial 3D printers [106], which also tend toward lower setup and running costs than conventional manufacturing systems [73, 77, 80]—presenting a faster and more cost-effective method of implant fabrication.

4.2.2 Template Manufacture

The general process involved in the generation of a cranioplasty implant through the rapid templating process is shown in Fig. 4 and is as follows:

Machine is initialized and calibrated—as per machine recommendations

Before any form of additive manufacturing can begin, machine setup is required to ensure that production proceeds as required. This can involve factors such as material loading, preparation of the print area and calibration of the print area and printing apparatus. While this setup process must be conducted for every machine, the specifics vary between systems and models. It is fundamental that

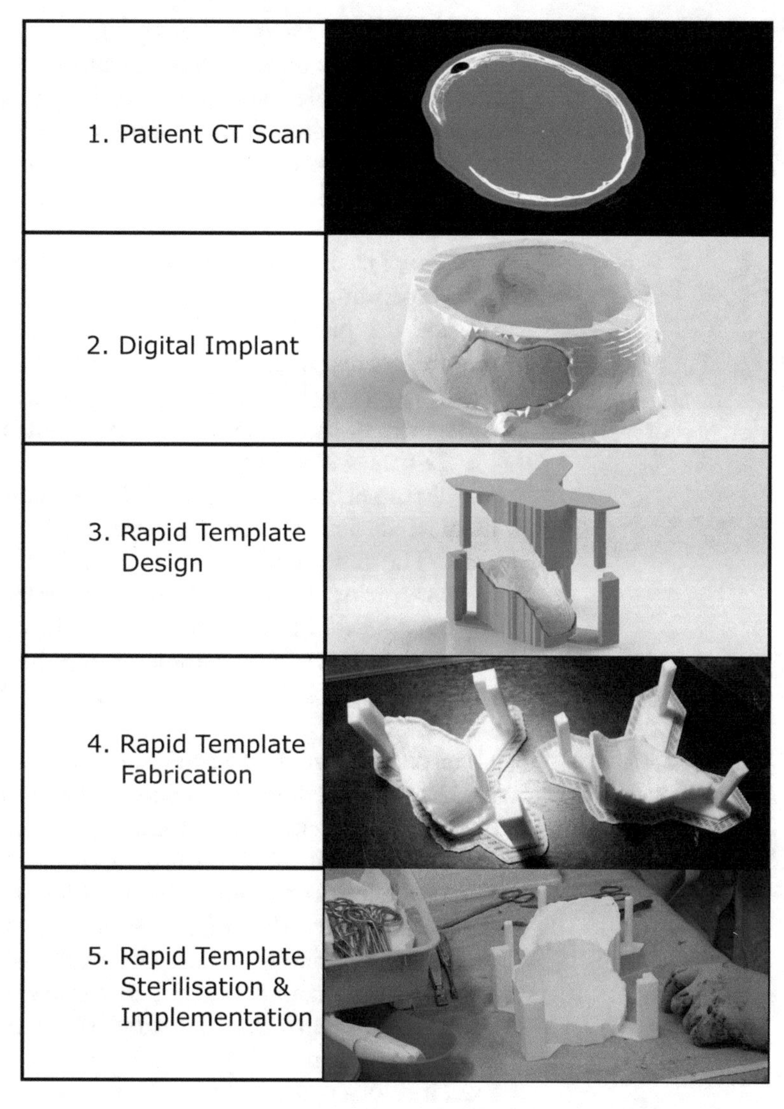

Fig. 4 Production process for custom cranioplasty generated using the rapid templating process

appropriate processes be followed to ensure effective machine function, optimal production outcome and to minimize failure risk.

Appropriate material is fitted within the machine

The nature of the rapid templating process as an intraoperative formation tool dictates that template formation materials used should be biocompatible to reduce patient risk—although this may be overcome using a biocompatible liner during the formation process. In addition, any material chosen must be capable of cleaning and sterilization to minimize infection risk. Clinical work with the rapid templating process has shown the use of FDM printers with biocompatible PLA material is effective in generating implants intraoperatively for large cranial defects.

Printing of the Rapid Template

The template is produced through additive manufacturing methods. While additive manufacturing technologies have increased reliability, it is recommended that manufacturing occurs with sufficient timeframes to allow production of additional components, should the initial prints fail to produce appropriate models.

Template is removed as cleaned/post-processed as necessary

Depending on the method of production and final purpose of use for the product model, a degree of post-processing may be required. This primarily involves the removal of support material, which can be either a manual or automated process depending on the nature of the machine, material used, and model geometry.

Template is sterilized

Terminal sterilization is a fundamental requirement for any medical device, to ensure patient safety and mitigate potential risks. The nature of the rapid templating process, intraoperative manufacturing, is such that both the template and the casted material must be sterilized prior to use in the operating theater. Clinical work has shown that templates can be effectively sterilized by participating hospitals using hydrogen peroxide (H_2O_2).

4.2.3 Intraoperative Implant Fabrication Using the Rapid Templating Process

Regardless of the injury profile, the surgical process of cranioplasty for large cranial defects remains relatively consistent. The patient is placed under general and local anaesthetic and the surgical team removes the skin of the scalp, exposing the bone defect and dura (which shelters the brain). The team prepares the edges of the bone defect to receive the implant—allowing the implant to be positioned appropriately within the skull flaw. The implant is then secured to the cranial bone. This can be achieved through numerous methods, including suturing the cranioplasty into place and fixing the implant with plates and screws [107]. Once the implant is fixed into place, the scalp is returned to its appropriate position and closed. Depending on the nature of the patient's injury additional features, such as a drain, will be implemented to ensure appropriate patient recovery.

The rapid templating process introduces an additional step where the implant is to be manufactured. This can be done in conjunction with preparation of the bone defect—minimizing surgical time and allowing implant tailoring based on characteristics only observable once the patient's scalp has been lifted.

In clinical examination of the technology implants were manufactured using PMMA bone cement. The process for forming the implant using rapid templating is shown in Fig. 5. This involved coating the template with sterile medical grade paraffin wax, to prevent the template halves from sticking together. This was done by rubbing the surface of the template with paraffin wax coated bandages. Bone cement was then mixed—with potential additives

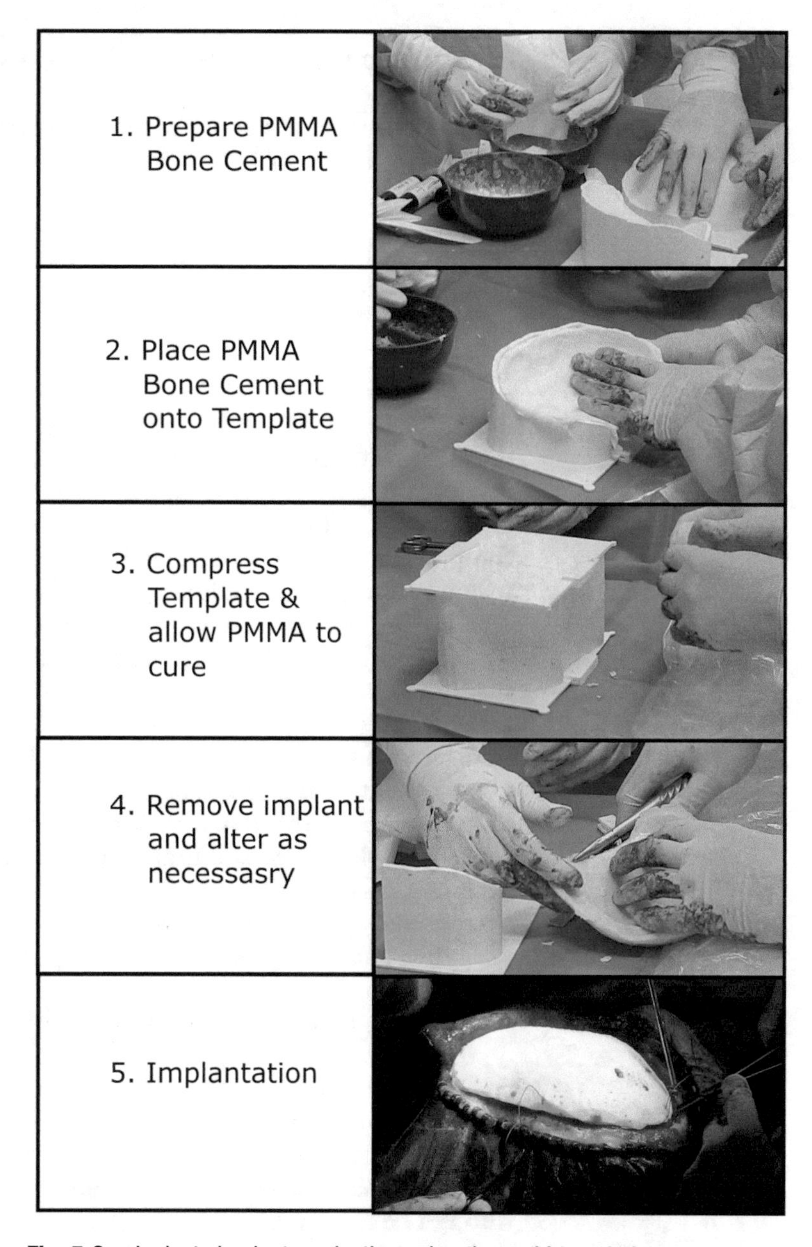

Fig. 5 Cranioplasty implant production using the rapid templating process

including antibiotics added at this stage—until it reached a dough-like consistency. At this stage, it was compressed into a flat shape and draped over the template. The two halves of the template were then compressed and the bone cement allowed to cure (taking approximately 10–15 min depending on brand, quantity, and additives). Once cured, the implant is removed from the template. It is prepared for implantation by removing any excess material (flashing) and drilling any necessary holes/slots. It is then ready for implantations.

5 Conclusions: Effectiveness and Utility of the Rapid Templating Process

The effectiveness and utility of the rapid templating process can be judged by the following factors:

Excellent Clinical Results:

Based on clinical work and literature analysis, the functional results, aesthetics, and stability of large cranial defect implants produced through the rapid templating process were deemed to be favorable [6, 45–48].

Rapid Turnaround:

Through the use of 3D printing technology the rapid templating process removes the tooling and prefabrication requirements of implant manufacture—minimizing the time from digital to physical product. These benefits both the patient and the medical industry—patient risk is minimized as receive their implant quicker and the process removes significant labor and cost from the implant manufacturing equation, leading to lower cost.

Ease of Use and Intraoperative formation/alteration:

This factor can be attributed to both the familiarity of the surgeon with the casting material (PMMA bone cement) and the capacity of the rapid template as a simplistic formation tool. Implant alteration can be completed with surgical tools. While not a benefit of the template, but rather the formation material, the presence of the template provides additional confidence in altering implants—as additional implants could be fabricated as necessary to ensure fit and form criteria are met. Intraoperative alteration allowed the implants to be tailored to factors not visible until the patient was open on the table, leading to better fit, form, and function.

Capacity for additives and alternative materials:

A benefit of the rapid templating process is the ability to complement the manufacturing material with bioactive additives—such as antibiotics. These can be added during the formation phase to provide enhanced properties to the final implant to ensure optimal patient outcomes. Using antibiotic-based bone cements through rapid templating is anticipated to provide increased infection resistance and comparable biocompatibility to conventional manufacturing methods and materials. Comprehensive in vitro infection studies that are in the process of being published were conducted by the authors to confirm the persisting risk mitigating potential of antibiotic eluting cement many days after the formation of implant material tokens compared to non-eluting controls.

The additives for use with the rapid templating process are not limited to antibiotics, just as the process is not limited to the use of PMMA bone cement. The versatility of the rapid templating process allows the formation of a wide variety of shape-setting materials into anatomic implants—metal, acrylic, and tissue-engineering based—along with the addition of additives such as bioactive materials supporting tissue attachment, growth factors, autologous pluripotent stem cells, and antibacterial or antibiotic agents.

The rapid templating process has been found to be an effective and practical method for reconstruction of large cranial defects by neurosurgical specialists. Work is now being progressed to apply the rapid templating process to spine and maxillofacial surgery applications in additional to surgical planning, practice, and education.

References

1. Rajendra PB et al (2009) Characteristics of associated craniofacial trauma in patients with head injuries: an experience with 100 cases. J Emerg Trauma Shock 2(2):89
2. Boffano P et al (2015) European Maxillofacial Trauma (EURMAT) project: a multicentre and prospective study. J Cranio-Maxillofac Surg 43(1):62–70
3. Ferreira PC et al (2015) Associated injuries in pediatric patients with facial fractures in Portugal: analysis of 1416 patients. J Cranio-Maxillofac Surg 43(4):437–443
4. Kraft A et al (2012) Craniomaxillofacial trauma: synopsis of 14,654 cases with 35,129 injuries in 15 years. Craniomaxillofac Trauma Reconstruct 5(1):41–50
5. Dujovny M et al (1997) Cranioplasty: cosmetic or therapeutic? Surg Neurol 47 (3):238–241
6. Marbacher S et al (2012) Intraoperative template-molded bone flap reconstruction for patient-specific cranioplasty. Neurosurg Rev 35(4):527–535
7. Annan M et al (2015) Sinking skin flap syndrome (or syndrome of the trephined): a review. Br J Neurosurg 29(3):314–318
8. Ashayeri K et al (2016) Syndrome of the trephined: a systematic review. Neurosurgery 79 (4):525–534
9. Abdou A et al (2015) Motor and neurocognitive recovery in the syndrome of the trephined: a case report. Ann Phys Rehabil Med 58(3):183
10. Hagan M, Bradley JP (2017) Syndrome of the trephined: functional improvement after reconstruction of large cranial vault defects. J Craniofac Surg 28(5):1129–1130
11. Tadros M, Costantino PD (2008) Advances in cranioplasty: a simplified algorithm to guide cranial reconstruction of acquired defects. Facial Plast Surg 24(01):135–145
12. Passalacqua NV, Rainwater CW (2015) Skeletal trauma analysis: case studies in context. Wiley, New York
13. Shah AM, Jung H, Skirboll S (2014) Materials used in cranioplasty: a history and analysis. Neurosurg Focus 36(4):E19
14. Harris DA et al (2014) History of synthetic materials in alloplastic cranioplasty. Neurosurg Focus 36(4):E20
15. Abhay S, Haines SJ (1997) Repairing holes in the head: a history of cranioplasty. Neurosurgery 40(3):588–603
16. Chibbaro S et al (2011) Decompressive craniectomy and early cranioplasty for the management of severe head injury: a prospective multicenter study on 147 patients. World Neurosurg 75(3):558–562
17. Liang W et al (2007) Cranioplasty of large cranial defect at an early stage after decompressive craniectomy performed for severe head trauma. J Craniofac Surg 18 (3):526–532
18. Bender A et al (2013) Early cranioplasty may improve outcome in neurological patients with decompressive craniectomy. Brain Inj 27(9):1073–1079
19. Song J et al (2014) Beneficial impact of early cranioplasty in patients with decompressive craniectomy: evidence from transcranial Doppler ultrasonography. Acta Neurochir 156(1):193–198
20. Cabraja M, Klein M, Lehmann T-N (2009) Long-term results following titanium cranioplasty of large skull defects. Neurosurg Focus 26(6):E10
21. Morina A et al (2011) Cranioplasty with subcutaneously preserved autologous bone grafts in abdominal wall—experience with 75 cases in a post-war country Kosova. Surg Neurol Int 2:72
22. Klinger DR et al (2014) Autologous and acrylic cranioplasty: a review of 10 years and 258 cases. World Neurosurg 82(3):e525–e530
23. Inamasu J, Kuramae T, Nakatsukasa M (2010) Does difference in the storage method of bone flaps after decompressive craniectomy affect the incidence of surgical site infection

after cranioplasty? Comparison between subcutaneous pocket and cryopreservation. J Trauma Acute Care Surg 68(1):183–187
24. Frazer RQ et al (2005) PMMA: an essential material in medicine and dentistry. J Long-Term Eff Med Implants 15(6):629–639
25. Kriegel RJ, Schaller C, Clusmann H (2007) Cranioplasty for large skull defects with PMMA (polymethylmethacrylate) or Tutoplast processed autogenic bone grafts. Zentralb Neurochir 68(4):182–189
26. Huang GJ et al (2015) Craniofacial reconstruction with poly (methyl methacrylate) customized cranial implants. J Craniofac Surg 26(1):64–70
27. Jaberi J et al (2013) Long-term clinical outcome analysis of poly-methyl-methacrylate cranioplasty for large skull defects. J Oral Maxillofac Surg 71(2):e81–e88
28. Marchac D, Greensmith A (2008) Long-term experience with methylmethacrylate cranioplasty in craniofacial surgery. J Plast Reconstr Aesthet Surg 61(7):744–752
29. Worm PV et al (2016) Polymethylmethacrylate imbedded with antibiotics cranioplasty: an infection solution for moderate and large defects reconstruction? Surg Neurol Int 7 (Suppl 28):S746
30. Hsu VM et al (2014) A preliminary report on the use of antibiotic-impregnated methyl methacrylate in salvage cranioplasty. J Craniofac Surg 25(2):393–396
31. Murray WR (1984) Use of antibiotic-containing bone cement. Clin Orthop Relat Res 190:89–95
32. Webb J, Spencer R (2007) The role of polymethylmethacrylate bone cement in modern orthopaedic surgery. Bone Joint J 89 (7):851–857
33. Golz T et al (2010) Temperature elevation during simulated polymethylmethacrylate (PMMA) cranioplasty in a cadaver model. J Clin Neurosci 17(5):617–622
34. Togawa D et al (2003) Histologic evaluation of human vertebral bodies after vertebral augmentation with polymethyl methacrylate. Spine 28(14):1521–1527
35. Pikis S, Goldstein J, Spektor S (2015) Potential neurotoxic effects of polymethylmethacrylate during cranioplasty. J Clin Neurosci 22(1):139–143
36. Matsuno A et al (2006) Analyses of the factors influencing bone graft infection after delayed cranioplasty. Acta Neurochir 148(5):535–540
37. Lee S-C et al (2009) Cranioplasty using polymethyl methacrylate prostheses. J Clin Neurosci 16(1):56–63
38. Rotaru H et al (2012) Cranioplasty with custom-made implants: analyzing the cases of 10 patients. J Oral Maxillofac Surg 70(2): e169–e176
39. Iaccarino C et al (2015) Preliminary results of a prospective study on methods of cranial reconstruction. J Oral Maxillofac Surg 73 (12):2375–2378
40. Turgut G, Özkaya Ö, Kayal MU (2012) Computer-aided design and manufacture and rapid prototyped polymethylmethacrylate reconstruction. J Craniofac Surg 23 (3):770–773
41. Lee C-H et al (2012) Analysis of the factors influencing bone graft infection after cranioplasty. J Trauma Acute Care Surg 73 (1):255–260
42. Bobinski L, Koskinen L-OD, Lindvall P (2013) Complications following cranioplasty using autologous bone or polymethylmethacrylate—retrospective experience from a single center. Clin Neurol Neurosurg 115 (9):1788–1791
43. Al-Tamimi YZ et al (2012) Comparison of acrylic and titanium cranioplasty. Br J Neurosurg 26(4):510–513
44. Piitulainen JM et al (2015) Outcomes of cranioplasty with synthetic materials and autologous bone grafts. World Neurosurg 83 (5):708–714
45. Stieglitz LH et al (2014) Intraoperative fabrication of patient-specific moulded implants for skull reconstruction: single-Centre experience of 28 cases. Acta Neurochir 156 (4):793–803
46. Kim B-J et al (2012) Customized cranioplasty implants using three-dimensional printers and polymethyl-methacrylate casting. J Kor Neurosurg Soc 52(6):541–546
47. Goh RC et al (2010) Customised fabricated implants after previous failed cranioplasty. J Plast Reconstr Aesthet Surg 63(9):1479–1484
48. Hay JA, Smayra T, Moussa R (2017) Customized polymethylmethacrylate cranioplasty implants using 3-dimensional printed polylactic acid molds: technical note with 2 illustrative cases. World Neurosurg 105:971–979.e1
49. Thien A et al (2015) Comparison of polyetheretherketone and titanium cranioplasty after decompressive craniectomy. World Neurosurg 83(2):176–180
50. Wiggins A et al (2013) Cranioplasty with custom-made titanium plates—14 years experience. Neurosurgery 72(2):248–256
51. Le Guéhennec L et al (2007) Surface treatments of titanium dental implants for rapid osseointegration. Dent Mater 23(7):844–854

52. Williams L, Fan K, Bentley R (2015) Custom-made titanium cranioplasty: early and late complications of 151 cranioplasties and review of the literature. Int J Oral Maxillofac Surg 44(5):599–608
53. Imanishi J, Choong PF (2015) Three-dimensional printed calcaneal prosthesis following total calcanectomy. Int J Surg Case Rep 10:83–87
54. Aranda JL et al (2015) Tridimensional titanium-printed custom-made prosthesis for sternocostal reconstruction. Eur J Cardio-Thorac Surg 48:e92–e94
55. Kurtz SM, Devine JN (2007) PEEK biomaterials in trauma, orthopedic, and spinal implants. Biomaterials 28(32):4845–4869
56. Ng ZY, Nawaz I (2014) Computer-designed PEEK implants: a peek into the future of cranioplasty? J Craniofac Surg 25(1):e55–e58
57. O'Reilly EB et al (2015) Computed-tomography modeled polyether ether ketone (PEEK) implants in revision cranioplasty. J Plast Reconstr Aesthet Surg 68(3):329–338
58. Gooch MR et al (2009) Complications of cranioplasty following decompressive craniectomy: analysis of 62 cases. Neurosurg Focus 26(6):E9
59. Cheng Y-K et al (2008) Factors affecting graft infection after cranioplasty. J Clin Neurosci 15 (10):1115–1119
60. Reddy S et al (2014) Clinical outcomes in cranioplasty: risk factors and choice of reconstructive material. Plast Reconstr Surg 133 (4):864–873
61. Mundinger GS et al (2016) Management of the repeatedly failed cranioplasty following large postdecompressive craniectomy: establishing the efficacy of staged free Latissimus Dorsi transfer/tissue expansion/custom polyetheretherketone implant reconstruction. J Craniofac Surg 27(8):1971–1977
62. Im S-H et al (2012) Long-term incidence and predicting factors of cranioplasty infection after decompressive craniectomy. J Kor Neurosurg Soc 52(4):396–403
63. Kwarcinski J et al (2017) Cranioplasty and craniofacial reconstruction: a review of implant material, manufacturing method and infection risk. Appl Sci 7(3):276
64. Lantada AD, Morgado PL (2012) Rapid prototyping for biomedical engineering: current capabilities and challenges. Annu Rev Biomed Eng 14:73–96
65. Tuomi J et al (2014) A novel classification and online platform for planning and documentation of medical applications of additive manufacturing. Surg Innov 21(6):553–559
66. Xu J et al (2015) Application of rapid prototyping pelvic model for patients with DDH to facilitate arthroplasty planning: a pilot study. J Arthroplast 30(11):1963–1970
67. Bagaria V et al (2011) Use of rapid prototyping and three-dimensional reconstruction modeling in the management of complex fractures. Eur J Radiol 80(3):814–820
68. Esses SJ et al (2011) Clinical applications of physical 3D models derived from MDCT data and created by rapid prototyping. Am J Roentgenol 196(6):W683–W688
69. Sherekar RM, Pawar AN (2014) Application of biomodels for surgical planning by using rapid prototyping: a review & case studies. Int J Innov Res Adv Eng 1(6):263–271
70. Waran V et al (2014) Utility of multimaterial 3D printers in creating models with pathological entities to enhance the training experience of neurosurgeons: technical note. J Neurosurg 120(2):489–492
71. Lipson H, Kurman M (2013) Fabricated: the new world of 3D printing. Wiley, New York
72. Grynol B (2014) Disruptive manufacturing: the effects of 3D printing. Deloitte, Canada
73. Huang SH et al (2013) Additive manufacturing and its societal impact: a literature review. Int J Adv Manuf Technol 67 (5–8):1191–1203
74. Ventola CL (2014) Medical applications for 3D printing: current and projected uses. Pharm Ther 39(10):704
75. Kietzmann J, Pitt L, Berthon P (2015) Disruptions, decisions, and destinations: enter the age of 3-D printing and additive manufacturing. Bus Horiz 58(2):209–215
76. Guo N, Leu MC (2013) Additive manufacturing: technology, applications and research needs. Front Mech Eng 8 (3):215–243
77. Thompson R (2007) Manufacturing processes for design professionals. Thames & Hudson, London
78. Gibson I, Rosen D, Stucker B (2015) Direct digital manufacturing. In: Additive manufacturing technologies. Springer, New York, pp 375–397
79. Chen D et al (2015) Direct digital manufacturing: definition, evolution, and sustainability implications. J Clean Prod 107:615–625
80. Gibson I, Rosen D, Stucker B (2014) Additive manufacturing technologies: 3D printing, rapid prototyping, and direct digital manufacturing. Springer, New York

81. Dudek P (2013) FDM 3D printing technology in manufacturing composite elements. Arch Metall Mater 58(4):1415–1418
82. Novakova-Marcincinova, L., et al. Special materials used in FDM rapid prototyping technology application. In: Intelligent Engineering Systems (INES), 2012 I.E. 16th International Conference on. 2012. IEEE
83. Singh S, Ramakrishna S, Singh R (2017) Material issues in additive manufacturing: a review. J Manuf Process 25:185–200
84. Wong JY, Pfahnl AC (2014) 3D printing of surgical instruments for long-duration space missions. Aviat Space Environ Med 85 (7):758–763
85. Hutmacher DW, Sittinger M, Risbud MV (2004) Scaffold-based tissue engineering: rationale for computer-aided design and solid free-form fabrication systems. Trends Biotechnol 22(7):354–362
86. Tarafder S et al (2013) Microwave-sintered 3D printed tricalcium phosphate scaffolds for bone tissue engineering. J Tissue Eng Regen Med 7(8):631–641
87. Hutmacher DW (2001) Scaffold design and fabrication technologies for engineering tissues—state of the art and future perspectives. J Biomater Sci Polym Ed 12(1):107–124
88. El-Hajje A et al (2014) Physical and mechanical characterisation of 3D-printed porous titanium for biomedical applications. J Mater Sci Mater Med 25(11):2471–2480
89. Pinkerton AJ (2016) Lasers in additive manufacturing. Opt Laser Technol 78:25–32
90. Kalpakjian S, Schmid SR, Kok C-W (2008) Manufacturing processes for engineering materials. Pearson-Prentice Hall, Upper Saddle River
91. de Treville S et al (2014) Valuing lead time. J Oper Manag 32(6):337–346
92. Huang SH et al (2013) Additive manufacturing and its societal impact: a literature review. Int J Adv Manuf Technol 67:1191–1203
93. Horn TJ, Harrysson OL (2012) Overview of current additive manufacturing technologies and selected applications. Sci Prog 95 (3):255–282
94. Chua CK, Leong KF, Liu ZH (2015) Rapid tooling in manufacturing. In: Nee AYC (ed) Handbook of manufacturing engineering and technology. Springer, New York, pp 2525–2549
95. Pham D, Dimov SS (2012) Rapid manufacturing: the technologies and applications of rapid prototyping and rapid tooling. Springer Science & Business Media, New York
96. Hieu L et al (2002) Design and manufacturing of cranioplasty implants by 3-axis cnc milling. Technol Health Care 10 (5):413–423
97. Delbeke D et al (2006) Procedure guideline for SPECT/CT imaging 1.0. J Nucl Med 47 (7):1227–1234
98. Brix G et al (2011) Dynamic contrast-enhanced CT studies: balancing patient exposure and image noise. Investig Radiol 46 (1):64–70
99. Miglioretti DL et al (2013) The use of computed tomography in pediatrics and the associated radiation exposure and estimated cancer risk. JAMA Pediatr 167(8):700–707
100. McMenamin PG et al (2014) The production of anatomical teaching resources using three-dimensional (3D) printing technology. Anat Sci Educ 7(6):479–486
101. An G et al (2017) Accuracy and efficiency of computer-aided anatomical analysis using 3D visualization software based on semi-automated and automated segmentations. Ann Anat 210:76–83
102. Pitas I (2000) Digital image processing algorithms and applications. Wiley, New York
103. Sahoo PK, Soltani S, Wong AK (1988) A survey of thresholding techniques. Comput Vis Graph Image Process 41(2):233–260
104. Patanè G, Giannini F, Attene M (2017) Computational methods for the morphological analysis and annotation of segmented 3D medical data. IMATI Report Series. nr. 17-08
105. Atherton PR, Caporael LR (1985) A subjective judgment study of polygon based curved surface imagery. In ACM SIGCHI Bulletin. ACM, New York
106. Klein GT, Lu Y, Wang MY (2013) 3D printing and neurosurgery—ready for prime time? World Neurosurg 80(3):233–235
107. Khader BA, Towler MR (2016) Materials and techniques used in cranioplasty fixation: a review. Mater Sci Eng C 66:315–322

Chapter 21

Computer-Assisted Approaches to Identify Functional Gene Networks Involved in Traumatic Brain Injury

Anthony San Lucas, John Redell, Pramod Dash, and Yin Liu

Abstract

Traumatic brain injury (TBI), a leading cause of death and disability in industrialized countries, is a result of an external force damaging brain tissue, accompanied with delayed pathogenic events which aggravate the injury. Molecular responses to TBI have not been well characterized, leaving molecular classification of TBI cases difficult. TBI subtype classification is an important step towards the development and selective application of novel treatments. To improve TBI classification, we have performed a network-based computational analysis on gene expression profiles from entire rat genome to identify functional gene subnetworks. The gene expression profiles are obtained from two experimental models of injury in rats: the controlled cortical impact (CCI; a focal brain injury) and fluid percussion injury (FPI; a diffuse brain injury). We demonstrate that the analysis of gene subnetworks is more suitable to classify the heterogeneous responses to different TBI models, compared to conventional analysis using an individual gene list. We therefore believe that effectively incorporating gene expression profiles into protein interaction information can improve the identification of functional subnetworks involved in TBI. The systems approach could lead to a better understanding of the underlying complexities of the molecular responses after TBI and the identified subnetworks could have important prognostic functions for patients who sustain mild TBIs.

Key words Gene networks, TBI subtype classification, Biomarkers, Protein interactions, Gene expression profiles, Greedy search algorithm

1 Introduction

Traumatic brain injury (TBI) results from an external force causing immediate damage to brain tissue, followed by secondary pathogenic events which ultimately give rise to neurodegeneration. Dependent on the context of the primary injury, different cell responses are initiated that can exacerbate the injury to varying degrees. Cell death resulting from the initial impact on the brain tissue is irreversible, so treatments normally focus on minimizing the secondary injury due to these cell responses [1]. To date, these secondary injury responses have been poorly characterized, leaving

Amit K. Srivastava and Charles S. Cox, Jr. (eds.), *Pre-Clinical and Clinical Methods in Brain Trauma Research*, Neuromethods, vol. 139, https://doi.org/10.1007/978-1-4939-8564-7_21,

molecular classification of TBI difficult [2, 3]. TBI remains a leading cause of death and disability in the industrialized countries and represents a growing health problem [4]. Thus, even a modest improvement in patient outcome could have significant public health benefits [5, 6]. Using computer-assisted bioinformatics approaches, we aim to improve the identification of biomarkers that can distinguish the CCI (controlled cortical impact) and FPI (fluid percussion injury) experimental models of TBI, both of which qualitatively recapitulate a number of functional deficits and pathological responses exhibited in human TBI cases. These biomarkers, if successfully identified, could be used to better direct treatments to TBI patients, and more optimistically they could be potential targets of novel treatments.

Recent years have witnessed that an increasing number of disease markers have been identified through computational analysis of genome-wide expression profiles. Typically, gene expression profiling studies are limited to focus on individual genes that are significantly differentially expressed between different classes of diseases. However, single-gene analyses have been criticized for several reasons [7, 8]. In the cases of TBI, if we only examined the differences in the expression levels of individual genes across different TBI models and simply neglected the genes that are not associated with a TBI model at a significance threshold, we would fail to account for the complexities and redundancies that arise from gene interactions inherent to the TBI responses. Discarded genes showing modest differential expression between TBI classes may represent false negatives and may be important biomarkers of TBI. We hypothesize that these genes may be identified within functional units of genes as subnetworks, which in aggregate have a significant association to a TBI class. To test this hypothesis, we have proposed a data-driven model and identified biomarkers not as individual genes but as gene subnetworks, by incorporating the gene expression profiles from injury models and the protein–protein interaction information from existing databases. The genes in each of the identified subnetworks are expected to be highly correlated and exhibit a coherent expression profile across samples, while others exist as background noise. It is also expected that the genes in a functional subnetwork exhibit high topological similarity with each other and should lead to a biologically meaningful sample classification. Here, with simulation and real data analysis, we show that our computational systems approach based on network theory performs better than individual gene analyses in TBI classification. The identified subnetworks can provide insights into the multifactorial relationships of genes and delineate the underlying complexities of the biological processes involved in different TBI classes.

2 Data Sources

Animal subjects and surgeries: Male Sprague-Dawley (SD) rats (275–300 g) were purchased from Charles River Laboratories (Wilmington, MA). All experimental procedures were approved by the local institutional animal care and use committee and were conducted in accord with recommendations provided in the Guide for the Care and Use of Laboratory Animals. Protocols were designed to minimize pain and discomfort during the injury procedure and recovery period.

mFPI and mCCI brain injury: Our injury models were described in [3]. After injury or sham operation, animals were placed in a warm chamber and allowed to completely recover from anesthesia, and then returned to their home cages.

Gene expression microarray: Total RNA from the rat ipsilateral cortex ($n = 6$/group) was isolated using the mirVana miRNA Isolation Kit (Invitrogen, Carlsbad, CA), following the manufacturers' recommended protocol, and amplified using the Illumina TotalPrep RNA Amplification Kit (Ambion, Austin, TX). RNA amplification and microarray hybridization was carried out by The University of Texas Health Science Center Houston Microarray Core Laboratory (Houston, TX). Briefly, first-strand complementary DNA (cDNA) was generated from total RNA by reverse transcription. Second-strand cDNA synthesis was initiated by the addition of RNase H/DNA polymerase mix. The complementary RNA (cRNA) was amplified by the in vitro transcription reaction (IVT). cRNA (750 ng) was loaded onto RatRefSeq-12 Illumina Sentrix Beadchip Arrays (Illumina, Inc., San Diego, CA), hybridized overnight, washed, and incubated with streptavidin-Cy3 to detect hybridized biotin-labeled cRNA probes. Arrays were dried and scanned with a BeadArray Reader (Illumina).

Protein–protein interaction (PPI) information: Experimentally detected PPIs were downloaded from BioGRID [9], DIP [10], and HPRD [11] databases. Since there are a limited number of experiments detecting PPIs in the rat genome, we also obtained predicted rat PPIs based on onthology, where orthologous interactions were generated by mapping experimentally detected PPIs in human or mouse genomes, to pairs of orthologs in rat, if such orthologs are available in HomoloGene database [12]. Given the resources of protein interactions as described above, a protein interaction network was constructed, resulting in a PPI network with 18,781 proteins and 207,829 edges.

3 Methods

3.1 Data Transformation

Because most raw gene expression values are not normally distributed but highly skewed, the Box-Cox transformation [13] was used to normalize the distribution for each gene expression value. The Kolmogorov–Smirnov test was used to test for normality of the transformed distribution at a 5% significance level.

3.2 A Weighted Network Constructed from Protein Interaction Information and Gene Expression Data

A gene network can be described by means of a graph $G = (V, E)$, where V is the vertex set representing genes, and E is the edge set representing functional relationship between genes. In this study, if $e_{ij} = 0$, there is no edge between two genes i and j based on the protein interaction network. Each edge of the interaction network is further weighted by overlaying gene expression information. Specifically, we have calculated the absolute value of the Pearson correlation as the edge weight $e_{ij} = \text{abs}(\text{cor}(x_i, x_j))$, where x_i and x_j represent the normalized gene expression vectors for genes i and j, respectively. Therefore, each edge of the protein interaction network is weighted by the level of co-expression between its two corresponding genes, and the weights lie between 0 and 1.

3.3 Scoring Subnetworks

We define the subnetwork scoring function S as a weighted sum of class relevance R and modularity M.

$$S = \beta M + R \tag{1}$$

Here M describes the subnetwork connectivity, and R is a measure of the discriminatory power of the subnetwork genes to differentiate classes. In addition, the parameter β allows us to trade off the effects of the gene expression information with the network modularity on the subnetwork score. To simplify the scoring algorithm, we set $\beta = 1$, assuming equal weights of network modularity and class relevance when calculating the network score.

To get a measure of how strongly the genes within a subnetwork are connected, the modularity M is calculated as the mean clustering coefficients C_i of the genes in a subnetwork,

$$M = \frac{\sum_i C_i}{n} \quad [\text{for } n \geq 3; \;\; 0 \text{ otherwise}] \tag{2}$$

where C_i is defined as in Dong and Horvath [14, 15]. C_i is the clustering coefficient for node i where nodes l and m are node neighbors of i, and e_{ij} represents the weight of the edge between nodes i and j in the subnetwork:

$$C_i = \frac{\sum_{l \neq i} \sum_{m \neq i, m \neq l} e_{il} e_{lm} e_{mi}}{\left(\sum_{l \neq i} e_{il}\right)^2 - \sum_{l \neq i} e_{il}^2} \tag{3}$$

Intuitively, C_i is a ratio of the weighted triangles that can be made with node i and its neighbors over the sum of the weighted possible triangles extending off of node i. For a general weighted network with e_{ij} between 0 and 1, the clustering coefficient of node i, C_i also lies between 0 and 1. C_i equals 1 if and only if all neighbors of node i are connected to each other.

The class relevance R is a measure of the ability for a subnetwork to distinguish two classes. To calculate this, the expression values g_{ij} are first normalized to z-scores, z_{ij}, which for each gene i has mean of 0 and s.d. of 1 over all samples j. The individual z-scores of each member gene in the subnetwork K are averaged into summarized expression v_{kj}, as $v_{kj} = \frac{\sum_{i=1}^{k} z_{ij}}{k}$. A t-test was then used to compare the expression values of subnetwork K for the samples of two classes, $R = T\text{-score}\ (v_{kj}, v_{k\hat{j}})$, where v_{kj} and $v_{k\hat{j}}$ represent the summarized expression scores of samples from classes j and classes $\hat{j}$, respectively. In this study, the two classes refer to CCI and FBI injury models.

3.4 Greedy Search for Identifying Significant Subnetworks

A framework demonstrating the steps for finding significant subnetworks is described in Fig. 1. We apply the greedy search algorithm in searching subnetworks [16]. To identify significant subnetworks that discriminate CCI and FPI, subnetworks are scored comparing two classes. For each comparison, individual differentially expressed genes are used as seeds for growing potential subnetworks. From each seed, two neighboring genes are iteratively added to the seed and the subnetwork score are recalculated. The pair of neighboring genes that give the biggest improvement in subnetwork score is added to the seed to form an initial subnetwork of three genes (i.e., an initial triangular subnetwork). Single

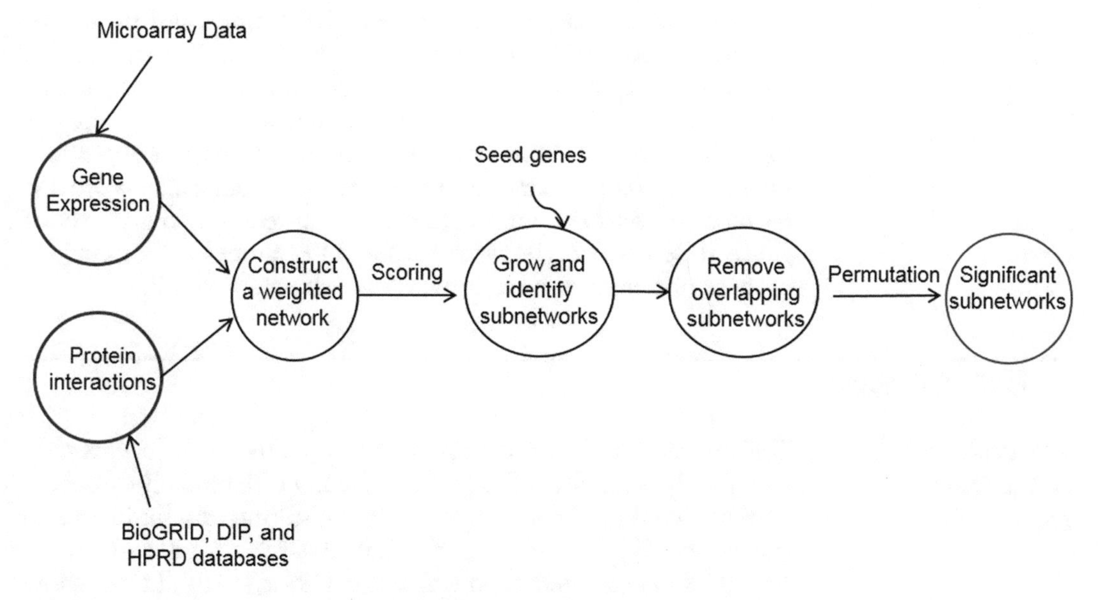

Fig. 1 An overview of subnetwork identification approach

neighbor nodes are then added iteratively until the subnetwork score could no longer be improved. The scoring method is described in Sect. 3.3. It is likely that genes are shared among different subnetworks, resulting in redundant subnetworks. The redundant subnetworks can be removed by the following steps:

1. Obtaining the scores of all subnetworks and sorting them in a descending order of scores.
2. Iterating through the list of subnetworks and checking for redundancy
   (a) If a subnetwork is contained within a higher-scoring subnetwork, discard the lower-scoring subnetwork.
   (b) If a subnetwork is a super set of a higher-scoring subnetwork, discard the super set.
   (c) If there is an overlap in genes between a lower-scoring and higher-scoring subnetwork:
      - If the overlap $\geq$50% (number of overlapping genes/total number of unique genes), discard lower-scoring subnetwork.
      - If the overlap <50%, keep both subnetworks (document them for manual inspection).

To select the significant subnetworks, we can calculate the empirical *p*-values of the identified subnetworks. We first generate the null distribution by permuting the expression vector of genes in the full network. This permutation test dissociates the relationship between protein interaction and gene expression information. We then run the same subnetwork identification procedure on the permuted data. This process is repeated 100 times and the scores of the resulted random subnetworks are recorded for each permutation. The empirical *p*-value for the real subnetwork score is calculated as the fraction of the random subnetworks having a higher score than that real subnetwork. Only those with empirical *p*-values smaller than 0.05 will be selected for further analysis. This process resulted in a set of subnetworks containing relatively unique set of genes, differentiating between CCI and FPI models. The examples of these discriminative subnetworks are shown in Fig. 2.

4 Method Evaluations

4.1 Examples of Functional Subnetworks

Unlike traditional expression profiling methods, our network-based analysis can identify genes that are not differentially expressed between classes. An example of the resulting discriminative is shown in Fig. 2. The genes mitogen-activated protein kinase 1 (MAPK1) and interleukin 6 family cytokine (LIF) did not show significant differentiated expression between CCI and FPI samples,

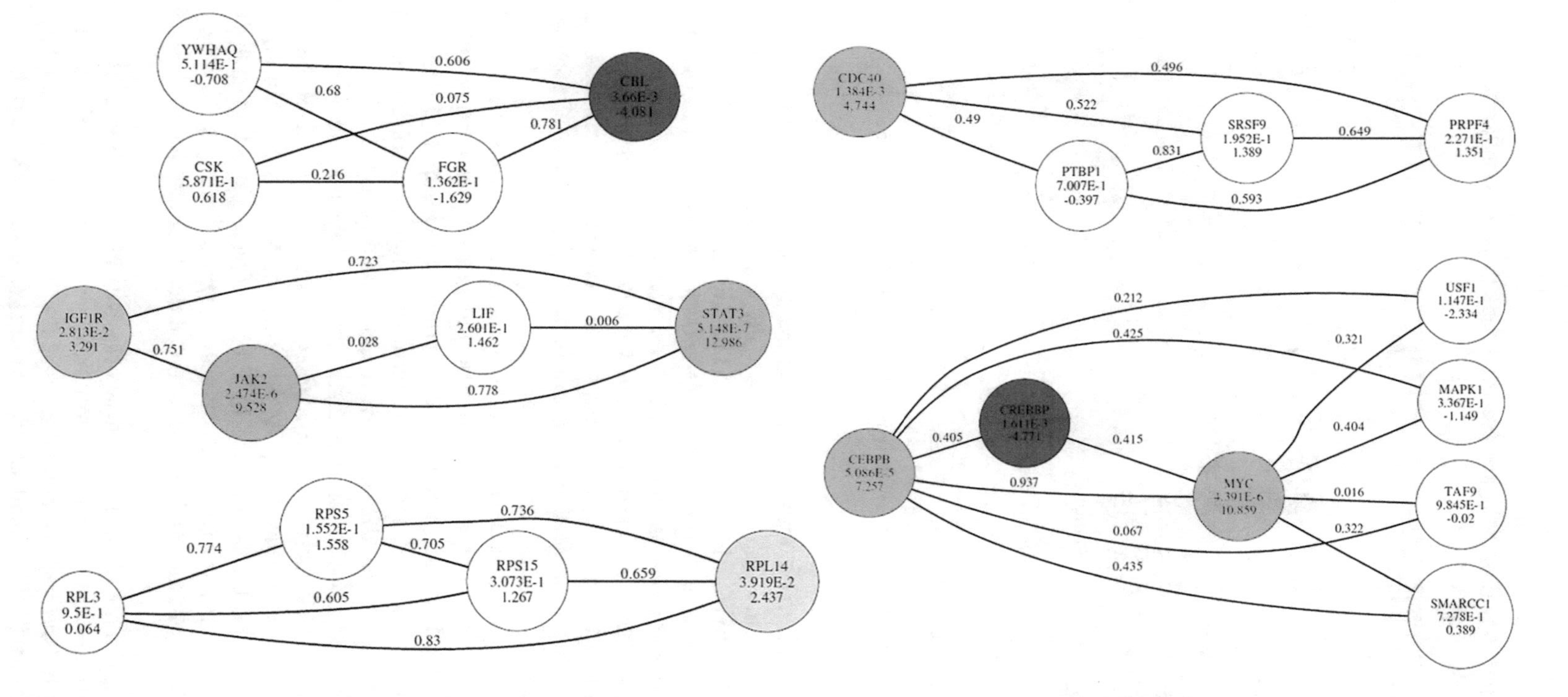

Fig. 2 Top five most significant subnetworks. Each subnetwork contains colored nodes. The colors on the nodes indicate how significantly that node differentiates the CCI and FPI classes. Green indicates that the *z*-scores for the CCI class have higher values compared to the FPI class. Red indicates that the *z*-scores for the CCI class have lower values compared to the FPI class. If a node is white, the corresponding gene does not significantly differentiate (p-value <0.05) the two sample groups. The intensity of the color corresponds to the level of significance. The numbers inside each node represent the discriminatory power of the gene, indicated by the *t*-score and the corresponding *p*-value. The numbers along each edge represent the edge weights

Table 1
Gene Ontology (GO) biological process annotations for significant subnetwork genes

GO biological process	FDR
Response to stress	8.36E−11
Response to chemical stimulus	1.96E−09
Metabolic process	8.75E−08
Positive regulation of cellular process	6.71E−07
Cell cycle	7.59E−07
Blood coagulation	1.11E−06
Regulation of body fluid levels	1.30E−06
Response to wounding	1.36E−06
Regulation of programmed cell death	5.49E−06
Cell proliferation	6.27E−06

The top ten most enriched GO biological process terms with their corresponding corrected *p*-values are listed. FDR, adjusted *p*-values for multiple testing by Benjamini and Hochberg's procedure

but they played an important role in the discriminative subnetworks by interconnecting many differentially expressed genes, such as CEBPB, MYC, JAK2, and STAT3. Given the fact that both MPAK1 and LIF genes are well-known players in the cytokine signaling pathway involved in inflammatory response, our results suggest they can serve as potential targets for intervention. To further investigate the functions of the identified subnetworks, we extract the biological process annotations from Gene Ontology database [17], and examine whether any Gene Ontology terms are overrepresented by the union of genes in the 50 most significant subnetworks, compared to an expected genome-wide representation [18]. Because there is a lot of redundancy in the GO tree, we use the "GO slim" terms to determine a high level of biological process categories the subnetwork genes belong to [19]. The GO enrichment *p*-values are calculated by the hyper-geometric test, followed by Benjamini and Hochberg's multiple hypotheses testing correction procedure [20]. As a result, the ten most significantly enriched GO terms are listed in Table 1. Overall, the GO analysis shows significant overrepresentation of genes belonging to stress, metabolic process, cell growth and cell death categories.

4.2 TBI Subtype Classification Evaluation

Now given the identified subnetworks, we can test their validity and performance in the classification problem. However, in this study, we only have experimental data available for 12 rats (six samples/ TBI model). The small sample size makes it difficult to train and test a classifier. Therefore, we have performed a simulation study to achieve an unbiased classification evaluation. The mean and standard deviation of each gene are estimated from the observed data. Given these parameters, we use the packages in R studio [21] to simulate gene expression datasets corresponding to CCI and FPI conditions, with 100 samples per condition. Given the simulated datasets, we performed a fivefold cross-validation to compute the classification accuracy. First, the simulated gene expression data corresponding to the identified subnetwork markers are used to encode features for a Support Vector Machine (SVM) classifier [22]. Then, we divide the CCI and FPI samples into five equal parts, respectively. We use four-fifth of the samples to train SVM and the remaining one-fifth of samples to test the trained classifier. Finally, we evaluate the classification results using an *F*-score as in [23], where $F = 2 \times \text{Precision} \times \text{Recall}/(\text{Precision} + \text{Recall})$. In this particular classification task, the precision is the proportion of classified CCI samples that are true CCIs, and the recall is the proportion of true CCI samples that are correctly classified by our method. We repeat the process for ten times to obtain an averaged *F*-score over ten iterations of cross-validation experiments. An overview of the classifier training, testing, and evaluation is depicted in Fig. 3.

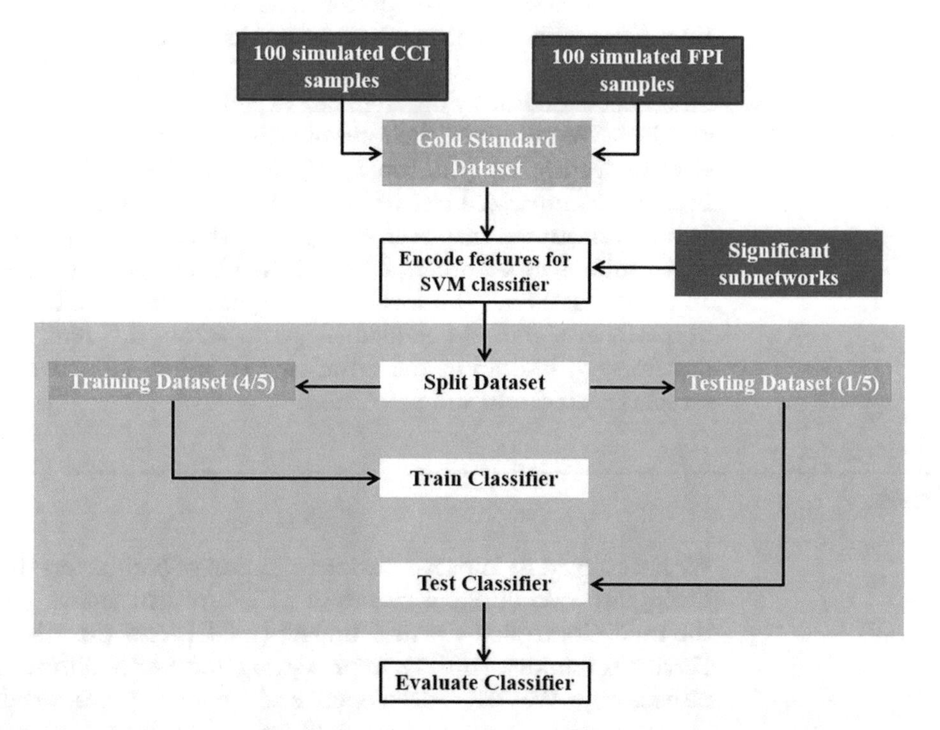

Fig. 3 Evaluating classification performance of the identified subnetworks in a simulation study. The activities of identified subnetworks calculated as the mean activities of its member genes (Sect. 3.3) are used as the features of the support vector machine (SVM) to classify the samples. The classifier performance is measured by the *F*-score

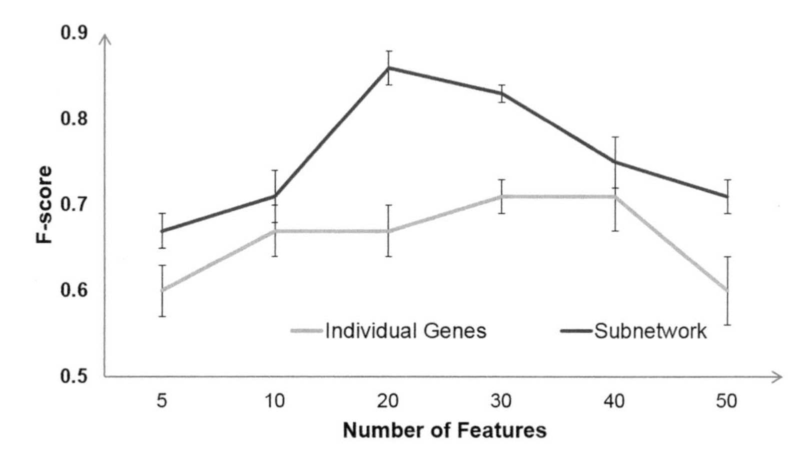

Fig. 4 Comparison of classification accuracy between subnetwork and individual genes. The classification performance of subnetwork and individual gene features are shown in blue and green, respectively. For each size of feature set, ten iterations of fivefold cross-validation are used to split the dataset, train, and evaluate the classifier. The curves show the median of classification performance, measured by the *F*-scores, and error bars indicate the standard deviation over ten cross-validation experiments

We also compared the performance of TBI subtype classification using either individual genes or identified subnetworks as features. The list of individual genes were obtained as the 100 most significantly differentially expressed genes between CCI and FPI. We examined their classification accuracy using different sizes of feature set (the top 5, 10, 20, 30, 40, or 50 features). Figure 4 summarized test results using different sizes of features. It is demonstrated that over all the tests, the SVM using 20 functional subnetwork features achieved the highest performance with an *F*-score of 0.85. We have also shown the functional subnetworks outperform significant individual genes across different sizes of feature sets, indicating the advantage of using subnetworks for sample classification and prediction.

5 Notes

We have aimed to improve the identification of biomarkers that can distinguish two different classes of TBI in rodent animal models: the mild Controlled Cortical Impact (mCCI) and the mild Fluid Percussion Injury (mFPI), representing focal and diffuse TBIs, respectively. We have developed and applied a network-based approach on gene expression profiles from the entire rat genome. Our network-based analysis can identify genes that are not differentially expressed but are essential for maintaining the integrity of a subnetwork whose overall expression is discriminative between samples. In addition, another advantage of our network-based analysis demonstrated from our preliminary study is that the list of identified significant gene subnetworks achieves higher sensitivity and specificity in classifying the heterogeneous responses corresponding to different classes of TBI, compared to a conventional analysis using an individual gene list. We therefore believe effectively incorporating gene expression profiles into protein interaction information can identify functional subnetworks that better classify different classes of TBI and are more reproducible across related studies than individual genes selected without network information. We understand that translating the knowledge gained from an animal model to molecular biomarkers identification in patients is practically challenging, simply because the brain tissue in TBI patients is rarely available, but the use of peripheral tissues such as lymphoblast or blood could be a potential solution. If successful, these identified biomarkers could be used to better direct the diagnosis and treatment to TBI patients, and more optimistically they could help to develop rationale-based therapies for treating the millions of Americans who suffer from TBI.

Acknowledgement

This work is supported in part by NIH grant R01 LM010022 and the seed grant from the University of Texas Health Science Center at Houston.

References

1. Yu Z, Morrison B 3rd (2010) Experimental mild traumatic brain injury induces functional alteration of the developing hippocampus. J Neurophysiol 103(1):499–510. https://doi.org/10.1152/jn.00775.2009
2. Grandhi R, Bonfield CM, Newman WC, Okonkwo DO (2014) Surgical management of traumatic brain injury: a review of guidelines, pathophysiology, neurophysiology, outcomes, and controversies. J Neurosurg Sci 58(4):249–259
3. Redell JB, Moore AN, Grill RJ, Johnson D, Zhao J, Liu Y, Dash PK (2013) Analysis of functional pathways altered after mild traumatic brain injury. J Neurotrauma 30(9):752–764. https://doi.org/10.1089/neu.2012.2437
4. Albert-Weissenberger C, Siren AL (2010) Experimental traumatic brain injury. Exp Transl Stroke Med 2(1):16. https://doi.org/10.1186/2040-7378-2-16
5. Jeter CB, Hergenroeder GW, Hylin MJ, Redell JB, Moore AN, Dash PK (2013) Biomarkers for the diagnosis and prognosis of mild traumatic brain injury/concussion. J Neurotrauma 30(8):657–670. https://doi.org/10.1089/neu.2012.2439
6. Mychasiuk R, Farran A, Esser MJ (2014) Assessment of an experimental rodent model of pediatric mild traumatic brain injury. J Neurotrauma 31(8):749–757. https://doi.org/10.1089/neu.2013.3132
7. Fortney K, Kotlyar M, Jurisica I (2010) Inferring the functions of longevity genes with modular subnetwork biomarkers of Caenorhabditis elegans aging. Genome Biol 11(2):R13. https://doi.org/10.1186/gb-2010-11-2-r13
8. Savarraj JPJ, Parsha K, Hergenroeder GW, Zhu L, Bajgur SS, Ahn S, Lee K, Chang T, Kim DH, Liu Y, Choi HA (2017) Systematic model of peripheral inflammation after subarachnoid hemorrhage. Neurology 88(16):1535–1545. https://doi.org/10.1212/WNL.0000000000003842
9. Chatr-Aryamontri A, Breitkreutz BJ, Oughtred R, Boucher L, Heinicke S, Chen D,

Stark C, Breitkreutz A, Kolas N, O'Donnell L, Reguly T, Nixon J, Ramage L, Winter A, Sellam A, Chang C, Hirschman J, Theesfeld C, Rust J, Livstone MS, Dolinski K, Tyers M (2015) The BioGRID interaction database: 2015 update. Nucleic Acids Res 43 (Database issue):D470–D478. https://doi.org/10.1093/nar/gku1204

10. Salwinski L, Miller CS, Smith AJ, Pettit FK, Bowie JU, Eisenberg D (2004) The database of interacting proteins: 2004 update. Nucleic Acids Res 32(Database issue):D449–D451. https://doi.org/10.1093/nar/gkh086
11. Goel R, Harsha HC, Pandey A, Prasad TS (2012) Human protein reference database and human proteinpedia as resources for phosphoproteome analysis. Mol BioSyst 8(2):453–463. https://doi.org/10.1039/c1mb05340j
12. NCBI_Resource_Coordinators (2016) Database resources of the National Center for Biotechnology Information. Nucleic Acids Res 44 (D1):D7–D19. https://doi.org/10.1093/nar/gkv1290
13. Sakia R (1992) The Box-Cox transformation technique: a review. Statistician 41:169–178
14. Zhang B, Horvath S (2005) A general framework for weighted gene co-expression network analysis. Stat Appl Genet Mol Biol 4:Article 17. https://doi.org/10.2202/1544-6115.1128
15. Dong J, Horvath S (2007) Understanding network concepts in modules. BMC Syst Biol 1:24. https://doi.org/10.1186/1752-0509-1-24
16. Carey VJ, Gentry J, Whalen E, Gentleman R (2005) Network structures and algorithms in bioconductor. Bioinformatics 21(1):135–136. https://doi.org/10.1093/bioinformatics/bth458
17. Huntley RP, Sawford T, Mutowo-Meullenet P, Shypitsyna A, Bonilla C, Martin MJ, O'Donovan C (2015) The GOA database: gene ontology annotation updates for 2015. Nucleic Acids Res 43(Database issue):D1057–D1063. https://doi.org/10.1093/nar/gku1113
18. Wang Z, Xu W, Liu Y (2015) Integrating full spectrum of sequence features into predicting functional microRNA-mRNA interactions. Bioinformatics 31(21):3529–3536. https://doi.org/10.1093/bioinformatics/btv392
19. Mi H, Huang X, Muruganujan A, Tang H, Mills C, Kang D, Thomas PD (2017) PANTHER version 11: expanded annotation data from gene ontology and reactome pathways, and data analysis tool enhancements. Nucleic Acids Res 45(D1):D183–D189. https://doi.org/10.1093/nar/gkw1138
20. Benjamini Y, Hochberg Y (1995) Controlling the false discovery rate: a practical and powerful approach to multiple testing. J R Stat Soc B 57 (1):289–300
21. Gentleman RC, Carey VJ, Bates DM, Bolstad B, Dettling M, Dudoit S, Ellis B, Gautier L, Ge Y, Gentry J, Hornik K, Hothorn T, Huber W, Iacus S, Irizarry R, Leisch F, Li C, Maechler M, Rossini AJ, Sawitzki G, Smith C, Smyth G, Tierney L, Yang JY, Zhang J (2004) Bioconductor: open software development for computational biology and bioinformatics. Genome Biol 5(10): R80. https://doi.org/10.1186/gb-2004-5-10-r80
22. Zhu Y, Shen X, Pan W (2009) Network-based support vector machine for classification of microarray samples. BMC Bioinformatics 10 (Suppl 1):S21. https://doi.org/10.1186/1471-2105-10-S1-S21
23. Wang Z, San Lucas FA, Qiu P, Liu Y (2014) Improving the sensitivity of sample clustering by leveraging gene co-expression networks in variable selection. BMC Bioinformatics 15:153. https://doi.org/10.1186/1471-2105-15-153

Chapter 22

Clinical Trials for Traumatic Brain Injury: Designs and Challenges

Juan Lu and Mirinda Gormley

Abstract

Traumatic brain injury (TBI) is a major public health problem affecting across all demographics. Despite its high impact on disability, mortality, and economics, currently there are no therapeutic drugs approved for the treatment of acute TBI. While many potential agents have shown effects in animal models, clinical trials in patients with TBI have shown little success and nearly 30 phase III clinical trials have failed to show efficacy for the treatment of TBI. This chapter discusses the challenges and designs of clinical trials along the critical path of developing new treatment for TBI. The overall objective is to promote effective clinical trial designs and optimize opportunities for developing new treatment for patients with TBI.

Key words Traumatic brain injury, Clinical trial design, Randomized controlled trials, Heterogeneity of the patient population, Statistical efficiency, Outcome measurement

1 Introduction

TBI is a major public health problem that has widespread impacts on all demographics globally. In the United States, TBI affects approximately 1.7–3.8 million individuals each year [1–4], and TBI-related injuries account for more than 50,000 deaths annually [3, 5], making up 30% of all injury-related deaths in the United States [3, 6].

Worldwide, TBI is the leading cause of disability for people under 40 years old [7]. The annual incidence of disability due to TBI is estimated at 100 per 100,000 population [7, 8]. The direct and indirect costs of TBI have been estimated between 60 and 76 billion dollars each year [3, 9–11]. Living with disability and addressing these costs poses significant strain and economic hardship on patients and their families, and society.

Despite the high mortality and economic costs associated with TBIs, currently there are no therapeutic drugs approved for the treatment of acute TBI [1, 12]. While many potential agents have shown effects in animal models, clinical trials in patients with TBI

Amit K. Srivastava and Charles S. Cox, Jr. (eds.), *Pre-Clinical and Clinical Methods in Brain Trauma Research*, Neuromethods, vol. 139, https://doi.org/10.1007/978-1-4939-8564-7_22,

have shown little success [4, 11–13] and nearly 30 phase III clinical trials have failed to show efficacy for the treatment of TBI [1, 14]. Thus far, many guidelines for treatment of TBI were built on Level 2 and 3 evidence [11].

Researchers often cite a variety of reasons that might have contributed to the failures of clinical trials in TBI, such as insufficient preclinical and earlier phase clinical data to inform multicenter phase III clinical trials [12–16], the heterogeneity of injury mechanism and patient population [2–4, 11, 13, 17–19], inefficient statistical approaches [13, 18, 20–22], lack of sensitive and effective outcome measures [1, 2, 4, 5, 13, 23], small sample sizes and recruitment challenges [11, 13, 15, 24], and lack of funding for large-scale, multicenter trials [11, 15]. To encourage better trials and treatments, researchers around the world have attempted to address the issues encountered with regard to clinical trial designs in order to optimize opportunities (Table 1).

Table 1
Acronyms

CIHR	Canadian Institute of Health Research
CDC	Centers for Disease Control
CENTER-TBI	Collaborative European Neurotrauma Effectiveness Research in TBI
CDE	Common Data Elements
DoD	Department of Defense
EC	European Commission
EU	European Union
FITBIR	Federal Interagency TBI Informatics System
FDA	Food and Drug Administration
GOS	Glasgow Outcome Scale
GOS-E	Glasgow Outcome Scale-Extended
IMPACT	International Mission on Prognosis and Clinical Trial Design in TBI
NINDS	National Institute of Neurological Disorders and Stroke
NIH	National Institutes of Health
OBTT	Operation Brain Trauma Therapy
PH	Psychological Health
RCT	Randomized Control Trials
TED	TBI Endpoints Development
TRACK-TBI	Transforming Research and Clinical Knowledge in TBI
TBI	Traumatic Brain Injury
WHO	World Health Organization

2 Methods

2.1 Initiatives and Activities for Advancing Clinical Research in TBI

Recognizing the importance of developing effective treatment and the methodological challenges in designing clinical trials in TBI, National Institute for Neurological Disorders and Stroke (NINDS) sponsored a workshop in May 2000 to discuss lessons learned from previously conducted trials that could be used to improve future studies. As a joint effort, experts from clinical, research, pharmaceutical, and the U.S. Food and Drug Administration (FDA) presented several recommendations regarding trial designs and conduct in the field [15]. Following the workshop, efforts and advances in the improvement of trial designs have been furthered. Table 2 shows a list of major initiatives and activities that have been

Table 2
Major initiatives and activities for advancing clinical research for TBI

Year	Activity	Description
2000	Expert Workshop [15]	NINDS sponsored workshop to review the design and conduct of clinical trials in TBI and develop recommendations.
2002	CDEs Project [25, 26]	NINDS sponsored initiative to develop data standards for clinical research within the neurological community.
2003	IMPACT Project [14]	NINDS sponsored research project to advance knowledge of prognosis, clinical trial design and treatment in TBI.
2009	Interagency Workshop: CDEs [25]	Federal interagency sponsored workshop, "Advancing Integrated Research in Psychological Health (PH) and TBI" to develop topic-driven CDEs for PH and TBI research.
2009	TRACK-TBI Project [27]	NINDS sponsored project to validate the feasibility of implementing the TBI-CDEs.
2011	InTBIR [28]	EC, CHIR, NIH co-sponsored international initiative to coordinate and leverage clinical research activities on TBI.
2012	FITBIR [29, 30]	DoD and NIH co-sponsored initiative which creates a platform for data sharing to accelerate TBI research
2012	White House Executive Order [31]	Executive order-improving access to mental health services for veterans, service members, and military families, to establish a national research action plan aimed at identifying strategies to establish biomarkers for early diagnosis and treatment of TBI and develop improved diagnostic criteria for TBI.
2013	TED Initiative [1, 32]	DoD sponsored to advance validation of endpoints acceptable to the FDA.
2013	CENTER-TBI Project [33]	EU sponsored project to better characterize TBI as a disease and identify the most effective clinical interventions for managing TBI.
2016	FDA Open Workshop [1, 34]	FDA sponsored workshop, "Advancing the Development of Biomarkers in TBI" to evaluate considerations for TBI biomarker development to improve clinical utility for TBI.

launched since 2000 to advance clinical research and search for effective treatment in TBI.

3 Methods

3.1 Registered TBI Clinical Trial Activities

Between January 1st 2000 and August 1st 2017, about 548 clinical trials were registered at ClinicalTrials.gov with a primary or secondary condition listed as TBI. ClinicalTrials.gov, a program funded by the NIH, is a web-based registry that contains information about public and privately funded human clinical trials conducted around the world [35]. Figure 1 illustrates the geographic distribution of the registered clinical trials during the search period for TBI throughout the world; the majority of these activities resided in the United States (61.9%) and Europe (19.3%). Figure 2 illustrates the number of interventional trials over the years. Since 2000, the number of trials has increased dramatically, then slowly decreased after peaking in 2013.

Among 548 registered trials, over two-thirds (395/548) were interventional trials, and the majority utilized randomized control design (87.0%). The common interventions under investigation were new treatment drugs (33.1%), behavioral intervention programs (21.9%), and new devices (19.1%). Consistent with the report by Li and colleagues [11], a large number of trials had very low planned enrollment; more than half of the trials (60.8%) had less than 100 participants, and nearly one-third (29.1%) reported enrollments of less than 24 patients (Table 3).

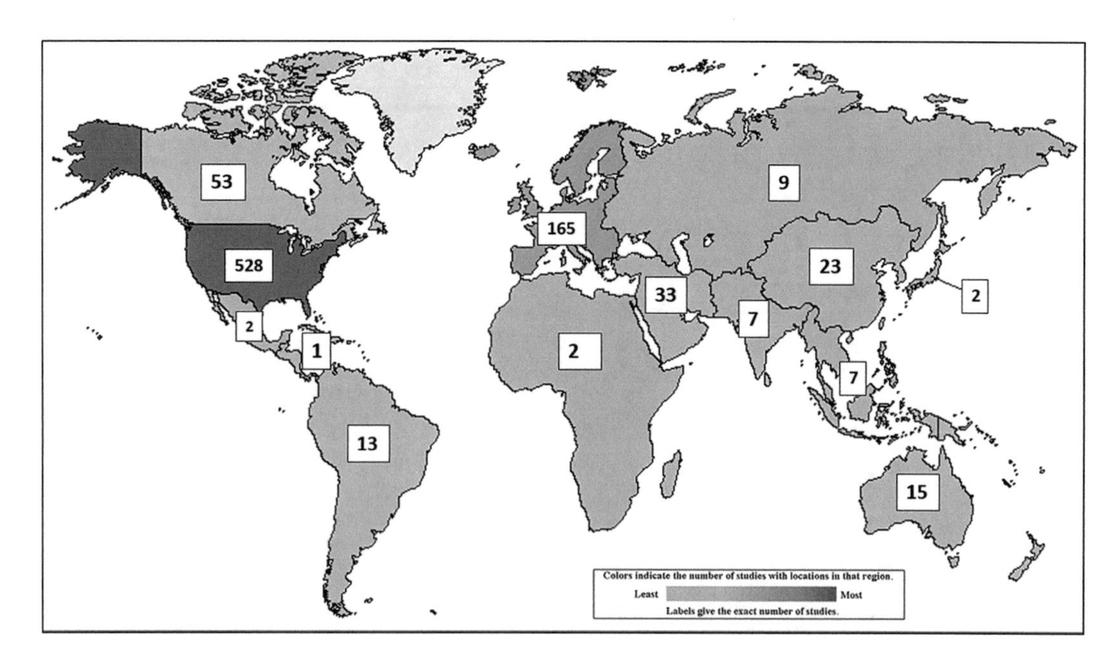

Fig. 1 Geographic distribution of clinical trials for TBI. Results as of search query conducted on August 1, 2017

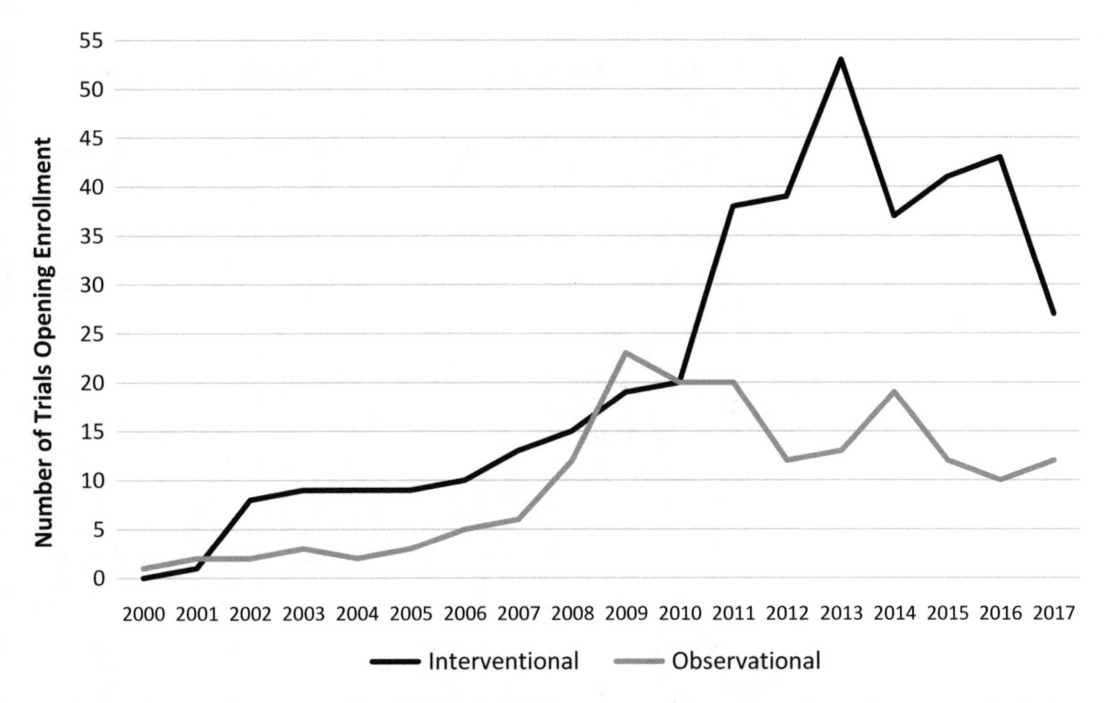

Fig. 2 Number of active or completed TBI clinical trials over years. Results as of search query conducted on August 1, 2017

4 Methods

4.1 Challenges and Design Considerations along the Critical Path of Developing New Treatment for TBI

The general critical path to developing a new intervention product encompasses the activities of basic research, prototype design or discovery, preclinical development, clinical development, and regulatory filing and approval (Fig. 3). Along the path, basic research is directed towards the fundamental understanding of biology and the disease process; translational research is directed towards moving basic discoveries from concept into clinical evaluation [36]. In the field of TBI research, therapeutic candidates identified through basic research were first tested through in vitro studies and animal models of TBI for implications in treatment of human TBI [37]. Therapeutic candidates then undergo a series of rigorously designed human clinical trials as they move from left to right along the path. As there is no FDA approved pharmacological treatment in acute TBI, all discussions in this chapter are focused on trial designs related to the development of a pharmacological agent.

4.2 Preclinical Studies

The objective of preclinical studies is to assess the safety, dosage, pharmacologic and biologic properties of a new treatment agent in preclinical models of TBI. However, the objective may not be achieved in a single study. Over the past several decades, preclinical studies have made extensive progress in understanding injury mechanisms, particularly the cascading of secondary insults

Table 3
Characteristics of the registered TBI interventional trials 2000–2017 ($N = 395$)[a]

Characteristic	N	%
Intervention		
Behavioral	70	21.9
Drug	106	33.1
Device	61	19.1
Procedure/protocol	22	6.9
Other	61	19.1
Missing	*75*	–
Phases		
Phase 1	32	17.6
Phase 1–2	19	10.4
Phase 2	63	34.6
Phase 2–3	8	4.4
Phase 3	32	17.6
Phase 4	28	15.4
Missing	*213*	–
Randomization		
Randomized	285	86.6
Non-randomized	44	13.4
Missing	*66*	–
Enrollment		
<25	115	29.1
25–50	62	15.7
51–75	58	14.7
76–99	33	8.4
100–500	108	27.3
500–1000	13	3.3
1000–10,000	3	0.8
>10,000	3	0.8
Recruitment		
Completed	140	35.4
Ongoing	150	38.0
Unknown status	44	11.1
Withdrawn	61	15.4

[a]Results as of search query conducted on August 1, 2017

following TBI. As a result, numerous potential therapeutic candidates progressed into human clinical trials, attempting to intervene in the cascade of secondary insults and reduce the morbidity and mortality rates for patients with TBIs [38]. However, the field is repeatedly perplexed by the inability to translate bench evidence to bedside. Among the reasons that contributed to these failures is the lack of compatible animal models to produce adequate preclinical data [39]. Researchers have suggested that our current animal models were relatively rigid and insufficient to represent the

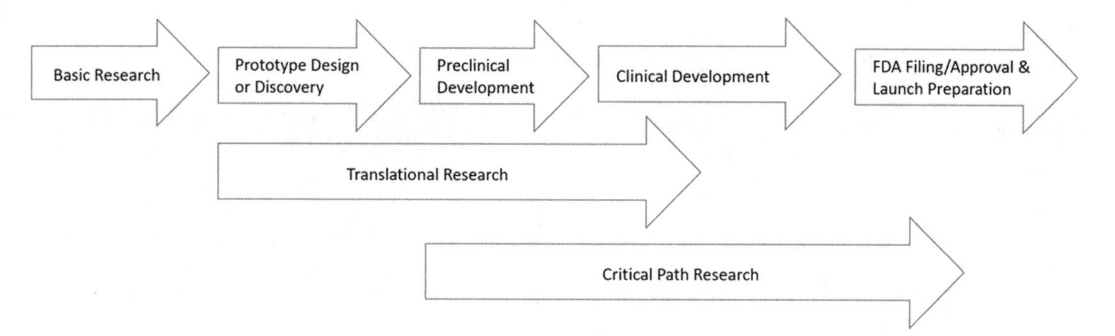

Fig. 3 Research support for product development [36]

heterogeneity of clinical TBI [37, 39–42]. First, all preclinical experiments were performed under a rigorous controlled environment and primarily driven by the treatment protocols that could not be reproduced in clinical studies [43]. Second, the endpoints in these preclinical studies were incompatible with human trials, where the treatment beneficial effects were mostly measured through "behavior outcomes" (e.g., Glasgow Outcome Scale) [44], and reflected the process of human functional recovery following injuries [15, 38].

To inform phase II TBI clinical trials, six theoretical requirements have been recommended by Muizelaar and Dore-Duffy [45]:

1. The treatment mechanism that targeted at the cellular damage should be confirmed both in animal models and in human studies.
2. Both the effect of intervention in blocking the cellular damage and improving clinical measures should be demonstrated in animal models.
3. The safety and tolerability of the intervention should be established in human TBIs.
4. The intervention drug should be able to penetrate the blood--brain barrier with adequate dosage for the targeted injury mechanism.
5. The intervention should be carried out during the effective "therapeutic window" corresponding to the targeted mechanism.
6. The outcome measures should be sensitive and capable of detecting meaningful intervention effect.

The limitations of current preclinical studies and the requirements for obtaining adequate clinical data have been discussed and reemphasized over the years [37–39, 43]. Numerous recommendations from the experts' discussions are highlighted in Table 4.

Table 4
Recommendations on designing preclinical studies in TBI

Recommendations
NINDS/Industrial Agencies Sponsored Workshop on Animal Models [37] • No single animal model of experimental traumatic brain injury can be embraced as a "gold standard" for the field. Multiple animal models should be used to replicate specific features of human TBI. • More attention should be placed on the language used to define the pathobiology of the traumatic condition. More care must be employed when using and applying terminology related to contusion, diffuse axonal injury, and secondary insult. • More attention must be given to the careful physiologic management and monitoring of all experimental animals, both before and following the induction of the traumatic brain insult. Such monitoring should include routine blood gas, blood glucose, blood pressure, and body temperature measurements. Further attention should be given to the age and sex of the chosen experimental animals. Attention to these details was viewed as essential to allow the interlaboratory comparisons necessary to validate a model fully and/or to test the efficacy of a specific drug. • Preclinical laboratory trials should include both acute and delayed dosing schedules both to demonstrate efficacy and to give some insight into the therapeutic window associated with the use of the chosen agent. • Any assessment of the recovery associated with TBI should consider both the acute and chronic phases of the recovery, as well as the rate of recovery. • The design and use of all chosen experimental animal models of TBI should include biomechanically relevant parameters.
NINDS Sponsored Workshop on Clinical Trials [15] • Test the lab model of injury severity clinically. • Establish animal intensive care units to study severe TBI. • Study pharmacokinetics in multiple animal models to establish the dose, how soon, how frequently, and for how long to give drug or initiate and continue therapy. • Test the intervention in at least two rodent models (such as weight drop or fluid percussion, or similar models in two rodent species) and in a larger animal if possible. • Test the intervention in more than one lab. • Try to model for both diffuse and focal injury, subarachnoid hemorrhage and ischemia. • Ensure adequate transport of a pharmaceutical agent into the cerebrospinal fluid and brain. • Study the time window of drug/intervention efficacy. • Establish a correlation between the window of opportunity in animals versus humans. • Establish dose–response curve. • Obtain as much toxicological data as possible.
DOD Sponsored Workshop on Pharmacotherapy [16] • Standardization of available animal models and introduce new models when scientifically necessary. • Reproduce important mechanisms of human TBI in animal models. • Design preclinical study using the same level of rigor as the ones in human RCTs and establish registration mechanism for preclinical studies. • Administration of therapeutic agents to mimic the timing, delivery route, and dosage feasible in humans in broad spectrum of animal models by multiple laboratories. • Select sensitive rodent behavior tasks that discriminate injury beyond 12 weeks of TBI.

In addition to inadequate animal models and design-related limitations, we need to recognize that failures to transform preclinical evidence to trials in TBI could be due to ineffective therapeutic agents or inappropriate target mechanisms. In an effort to address the need for effective novel therapies and selection of the most promising therapeutic agents, a multicentered preclinical drug screening consortium for acute therapies in severe TBI, operation brain trauma therapy (OBTT), was recently established as a therapy screening entity [46, 47]. The objective of OBTT is to include a spectrum of established TBI models at experienced centers and assess the effect of promising therapies on both conventional outcomes and serum biomarker levels [47–49]. The experience of OBTT in TBI modeling, evaluation of therapies, drug selection, and biomarker assessment on the first five therapies was reported in a special issue in the Journal of Neurotrauma in March, 2016 [48, 50–55]. This collaborative work provided an excellent example for advancing preclinical studies in the field.

4.3 Clinical Trials

The overall goal of a clinical trial is to efficiently develop a new, safe, and effective treatment agent. As indicated in the critical path by the FDA, the early phase of clinical development mostly concerns the safeties of the product; the later phase typically concerns the efficacies and roles in clinical practice. All clinical trials performed in TBI followed the classic path of preclinical development in regard to safety and efficacy evaluation phases. Following the preclinical studies, phase I trials assess the safety, dosage, pharmacologic and biologic properties of the agent; phases II trials document evidence of drug efficacy and side effects; and phase III trials evaluate efficacy and monitor adverse effects. After each phase, the results should be reviewed thoroughly to determine whether the evidence provides sufficient information necessary for future studies.

As an extremely heterogeneous disease, TBI poses significant methodological challenges in clinical trial designs, particularly in late phase (III or IV) studies [13, 15, 43]. Thus far, a majority of large, multicentered phase III trials failed to demonstrate significant improvement in clinical outcomes for TBI patients [14–16, 18, 56, 57]. A key challenge is the inability to replicate the positive evidence from preclinical/early phase human studies in large size multicenter trials. This led many efforts to improve the existing trial designs in the field. Several areas of challenges and related recommendations in designs of TBI trials were discussed below: (1) dealing with heterogeneous patient population, (2) improving statistical efficiencies in sample size estimation and primary outcome analysis, and (3) selecting more sensitive endpoint measurement.

4.3.1 Dealing with Patient Heterogeneity at Planning and Analytic Phases

IMPACT Project

In 2003, the International Mission on Prognosis and Clinical Trial Design in TBI (IMPACT) project was initiated to develop methodologies to improve the design and analysis of clinical trials of TBI [13]. Funded by NIH, the project was an international collaboration linking clinical, epidemiological, and statistical investigators from Belgium, the Netherlands, the United Kingdom, and the United States. The project was granted access to 11 TBI studies for establishing a database and exploring concepts to improve the design of clinical trials in TBI. The initial phase of the database contains clinical data on 9205 individual patients with moderate or severe TBI, 6535 from RCTs, and 2670 from observational studies [58]. During the continuation funding period, the number of studies was expanded, and more than 40,000 individuals with TBI were included in the study [59]. Within this large dataset, IMPACT investigators performed extensive simulation studies to explore different approaches in the improvement of trial designs for TBI. Several key findings and recommendations are discussed below, including the approaches in dealing with the heterogeneity of the patient population (e.g., baseline imbalance) and improving statistical efficiency in TBI trials [14, 59, 60].

Heterogeneity of the Patient Population and Baseline Imbalance

In a randomized controlled trial, randomization ensures the treatment allocation purely due to chance, though an imbalance of baseline characteristics could still exist. In trials of TBI, imbalance in baseline characteristics that influence trial outcomes, such as patient age or injury severity, can bias outcome estimation. Some strategies are commonly used to protect against the imbalance, such as applying stratified randomization, minimization or stringent inclusion criteria at the planning stage, and adjusting baseline covariates in the analytic phase [61, 62]. Both stratified randomization and minimization on important prognostic factors address the issue of imbalance and were often integrated in trial designs, but both approaches have some practical shortcomings.

Dealing with Imbalance at Planning Phase

In trials of TBI, more stringent enrollment criteria have been commonly employed in the past to decrease patient heterogeneity and select subgroups of patients that were more likely to respond to the mechanisms under investigation that contribute to the outcomes [60]. However, simulation studies using data within the IMPACT database showed that although stringent enrollment criteria could reduce the heterogeneity of the patient characteristics and increase statistical efficiency (i.e., the percentage reductions in sample size, which can be achieved without loss of statistical power relative to a different approach, e.g., utilizing less stringent enrollment criteria), the approach also led to alteration of the outcome distribution and substantial reductions in recruitment rate. The results showed that the more stringent enrollment criteria could lead to 65% and 41% reduction in study recruitment for the observational studies and RCTs in the IMPACT database, respectively [63]. Weighing the pros and cons of such an approach, the current recommendation advises keeping less stringent inclusion criteria. Since the approach not only maximizes the recruitment rates and enhances the generalizability of the results, it also increases the likelihood of the current understanding of mechanisms under investigation [14].

Dealing with Imbalance at Analytic Phase

As an alternative approach to minimize imbalances, covariate adjustment at a trial's analytic phase was further explored by the IMPACT investigators. Covariate adjustment is a procedure commonly used for controlling imbalanced prognostic factors and minimizing biased estimates of the treatment effect. The procedure is not only suitable under the context of observational studies, but also appropriate for RCTs which have moderate-to-severe imbalances in baseline characteristics for TBI [61, 64]. Utilizing logistic regression analysis to estimate the treatment effect with seven covariates, the IMPACT simulation studies demonstrated a 25% gain in statistical efficiency, compared with analysis without adjustment for covariates [65]. The subsequent study demonstrated that the effectiveness of covariates adjustment in gains of statistical efficiency is related to the degree of baseline imbalances. With greater heterogeneous patient populations, the greater the statistical efficiency could be gained. For those observational studies in the IMPACT database, the covariate adjustment resulted in a 30% gain of the statistical efficiency; whereas for those RCTs, the same covariate adjustment led to 16% gain in efficiency [63].

To carry out covariate adjustment analysis in RCTs, a couple of principles should be followed. First, the choice of baseline characteristics, by which an analysis is adjusted, should be determined by prior knowledge of influence on outcome rather than evidence of imbalance between treatment groups in the trial. Second, such information should be included in trial protocols and reported with analysis details [61, 66]. Both guidance on statistical principles by the International Conference of Harmonisation [67] and guidelines on adjustment for baseline covariates by the European Medicines Agency supported this notion and later provided very detailed guidance for practice [64].

In the case of TBI, core prognostic predictors of patients' age and baseline injury severity index (Glasgow Coma Scale motor score, pupillary reactivity) on 6-month mortality and functional outcome following moderate and severe TBI were further confirmed by the IMPACT study [68, 69]. In addition to these core predictors, the baseline brain CT classification, second insults (hypoxia and hypotension), subarachnoid hemorrhage, epidural hematoma, and certain lab values (glucose, hemoglobin) were identified as significant predictors [70–75]. The prognostic models based on these variables have been both internally and externally validated [70, 76–78]. Moreover, these prognostic models not only have been applicable in moderate and severe TBI populations of different settings, but also applicable for mild TBI [79].

4.3.2 Improving Statistical Efficiencies in Designs

Sample size estimation

For a fixed design, the trial sample size estimation is usually based on information obtained from prior trials. It is important to note that this information merely reflects the realistic expectations which the current trial wishes to aim for and does not directly relate to the eventual trial results. Moreover, in practice, the sample size estimation does not usually account for patient factors which might influence prognosis. In phase III TBI trials, the sample size estimations have traditionally been aimed at detecting a 10% absolute difference in the dichotomous Glasgow Outcome Scale (death/vegetative status/severe disability vs. moderate disability/good recovery) [44] between the two treatment arms, presuming that there will be an approximately 50% split of the dichotomous outcome distribution in the patient population. As indicated in many discussions, both expectations of effect size [15] and outcome distribution [60] may not be realistic in practice. In recent decades, the outcome distribution in severe TBI has been shifting towards positive outcomes and decreased mortality. For example, several recently completed trials have shown that the 6-month mortality following severe TBI ranges from 17.0% to 27.6% [80–82]. In placebo-treated patients, a mortality rate drop below 20% does not allow for a 10% shift in favorable outcome unless the drug reduces the mortality rate by greater than 50%. As a result, it is very difficult to obtain a 10% effect size [15]. Although the problem can be potentially overcome if a study restricts patient population with poor prognosis, such restriction could be offset with much slower enrollment and prolonged study period [63].

To ensure adequate power for future trials, several design principles have been recommended. For example, the expert workshop sponsored by NINDS has recommended to have reasonable expectations for effect size (e.g., 5–7.5%) and consider alternative designs, such as adaptive approaches or large simple trial designs where possible [15]. The IMPACT study further recommended to include important covariates and apply the ordinal model in primary outcome analysis to improve the statistical efficiency and statistical power in TBI trials [21].

Primary efficacy analysis

The Glasgow Outcome Scale (GOS) [44] is a well-established outcome measure for most phase III TBI trials. Conventionally, trials have collapsed the 5-category GOS as an unfavorable (dead, vegetative status, and severe disability) versus favorable (moderate disability and good recovery) outcome. Such an approach has been criticized for discarding valuable information among those collapsed categories, therefore resulting in statistical inefficiency in past trials [83, 84]. Several recent studies have explored the ordinal analysis of the GOS [83, 85, 86] and other ordinal outcomes commonly used in TBI and stroke clinical trials. These methodological works demonstrated that analyzing ordinal outcomes as ordinal could yield substantial gains over conventional dichotomized outcomes. In a simulation study, IMPACT study showed that using ordinal technique to analyze 5-category GOS could improve statistical efficiency in trials with moderate-to-severe patient populations. The ordinal approach of sliding dichotomy could decrease sample size by 40% without compromise of statistical power. Another ordinal approach of proportional odds model could also reduce similar percentages of sample size without affecting statistical power [83]. The subsequent IMPACT study showed that comparing with the use of 5-category GOS, utilization of the 8-category Extended Glasgow Outcome Scale (GOS-E) could further improve the statistical efficiency by 3–5% [87]. Therefore, it is recommended that future TBI trials use the ordinal outcomes and the methods of ordinal analyses.

4.3.3 Identifying Appropriate Outcome Measures

As with all trial designs, the validity of the clinical trials is dependent on the relevance of the chosen endpoint. Endpoints that do not correspond to the targeted treatment mechanism will yield biased conclusions. During the past several decades, GOS or GOS-E has been served as an FDA-accepted endpoint for late phase TBI trials. As a global outcome measure, GOS was introduced to measure the overall functional recovery, particularly to determine an individual's ability to return to pre-injury lifestyle following TBI [44]. To increase the outcome sensitivity in clinical trials, the 5-category GOS was later extended to 8-category, by subdividing the categories of severe, moderate, and good recovery into upper or lower degrees [88]. Although these measures are not a gold standard, they were widely used to compare the practice against the history in the field [15]. However, several limitations of using such measures as primary outcome in TBI trials must be recognized. First, GOS and GOS-E are specific to measure disability and handicap following TBI, but less appropriate to measure physical impairment [44]. Second, these measures are not sensitive to capture deficiencies and recoveries in the areas of cognitive or psychological health that a new intervention might target [1]. Third, both measures are subjective to misclassifications which could subsequently reduce the statistical power and bias the efficacy estimation [89–91]. To address this important issue, the TED Initiative was recently launched to support the collaborative effort of searching and validating the endpoints that will be acceptable to the U.S. FDA for use in future trials of TBI. This effort is in progress along with other efforts in the field for continuing the search of effective treatment for patients with TBI [1].

Acknowledgement

We would like to thank Monika Devanaboyina, an undergraduate student at Virginia Commonwealth University, for her assistance with validating references, working with figures, tables, and edits for the chapter.

References

1. Manley GT et al (2017) The traumatic brain injury endpoints development (TED) initiative: progress on a public-private regulatory collaboration to accelerate diagnosis and treatment of traumatic brain injury. J Neurotrauma 34(19):2721–2730. https://doi.org/10.1089/neu.2016.4729
2. Hoffer ME, Szczupak M, Balaban C (2016) Clinical trials in mild traumatic brain injury. J Neurosci Methods 272:77–81. S0165-0270(16)30073-5 [pii]
3. Yue JK et al (2017) Temporal profile of care following mild traumatic brain injury: predictors of hospital admission, follow-up referral and six-month outcome. Brain Inj 31(13–14):1820–1829. https://doi.org/10.1080/02699052.2017.1351000
4. Miller G (2010) Neuroscience. New guidelines aim to improve studies of traumatic brain injury. Science 328(5976):297. https://doi.org/10.1126/science.328.5976.297
5. Centers for Disease Control and Prevention (CDC) (2015) Traumatic brain injury in the United States: epidemiology and rehabilitation. National Center for Injury Prevention and Control; Division of Unintentional Injury Prevention, Atlanta, GA. CS261967-A
6. Taylor CA et al (2017) Traumatic brain injury-related emergency department visits, hospitalizations, and deaths—United States, 2007 and 2013. MMWR Surveill Summ 66(9):1–16. https://doi.org/10.15585/mmwr.ss6609a1
7. World Health Organization (2006) 3.6 Neurological disorders associated with malnutrition. In: Anonymous neurological disorders public health challenges. WHO Press, Geneva Switzerland, pp 111–175
8. Thornhill S et al (2000) Disability in young people and adults one year after head injury: prospective cohort study. BMJ 320(7250):1631–1635
9. Berg J, Tagliaferri F, Servadei F (2005) Cost of trauma in Europe. Eur J Neurol 12(Suppl 1):85–90. ENE1200 [pii]
10. Nelson LD et al (2017) Validating multidimensional outcome assessment using the TBI common data elements: an analysis of the TRACK-TBI pilot sample. J Neurotrauma 34(22):3158–3172. https://doi.org/10.1089/neu.2017.5139
11. Li LM, Menon DK, Janowitz T (2014) Cross-sectional analysis of data from the U.S. clinical trials database reveals poor translational clinical trial effort for traumatic brain injury, compared with stroke. PLoS One 9(1):e84336. https://doi.org/10.1371/journal.pone.0084336
12. Janowitz T, Menon DK (2010) Exploring new routes for neuroprotective drug development in traumatic brain injury. Sci Transl Med 2(27):27rv1. https://doi.org/10.1126/scitranslmed.3000330
13. Maas AI et al (2007) Prognosis and clinical trial design in traumatic brain injury: the IMPACT study. J Neurotrauma 24(2):232–238. https://doi.org/10.1089/neu.2006.0024
14. Maas AI et al (2010) IMPACT recommendations for improving the design and analysis of clinical trials in moderate to severe traumatic brain injury. Neurotherapeutics 7(1):127–134. https://doi.org/10.1016/j.nurt.2009.10.020
15. Narayan RK et al (2002) Clinical trials in head injury. J Neurotrauma 19(5):503–557. https://doi.org/10.1089/089771502753754037
16. Diaz-Arrastia R et al (2014) Pharmacotherapy of traumatic brain injury: state of the science and the road forward: report of the Department of Defense Neurotrauma Pharmacology Workgroup. J Neurotrauma 31(2):135–158. https://doi.org/10.1089/neu.2013.3019
17. Bagiella E et al (2010) Measuring outcome in traumatic brain injury treatment trials: recommendations from the traumatic brain injury clinical trials network. J Head Trauma Rehabil 25(5):375–382. https://doi.org/10.1097/HTR.0b013e3181d27fe3
18. Bragge P et al (2016) A state-of-the-science overview of randomized controlled trials evaluating acute management of moderate-to-severe traumatic brain injury. J Neurotrauma 33(16):1461–1478. https://doi.org/10.1089/neu.2015.4233

19. Roozenbeek B et al (2011) The added value of ordinal analysis in clinical trials: an example in traumatic brain injury. Crit Care 15(3):R127. https://doi.org/10.1186/cc10240
20. Maas AI et al (2012) Re-orientation of clinical research in traumatic brain injury: report of an international workshop on comparative effectiveness research. J Neurotrauma 29(1):32–46. https://doi.org/10.1089/neu.2010.1599
21. Maas AI, Roozenbeek B, Manley GT (2010) Clinical trials in traumatic brain injury: past experience and current developments. Neurotherapeutics 7(1):115–126. https://doi.org/10.1016/j.nurt.2009.10.022
22. Manley GT, Maas AI (2013) Traumatic brain injury: an international knowledge-based approach. JAMA 310(5):473–474. https://doi.org/10.1001/jama.2013.169158
23. Wilde EA et al (2010) Recommendations for the use of common outcome measures in traumatic brain injury research. Arch Phys Med Rehabil 91(11):1650–1660.e17. https://doi.org/10.1016/j.apmr.2010.06.033
24. Lingsma HF et al (2011) Between-centre differences and treatment effects in randomized controlled trials: a case study in traumatic brain injury. Trials 12:201. https://doi.org/10.1186/1745-6215-12-201
25. Thurmond VA et al (2010) Advancing integrated research in psychological health and traumatic brain injury: common data elements. Arch Phys Med Rehabil 91(11):1633–1636. https://doi.org/10.1016/j.apmr.2010.06.034
26. National Institute of Neurological Disorders and Stroke (2017) Project overview. NINDS common data elements. Available via https://www.commondataelements.ninds.nih.gov/ProjReview.aspx#tab=Introduction. Accessed 5 Aug 2017
27. University of California, San Francisco (2017) Transforming research and clinical knowledge in TBI (TRACK-TBI). Available via https://tracktbi.ucsf.edu/. Accessed 30 Aug 2017
28. International Initiative for Traumatic Brain Injury Research (2017) InTBIR mission. Available via https://intbir.nih.gov/mission. Accessed 30 Aug 2017
29. Hicks R et al (2013) Progress in developing common data elements for traumatic brain injury research: version two—the end of the beginning. J Neurotrauma 30(22):1852–1861. https://doi.org/10.1089/neu.2013.2938
30. Thompson HJ, Vavilala MS, Rivara FP (2015) Common data elements and federal interagency traumatic brain injury research informatics system for TBI research. Annu Rev Nurs Res 33(1):1–11. https://doi.org/10.1891/0739-6686.33.1
31. White House (2012) Executive order—improving access to mental health services for veterans, service members, and military families
32. University of California, San Francisco (2017) Welcome to the TED initiative. TBI Endpoints Development (TED) Initiative. Available via https://tbiendpoints.ucsf.edu/. Accessed 31 Aug 2017
33. CENTER-TBI (2017) Welcome to the CENTER-TBI website. CENTER-TBI. Available via https://www.center-tbi.eu/. Accessed 30 Aug 2017
34. US Food and Drug Administration (2016) Public workshop–advancing the development of biomarkers in traumatic brain injury, March 3, 2016. Available via https://www.fda.gov/MedicalDevices/NewsEvents/WorkshopsConferences/ucm483551.htm. Accessed 28 Aug 2017.
35. National Institute of Health (2017) Studies found for: Traumatic Brain Injury. Available via https://www.clinicaltrials.gov/ct2/results?cond=Traumatic+Brain+Injury&term=&cntry1=&state1=&recrs=. Accessed 1 Aug 2017
36. US Food and Drug Administration (2016) Innovation or stagnation: challenge and opportunity on the critical path to new medical products. Available via https://www.fda.gov/ScienceResearch/SpecialTopics/CriticalPathInitiative/CriticalPathOpportunitiesReports/ucm077262.htm#fig4. Accessed 7 Aug 2017
37. Povlishock JT et al (1994) Workshop on animal models of traumatic brain injury. J Neurotrauma 11(6):723–732. https://doi.org/10.1089/neu.1994.11.723
38. Bullock MR, Lyeth BG, Muizelaar JP (1999) Current status of neuroprotection trials for traumatic brain injury: lessons from animal models and clinical studies. Neurosurgery 45(2):207–217. discussion 217–220
39. Statler KD et al (2001) The simple model versus the super model: translating experimental traumatic brain injury research to the bedside. J Neurotrauma 18(11):1195–1206. https://doi.org/10.1089/089771501317095232
40. Morganti-Kossmann MC, Yan E, Bye N (2010) Animal models of traumatic brain injury: is there an optimal model to reproduce human brain injury in the laboratory? Injury 41(Suppl 1):S10–S13. https://doi.org/10.1016/j.injury.2010.03.032
41. O'Connor WT, Smyth A, Gilchrist MD (2011) Animal models of traumatic brain injury: a

critical evaluation. Pharmacol Ther 130 (2):106–113. https://doi.org/10.1016/j.pharmthera.2011.01.001
42. Nyanzu M et al (2017) Improving on laboratory traumatic brain injury models to achieve better results. Int J Med Sci 14(5):494–505. https://doi.org/10.7150/ijms.18075
43. Tolias CM, Bullock MR (2004) Critical appraisal of neuroprotection trials in head injury: what have we learned? NeuroRx 1(1):71–79. https://doi.org/10.1602/neurorx.1.1.71
44. Jennett B, Bond M (1975) Assessment of outcome after severe brain damage. Lancet 1 (7905):480–484. S0140-6736(75)92830-5 [pii]
45. Muizelaar JP, Dore-Duffy P (1996) Advances in experimental research and clinical intervention for catastrophic brain injury: the early 21st century. In: Levin HS, Benton AL, Paul Muizelaar J, Eisenberg HM (eds) Catastrophic brain injury. Oxford University Press, Oxford, UK, pp 233–252
46. Kochanek PM et al (2011) A novel multicenter preclinical drug screening and biomarker consortium for experimental traumatic brain injury: operation brain trauma therapy. J Trauma 71(1 Suppl):S15–S24. https://doi.org/10.1097/TA.0b013e31822117fe
47. Kochanek PM et al (2016) Approach to modeling, therapy evaluation, drug selection, and biomarker assessments for a Multicenter pre-clinical drug screening consortium for acute therapies in severe traumatic brain injury: operation brain trauma therapy. J Neurotrauma 33(6):513–522. https://doi.org/10.1089/neu.2015.4113
48. Kochanek PM et al (2016) Synthesis of findings, current investigations, and future directions: operation brain trauma therapy. J Neurotrauma 33(6):606–614. https://doi.org/10.1089/neu.2015.4133
49. Rasmussen TE, Crowder AT (2016) Synergy in science and resources. J Neurotrauma 33 (6):511–512. https://doi.org/10.1089/neu.2016.29007.ter
50. Dixon CE et al (2016) Cyclosporine treatment in traumatic brain injury: operation brain trauma therapy. J Neurotrauma 33(6):553–566. https://doi.org/10.1089/neu.2015.4122
51. Shear DA et al (2016) Nicotinamide treatment in traumatic brain injury: operation brain trauma therapy. J Neurotrauma 33 (6):523–537. https://doi.org/10.1089/neu.2015.4115
52. Bramlett HM et al (2016) Erythropoietin treatment in traumatic brain injury: operation brain trauma therapy. J Neurotrauma 33 (6):538–552. https://doi.org/10.1089/neu.2015.4116
53. Mountney A et al (2016) Simvastatin treatment in traumatic brain injury: operation brain trauma therapy. J Neurotrauma 33 (6):567–580. https://doi.org/10.1089/neu.2015.4130
54. Browning M et al (2016) Levetiracetam treatment in traumatic brain injury: operation brain trauma therapy. J Neurotrauma 33 (6):581–594. https://doi.org/10.1089/neu.2015.4131
55. Mondello S et al (2016) Insight into pre-clinical models of traumatic brain injury using circulating brain damage biomarkers: operation brain trauma therapy. J Neurotrauma 33(6):595–605. https://doi.org/10.1089/neu.2015.4132
56. Marklund N et al (2006) Evaluation of pharmacological treatment strategies in traumatic brain injury. Curr Pharm Des 12 (13):1645–1680
57. Lu J et al (2012) Randomized controlled trials in adult traumatic brain injury. Brain Inj 26 (13–14):1523–1548. https://doi.org/10.3109/02699052.2012.722257
58. Marmarou A et al (2007) IMPACT database of traumatic brain injury: design and description. J Neurotrauma 24(2):239–250. https://doi.org/10.1089/neu.2006.0036
59. Maas AI et al (2013) Advancing the care for traumatic brain injury: summary results from the IMPACT studies and perspectives on future research. Lancet Neurol 12(12):1200–1210
60. Roozenbeek B, Lingsma HF, Maas AI (2012) New considerations in the design of clinical trials for traumatic brain injury. Clin Investig (Lond) 2(2):153–162. https://doi.org/10.4155/cli.11.179
61. Roberts C, Torgerson DJ (1999) Understanding controlled trials: baseline imbalance in randomised controlled trials. BMJ 319(7203):185
62. Pocock SJ, Simon R (1975) Sequential treatment assignment with balancing for prognostic factors in the controlled clinical trial. Biometrics 31(1):103–115
63. Roozenbeek B et al (2009) The influence of enrollment criteria on recruitment and outcome distribution in traumatic brain injury studies: results from the impact study. J Neurotrauma 26(7):1069–1075. https://doi.org/10.1089/neu.2008.0569
64. European Medicines Agency (2013) Guideline on adjustment for baseline covariates EMA/295050/2013
65. Hernandez AV, Steyerberg EW, Habbema JD (2004) Covariate adjustment in randomized

controlled trials with dichotomous outcomes increases statistical power and reduces sample size requirements. J Clin Epidemiol 57 (5):454–460. https://doi.org/10.1016/j.jclinepi.2003.09.014

66. Senn S (1994) Testing for baseline balance in clinical trials. Stat Med 13:1715–1726
67. ICH (1999) ICH harmonised tripartite guideline. Statistical principles for clinical trials. International conference on harmonisation E9 expert working group. Stat Med 18 (15):1905–1942
68. Mushkudiani NA et al (2007) Prognostic value of demographic characteristics in traumatic brain injury: results from the IMPACT study. J Neurotrauma 24(2):259–269. https://doi.org/10.1089/neu.2006.0028
69. Marmarou A et al (2007) Prognostic value of the Glasgow Coma Scale and pupil reactivity in traumatic brain injury assessed pre-hospital and on enrollment: an IMPACT analysis. J Neurotrauma 24(2):270–280. https://doi.org/10.1089/neu.2006.0029
70. Steyerberg EW et al (2008) Predicting outcome after traumatic brain injury: development and international validation of prognostic scores based on admission characteristics. PLoS Med 5(8):e165.; discussion e165. https://doi.org/10.1371/journal.pmed.0050165
71. Maas AI et al (2007) Prognostic value of computerized tomography scan characteristics in traumatic brain injury: results from the IMPACT study. J Neurotrauma 24 (2):303–314. https://doi.org/10.1089/neu.2006.0033
72. McHugh GS et al (2007) Prognostic value of secondary insults in traumatic brain injury: results from the IMPACT study. J Neurotrauma 24(2):287–293. https://doi.org/10.1089/neu.2006.0031
73. Butcher I et al (2007) Prognostic value of admission blood pressure in traumatic brain injury: results from the IMPACT study. J Neurotrauma 24(2):294–302. https://doi.org/10.1089/neu.2006.0032
74. Van Beek JG et al (2007) Prognostic value of admission laboratory parameters in traumatic brain injury: results from the IMPACT study. J Neurotrauma 24(2):315–328. https://doi.org/10.1089/neu.2006.0034
75. Murray GD et al (2007) Multivariable prognostic analysis in traumatic brain injury: results from the IMPACT study. J Neurotrauma 24 (2):329–337. https://doi.org/10.1089/neu.2006.0035
76. Perel P, Arango M (2008) Statistical reanalysis of functional outcomes in stroke trials. BMJ 336:425–429
77. Roozenbeek B et al (2012) Prediction of outcome after moderate and severe traumatic brain injury: external validation of the international mission on prognosis and analysis of clinical trials (IMPACT) and corticoid randomisation after significant head injury (CRASH) prognostic models. Crit Care Med 40 (5):1609–1617. https://doi.org/10.1097/CCM.0b013e31824519ce
78. Lingsma H et al (2013) Prognosis in moderate and severe traumatic brain injury: external validation of the IMPACT models and the role of extracranial injuries. J Trauma Acute Care Surg 74(2):639–646. https://doi.org/10.1097/TA.0b013e31827d602e
79. Jacobs B et al (2010) Outcome prediction in mild traumatic brain injury: age and clinical variables are stronger predictors than CT abnormalities. J Neurotrauma 27(4):655–668. https://doi.org/10.1089/neu.2009.1059
80. Maas AI et al (2006) Efficacy and safety of dexanabinol in severe traumatic brain injury: results of a phase III randomised, placebo-controlled, clinical trial. Lancet Neurol 5 (1):38–45. S1474-4422(05)70253-2 [pii]
81. Wright DW et al (2014) Very early administration of progesterone for acute traumatic brain injury. N Engl J Med 371(26):2457–2466. https://doi.org/10.1056/NEJMoa1404304
82. Skolnick BE et al (2014) A clinical trial of progesterone for severe traumatic brain injury. N Engl J Med 371(26):2467–2476. https://doi.org/10.1056/NEJMoa1411090
83. McHugh GS et al (2010) A simulation study evaluating approaches to the analysis of ordinal outcome data in randomized controlled trials in traumatic brain injury: results from the IMPACT Project. Clin Trials 7(1):44–57. https://doi.org/10.1177/1740774509356580
84. Altman DG, Dore CJ (1990) Randomisation and baseline comparisons in clinical trials. Lancet 335(8682):149–153. 0140-6736(90)90014-V [pii]
85. Bath PM et al (2008) Use of ordinal outcomes in vascular prevention trials: comparison with binary outcomes in published trials. Stroke 39 (10):2817–2823. https://doi.org/10.1161/STROKEAHA.107.509893
86. Optimising Analysis of Stroke Trials (OAST) Collaboration et al (2007) Can we improve the statistical analysis of stroke trials? Statistical reanalysis of functional outcomes in stroke

trials. Stroke 38(6):1911–1915. STROKEAHA.106.474080 [pii]

87. Weir J et al (2012) Does the extended Glasgow Outcome Scale add value to the conventional Glasgow Outcome Scale? J Neurotrauma 29 (1):53–58. https://doi.org/10.1089/neu.2011.2137
88. Wilson JT, Pettigrew LE, Teasdale GM (1998) Structured interviews for the Glasgow Outcome Scale and the extended Glasgow Outcome Scale: guidelines for their use. J Neurotrauma 15(8):573–585. https://doi.org/10.1089/neu.1998.15.573
89. Lu J et al (2008) Effects of Glasgow Outcome Scale misclassification on traumatic brain injury clinical trials. J Neurotrauma 25(6):641–651. https://doi.org/10.1089/neu.2007.0510
90. Lu J et al (2012) Impact of GOS misclassification on ordinal outcome analysis of traumatic brain injury clinical trials. J Neurotrauma 29 (5):719–726. https://doi.org/10.1089/neu.2010.1746
91. Choi SC et al (2002) Misclassification and treatment effect on primary outcome measures in clinical trials of severe neurotrauma. J Neurotrauma 19(1):17–24

INDEX

Amit K. Srivastava and Charles S. Cox, Jr. (eds.), *Pre-Clinical and Clinical Methods in Brain Trauma Research*, Neuromethods, vol. 139, https://doi.org/10.1007/978-1-4939-8564-7, © Springer Science+Business Media, LLC, part of Springer Nature 2018

G

H

I

L

M

N

O

P

R

S

T

Printed in the United States
By Bookmasters